New Aspects *Of* Plasma Physics

Proceedings of the 2007 ICTP Summer College on Plasma Physics

New Aspects *Of* Plasma Physics

Proceedings of the 2007 ICTP Summer College on Plasma Physics

edited by

Padma K Shukla

Ruhr-Universität Bochum, Germany

Lennart Stenflo & Bengt Eliasson

Umeå University, Sweden

World Scientific

NEW JERSEY • LONDON • SINGAPORE • BEIJING • SHANGHAI • HONG KONG • TAIPEI • CHENNAI

Published by

World Scientific Publishing Co. Pte. Ltd.
5 Toh Tuck Link, Singapore 596224
USA office: 27 Warren Street, Suite 401-402, Hackensack, NJ 07601
UK office: 57 Shelton Street, Covent Garden, London WC2H 9HE

British Library Cataloguing-in-Publication Data
A catalogue record for this book is available from the British Library.

NEW ASPECTS OF PLASMA PHYSICS
Proceedings of the 2007 ICTP Summer College on Plasma Physics

ISBN-13 978-981-279-977-7
ISBN-10 981-279-977-X

Printed in Singapore by B & JO Enterprise

FOREWORD

The "2007 ICTP Summer College on Plasma Physics" was held at the Abdus Salam International Centre for Theoretical Physics (ICTP), Trieste, Italy, during the period 30 July to 24 August 2007. The summer college was organized by S. M. Mahajan, P. K. Shukla, R. Bingham, L. Stenflo and Z. Yoshida. The College on Plasma Physics is a permanent feature of ICTP.

The purpose of the summer college was to provide training for young scientists from all over the world, mainly from third world countries, and to give them the opportunity to interact with the senior scientists in an informal manner. The first part of the summer college consisted of classroom teaching and seminars in preparation for the fourth week, when a plasma physics workshop was held with a large number of talks by invited speakers and experts. The summer college was attended by approximately one hundred and twenty participants from the developing countries and industrial nations.

The main focus of the scientific program was on magnetic confinement fusion and tokamak physics, intense laser-plasma interactions and plasma-based particle acceleration, turbulence, dusty plasmas, and quantum plasmas. A selected number of papers in these areas appears in this book.

The editors express their sincere gratitude to the ICTP director Professor K R Sreenivasan and Professors S. M. Mahajan, R. Bingham and Z. Yoshida for their wholehearted support to the 2007 Summer College on Plasma Physics. We would also like to thank the staff at the ICTP for their excellent support and help during the activity. In addition, the organizers thank the speakers and attendees for their contributions which have resulted in the success of the Summer College. Specifically, we appreciate the speakers for delivering excellent lectures and talks, and for supplying well-prepared manuscripts for publication in the present book.

P. K. Shukla, L. Stenflo and B. Eliasson
Ruhr-University Bochum, Germany, and Umeå University, Sweden

CONTENTS

Foreword v

Nonlinear Collective Processes in Very Dense Plasmas
P. K. Shukla, B. Eliasson and D. Shaikh 1

Quantum, Spin and QED Effects in Plasmas
G. Brodin and M. Marklund 26

Spin Quantum Plasmas — New Aspects of Collective Dynamics
M. Marklund and G. Brodin 35

Revised Quantum Electrodynamics with Fundamental Applications
B. Lehnert 52

Quantum Methodologies in Beam, Fluid and Plasma Physics
R. Fedele 87

Plasma Effects in Cold Atom Physics
J. T. Mendonça, J. Loureiro, H. Terças and R. Kaiser 133

General Properties of the Rayleigh-Taylor Instability in Different Plasma Configurations: The Plasma Foil Model
F. Pegoraro and S. V. Bulanov 152

The Rayleigh-Taylor Instability of a Plasma Foil Accelerated by the Radiation Pressure of an Ultra Intense Laser Pulse
F. Pegoraro and S. V. Bulanov 162

Generation of Galactic Seed Magnetic Fields
H. Saleem 174

Nonlinear Dynamics of Mirror Waves in Non-Maxwellian Plasmas
O. A. Pokhotelov et al. 195

Formation of Mirror Structures Near Instability Threshold
E. A. Kuznetsov, T. Passot and P. L. Sulem 221

Nonlinear Dispersive Alfvén Waves in Magnetoplasmas
P. K. Shukla, B. Eliasson, L. Stenflo and R. Bingham 232

Properties of Drift and Alfvén Waves in Collisional Plasmas
J. Vranjes, S. Poedts and B. P. Pandey 256

Current Driven Acoustic Perturbations in Partially Ionized Collisional Plasmas
J. Vranjes, S. Poedts, M. Y. Tanaka and B. P. Pandey 285

Multifluid Theory of Solitons
F. Verheest 316

Nonlinear Wavepackets in Pair-Ion and Electron-Positron-Ion Plasmas
I. Kourakis et al. 355

Electro-Acoustic Solitary Waves in Dusty Plasmas
A. A. Mamun and P. K. Shukla 374

Physics of Dust in Magnetic Fusion Devices
Z. Wang et al. 394

Short Wavelength Ballooning Mode in Tokamaks
A. Hirose and N. Joiner 476

Effects of Perpendicular Shear Superposition and Hybrid Ions Intruduction on Parallel Shear Driven Plasma Instabilities
T. Kaneko and R. Hatakeyama 488

NONLINEAR COLLECTIVE PROCESSES IN VERY DENSE PLASMAS

P. K. SHUKLA* and B. ELIASSON

Institut für Theoretische Physik IV, Fakultät für Physik und Astronomie, Ruhr-Universität Bochum, D-44780 Bochum, Germany
**E-mail: ps@tp4.rub.de*
www.tp4.rub.de

D. SHAIKH

Institute of Geophysics and Planetary Physics, University of Californina, Riverside, CA 92521, USA

We present simulation studies of the formation and dynamics of dark solitons and vortices, and of nonlinear interactions between intense circularly polarized electromagnetic (CPEM) waves and electron plasma oscillations (EPOs) dense in quantum electron plasmas. The electron dynamics in the latter is governed by a pair of equations comprising the nonlinear Schrödinger and Poisson system of equations, which conserves electrons and their momentum and energy. Nonlinear fluid simulations are carried out to investigate the properties of fully developed two-dimensional (2D) electron fluid turbulence in a dense Fermi (quantum) plasma. We report several distinguished features that have resulted from our 2D computer simulations of the nonlinear equations which govern the dynamics of nonlinearly interacting electron plasma oscillations (EPOs) in the Fermi plasma. We find that a 2D quantum electron plasma exhibits dual cascades, in which the electron number density cascades towards smaller turbulent scales, while the electrostatic potential forms larger scale eddies. The characteristic turbulent spectrum associated with the nonlinear electron plasma oscillations determined critically by quantum tunneling effect. The turbulent transport corresponding to the large-scale potential distribution is predominant in comparison with the small-scale electron number density variation, a result that is consistent with the classical diffusion theory. The dynamics of the CPEM waves is also governed by a nonlinear schrödinger equation, which is nonlinearly coupled with the nonlinear Schrödinger equation of the EPOs via the relativistic ponderomotive force, the relativistic electron mass increase in the CPEM field, and the electron density fluctuations. The present governing equations in one spatial dimension admit stationary solutions in the form a dark envelope soliton. The dynamics of the latter reveals its robustness. Furthermore, we numerically demonstrate the existence of cylindrically symmetric two-dimensional quantum electron vortices, which survive during collisions. The nonlinear equations admit the modulational instability of an intense CPEM pump wave against EPOs, leading to the formation and trapping of localized CPEM wave pipes in the electron density hole that is associated with a positive potential distribution in our dense plasma.

1. Introduction

About forty five years ago, Pines[1] had laid down foundations for quantum plasma physics through his studies of the properties of electron plasma oscillations (EPOs) in a dense Fermi plasma. The high-density, low-temperature quantum Fermi plasma is significantly different from the low-density, high-temperature "classical plasma" obeying the Maxwell-Boltzmann distribution. In a very dense quantum plasma, there are new equations of state[2–4] associated with the Fermi-Dirac plasma particle distribution function and there are new quantum forces involving the quantum Bohm potential[5] and the electron-1/2 spin effect[6] due to magnetization. It should be noted that very dense quantum plasmas exist in intense laser-solid density plasma interaction experiments,[7–10] in laser-based inertial fusion,[11] in astrophysical and cosmological environments,[12–15] and in quantum diodes.[16–18]

During the last decade, there has been a growing interest in investigating new aspects of dense quantum plasmas by developing the quantum hydrodynamic (QHD) equations[5] by incorporating the quantum force associated with the Bohm potential.[5] The the Wigner-Poisson (WP) model[19,20] has been used to derive a set of quantum hydrodynamic (QHD) equations[2,3] for a dense electron plasma. The QHD equations include the continuity, momentum and Poisson equations. The quantum nature[2] appears in the electron momentum equation through the pressure term, which requires the knowledge of the Wigner distribution for a quantum mixture of electron wave functions, each characterized by an occupation probability. The quantum part of the electron pressure is represented as a quantum force[2,5] $-\nabla\phi_B$, where $\phi_B = -(\hbar^2/2m_e\sqrt{n_e})\nabla^2\sqrt{n_e}$, $\hbar$ is the Planck constant divided by 2π, m_e is the electron mass, and n_e is the electron number density. Defining the effective wave function $\psi = \sqrt{n_e(\mathbf{r},t)}\exp[iS(\mathbf{r},t)/\hbar]$, where $\nabla S(\mathbf{r},t) = m_e\mathbf{u}_e(\mathbf{r},t)$ and $\mathbf{u}_e(\mathbf{r},t)$ is the electron velocity, the electron momentum equation can be represented as an effective nonlinear Schrödinger (NLS) equation,[2–4] in which there appears a coupling between the wave function and the electrostatic potential associated with the EPOs. The electrostatic potential is determined from the Poisson equation. We thus have the coupled NLS and Poisson equations, which govern the dynamics of nonlinearly interacting EPOs is a dense quantum plasmas. This mean-field model of Ref.[2,3] is valid to the lowest order in the correlation parameter, and it neglects correlations between electrons. The QHD equations is useful for deriving the Child-Langmuir law in the quantum regime[17,18] and for studying numerous collective effects[2–4,21–24] involving different quantum forces (e.g. due to the Bohm potential[5] and the pressure law[2,3] for the Fermi plasma, as well as the potential energy of the electron$-1/2$ spin magnetic moment in a magnetic field[44]). In dense plasmas, quantum mechanical effects (e.g. tunnelling) are important since the de Broglie length of the charge carriers (e.g. electrons and holes/positrons) is comparable to the dimensions of the system. Studies of collective interactions in dense quantum plasmas are relevant for the next generation intense laser-solid density plasma experiments,[8,10,25] for superdense astrophysical bodies[12,14,15,26] (e.g. the interior of white dwarfs and neutron stars), as well as for micro and nano-scale objects (e.g. quantum diodes,[17,18] quantum dots and nanowires,[27] nano-photonics,[28,29] ultra-small electronic devices[30]) and micro-plasmas.[31] Quantum transport models similar to the QHD plasma model has also been used in superfluidity[32] and superconductivity,[33] as well as the study of metal clusters and nanoparticles, where they are referred to as nonstationary Thomas-Fermi models.[34] The density functional theory[35–37] incorporates electron-electron correlations, which are neglected in the present paper.

It has been recently recognized[25,38,39] that quantum mechanical effects play an important role in intense laser-solid density plasma interaction experiments. In the latter, there are nonlinearities[40] associated with the electron mass increase in the electromagnetic (EM) fields and the modification of the electron number density by the relativistic ponderomotive force. Relativistic nonlinear effects in a classical plasma is very important, because they provide the possibility of the compression and localization of intense electromagnetic waves. In this Letter, we consider nonlinear interactions between intense CPEM waves and EPOs in dense quantum plasmas, which are relevant for a variety of applications in laboratories.[9,10]

In this paper, we investigate, by means of computer simulations, the formation and dynamics of dark/gray envelope solitons and vortices in quantum electron plasmas with fixed ion background. The results are relevant for the transport of information at quantum scales in micro-plasmas as well as in micro-mechanical systems and microelectronics. For our purposes, we shall use an effective Schrödinger-Poisson model,[2,21–24] which was developed by employing the Wigner-Poisson phase space formalism on the Vlasov equation coupled with the Poisson equation for the electric potential. Such a model was originally derived by Hartree in the context of atomic physics for studying the self-consistent effect of atomic electrons on the Coulomb potential of the nucleus. The properties of 2D electron fluid turbulence and associated electron transport in quantum plasmas are investigated numerically by simulations. We find that the nonlinear coupling between the EPOs of different scale sizes gives rise to small-scale electron density structures, while the electrostatic potential cascades towards large-scales. Finally, we present theoretical and simulation studies of the CPEM wave modulational instability against EPOs, as well as the trapping of localized CPEM waves into a quantum electron hole in very dense quantum plasmas, which may be relevant for the next generation intense laser-plasma interaction experiments.

2. Dark solitons and vortices in a dense quantum plasma

In this section, we discuss the nonlinear properties and dynamics of dark solitons and vortices in a quantum plasma.[4] Generalizing the one-dimensional Schrödinger-Poisson system of equations[2] to multi-space dimensions, we have

$$i\frac{\partial \Psi}{\partial t} + A\nabla^2\Psi + \varphi\Psi - |\Psi|^4\Psi = 0, \tag{1}$$

and

$$\nabla^2\varphi = |\Psi|^2 - 1, \tag{2}$$

where the wave function Ψ is normalized by $\sqrt{n_0}$, the electrostatic potential φ by $k_B T_F/e$, the time t by $\hbar K_B T_F$ and the space $\mathbf{r}$ by λ_D. We have introduced the notations $\lambda_D = (k_B T_F/4\pi n_0 e^2)^{1/2}$ and $A = \Gamma_Q/2$, where the quantum coupling parameter $\Gamma_Q = 4\pi e^2 m/\hbar n_0^{1/3}$ can be both smaller and larger than unity for typical metallic electrons.[2] Here n_0 is the equilibrium electron particle density, $k_B T_F \simeq \hbar n_0^{2/3}/m_e$ is the Fermi temperature, m_e is the electron mass, e is the magnitude of the electron charge, k_B is Boltzmann's constant, and $\hbar$ is the Planck constant divided by 2π. Strictly speaking, the nonlinearity $|\Psi|^4$ in the last term in the left-hand side of Eq. (1) was derived for the one-dimensional model[2] and takes the form $|\Psi|^{4/D}$ in D dimensions. However, our numerical investigations of the profiles of dark solitons and vortices have shown very small differences if we use $D = 2$ (for two dimensions) instead of $D = 1$; henceforth, we will keep Eqs. (1) and (2) in the present form. The system (1) and (2) is supplemented by the Maxwell equation

$$\partial \mathbf{E}/\partial t = iA\left(\Psi\nabla\Psi^* - \Psi^*\nabla\Psi\right), \tag{3}$$

where the electric field $\mathbf{E} = -\nabla\varphi$. The system of equations (1)–(3) conserves the number of electrons $N = \int |\Psi|\, d^3x$, the electron momentum $\mathbf{P} = -i\int \Psi^*\nabla\Psi\, d^3x$, the electron angular momentum $\mathbf{L} = -i\int \Psi^*\mathbf{r}\times\nabla\Psi\, d^3x$, and the total energy $\mathcal{E} = \int(-\Psi^* A\nabla^2\Psi + |\nabla\varphi|^2/2 + |\Psi|^6/3)\, d^3x$. We note that one-dimensional version of Eq. (1) without the φ-term has also been used to describe the behaviour of a Bose-Einstein condensate.[41]

Let us first consider a quasi-stationary, one-dimensional structure moving with a constant speed v_0, and make the ansatz $\Psi = W(\xi)\exp(iKx - i\Omega t)$, where W is a complex-valued function of the argument $\xi = x - v_0 t$, and K and Ω are a constant wavenumber and frequency shift, respectively. By the choice $K = v_0/2A$, we can then write the coupled system of equations as

$$\frac{d^2W}{d\xi^2} + \lambda W + \frac{\varphi W}{A} - \frac{|W|^4 W}{A} = 0, \tag{4}$$

and

$$\frac{d^2\varphi}{d\xi^2} = |W|^2 - 1, \tag{5}$$

where $\lambda = \Omega/A - v_0^2/4A^2$ is an eigenvalue of the system. From the boundary conditions $|W| = 1$ and $\varphi = 0$ at $|\xi| = \infty$, we determine $\lambda = 1/A$ and $\Omega = 1 + v_0^2/4A$. The system of Eqs. (4) and (5) supports a first integral in the form

$$\begin{aligned} H = & A\left|\frac{dW}{d\xi}\right|^2 - \frac{1}{2}\left(\frac{d\varphi}{d\xi}\right)^2 + |W|^2 \\ & - \frac{|W|^6}{3} + \varphi|W|^2 - \varphi - \frac{2}{3} = 0, \end{aligned} \tag{6}$$

where we have used the boundary conditions $|W| = 1$ and $\varphi = 0$ at $|\xi| = \infty$.

We have solved (4) and (5) numerically and have presented the results in Fig. 1. Here we have plotted the profiles of W^2 and φ for a few values of A, where W was set to -1 on the left boundary and to $+1$ on the right boundary, i.e. the phase shift is 180 degrees between the two boundaries. We see that we have solutions in the form of a dark soliton, with a localized depletion of the electron density $N_e = |W|^2$, and where W has different sign on different sides of the solitary structure. The local depletion of the electron density is associated with a positive potential. Larger values of the parameter A give rise to larger-amplitude and wider dark solitons. Unlike a dark soliton associated with usual cubic Schrödinger equation in which the group dispersion and the nonlinearity coefficient have opposite sign, the

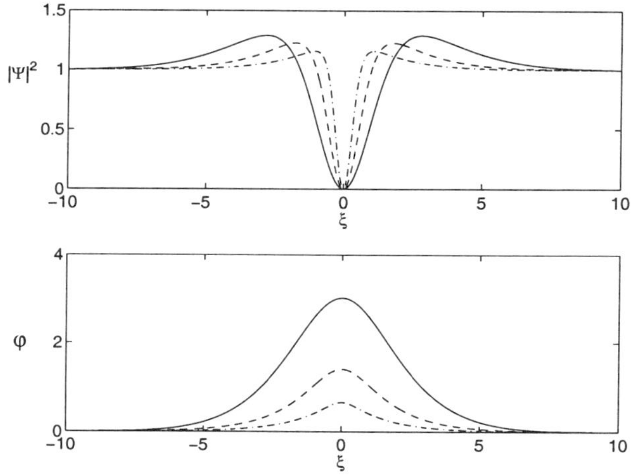

Fig. 1. The electron density $|\Psi|^2$ (the upper panel) and electrostatic potential φ (the lower panel) associated with a dark soliton supported by the system of equations (4) and (5), for $A = 5$ (solid lines), $A = 1$ (dashed lines), and $A = 0.2$ (dash-dotted line). After Ref. 4.

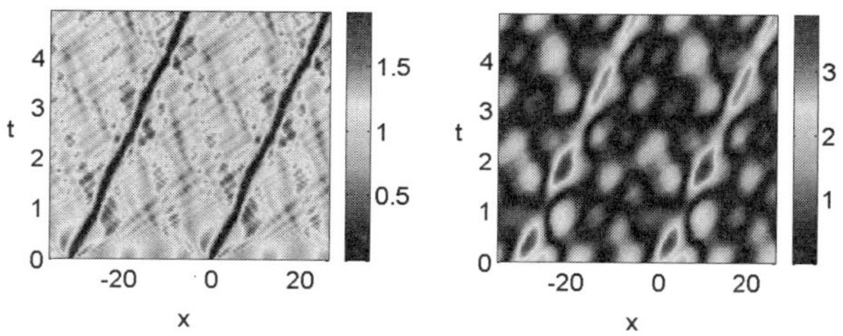

Fig. 2. The time-development of the electron density $|\Psi|^2$ (left-hand panel) and electrostatic potential φ (the right-hand panel), obtained from a simulation of the system of equations (1) and (2). The initial condition is $\Psi = 0.18 + \tanh[20\sin(x/10)]\exp(iKx)$, with $K = v_0/2A$, $A = 5$ and $v_0 = 5$. After Ref. 4.

modulus of the wave function in the present work has localized maxima on both sides of the density depletion. If the boundary conditions are shifted below 180 degrees (i.e. by a complex number), we have a "grey soliton" which is characterized by a non-zero density at the center of the soliton. In order to assess the dynamics and stability of the dark soliton, we have solved the time-dependent system of Eqs. (1) and (2) numerically, and have displayed the result in Fig. 2. The initial condition is $\Psi = 0.18 + \tanh[20\sin(x/10)]\exp(iKx)$, where $K = v_0/2A$, $A = 5$ and $v_0 = 5$. We clearly see oscillations and wave turbulence in the time-dependent solution presented in Fig. 2. Two very clear and long-lived dark solitons are visible, associated with a positive potential of $\varphi \approx 3$, which is consistent with the quasi-stationary solution of Fig. 1 for $A = 5$. Hence, the dark solitons seem to be robust structures that can withstand perturbations and turbulence during a considerable time.

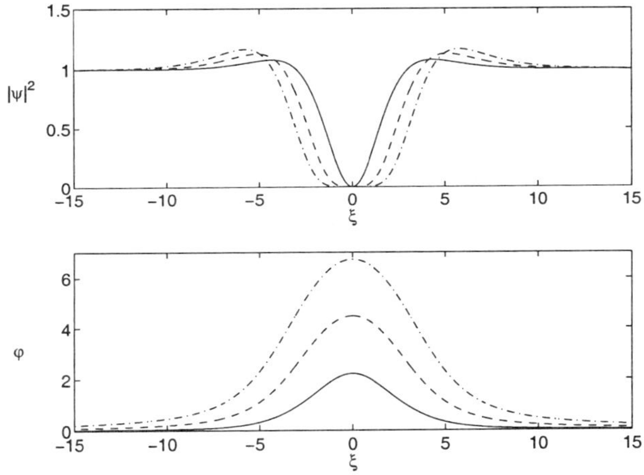

Fig. 3. The electron density $|\Psi|^2$ (upper panel) and electrostatic potential φ (lower panel) associated with a two-dimensional vortex supported by the system (7) and (8), for the charge states $n = 1$ (solid lines), $n = 2$ (dashed lines) and $n = 3$ (dash-dotted lines). We used $A = 5$ in all cases. After Ref. 4.

We next consider two-dimensional vortex structures of the form $\Psi = \psi(r)\exp(in\theta - i\Omega t)$, where r and θ are the polar coordinates defined via $x = r\cos(\theta)$ and $y = r\sin(\theta)$, Ω is a constant frequency shift, and $n = 0, \pm 1, \pm 2, \ldots$ for different excited states (charge states). With this, we can write Eqs. (1) and (2) in the form

$$\Omega\psi + A\left(\frac{d^2}{dr^2} + \frac{1}{r}\frac{d}{dr} - \frac{n^2}{r^2}\right)\psi + \varphi\psi - |\psi|^4\psi = 0, \tag{7}$$

and

$$\left(\frac{d^2}{dr^2} + \frac{1}{r}\frac{d}{dr}\right)\varphi = |\psi|^2 - 1, \tag{8}$$

respectively, where the boundary conditions $\psi = 1$ and $\varphi = d\psi/dr = 0$ at $r = \infty$ determine $\Omega = 1$. Different signs of n describe different rotation

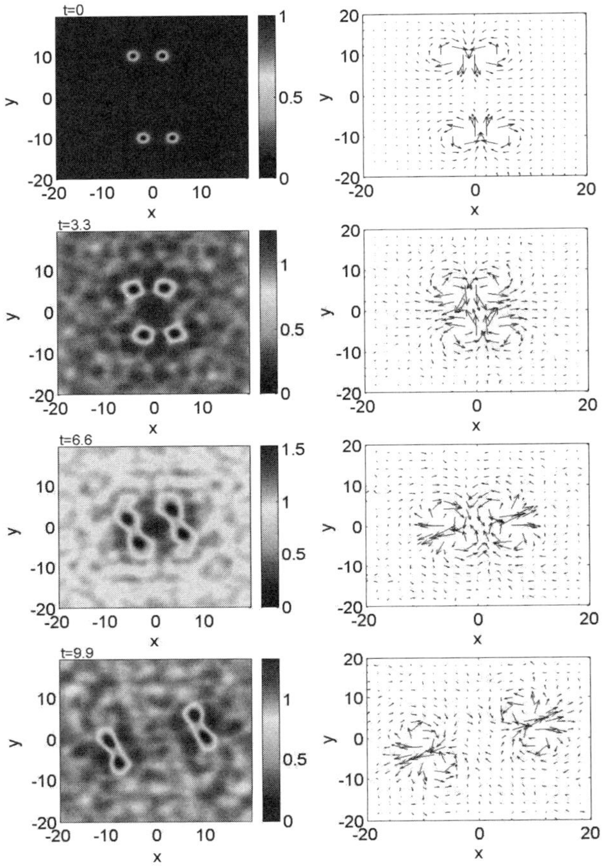

Fig. 4. The electron density $|\Psi|^2$ (left panel) and an arrow plot of the electron current $i\,(\Psi\nabla\Psi^* - \Psi^*\nabla\Psi)$ (right panel) associated with singly charged ($n = 1$) two-dimensional vortices, obtained from a simulation of the time-dependent system of equations (1) and (2), at times $t = 0$, $t = 3.3$, $t = 6.6$ and $t = 9.9$ (upper to lower panels). We used $A = 5$. The singly charged vortices form pairs and keep their identities. After Ref. 4.

directions of the vortex. For $n \neq 0$, we must have $\psi = 0$ at $r = 0$, and from symmetry considerations we have $d\varphi/dr = 0$ at $r = 0$. In Fig. 3, we display numerical solutions of Eqs. (7) and (8) for different charge states n and for $A = 5$. We see that the vortex is characterized by a complete depletion of the electron density at the core of the vortex, and is associated with a positive electrostatic potential. In order to assess the stability of the vortices, we have numerically solved the time-dependent system of Eqs. (1)

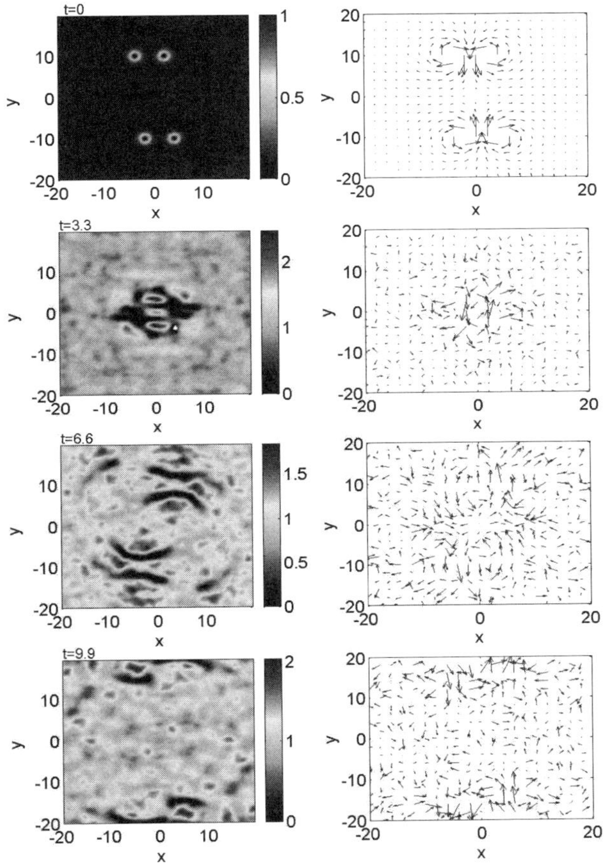

Fig. 5. The electron density $|\Psi|^2$ (left panel) and an arrow plot of the electron current $i\,(\Psi\nabla\Psi^* - \Psi^*\nabla\Psi)$ (right panel) associated with double charged ($n = 2$) two-dimensional vortices, obtained from a simulation of the time-dependent system of Eqs. (1) and (2), at times $t = 0$, $t = 3.3$, $t = 6.6$ and $t = 9.9$ (upper to lower panels). We used $A = 5$. The doubly charged vortices dissolve into nonlinear structures and wave turbulence. After Ref. 4.

and (2) in two-space dimensions for singly charged vortices and presented our results in Fig. 4. We have placed four vortex-like structures at some distance from each other, by the initial condition $\Psi = f_1 f_2 f_3 f_4$, where $f_j = \tanh[\sqrt{(x-x_j)^2 + (y-y_j)^2}]\exp[+in\arg(x-x_j, y-y_j)]$. Here $(x_1, y_1) = (-4, 10)$, $(x_2, y_2) = (2, 10)$, $(x_3, y_3) = (-2, -10)$, and $(x_4, y_4) = (4, -10)$. The function $\arg(x, y)$ denotes the angle between the x axis and the point

(x, y), and it takes values between $-\pi$ and π. The initial conditions are such that the vortices are organized in two vortex pairs, as seen in the upper panels of Fig. 4. The vortices in the pairs have opposite polarity on the rotation, as seen in the electron fluid rotation direction in the upper right panel. The time-development of the system exhibits that the "partners" in the vortex pairs attract each other and propagate together with a constant velocity. When the two vortex pairs collide and interact (see the second and third pairs of panels in Fig. 4), the vortices keep their identities and change partners in a manner of asymptotic freedom, resulting into two new vortex pairs which propagate obliquely to the original propagation direction. For vortices that are multiply charged ($|n| > 1$), we have a breakup of the vortices and the formation of quasi one-dimensional dark solitons and pairs of vortices with single charge states. One such example is shown in Fig. 5, where we have simulated the system of Eqs. (1) and (2), with the same initial condition as the one in Fig. 4, except that we here have taken $n = 2$ to make the vortices doubly charged. The second row of panels in Fig. 5 reveals that the vortex pairs keep their identities for some time, while a quasi one-dimensional density cavity is formed between the two vortex pairs. At a later stage, the four vortices dissolve into complicated nonlinear structures and wave turbulence. Hence, the nonlinear dynamics is very different between singly and multiply charged solitons, where only singly charged vortices are long-lived and keep their identities. This is in line with previous results on the nonlinear Schrödinger equation, where it was noted that vortices with higher charge states are unstable.[42] In the numerical simulations of Eqs. (1) and (2), we used a pseudo-spectral method to approximate the x and y derivatives and a fourth-order Runge-Kutta scheme for the time-stepping. The numerical simulations confirmed the conservation laws of the electron number, momentum and energy up to the accuracy of the numerical scheme. The numerical solutions of the time-independent systems (4)–(5) and (7)–(8) were obtained by using the Newton method, where the ξ derivatives were approximated with a second-order centered difference scheme with appropriate boundary conditions on Ψ and φ.

3. Turbulence in quantum plasmas

In this Section, we use the coupled NLS and Poisson equations for investigating, by means of computer simulations, the properties of 2D electron fluid turbulence and associated electron transport in quantum plasmas.[43] We find that the nonlinear coupling between the EPOs of different scale sizes gives rise to small-scale electron density structures, while the electrostatic potential cascades towards large-scales. The total energy associated with our quantum electron plasma turbulence, nonetheless, processes a characteristic spectrum, which is a *non-* Kolmogorov-like. The electron diffusion caused by the electron fluid turbulence is consistent with the dynamical evolution of turbulent mode structures.

For our 2D turbulence studies, we use the nonlinear Schrödinger-Poisson equations[2,4]

$$i\sqrt{2H}\frac{\partial\Psi}{\partial t} + H\nabla^2\Psi + \varphi\Psi - |\Psi|^2\Psi = 0, \tag{9}$$

and

$$\nabla^2\varphi = |\Psi|^2 - 1, \tag{10}$$

which are valid at zero electron temperature for the Fermi-Dirac equilibrium distribution, and which govern the dynamics of nonlinearly interacting EPOs of different wavelengths. In Eqs. (9) and (10) the wave function Ψ is normalized by $\sqrt{n_0}$, the electrostatic potential φ by k_BT_F/e, the time t by the electron plasma period ω_{pe}^{-1}, and the space $\mathbf{r}$ by the Fermi Debye radius λ_D. We have introduced the notations $\lambda_D = (k_BT_F/4\pi n_0e^2)^{1/2} \equiv V_F/\omega_{pe}$ and $\sqrt{H} = \hbar\omega_{pe}/\sqrt{2}k_BT_F$, where the Fermi electron temperature $k_BT_F = (\hbar^2/2m_e)(3\pi^2)^{1/3}n_0^{2/3}$, e is magnitude of the electron charge, and $\omega_{pe} = (4\pi n_0e^2/m_e)^{1/2}$ is the electron plasma frequency. The origin of the various terms in Eq. (9) is obvious. The first term is due to the electron inertia, the H-term in (9) is associated from the quantum tunneling involving the Bohm potential, $\varphi\Psi$ comes from the nonlinear coupling between the scalar potential (due to the space charge electric field) and the electron wave function, and the cubic nonlinear term is the contribution of the electron pressure[2] for the Fermi plasma that has a quantum statistical equation of state.

Equations (9) and (10) admit a set of conservation laws,[44] including the number of electrons $N = \int \Psi^2 dxdy$, the electron momentum $\mathbf{P} = -i\int \Psi^*\nabla\Psi dxdy$, the electron angular momentum $\mathbf{L} = -i\int \Psi^*\mathbf{r}\times\nabla\Psi dxdy$, and the total energy $\mathcal{E} = \int[-\Psi^*H\nabla^2\Psi + |\nabla\varphi|^2/2 + |\Psi|^3/2]dxdy$. In obtaining the total energy $\mathcal{E}$, we have used the relation $\partial\mathbf{E}/\partial t = iH(\Psi\nabla\Psi^* - \Psi^*\nabla\Psi)$, where the electric field $\mathbf{E} = -\nabla\varphi$. The conservations laws are used to maintain the accuracy of the numerical integration of Eqs. (9) and (10), which hold for quantum electron-ion plasmas with fixed ion background. The assumption of immobile ions is valid, since the EPOs (given by the dispersion relation[2,3] $\omega^2 = \omega_{pe}^2 + k^2V_F^2 + \hbar^2k^4/4m_e^2$) occur on the electron plasma period, which is much shorter than the ion plasma period ω_{pi}^{-1}. Here

ω and k are the frequency and the wave-number, respectively. The ion dynamics, which may become important in the nonlinear phase on a longer timescale (say of the order of ω_{pi}^{-1}), in our investigation can easily be incorporated by replacing 1 in Eq. (10) by n_i, where the normalized (by n_0) ion density n_i is determined from $d_t n_i + n_i \nabla \cdot \mathbf{u}_i = 0$ and $d_t \mathbf{u}_i = -C_s^2 \nabla \varphi$, where $d_t = (\partial/\partial t) + \mathbf{u}_i \cdot \nabla$, $\mathbf{u}_i$ is the ion velocity, $C_s = (T_F/m_i)^{1/2}$ is the ion sound speed, and m_i is the ion mass.

The nonlinear mode coupling interaction studies are performed to investigate the multi-scale evolution of a decaying 2D electron fluid turbulence, which is described by Eqs. (9) and (10). All the fluctuations are initialized isotropically (no mean fields are assumed) with random phases and amplitudes in Fourier space, and evolved further by the integration of Eqs. (9) and (10), using a fully de-aliased pseudospectral numerical scheme[45] based on the Fourier spectral methods. The spatial discretization in our 2D simulations uses a discrete Fourier representation of turbulent fluctuations. The numerical algorithm employed here conserves energy in terms of the dynamical fluid variables and not due to a separate energy equation written in a conservative form. The evolution variables use periodic boundary conditions. The initial isotropic turbulent spectrum was chosen close to k^{-2}, with random phases in all three directions. The choice of such (or even a flatter than -2) spectrum treats the turbulent fluctuations on an equal footing and avoids any influence on the dynamical evolution that may be due to the initial spectral non-symmetry. The equations are advanced in time using a second-order predictor-corrector scheme. The code is made stable by a proper de-aliasing of spurious Fourier modes, and by choosing a relatively small time step in the simulations. Our code is massively parallelized using Message Passing Interface (MPI) libraries to facilitate higher resolution in a 2D computational box, with a resolution of 512^2 grid points.

We study the properties of 2D fluid turbulence, composed of nonlinearly interacting EPOs, for two specific physical systems. These are the dense plasmas in the next generation laser-based plasma compression (LBPC) schemes[10] as well as in superdense astrophysical objects[14,15,26] (e.g. white dwarfs). It is expected that in LBPC schemes, the electron number density may reach 10^{27} cm^{-3} and beyond. Hence, we have $\omega_{pe} = 1.76 \times 10^{18}$ s^{-1}, $k_B T_F = 1.7 \times 10^{-9}$ erg, $\hbar\omega_{pe} = 1.7 \times 10^{-9}$ erg, and $H = 1$. The Fermi Debye length $\lambda_D = 0.1$ Å. On the other hand, in the interior of white dwarfs, we typically have[46] $n_0 \sim 10^{30}$ cm^{-3} (such values are also common in dense neutron stars and supernovae), yielding $\omega_{pe} = 5.64 \times 10^{19}$ s^{-1}, $k_B T_F = 1.7 \times 10^{-7}$ erg, $\hbar\omega_{pe} = 5.64 \times 10^{-8}$ erg, $H \approx 0.3$, and $\lambda_D = 0.025$ Å.

The numerical solutions of Eqs. (9) and (10) for $H = 1$ and $H = 0.025$ (corresponding to $n_0 = 10^{27}$ cm^{-3} and $n_0 = 10^{30}$ cm^{-3}, respectively) are displayed in Figs. 6 and 7, respectively, which are the electron number density and electrostatic (ES) potential distributions in the (x, y)-plane.

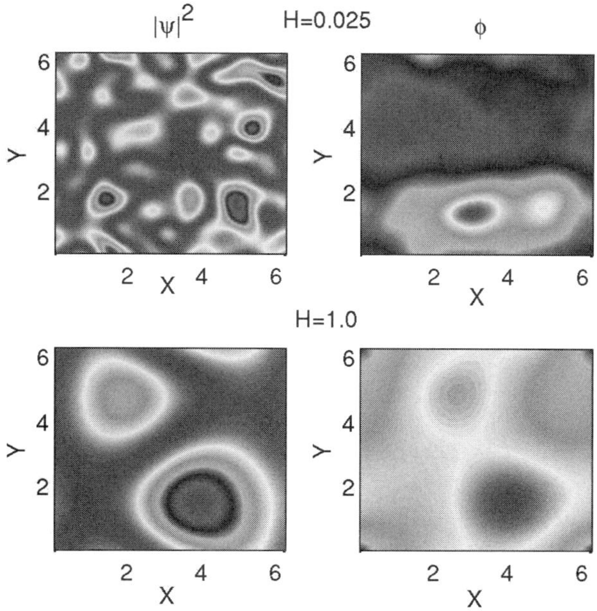

Fig. 6. Small scale fluctuations in the electron density resulted from a steady turbulence simulations of our 2D electron plasma. Forward cascades are responsible for the generation of small-scale fluctuations. Large scale structures are present in the electrostatic potential, essentially resulting from an inverse cascade. The 2D electron fluid turbulence interestingly relaxes towards an Iroshnikov-Kraichnan (IK) type $k^{-3/2}$ spectrum in a dense plasma for $H = 1$ as shown in the next figure. After Ref. 43.

Figures 6 and 7 reveal that the electron density distribution has a tendency to generate smaller length-scale structures, while the ES potential cascades towards larger scales. The co-existence of the small and larger scale structures in turbulence is a ubiquitous feature of various 2D turbulence systems. For example, in 2D hydrodynamic turbulence, the incompressible fluid admits two invariants, namely the energy and the mean squared vorticity. The two invariants, under the action of an external forcing, cascade simultaneously in turbulence, thereby leading to a dual cascade phenomena. In these processes, the energy cascades towards longer length-scales, while

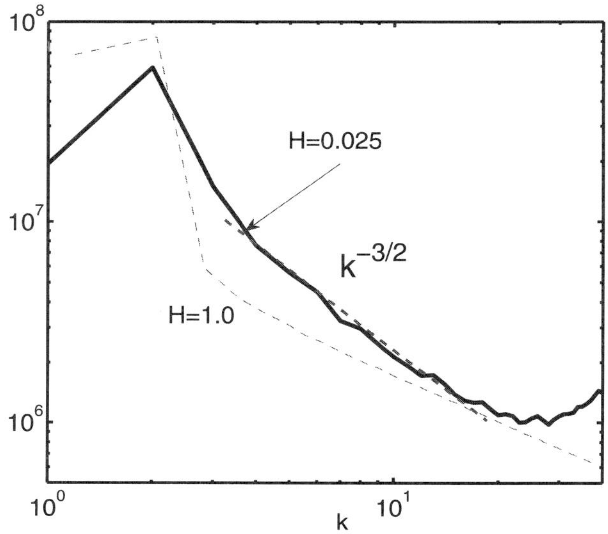

Fig. 7. The 2D electron fluid turbulence interestingly relaxes towards an Iroshnikov-Kraichnan (IK) type $k^{-3/2}$ spectrum in a dense plasma for $H = 1$. $H = 0.025$ results in a flat spectrum. After Ref. 43.

the fluid vorticity transfers spectral power towards shorter length-scales. Usually, a dual cascade is observed in a driven turbulence simulation, in which certain modes are excited externally through random turbulent forces in spectral space. The randomly excited Fourier modes transfer the spectral energy by conserving the constants of motion in k-space. On the other hand, in freely decaying turbulence, the energy contained in the large-scale eddies is transferred to the smaller scales, leading to a statistically stationary inertial regime associated with the forward cascades of one of the invariants. Decaying turbulence often leads to the formation of coherent structures as turbulence relaxes, thus making the nonlinear interactions rather inefficient when they are saturated. The power spectrum exhibits an interesting feature in our 2D electron plasma system, unlike the 2D hydrodynamic turbulence.[47–49] The spectral slope in the 2D quantum electron fluid turbulence is close to the Iroshnikov-Kraichnan power law[50,51] $k^{-3/2}$, rather than the usual Kolomogrov power law[47] $k^{-5/3}$. We further find that this scaling is not universal and is determined critically by the quantum tunneling effect. For instance, for a higher value of H=1.0 the spectrum becomes more flat (see Fig 7). Physically, the flatness (or deviation from the $k^{-5/3}$), results from the short wavelength part of the EPOs spectrum which is controlled by the quantum tunneling effect associated with the Bohm potential. The peak in the energy spectrum can be attributed to the higher turbulent power residing in the EPO potential, which eventually leads to the generation of larger scale structures, as the total energy encompasses both the electrostatic potential and electron density components. In our dual cascade process, there is a delicate competition between the EPO dispersions caused by the statistical pressure law (giving the $k^2V_F^2$ term, which dominates at longer scales) and the quantum Bohm potential (giving the $\hbar^2k^4/4m_e^2$ term, which dominates at shorter scales with respect to a source). Furthermore, it is interesting to note that exponents other than $k^{-5/3}$ have also been observed in numerical simulations[52,53] of the Charney and 2D incompressible Navier-Stokes equations.

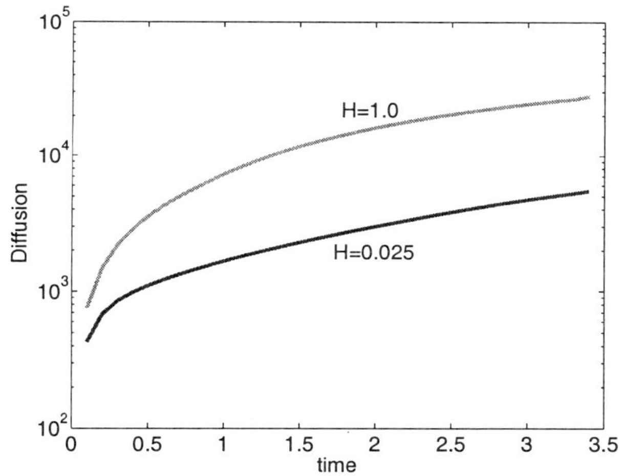

Fig. 8. Time evolution of an effective electron diffusion coefficient associated with the large-scale electrostatic potential and the small-scale electron density. Here a comparison between $H = 1$ and $H = 0.025$ is shown. After Ref. 43.

We finally estimate the electron diffusion coefficient in the presence of small and large scale turbulent EPOs in our quantum plasma. An effective electron diffusion coefficient caused by the momentum transfer can be calculated from $D_{eff} = \int_0^\infty \langle \mathbf{P}(\mathbf{r},t) \cdot \mathbf{P}(\mathbf{r},t+t') \rangle dt'$, where $\mathbf{P}$ is electron momentum and the angular bracket denotes spatial averages and the ensemble averages are normalized to unit mass. Since the 2D structures are confined to a $x-y$ plane, the effective electron diffusion coefficient, D_{eff}, essentially relates the diffusion processes associated with random translational motions of the electrons in nonlinear plasmonic fields. We compute D_{eff} in our simulations, to measure the turbulent electron transport that is associated with the turbulent structures that we have reported herein. It is observed that the effective electron diffusion is lower when the field perturbations are Gaussian. On the other hand, the electron diffusion increases rapidly with the eventual formation of longer length-scale structures, as shown in Fig. 8. The electron diffusion due to large scale potential distributions in quantum plasmas dominates substantially, as depicted by the solid-curve in Fig. 8. Furthermore, in the steady-state, nonlinearly coupled EPOs form stationary structures, and D_{eff} saturates eventually. Thus, remarkably an enhanced electron diffusion results primarily due to the emergence of large-scale potential structures in our 2D quantum plasma.

4. Interaction between intense electromagnetic waves and quantum plasma oscillations

In this section, we discuss the nonlinear interaction between intense electromagnetic radiation and quantum plasma oscillations.[54] We consider a one-dimensional geometry of an unmagnetized dense electron-ion plasma, in which immobile ions form the neutralizing background. Thus, we are investigating the phenomena on a timescale shorter than the ion plasma period. Our dense quantum plasma contains an intense circularly polarized electromagnetic (CPEM) plane wave that nonlinearly interacts with EPOs. The nonlinear interaction between intense CPEM waves and EPOs gives rise to an envelope of the CPEM vector potential $\mathbf{A}_\perp = A_\perp(\hat{\mathbf{x}} + i\hat{\mathbf{y}})\exp(-i\omega_0 t + ik_0 z)$, which obeys the nonlinear Schrödinger equation[40]

$$2i\Omega_0\left(\frac{\partial}{\partial t} + V_g\frac{\partial}{\partial z}\right)A_\perp + \frac{\partial^2 A_\perp}{\partial z^2} - \left(\frac{|\psi|^2}{\gamma} - 1\right)A_\perp = 0, \tag{11}$$

where the electron wave function ψ and the scalar potential are governed by, respectively,

$$iH_e\frac{\partial\psi}{\partial t} + \frac{H_e^2}{2}\frac{\partial^2\psi}{\partial z^2} + (\phi - \gamma + 1)\psi = 0, \tag{12}$$

and

$$\frac{\partial^2\phi}{\partial z^2} = |\psi|^2 - 1, \tag{13}$$

where $\Omega_0 = \omega_0/\omega_{pe}$, $V_g = v_g/c$, $H_e = \hbar\omega_{pe}/mc^2$, $v_g = k_0c^2/\omega_0$ is the group velocity of the CPEM waves, and $\gamma = (1 + |A_\perp|^2)^{1/2}$ is the relativistic gamma factor due to the electron quiver velocity in the CPEM wave fields. Furthermore, $\omega_0 = (k_0^2 c^2 + \omega_{pe}^2)^{1/2}$ is the CPEM wave frequency, k_0 is the wavenumber, c is the speed of light in vacuum, $\omega_{pe} = (4\pi n_0 e^2/m)^{1/2}$ is the electron plasma frequency, e is the magnitude of the electron charge, n_0 is the equilibrium electron number density, and m is the electron rest mass. In (11)–(13) the time and space variables are normalized by the inverse electron plasma frequency ω_{pe}^{-1} and skin depth $\lambda_e = c/\omega_{pe}$, respectively, the scalar potential ϕ by mc^2/e, the vector potential $A_\perp$ by mc^2/e, and the electron wave function $\psi(z,t)$ by $n_0^{1/2}$. The nonlinear coupling between intense CPEM waves and EPOs comes about due to the nonlinear current density, which is represented by the term $|\psi|^2 A_\perp/\gamma$ in Eq. (11). The electron number density is defined as $n_e = \psi\psi^* = |\psi|^2$, where the asterisk denotes the complex conjugate. In Eq. (12), $1-\gamma$ is the relativistic ponderomotive potential,[40] which arises due to the cross-coupling between the CPEM wave-induced electron quiver velocity and the CPEM wave magnetic field. The second term in the left-hand side in (12) is associated with the quantum Bohm potential.[5]

It is well known[55] that a relativistically strong electromagnetic wave in a classical electron plasma is subjected to the Raman scattering and modulational instabilities. At quantum scales, these instabilities will be modified by the dispersive effects caused by the tunnelling of the electrons. In order to investigate the quantum mechanical effects on the relativistic parametric instabilities in a dense quantum plasma in the presence of a relativistically strong CPEM pump wave, we let $\phi(z,t) = \phi_1(z,t)$, $A_\perp(z,t) = [A_0 + A_1(z,t)]\exp(-i\alpha_0 t)$ and $\psi(z,t) = [1+\psi_1(z,t)]\exp(-i\beta_0 t)$, where A_0 is the large-amplitude CPEM pump and A_1 is the small-amplitude fluctuations of the CPEM wave amplitude due to the nonlinear coupling between CPEM waves and EPOs, i.e. $|A_1| \ll |A_0|$, and ψ_1 ($\ll 1$) is the small-amplitude perturbations in the electron wave function. The constant frequency shifts, determined from Eqs. (11) and (12), are $\alpha_0 = (1/\gamma_0 - 1)/(2\Omega_0)$ and $\beta_0 = (1-\gamma_0)/H_e$, where $\gamma_0 = (1+|A_0|^2)^{1/2}$. The first-order perturbations in the electromagnetic vector potential and the electron wave function are expanded into their respective sidebands as $A_1(z,t) = A_+ \exp(iKz - i\Omega t) + A_- \exp(-iKz + i\Omega t)$ and $\psi_1(z,t) = \psi_+ \exp(iKz - i\Omega t) + \psi_- \exp(-iKz + i\Omega t)$, while the potential is expanded as $\phi(z,t) = \widehat{\phi}\exp(iKz - i\Omega t) + \widehat{\phi}^* \exp(-iKz + i\Omega t)$, where Ω and K are the frequency and wave number of the electron plasma oscillations, respectively. Inserting the above mentioned Fourier ansatz into Eqs. (11)–(13), linearizing the resultant system of equations, and sorting into equations for different Fourier modes, we obtain the nonlinear dispersion relation

$$1-\left(\frac{1}{D_+}+\frac{1}{D_-}\right)\left(1+\frac{K^2}{D_L}\right)\frac{|A_0|^2}{2\gamma_0^3}=0, \tag{14}$$

where $D_\pm = \mp 2\Omega_0(\Omega - V_g K) + K^2$ and $D_L = 1 + H_e^2 K^4/4 - \Omega^2$. We note that $D_L = 0$ yields the linear dispersion relation $\Omega^2 = 1 + H_e^2 K^4/4$ for the EPOs in a dense quantum plasma.[1] For $H_e \to 0$ we recover from (14) the nonlinear dispersion relation for relativistically large amplitude electromagnetic waves in a classical electron plasma.[55] The dispersion relation (14) governs the Raman backward and forward scattering instabilities, as well as the modulational instability. In the long wavelength limit $V_g \ll 1$, $\Omega_0 \approx 1$ we introduce the ansatz $\Omega = i\Gamma$, where the normalized (by ω_{pe}) growth rate $\Gamma \ll 1$, and obtain from Eq. (14) the growth rate $\Gamma = (1/2)|K|\{(|A_0|^2/\gamma_0^3)[1 + K^2/(1 + H_e^2 K^4/4)] - K^2\}^{1/2}$ of the modulational instability. For $|K| < 1$ and $H_e < 1$, the linear growth rate is only weakly depending on the quantum parameter H_e. However, possible nonlinear saturation of the modulational instability may lead to localized CPEM wave packets, which are trapped in a quantum electron hole. Such localized electromagnetic wavepackets would have length scales much shorter than those involved in the modulational instability process. Here quantum diffraction effects (associated with the quantum Bohm potential) become very important. In order to investigate the quantum diffraction effect on such localized electromagnetic pulses, we consider a steady state structure moving with a constant speed V_g. Inserting the ansatz $A_\perp = W(\xi)\exp(-i\Omega t)$, $\psi = P(\xi)\exp(ikx - i\omega t)$ and $\phi = \phi(\xi)$ into Eqs. (11)–(13), where $\xi = z - V_g t$, $k = V_g/H_e$ and $\omega = V_g^2/2H_e$, and where $W(\xi)$ and $P(\xi)$ are real, we obtain from (11)-(13) the coupled system of equations

$$\frac{\partial^2 W}{\partial \xi^2}+\left(\lambda-\frac{P^2}{\gamma}+1\right)W=0, \tag{15}$$

$$\frac{H_e^2}{2}\frac{\partial^2 P}{\partial \xi^2}+(\phi-\gamma+1)P=0, \tag{16}$$

where $\gamma = (1+W^2)^{1/2}$, and

$$\frac{\partial^2 \phi}{\partial \xi^2}=P^2-1, \tag{17}$$

with the boundary conditions $W = \Phi = 0$ and $P^2 = 1$ at $|\xi| = \infty$. In Eq. (15), $\lambda = 2\Omega_0\Omega$ represents a nonlinear frequency shift of the CPEM wave.

In the limit $H_e \to 0$, we have from (16) $\phi = \gamma - 1$, where $P \neq 0$, and we recover the classical (non-quantum) case of the relativistic solitary waves in a cold plasma.[56] We note that the system of equations (15)–(17) admits a Hamiltonian

$$\begin{aligned} Q_H =& \frac{1}{2}\left(\frac{\partial W}{\partial \xi}\right)^2 + \frac{H_e^2}{2}\left(\frac{\partial P}{\partial \xi}\right)^2 - \frac{1}{2}\left(\frac{\partial \phi}{\partial \xi}\right)^2 \\ &+ \frac{1}{2}(\lambda + 1)W^2 + P^2 - \gamma P^2 + \phi P^2 - \phi = 0, \end{aligned} \tag{18}$$

where we have used the boundary conditions $\partial/\partial\xi = 0$, $W = \phi = 0$ and $|P| = 1$ at $|\xi| = \infty$.

In order to asses the importance of our investigation, we now present numerical solutions of (8)–(13) and (15)–(17), ensuring that (18) is conserved. We chose parameters that are representative of the next generation laser-based plasma compression (LBPC) schemes.[10,11] The formula[40] $eA_\perp/mc^2 = 6 \times 10^{-10}\lambda_s\sqrt{I}$ will determine the normalized vector potential, provided that the CPEM wavelength λ_s (in microns) and intensity I (in W/cm^2) are known. It is expected that in LBPC schemes, the electron number density n_0 may reach 10^{27} cm^{-3} and beyond, and the peak values of $eA_\perp/mc^2$ may be in the range 1-2 (e.g. for focused EM pulses with $\lambda_s \sim 0.15$ nm and $I \sim 5 \times 10^{27}$ W/cm^2). For $\omega_{pe} = 1.76 \times 10^{18}$ s^{-1}, we have $\hbar\omega_{pe} = 1.76 \times 10^{-9}$ erg and $H_e = 0.002$, since $mc^2 = 8.1 \times 10^{-7}$ erg. The electron skin depth $\lambda_e \sim 1.7$ Å. On the other hand, a higher value of $H_e = 0.007$ is achieved for $\omega_{pe} = 5.64 \times 10^{18}$ s^{-1}. Thus, our numerical solutions below, based on these two values of H_e, have focused on scenarios that are relevant for the next generation intense laser-solid density plasma interaction experiments.[10]

We first numerically solved Eqs. (15)–(17) for several values of H_e. Here, we solved the nonlinear boundary value problem with the boundary conditions $W = \phi = 0$ and $P = 1$ at the boundaries at $\xi = \pm 10$. We used centered second-order approximations for the second derivatives and solved the obtained nonlinear system of equations numerically by using the Newton method. The results are displayed in Figs. 9 and 10. We see that the solitary envelope pulse is composed of a single maximum of the localized vector potential W and a local depletion of the electron density P^2, and a localized positive potential ϕ at the center of the solitary pulse. The latter has a continuous spectrum in λ, where larger values of negative λ are associated with larger amplitude solitary EM pulses. At the center of the solitary EM pulse, the electron density is partially depleted, as in panels a) of Fig. 9, and for larger amplitudes of the EM waves we have stronger depletion of the electron density, as shown in panels b) and c) of Fig. 9. For cases where the electron density goes to almost zero in the classical case,[56] one important quantum effect is that the electrons can tunnel into the depleted region. This is seen in Fig. 10, where the electron density remains nonzero for the larger value of H_e in panels a), while the density shrinks to zero for the smaller value of H_e in panel b).

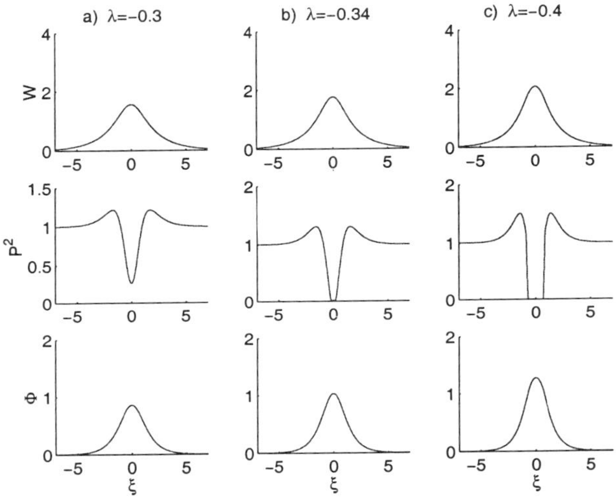

Fig. 9. The profiles of the CPEM vector potential $A_\perp$, the electron number density and the scalar potential (upper to lower rows of panels) for $\lambda = -0.3$, $\lambda = -3.4$ and $\lambda = -0.4$, with $H_e = 0.002$. After Ref. 54.

In order to investigate the quantum diffraction effects on the dynamics of localized CPEM wavepackets, we have solved the system of Eqs. (11)–(13) numerically. We considered the long-wavelength limit $\omega_0 \approx 1$ and $V_g \approx 0$. In the initial conditions, we use an EM pump with a constant amplitude $A_\perp = A_0 = 1$ and a uniform plasma density $\psi = 1$. A small amplitude noise (random numbers) of order 10^{-2} is added to $A_\perp$ to give a seed for any instability. The numerical results are displayed in Figs. 11 and 4 for $H_e = 0.002$ and $H_e = 0.007$, respectively. In both cases, we see an initial linear growth phase and a wave collapse at $t \approx 70$, in which almost all the CPEM wave energy is contracted into a few well separated localized CPEM wave pipes. These are characterized by a large bell-shaped amplitude of the CPEM wave, an almost complete depletion of the electron number density at the center of the CPEM wavepacket, and a large-amplitude positive

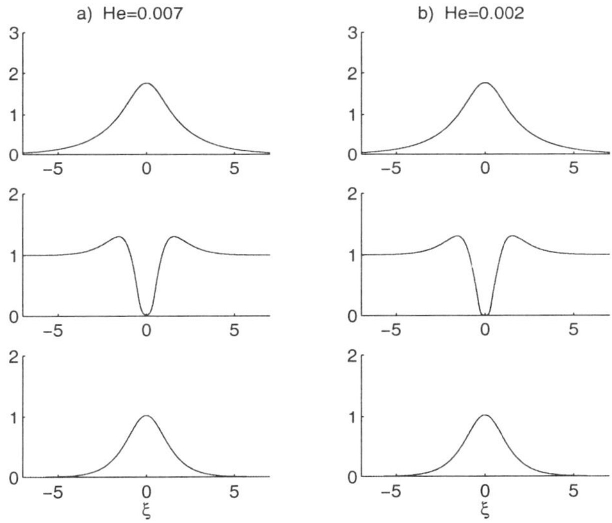

Fig. 10. The profiles of the CPEM vector potential $A_\perp$, the electron number density and the scalar potential (upper to lower rows of panels) for $H_e = 0.007$ and $H_e = 0.002$, with $\lambda = -0.34$. After Ref. 54.

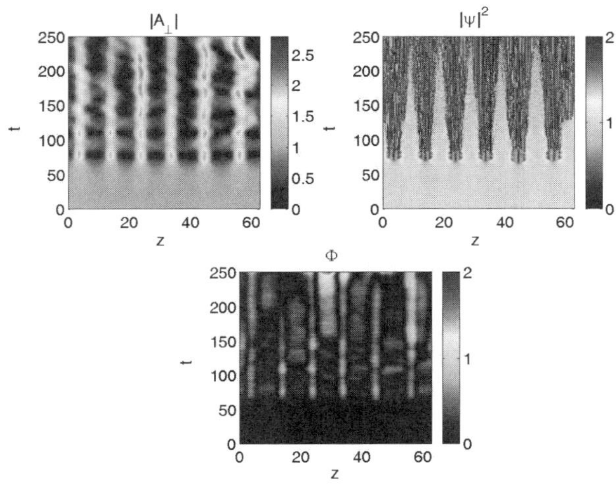

Fig. 11. The dynamics of the CPEM vector potential $A_\perp$ and the electron number density $|\psi|^2$ (upper panels) and of the electrostatic potential Φ (lower panel) for $H_e = 0.002$. After Ref. 54.

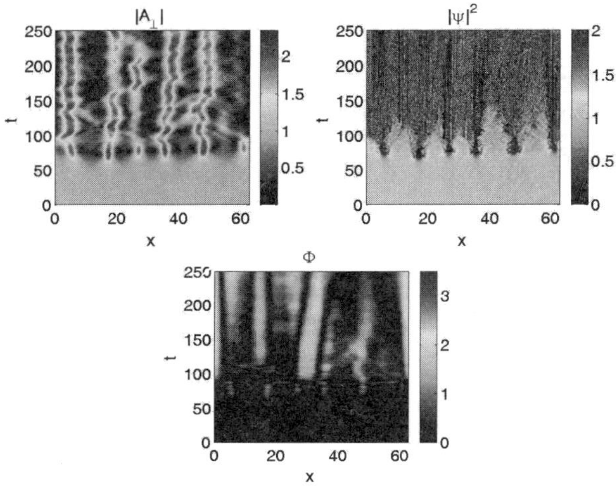

Fig. 12. The dynamics of the CPEM vector potential $A_\perp$ and the electron number density $|\psi|^2$ (upper panels) and the electrostatic potential ϕ (lower panel) for $H_e = 0.007$. After Ref. 54.

electrostatic potential. Comparing Fig. 11 with Fig. 12, we see that there is a more complex dynamics in the interaction between the CPEM wavepackets for the larger $H_e = 0.007$, shown in Fig. 12, in comparison with $H_e = 0.002$, shown in Fig. 11, where the wavepackets are almost stationary when they are fully developed. We have here neglected the effects of the ion dynamics. The latter may be important for the development of expanding plasma bubbles (cavities) on longer timescales (e.g. the ion plasma period).[57]

5. Conclusions

In summary, we have demonstrated the existence of localized nonlinear structures in quantum electron plasmas. The electron dynamics in the latter is governed by a coupled nonlinear Schrödinger and Poisson system of equations, which admit a set of conserved quantities (the total number of electrons, the electron momentum, the electron angular momentum, and the electron energy). The latter were checked numerically. Quasi-stationary, localized structures in the form of one-dimensional dark solitons and two-dimensional vortices were found by solving the time-independent coupled system of equations numerically. These structures are associated with a local depletion of the electron density associated with positive electrostatic potential, and are parameterised by the quantum coupling parameter only.

In the two-dimensional geometry, we have a class of vortices of different excited states (charge states) associated with a complete depletion of the electron density and an associated positive potential. The numerical simulation of the time-dependent system of equations shows the formation of stable dark solitons in one-space dimension with an amplitude consistent with the one found from the time-independent solutions. In two-space dimensions, the dark solitons of the first excited state were found to be stable and the preferred nonlinear state was in the form of vortex pairs of vortices with different polarities. One-dimensional dark solitons and singly charge two-dimensional vortices are thus long-lived nonlinear structures, which can transport information at quantum scales in micro-mechanical systems and dense laboratory plasmas. We have presented computer simulation studies of 2D fluid turbulence in a dense quantum plasma. Our simulations, for the parameters that are representative of the next generation intense laser-solid density plasma experiments as well as of the superdense astrophysical bodies, reveal new features of the dual cascade in a fully developed 2D electron fluid turbulence. Specifically, we find that the power spectrum associated with nonlinearly interacting EPOs in quantum plasmas follow a non-Kolmogorov-like spectrum. The deviation from a Kolmogorov-like spectrum resulting from the flattening of the spectrum is mediated essentially by the nonlinear EPOs interactions in the inertial range (basically controlled by the electron plasma wave dispersion effect represented by $\hbar^2 k^4/4m_e^2$), which impedes the spectral transfer of the turbulent power associated with the short scale Fourier modes. In the nonlinear regime, the inhibition of the spectral transfer is caused by short scale EPOs that are nonlinearly excited by the mode coupling of the EPOs in the forward cascade regime, which then grow, acquire nonlinear amplitudes, and eventually saturate in the nonlinear phase. We have also presented theoretical and computer simulation studies of nonlinearly interacting intense CPEM waves and EPOs in very dense quantum plasmas. The localized dark solitons, vortices, and CPEM wave structures, as discussed here, may be useful for information transfer as well as for electron acceleration in dense quantum plasmas.

References

1. D. Pines, *J. Nucl. Energy: Part C: Plasma Phys.* **2**, 5 (1961).
2. G. Manfredi and F. Haas, *Phys. Rev. B* **64**, 075316 (2001).
3. G. Manfredi, *Fields Inst. Commun.* **46**, 263 (2005).
4. P. K. Shukla and B. Eliasson, *Phys. Rev. Lett.* **96** 245001 (2006).
5. C. L. Gardner and C. Ringhofer, *Phys. Rev. E* **53**, 157 (1996).

6. M. Marklund and G. Brodin, *Phys. Rev. Lett.* **98**, 025001 (2007).
7. S. X. Hu and C. H. Keitel, *Phys. Rev. Lett.* **83**, 4709 (1999).
8. Y. A. Salamin *et al.*, *Phys. Rep.* **427**, 41 (2006).
9. S. H. Glenzer *et al.*, *Phys. Rev. Lett.* **98**, 065002 (2007).
10. V. M. Malkin *et al.*, *Phys. Rev. E* **75**, 026404 (2007).
11. H. Azechi *et al.*, *Plasma Phys. Control. Fusion* **48**, B267 (2006).
12. M. Opher *et al.*, *Phys. Plasmas* **8**, 2454 (2001).
13. O. G. Benvenuto and M. A. De Vito, *Mon. Not. R. Astron. Soc.* **362**, 891 (2005).
14. G. Chabrier *et al.*, *J. Phys.: Condens. Matter* **14**, 9133 (2002).
15. G. Chabrier *et al.*, *J. Phys. A: Math. Gen.* **39**, 4411 (2006).
16. Y.Y. Lau *et al.*, *Phys. Rev. Lett.* **66**, 1446 (1991).
17. L. K. Ang *et al.*, *Phys. Rev. Lett.* **91**, 208303 (2003).
18. L. K. Ang and P. Zhang, *Phys. Rev. Lett.* **98**, 164802 (2007).
19. E. P. Wigner, *Phys. Rev.* **40**, 749 (1932).
20. M. Hillery *et al.*, *Phys. Rep.* **106**, 121 (1984).
21. F. Haas, G. Manfredi, and M. Feix, *Phys. Rev. E* **62**, 2763 (2000).
22. D. Anderson *et al.*, *Phys. Rev. E* **65**, 046417 (2002).
23. F. Haas, L. G. Garcia, J. Goedert, and G. Manfredi, *Phys. Plasmas* **10**, 3858 (2003).
24. F. Haas, *Phys. Plasmas* **12**, 062117 (2005).
25. G. Mourou *et al.*, *Rev. Mod. Phys.* **78**, 309 (2006).
26. A. K. Harding and D. Lai, *Rep. Prog. Phys.* **69**, 2631 (2006).
27. G. V. Shpatakovskaya, *J. Exp. Teor. Phys.* **102**, 466 (2006).
28. W. L. Barnes *et al.*, *Nature (London)* **424**, 824 (2003).
29. D. E. Chang *et al.*, *Phys. Rev. Lett.* **97**, 053002 (2006).
30. P. A. Markowich *et al.*, *Semiconductor Equations* (Springer, Berlin, 1990).
31. K. H. Becker, K. H. Schoenbach, and J. G. Eden, *J. Phys. D: Appl. Phys.* **39**, R55 (2006).
32. M. Loffredo and L. Morato, *Nuovo Cimento Soc. Ital Fis. B* **108B**, 205 (1993).
33. R. Feynman, *Statistical Mechanics, A Set of of Lectures* (Benjamin, Reading, 1972).
34. A. Domps *et al.*, *Phys. Rev. Lett.* **80**, 5520 (1998).
35. P. Hohenberg and W. Kohn, *Phys. Rev.* **136**, B864 (1964).
36. W. Kohn and L. J. Sham, *Phys. Rev.* **140**, A1133 (1965).
37. L. Brey *et al.*, *Phys. Rev. B* **42**, 1240 (1990).
38. A. V. Andreev, *JETP Lett.* **72**, 238 (2000).
39. M. Marklund and P. K. Shukla, *Rev. Mod. Phys.* **78**, 591 (2006).
40. P. K. Shukla *et al.*, *Phys. Rep.* **138**, 1 (1986).
41. E. B. Kolomeisky *et al.*, *Phys. Rev. Lett.* **85**, 1146 (2000).
42. I. A. Ivonin, V. P. Pavlenko, and H. Persson, *Phys. Rev. E* **60**, 492 (1999).
43. D. Shaikh and P. K. Shukla, *Phys. Rev. Lett.* **99**, 125002 (2007).
44. M. Marklund and G. Brodin, *Phys. Rev. Lett.* **98**, 025001 (2007).
45. D. Gottlieb and S. A. Orszag, *Numerical Analysis of Spectral Methods* (SIAM, Philadelphia, 1977).

46. I. Iben Jr. and A. V. Tutukov, *Astrophys. J.* **282**, 615 (1984).
47. A. N. Kolmogorov, *C. R. Acad. Sci. USSR* **30**, 301 (1941).
48. M. Lesieur, *Turbulence in Fluids* (Kluwer, Dordrecht, 1990).
49. U. Frisch, *Turbulence* (Cambridge University Press, Cambridge, England, 1995).
50. P. Iroshnikov, *Sov. Astron.* **7**, 566 (1963).
51. R. H. Kraichnan, *Phys. Fluids* **8**, 1385 (1965).
52. V. D. Larichev and J. C. McWilliams, *Phys. Fluids A* **3** 938 (1991).
53. R. K. Scott, *Phys. Rev. E* **75**, 046301 (2007).
54. P. K. Shukla and B. Eliasson, *Phys. Rev. Lett.* **99**, 096401 (2007).
55. C. J. McKinstrie and R. Bingham, *Phys. Fluids B* **4**, 2626 (1992).
56. J. H. Marburger and R. F. Tooper, *Phys. Rev. Lett.* **35**, 1001 (1975).
57. M. Borghesi *et al.*, *Phys. Rev. Lett.* **88**, 135002 (2002).

QUANTUM, SPIN AND QED EFFECTS N PLASMAS

G. Brodin and M. Marklund

Department of Physics, Umeå University
SE-901 87 Umeå, Sweden
E-mail: gert.brodin@physics.umu.se

Plasmas are usually described using classical equations. While this is often a good approximation, where are situations when a quantum description is motivated. In this paper we will include several quantum effects, ranging from particle dispersion, which give raise to the so called Bohm potential, to spin effects, and to quantum electrodynamical effects. The later effects appears when the field strength approaches the Schwinger critical field, which may occur in for example astrophysical systems. Examples of how to model such quantum effects will be presented, and the phenomena resulting from these models will be discussed.

Keywords: Quantum plasmas; QED effects, Spin effects

1. Introduction

A characteristic feature of standard plasmas is the domination of collective forces over single particle forces. This scaling is equivalent to saying that there should be large number of particles in a Debye sphere. For this regime, as opposed to strongly coupled plasmas where the opposite condition holds, classical equations of motion is usually thought to be adequate. While this certainly applies to some of the wellknown quantum effects, it is not true in general, however. In this article we will focus on plasmas that are classical in the sense that there are large numbers of particles in a Debye sphere, which implies the importance of collective processes. But at the same time we will include a number of different quantum plasma effects, that have been of much interest recently see e.g. Refs.[1–13] The interest in quantum plasma effects has several different origins, for example recent progress in nanoscale technology,[14] various astrophysical applications,[15–17] high intensity effects made relevant by the continuous increase of laser powers,[18,19] as well as a general theoretical motives.[20–24] As indicated above, the combined focus on collective and quantum plasma effects are to some extent contradictory, as in many cases these effects are important in different regimes. Nevertheless there are several important reasons to treat them simultaneously:

(1) *Unification:* Before a detailed calculation has been done, it can be difficult to know whether quantum or collective effects will be dominant in a specific problem. In this case it is useful to be able to start from a set of equations including both types of phenomena.
(2) *Different scalings:* Certain quantum effects, in particular those due quantum electrodynamics (QED) and particle spin does not necessarily become insignificant even if there is a large number of particles in a Debye sphere. For such plasmas, collective and quantum effects can simultaneously be important.
(3) *Symmetry dependent effects:* The symmetry properties of the standard and quantum terms differ to some extent in the equations of motion. Thus for a problem with a specific geometry, a classical effect may sometimes vanish due to a symmetry, while a small quantum effect survives and dominate the dynamical picture.
(4) *Extreme regimes:* In certain extreme regimes, as for example found in astrophysics, the formal conditions for collective and quantum effects to be important simultaneously can be fulfilled.

In this paper we will describe a number of different quantum phenomena that can be fit into the standard Maxwell-Fluid model by adding various terms to the classical equations. In particular we will be dealing with particle dispersion,[1–13] giving raise to the so called Bohm potential, effects related to the spin-half properties of the particles,[20,21] as well as QED effects such as vacuum polarization and magnetization,[18,19] which becomes important for field strengths approaching the Schwinger critical field, $E_{\text{crit}} \equiv m_e^2 c^3/\hbar e \approx 10^{16}$ V/cm^{-1}. Here m_e is the electron mass, c is the speed of light in vacuum, e is the elementary charge and $\hbar$ is Plank's constant. The applicability of the presented models will be discussed, and a number of phenomena induced by the quantum terms will be shown. Finally, various applications to specific plasmas will be pointed out.

2. Particle dispersion and Fermi pressure

Naturally the most basic quantum effect is that particles are described by wave functions rather than classical point particles. Following e.g. Ref.,[14] the particles are described by the statistical mixture of N states ψ_i, $i = 1, 2, \ldots, N$ where the index i sums over all particles independent of species. We then take each ψ_i to satisfy a single particle Schrdinger equation where the potentials $(\mathbf{A}, \phi)$ is due to the collective charge and current densities, i.e. $\varepsilon_0 \nabla^2 \phi = -\sum_{i=1}^{N} q_i p_i |\psi_i|^2$, etc., where p_i is the occupation probability of state ψ_i.This model amounts to assume that all entanglement between particles are neglected. To derive a fluid description we make the ansatz $\psi_i = \sqrt{n_i} \exp(iS_i/\hbar)$ where n_i is the particle density due to particle i, S_i is real, and the velocity of the i'th particle is $\mathbf{u}_i = \nabla S_i/m_i - (q_i/m_i c)\mathbf{A}$, where q_i and m_i is the charge and mass. Next we define the global density and velocity as $n = \sum_j p_j n_j$ and $\mathbf{u} = \sum_j p_j n_j \mathbf{u_j}/n$, where j runs over all particles. Separating the real and the imaginary part in the Schrdinger equation, we obtain the continuity equation

$$\frac{\partial n}{\partial t} + \nabla \cdot (n\mathbf{u}) = 0, \tag{1}$$

and the momentum equation

$$\frac{\partial \mathbf{u}}{\partial t} + (\mathbf{u} \cdot \nabla)\, \mathbf{u} = \frac{q}{m} (\mathbf{E} + \mathbf{u} \times \mathbf{B}) - \frac{1}{mn} \nabla P + \frac{\hbar^2}{2m^2} \nabla \left(\frac{1}{\sqrt{n}} \nabla^2 \sqrt{n} \right). \tag{2}$$

The last term is the gradient of the so called Bohm potential, and the tendency to smoothen a density profile naturally reflects the dispersive tendencies of a localized wave packet. Furthermore, we stress that the pressure term contains both the fermion pressure, P_F, and the thermal pressure, P_t. For low temperature plasmas, where the Fermi pressure is of most significance, P_F can be written as $P_F = (4\pi^2\hbar^2/5m)(3/8\pi)^{2/3} n^{5/3}$.

As a simple illustration of some effects due to the quantum terms we can study linear wave propagation in a homogeneous plasma that may - or may not - be magnetized. Since both the thermal pressure, the Fermi pressure and the Bohm potential becomes proportional to ∇n_1, where n_1 is th density perturbation of the total density $n = n_0 + n_1$, it turns out[25] that the above quantum effects can be captured by making the following simple substitutions

$$v_t^2 \to v_t^2 + \frac{3}{5} v_F^2 + \frac{\hbar^2 k^2}{4m^2}. \tag{3}$$

in any classical linear dispersion relations. Here v_t is the thermal velocity, v_F the Fermi velocity and k the wavenumber of the perturbation. As a specific example we may consider Langmuir waves in which case the quantum version of the dispersion relation thus becomes

$$\omega^2 = \omega_p^2 + k^2 \left(v_t^2 + \frac{3}{5} v_F^2 + \frac{\hbar^2 k^2}{4m^2} \right) \tag{4}$$

where ω_p is the plasma frequency. The dispersion relation (4) was recently experimentally verified in X-ray scattering experiments made in Laser produced plasmas.[26]

3. Particle spin

The treatment of the previous section can be generalized to include the effects of particle spin. The following modifications are then necessary:[20,21]

(1) Replace the Schrdinger equation for a scalar wave function with the Pauli equation for the spinors.
(2) Decompose the spinors according to $\psi_i = \sqrt{n_i}\exp(iS_i/\hbar)\varphi_i$, where φ_i is a normalized two spinor.
(3) Introduce the velocity $\mathbf{u}_i$ and the spin vector $\mathbf{s}_i$ as $\mathbf{u}_i = (1/m)(\nabla S_i - i\hbar\varphi_i^\dagger\nabla\varphi_i) - (q_i/m_i c)\mathbf{A}$ and $\mathbf{s}_i = (\hbar/2)\varphi_i^\dagger\sigma\varphi_i$, where $\sigma = (\sigma_1, \sigma_2, \sigma_3)$ and $\sigma_{1,2,3}$ are the Pauli spin matrices.

It is no surprise that the resulting equations are considerably more complicated than the spinless equations in the preceding section. Rather than presenting the full theory (see Refs.[20,21]) here, we will focus on the leading contributions where a number of terms of higher order in $\hbar$ are neglected. The spin effects can then be captured by a spin force $\mathbf{F}_{sp}$ that is added to the momentum equation, and a magnetization current $\mathbf{j}_m$ associated with the spins. These expressions in turn depend on a macroscopic spin vector $\mathbf{s} = \sum_i \mathbf{s}_i/n$, that is described by a separate evolution equation complementing the Maxwell-fluid system. The results for the electrons, denoted by index e are

$$\mathbf{F}_{sp} = \frac{2\mu_B n_e}{\hbar} s^j \nabla B_j \tag{5}$$

$$\mathbf{j}_m = \nabla \times \mathbf{M} = \nabla \times \left(\frac{2n_e\mu_B\mathbf{s}}{\hbar}\right) \tag{6}$$

$$\left(\frac{\partial}{\partial t} + \mathbf{u}\cdot\nabla\right)\mathbf{s} = \frac{2\mu_B}{\hbar}\mathbf{B}\times\mathbf{s} \tag{7}$$

where $\mu_B = e\hbar/2m_e$ is the Bohr magneton, $\mathbf{B}$ is the magnetic field, $\mathbf{M}$ is the magnetization vector and we use the Einstein summation convention in Eq. (5). The spin effects associated with the ions is usually smaller due to their larger mass. For a generalization including the spin contribution for an arbitrary particle species, see Refs.[20,21]

There is a rich variety of new dynamical effects associated with the spins, as described by Eqs. (5)-(7). However, in order to start exploring the dynamics, we must first have an expression for the spin vector in thermodynamic equilibrium. The result for spin half particles is[20,21] s $=(\hbar/2)\tanh(\mu_B B_0/T)$, where B_0 is the unperturbed magnetic field and T is the temperature given in energy units. A simple example of the results that can be derived from the Maxwell-Fluid results complemented by (5)-(7) is the modification of the Alfvn velocity. Taking the MHD limit, it turns out that the Alfvn velocity $C_A = \left(B_0^2/\mu_0\rho_0\right)^{1/2}$ is modified according to[27]

$$C_A \to \frac{C_A}{(1+(\hbar\omega_{pe}^2/2mc^2\omega_{ce}^{(0)})\tanh(\mu_B B_0/T))^{1/2}} \tag{8}$$

where $\omega_{ce}^{(0)} = eB_{0\text{ext}}/m_e$ is the electron cyclotron frequency due to the external field $B_{0\text{ext}}$ only, i.e. the contribution from the zero order spin magnetization is excluded. The substitution (8) applies for linear homogeneous MHD theory in general, i.e. both for the shear Alfvn mode as well as for the fast and slow magnetosonic modes.

4. High field and short wavelength QED effects

The first order QED effects can effectively be modeled through the Heisenberg-Euler Lagrangian density.[28,29] This Lagrangian describes a vacuum perturbed by a slowly varying electromagnetic field. The effect of rapidly varying fields can be accounted for by adding a derivative correction to the Lagrangian.[30] This correction is referred to as the derivative QED correction or the short wavelength QED correction. The Heisenberg-Euler Lagrangian density with the derivative correction reads

$$\mathcal{L} = \mathcal{L}_0 + \mathcal{L}_{HE} + \mathcal{L}_D = \frac{\varepsilon_0}{4}F_{ab}F^{ab} + \frac{\varepsilon_0^2\kappa}{16}\left[4\left(F_{ab}F^{ab}\right)^2 + 7\left(F_{ab}\widehat{F}^{ab}\right)^2\right] \\ +\sigma\varepsilon_0\left[\left(\partial_a F^{ab}\right)\left(\partial_c F^{c}_{\ b}\right) - F_{ab}\Box F^{ab}\right], \tag{9}$$

where $\mathcal{L}_0$ is the classical Lagrangian density, while $\mathcal{L}_{HE}$ represents the Heisenberg-Euler correction due to first order strong field QED effects, $\mathcal{L}_D$ is the derivative correction, $\Box = \partial_a\partial^a$ is the d'Alembertian, F^{ab} is the electromagnetic field tensor and $\widehat{F}^{ab} = \epsilon^{abcd}F_{cd}/2$ where ϵ^{abcd} is the totally antisymmetric tensor. The parameter $\kappa = 2\alpha^2\hbar^3/45m^4c^5$ gives the nonlinear coupling, $\sigma = (2/15)\alpha c^2/\omega_e^2$ is the coefficient of the derivative correction and $\alpha = e^2/4\pi\hbar c\varepsilon_0$ is the fine structure constant, where ε_0 is the free space permittivity. We obtain the field equations from the Euler-Lagrange equations $\partial_b \left[\partial\mathcal{L}/\partial F_{ab}\right] = \mu_0 j^a$,

$$(1 + 2\sigma\Box)\,\partial_a F^{ab} = 2\varepsilon_0\kappa\partial_a \left[\left(F_{cd}F^{cd}\right)F^{ab} + \tfrac{7}{4}\left(F_{cd}\widehat{F}^{cd}\right)\widehat{F}^{ab}\right] + \mu_0 j^b, \quad (10)$$

where j^a is the four-current and μ_0 is the free space permeability.

Using three-vector notation, the corresponding sourced Maxwell equations resulting from the derivative corrected field equation thus become

$$\left[1 + 2\sigma\left(-\frac{1}{c^2}\frac{\partial^2}{\partial t^2} + \nabla^2\right)\right]\nabla\cdot\mathbf{E} = \frac{\rho + \rho_{\text{vac}}}{\varepsilon_0}, \quad (11)$$

$$\left[1 + 2\sigma\left(-\frac{1}{c^2}\frac{\partial^2}{\partial t^2} + \nabla^2\right)\right]\left(\nabla\times\mathbf{B} - \frac{1}{c^2}\frac{\partial\mathbf{E}}{\partial t}\right) = \mu_0(\mathbf{j} + \mathbf{j}_{\text{vac}}), \quad (12)$$

where the vacuum charge density is $\rho_{\text{vac}} = -\nabla\cdot\mathbf{P}$ and the vacuum current density is $\mathbf{j}_{\text{vac}} = \partial\mathbf{P}/\partial t - \nabla\times\mathbf{M}$ with the vacuum polarization and magnetization given by

$$\mathbf{P} = 2\epsilon_0^2\kappa[2(E^2 - c^2B^2)\mathbf{E} + 7c^2(\mathbf{E}\cdot\mathbf{B})\mathbf{B}] \quad (13)$$

and

$$\mathbf{M} = 2c^2\epsilon_0^2\kappa[-2(E^2 - c^2B^2)\mathbf{B} + 7(\mathbf{E}\cdot\mathbf{B})\mathbf{E}] \quad (14)$$

respectively. The source free Maxwell equations are $\nabla\cdot\mathbf{B} = 0$ and

$$\nabla\times\mathbf{E} = -\frac{\partial\mathbf{B}}{\partial t}. \quad (15)$$

The QED vacuum contribution can give raise to a large number of physical effects.[19] For example the nonlinear vacuum terms implies processes such as photon-photon scattering. However, in order to keep our examples simple, we here consider just linear wave propagation, and also treat the high field effects proportional to κ and the short wavelength effects proportional to σ separately from now on.

As our first example we consider short wavelength linear wave propagation in a magnetized plasma, and keep only the QED terms proportional to σ. It turns out that the effects due to a finite σ can be included in a very simple manner, by making the substitution

$$\omega_{ps}^2 \to \frac{\omega_{ps}^2}{1-\zeta}. \quad (16)$$

everywhere in the susceptibility tensor of a plasma,[25] where $\zeta = 2\sigma(\omega^2/c^2 - k^2)$. In most cases the short wavelength QED corrections is a very small effect. The possibility to confirm such effects in laboratory has been discussed in some detail by Ref.[25]

As our second QED example we consider the effects of the vacuum polarization a magnetization due to a strong external magnetic field $\mathbf{B} = B_0\widehat{\mathbf{z}}$. We note that the term proportional to κ contributes with terms that are linear in the wave field and quadratic in B_0. Following Ref.[31] we find that the susceptibility tensor can be modified to include the QED effect of strong magnetic fields by adding the correction

$$\chi_{\text{QED}} = -4\xi \begin{pmatrix} 1 - n_{\parallel}^2 & 0 & n_{\perp}n_{\parallel} \\ 0 & 1 - n^2 - 2n_{\perp} & 0 \\ n_{\perp}n_{\parallel} & 0 & -\frac{5}{2} - n_{\perp}^2 \end{pmatrix}, \tag{17}$$

where $\xi \equiv \kappa\varepsilon_0 c^2 B_0^2 = (\alpha/90\pi)(cB_0/E_{\text{crit}})^2$, $n_{\perp} = k_{\perp}c/\omega$, $n_{\parallel} = k_{\parallel}c/\omega$ and $n = kc/\omega$. Here the indices $\perp$ and $\parallel$ denote the directions perpendicular and parallel to the external magnetic field respectively, and we have chosen the wavevector to lie in the xz-plane. In magnetar environments with extreme magnetic fields, the parameter ξ can approach unity.[17] As a consequence, wave propagation in the electron-positron plasma surrounding magnetars are likely to be significantly affected by strong field QED effects.

5. Concluding remarks

In this paper we have given a brief review of how the plasma dynamics is modified by various quantum, spin and QED effects. The approach has been to modify the Maxwell-Fluid equations, in order to keep contact with the theoretical results and methods developed for classical systems. In order to illustrate the usefulness of the modified equations, we have presented some simple results for linear wave propagation. It should be stressed, however, that much of the work dealing with quantum and QED effects focuses on nonlinear phenomena.[19] The set of plasmas where the new phenomena tend to be important can be briefly described as follows:

(1) *The Bohm potential and Fermi pressure:* Low temperature and/or high density plasmas. This includes solid state plasmas, white dwarf stars and to some extent laser produced plasmas. Ultra-cold plasmas generated from Rydberg states are also of interest in this context.
(2) *Spin effects:* Due to the complexity of the spin dynamics, it is difficult to give simple conditions when these effects are important. However, a few simple rules of thumb can be given: Spin effects are important if the energy difference between the two spin states is larger than the thermal energy. This applies to plasmas in the vicinity of magnetars, and possibly also to ultracold plasmas. Furthermore, spin effects can

be important in low-temperature high density plasmas, similarly to the ones described in point 1. Finally spin effects can be important if $C_A^2 \lesssim \mu_B B_0/m_i$, which contrary to the first conditions tend to be fulfilled in plasmas embedded in a rather weak external magnetic field.

(3) *High field QED effects:* The characteristic scale for this phenomena is the Schwinger critical field. However, since qualitatively new phenomena (i.e. photon-photon scattering in vacuum) occur due to these terms, there is a hope to see such QED effects even before the laser intensities reach this extreme scale. Furthermore, astrophysical plasmas in the vicinity of magnetars can be subject to magnetic field strengths exceeding the critical field.

(4) *Short wavelength QED effects:* These are important when the wavelength approach the Compton wavelength, provided the plasma density is very high, such that the c/ω_p is comparable or smaller than the Compton wavelength.

The above list should not be taken too literally, as quantum effects certainly can be more important whenever a classical effect vanishes due to some symmetry obeyed by the classical terms only. We conclude this paper by pointing out that the field of quantum plasmas is a very rich one, and that many important aspects remain to be discovered.

References

1. F. Haas, G. Manfredi, and M. R. Feix, Phys. Rev. E **62**, 2763 (2000).
2. D. Anderson, B. Hall, M. Lisak, and M. Marklund, Phys. Rev. E **65**, 046417 (2002).
3. F. Haas, L. G. Garcia, J. Goedert, and G. Manfredi, Phys. Plasmas **10**, 3858 (2003).
4. F. Haas, Phys. Plasmas **12**, 062117 (2005).
5. L. G. Garcia, F. Haas, L. P. L. de Oliviera, and J. Goedert, Phys. Plasmas **12**, 012302 (2005).
6. M. Marklund, Phys. Plasmas **12**, 082110 (2005).
7. P. K. Shukla and L. Stenflo, Phys. Lett. A **355**, 378 (2006).
8. P. K. Shukla, Phys. Lett. A **357**, 229 (2006).
9. P. K. Shukla, L. Stenflo, and R. Bingham, Phys. Lett. A **359**, 218 (2006).
10. P. K. Shukla, Phys. Lett. A **352**, 242 (2006).
11. P. K. Shukla and B. Eliasson, Phys. Rev. Lett. **96**, 245001 (2006).
12. P. K. Shukla, S. Ali, L. Stenflo, and M. Marklund, Phys. Plasmas **13**, 112111 (2006).
13. F. Haas, Europhys. Lett. **44**, 45004 (2007).
14. G. Manfredi, Fields Inst. Commun. **46**, 263 (2005).

15. G. Brodin, M. Marklund, B. Eliasson, and P. K. Shukla, Phys. Rev. Lett. **98**, 125001 (2007).
16. M. G. Baring and A. K. Harding, Astrophys. J. **547**, 929 (2001).
17. V. S. Beskin, A. V. Gurevich and Ya. N. Istomin, *Physics of the Pulsar Magnetosphere* (Cambridge university press, Cambridge, 1993).
18. E. Lundstrm *et al.*, Phys. Rev. Lett. **96**, 083602 (2006).
19. M. Marklund and P. K. Shukla, Rev. Mod. Phys. **78**, 591 (2006).
20. M. Marklund and G. Brodin, Phys. Rev. Lett. **98**, 025001 (2007).
21. G. Brodin and M. Marklund, New J. Phys. **9**, 277, (2007).
22. W. Dittrich and H. Gies, *Probing the Quantum Vacuum* (Springer–Verlag, Berlin, 2000).
23. G. Brodin, M. Marklund, and L. Stenflo, Phys. Rev. Lett. **87**, 171801 (2001).
24. P. R. Holland, *The Quantum Theory of Motion* (Cambridge University Press, Cambridge, 1993).
25. J. Lundin *et al.*, Phys. Plasmas **14**, 062112 (2007).
26. S. H. Glenzer *et al.*, Phys. Rev. Lett, **98**, 065002, 2007
27. G. Brodin and M. Marklund, Phys. Rev. E. In press.
28. W. Heisenberg and H. Euler, Z. Physik **98**, 714 (1936).
29. J. Schwinger, Phys. Rev. **82**, 664 (1951).
30. S. G. Mamaev, V. M. Mostepanenko, and M. I. Eǐdes, Sov. J. Nucl. Phys. **33**, 569 (1981).
31. G. Brodin, M. Marklund, L. Stenflo and P. K. Shukla, New J. Phys. **8**, 16 (2006).

SPIN QUANTUM PLASMAS — NEW ASPECTS OF COLLECTIVE DYNAMICS

M. MARKLUND* and G. BRODIN

Department of Physics, Umeå University, SE-901 87 Umeå, Sweden
**E–mail: mattias.marklund@physics.umu.se*

Quantum plasmas is a rapidly expanding field of research, with applications ranging from nanoelectronics, nanoscale devices and ultracold plasmas, to inertial confinement fusion and astrophysics. Here we give a short systematic overview of quantum plasmas. In particular, we analyze the collective effects due to spin using fluid models. The introduction of an intrinsic magnetization due to the plasma electron (or positron) spin properties in the magnetohydrodynamic limit is discussed. Finally, a discussion of the theory and examples of applications is given.

1. Introduction

The field of quantum plasmas is a rapidly growing field of research. From the non-relativistic domain, with its basic description in terms of the Schrödinger equation, to the strongly relativistic regime, with its natural connection to quantum field theory, quantum plasma physics provides promises of highly interesting and important application, fundamental connections between different areas of science, as well as difficult challenges from a computational perspective. The necessity to thoroughly understand such plasmas motivates a reductive principle of research, for which we successively build more complex models based on previous results. The simplest lower order effect due to relativistic quantum mechanics is the introduction of spin, and as such thus provides a first step towards a partial description of relativistic quantum plasmas.

Already in the 1960's, Pines studied the excitation spectrum of quantum plasmas,[1,2] for which we have a high density and a low temperature as compared to normal plasmas. In such systems, the finite width of the electron wave function makes quantum tunnelling effects crucial, leading to an altered dispersion relation. Since the pioneering work by Pines, a number or theoretical studies of quantum statistical properties of plasmas has been done (see, *e.g.*, Ref. 3 and references therein). For example, Bezzerides & DuBois presented a kinetic theory for the quantum electrodynamical properties of nonthermal plasmas,[4] while Hakim & Heyvaerts presented a covariant Wigner function approach for relativistic quantum plasmas.[5] Recently there has been an increased interest in the properties of quantum plasmas.[6–25] The studies has been motivated by the development in nanostructured materials[26] and quantum wells,[27] the discovery of ultracold plasmas[28] (see Ref. 29 for an experimental demonstration of quantum plasma oscillations in Rydberg systems), astrophysical applications,[30] or a general theoretical interest. Moreover, it has recently been experimentally shown that quantum dispersive effects are important in inertial confinement plasmas.[31] The list of quantum mechanical effects that can be included in a fluid picture includes the dispersive particle properties accounted for by the Bohm potential,[6–18] the zero temperature Fermi pressure,[6–10] spin properties[19–21] as well as certain quantum electrodynamical effects.[32–35] Within such descriptions,[6–10,19,20,33–35] quantum and classical collective effects can be described within a unified picture.

2. The microscopic equations: Schrödinger and Pauli dynamics

2.1. *The Schrödinger description*

The basic equation of nonrelativistic quantum mechanics is the Schrödinger equation. The dynamics of an electron, represented by its wave function ψ, in an external electromagnetic potential ϕ is governed by

$$i\hbar\frac{\partial\psi}{\partial t} + \frac{\hbar}{2m_e}\nabla^2\psi + e\phi\psi = 0, \tag{1}$$

where $\hbar$ is Planck's constant, m_e is the electron mass, and e is the magnitude of the electron charge. This complex equation may be written as two real equations, writing $\psi = \sqrt{n}\exp iS/\hbar$, where n is the amplitude and S the phase of the wave function, respectively.[36] Such a decomposition was presented by de Broglie and Bohm in order to understand the dynamics of the electron wave packet in terms of classical variables. Using this decomposition in Eq. (1), we obtain

$$\frac{\partial n}{\partial t} + \nabla\cdot(n\mathbf{v}) = 0, \tag{2}$$

and

$$m_e\frac{d\mathbf{v}}{dt} = e\nabla\phi + \frac{\hbar^2}{2m_e}\nabla\left(\frac{\nabla^2\sqrt{n}}{\sqrt{n}}\right), \tag{3}$$

where the velocity is defined by $\mathbf{v} = \nabla S/m_e$. The last term of Eq. (3) is the gradient of the Bohm–de Broglie potential, and is due to the effect of wave function spreading, giving rise to a dispersive-like term. We also note the striking resemblance of Eqs. (2) and (3) to the classical fluid equations.

2.2. *The Pauli description*

In relativistic quantum mechanics, the spin of the electron (and positron) is rigorously introduced through the Dirac Hamiltonian

$$H = c\alpha \cdot (\mathbf{p} + e\mathbf{A}) - e\phi + \beta m_e c^2, \tag{4}$$

where $\alpha = (\alpha_1, \alpha_2, \alpha_3)$, e is the magnitude of the electron charge, c is the speed of light, $\mathbf{A}$ is the vector potential, ϕ is the electrostatic potential, and the relevant matrices are given by

$$\alpha = \begin{pmatrix} 0 & \sigma \\ \sigma & 0 \end{pmatrix}, \qquad \beta = \begin{pmatrix} \mathsf{I} & 0 \\ 0 & -\mathsf{I} \end{pmatrix}. \tag{5}$$

Here I is the unit 2×2 matrix and $\sigma = (\sigma_1, \sigma_2, \sigma_3)$, where we have the Pauli spin matrices

$$\sigma_1 = \begin{pmatrix} 0 & 1 \\ 1 & 0 \end{pmatrix}, \sigma_2 = \begin{pmatrix} 0 & -i \\ i & 0 \end{pmatrix}, \text{ and } \sigma_3 = \begin{pmatrix} 1 & 0 \\ 0 & -1 \end{pmatrix}. \tag{6}$$

From the Hamiltonian (4), a nonrelativistic counterpart may be obtained, taking the form

$$H = \frac{1}{2m_e}(\mathbf{p} + e\mathbf{A})^2 + \frac{e\hbar}{2m_e}\mathbf{B} \cdot \sigma - e\phi. \tag{7}$$

Thus, the electron possesses a magnetic moment $\mathbf{m} = -\mu_B \langle\psi|\sigma|\psi\rangle / \langle\psi|\psi\rangle$, where $\mu_B = e\hbar/2m_e$ is the Bohr magneton, giving a contribution $-\mathbf{B} \cdot \mathbf{m}$ to the energy. The latter shows the paramagnetic property of the electron, where the spin vector is anti-parallel to the magnetic field in order to minimize the energy of the magnetized system. According to (7) and the relation $dF/dt = \partial F/\partial t + (1/i\hbar)[F, H]$, where F is some operator and $[,]$ is the Poisson bracket, we have the following evolution equations for the position and momentum in the Heisenberg picture[36–38]

$$\mathbf{v} \equiv \frac{d\mathbf{x}}{dt} = \frac{1}{m_e}(\mathbf{p} + e\mathbf{A}), \tag{8}$$

$$m_e \frac{d\mathbf{v}}{dt} = -e(\mathbf{E} + \mathbf{v} \times \mathbf{B}) - \frac{2}{\hbar}\mu_B \nabla(\mathbf{B} \cdot \mathbf{s}), \tag{9}$$

while the spin evolution is given by

$$\frac{d\mathbf{s}}{dt} = \frac{2}{\hbar}\mu_B \mathbf{B} \times \mathbf{s}, \tag{10}$$

where the spin operator is given by

$$\mathbf{s} = \frac{\hbar}{2}\sigma. \tag{11}$$

The above equations thus gives the quantum operator equivalents of the equations of motion for a classical particle, including the evolution of the spin in a magnetic field.

The non-relativistic evolution of spin $\frac{1}{2}$ particles, as described by the two-component spinor $\Psi_{(\alpha)}$, is given by the Pauli equation (see, *e.g.*, 36)

$$i\hbar\frac{\partial\psi}{\partial t} + \left[\frac{\hbar^2}{2m_e}\left(\nabla + \frac{ie}{\hbar}\mathbf{A}\right)^2 - \mu_B\mathbf{B}\cdot\sigma + e\phi\right]\psi = 0, \tag{12}$$

where $\mathbf{A}$ is the vector potential, $\mu_B = e\hbar/2m_e$ is the Bohr magneton, and $\sigma = (\sigma_1, \sigma_2, \sigma_3)$ is the Pauli spin vector.

Now, in the same way as in the Schrödinger case, we may decompose the electron wave function ψ into its amplitude and phase. However, as the electron has spin, the wave function is now represented by a 2-spinor instead of a c-number. Thus, we may use $\psi = \sqrt{n}\exp(iS/\hbar)\varphi$, where φ, normalized such that $\varphi^\dagger\varphi = 1$, now gives the spin part of the wave function. Multiplying the Pauli equation (12) by $\psi^\dagger$, inserting the above wave function decomposition and taking the gradient of the resulting phase evolution equation, we obtain the conservation equations

$$\frac{\partial n}{\partial t} + \nabla\cdot(n\mathbf{v}) = 0 \tag{13}$$

and

$$m_e\frac{d\mathbf{v}}{dt} = -e(\mathbf{E} + \mathbf{v}\times\mathbf{B}) + \frac{\hbar^2}{2m_e}\nabla\left(\frac{\nabla^2\sqrt{n}}{\sqrt{n}}\right) - \frac{2\mu_B}{\hbar}(\nabla\otimes\mathbf{B})\cdot\mathbf{s} - \frac{1}{m_e n}\nabla\cdot(n\mathbf{\Sigma}) \tag{14}$$

respectively. The spin contribution to Eq. (14) is consistent with the results of Ref. 39. Here the velocity is defined by

$$\mathbf{v} = \frac{1}{m_e}\left(\nabla S - i\hbar\varphi^\dagger\nabla\varphi\right) + \frac{e\mathbf{A}}{m_e c}, \tag{15}$$

the spin density vector is

$$\mathbf{s} = \frac{\hbar}{2}\varphi^\dagger\sigma\varphi, \tag{16}$$

which is normalized according to

$$|\mathbf{s}| = \hbar/2, \tag{17}$$

and we have defined the symmetric gradient spin tensor

$$\mathbf{\Sigma} = (\nabla s_a) \otimes (\nabla s^a). \tag{18}$$

Moreover, contracting Eq. (12) by $\psi^\dagger \sigma$, we obtain the spin evolution equation

$$\frac{d\mathbf{s}}{dt} = \left\{ \frac{2\mu_B}{\hbar}\mathbf{B} - \frac{1}{m_e n}\left[\partial_a(n\partial^a \mathbf{s})\right] \right\} \times \mathbf{s}. \tag{19}$$

We note that the last equation allows for the introduction of an effective magnetic field $\mathbf{B}_{\text{eff}} \equiv (2\mu_B/\hbar)\mathbf{B} - (m_e n)^{-1}\left[\partial_a(n\partial^a \mathbf{s})\right]$. However, this will not pursued further here (for a discussion, see Ref. 36).

Comparing the effects due to spin from the Pauli dynamics with the Schrödinger theory, we see a significant increase in the complexity of the fluid like equations due the presence of spin. The fact that the spin couples linearly to the magnetic field makes the dynamical aspects of such Pauli systems very rich. Moreover, when going over to the collective regime, the back reaction through Maxwell's equation can yield interesting new properties of such spin plasmas. In fact, the introduction of an intrinsic magnetization can give rise to linear instability regimes, much like the Jeans instability (see Sec. 5.2.).

3. Collective plasma dynamics

As pointed out in the previous section, the route from single wavefunction dynamics to collective effects introduces a new complexity into the system. At the classical level, the ordinary pressure is such an effect. In the quantum case, a similar term, based on the thermal distribution of spins, will be introduced.

3.1. *Multistream model*

The multistream model of classical plasmas was successfully introduced by Dawson.[40] Here we will focus on the electrostatic interaction between a multistream quantum plasma described within the Schrödiner model, a system first investigated in Ref. 7 (where also the stationary regime was probed). Thus, we have the governing equations (2) and (3) but for N beams of electrons on a stationary ion background, *i.e.*, Using this decomposition in Eq. (1), we obtain

$$\frac{\partial n_\alpha}{\partial t} + \nabla \cdot (n_\alpha \mathbf{v}_\alpha) = 0, \tag{20}$$

and

$$m_e \frac{d\mathbf{v}_\alpha}{dt} = e\nabla\phi + \frac{\hbar^2}{2m_e}\nabla\left(\frac{\nabla^2\sqrt{n_\alpha}}{\sqrt{n_\alpha}}\right), \tag{21}$$

now coupled through the self-consistent electrostatic potential governed by

$$\nabla^2\phi = \frac{e}{\epsilon_0}\sum_{\alpha=1}^{N}(n_\alpha - n_0). \tag{22}$$

Here, $d/dt = \partial_t + \mathbf{v}_\alpha \cdot \nabla$ and n_0 is the density of the stationary ion background.

In the one-stream case ($\alpha = 1$), we have the equilibrium solution $\mathbf{v} = \mathbf{v}_0$ (a constant drift relative the stationary ion background) and the constant electron density $n = n_0$ (such that $\phi = 0$). Perturbing this system a Fourier decomposing the perturbations, such that $n = n_0 + \delta n \exp[i(\mathbf{k}\cdot\mathbf{x} - \omega t)]$), $\mathbf{v} = \mathbf{v}_0 + \delta\mathbf{v}\exp[i(\mathbf{k}\cdot\mathbf{x} - \omega t)]$, and $\phi = \delta\phi\exp[i(\mathbf{k}\cdot\mathbf{x} - \omega t)]$, we obtain[1,7]

$$(\omega - \mathbf{k}\cdot\mathbf{v}_0)^2 = \omega_p^2 + \frac{\hbar^2 k^4}{4m_e^2}, \tag{23}$$

where the last term is the Bohm–de Broglie correction to the dispersion relation. Here we have the electron plasma frequency $\omega_p = (e^2 n_0/\epsilon_0 m_e)^{1/2}$.

Similarly to the one-stream case, we obtain the dispersion relation[7,41]

$$1 = \frac{\omega_{p1}^2}{(\omega - \mathbf{v}_{01}\cdot\mathbf{k})^2 - \hbar^2 k^4/4m_e^2} + \frac{\omega_{p2}^2}{(\omega - \mathbf{v}_{02}\cdot\mathbf{k})^2 - \hbar^2 k^4/4m_e^2}, \tag{24}$$

for two propagating electron beams (with velocities $\mathbf{v}_{01}$ and $\mathbf{v}_{02}$) with background densities n_{01} and n_{02}. The quantum effect has a subtle influence on the stability of the perturbed plasma. For the case $n_{01} = n_{02} = n_0/2$ and $\mathbf{v}_{01} = -\mathbf{v}_{02} = \mathbf{v}_0$, we have the instability condition

$$\frac{4}{K^2}\left(1 - \frac{1}{K^2}\right) < H^2 < \frac{4}{K^2}, \tag{25}$$

in terms of the normalized wavenumber $K = kv_0/\omega_p$ and the quantum parameter $H = \hbar\omega_p/m_e v_0^2$ (see Fig. 1).[7,41] We see that when $H = 0$, we

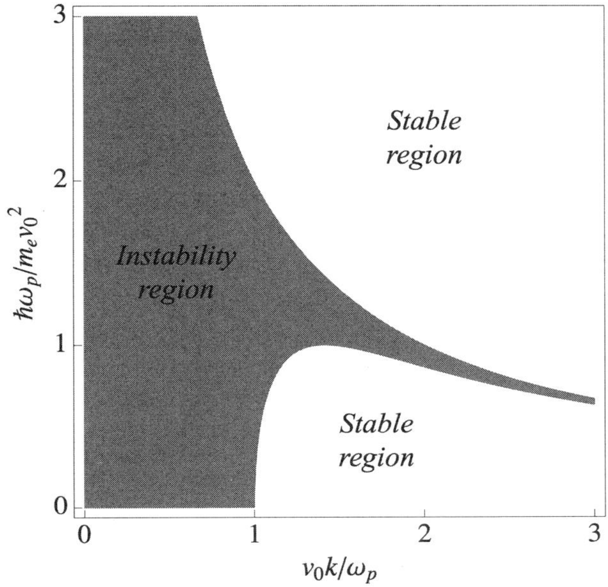

Fig. 1. The regions of stability and instability in the case of the quantum two-stream interaction.[7,41]

have unstable perturbations for $0 < K < 1$, but when $H \neq 0$ a considerably more complex instability region develops.

A model for treating partial coherence in such systems, based on the Wigner transform technique,[42–45] can also be developed[41] (see also Ref. 46). Moreover, using the equations (13) and (14), a similar framework may be set up for electron streams with spin properties.

3.2. *Fluid model*

3.2.1. *Plasmas based on the Schrödinger model*

Suppose that we have N electron wavefunctions, and that the total system wave function can be described by the factorization $\psi(\mathbf{x}_1, \mathbf{x}_2, \ldots \mathbf{x}_N) = \psi_1\psi_2\ldots\psi_N$. For each wave function ψ_α, we have a corresponding probability $\mathcal{P}_\alpha$. From this, we first define $\psi_\alpha = n_\alpha \exp(iS_\alpha/\hbar)$ and follow the steps leading to Eqs. (2) and (3). We now have N such equations the wave functions $\{\psi_\alpha\}$. Defining[6]

$$n \equiv \sum_{\alpha=1}^{N} \mathcal{P}_\alpha n_\alpha \tag{26}$$

and

$$\mathbf{v} \equiv \langle \mathbf{v}_\alpha \rangle = \sum_{\alpha=1}^{N} \frac{\mathcal{P}_\alpha n_\alpha \mathbf{v}_\alpha}{n}, \tag{27}$$

we can define the deviation from the mean flow according to

$$\mathbf{w}_\alpha = \mathbf{v}_\alpha - \mathbf{v}. \tag{28}$$

Taking the average, as defined by (27), of Eqs. (2) and (3) and using the above variables, we obtain the quantum fluid equation

$$\frac{\partial n}{\partial t} + \nabla \cdot (n\mathbf{v}) = 0 \tag{29}$$

and

$$m_e n \left(\frac{\partial}{\partial t} + \mathbf{v} \cdot \nabla \right) \mathbf{v} = en\nabla\phi - \nabla p + \frac{\hbar^2 n}{2m_e} \nabla \left\langle \left(\frac{\nabla^2 \sqrt{n_\alpha}}{\sqrt{n_\alpha}} \right) \right\rangle, \tag{30}$$

where we have assumed that the average produces an isotropic pressure $p = m_e n \langle |\mathbf{w}_\alpha|^2 \rangle$ We note that the above equations still contain an explicit sum over the electron wave functions. For typical scale lengths larger than the Fermi wavelength λ_F, we may approximate the last term by the Bohm–de Broglie potential[6]

$$\left\langle \frac{\nabla^2 \sqrt{n_\alpha}}{\sqrt{n_\alpha}} \right\rangle \approx \frac{\nabla^2 \sqrt{n}}{\sqrt{n}}. \tag{31}$$

Using a classical or quantum model for the pressure term, we finally have a quantum fluid system of equations. For a self-consistent potential ϕ we furthermore have

$$\nabla^2 \phi = \frac{e}{\epsilon_0}(n - n_i). \tag{32}$$

3.2.2. *Spin plasmas*

The collective dynamics of electrons with spin and some of the spin modifications of the classical dispersion relation was presented in Ref. 19. Here we will follow Refs. 19 and 20 for the derivation of the governing equations. Suppose that we have N wave functions for the electrons with magnetic moment $\mu_e = -\mu_B$, and that, as in the case of the Schrödinger description, the total system wave function can be described by the factorization $\psi = \psi_1 \psi_2 \ldots \psi_N$. Then the density is defined as in Eq. (26) and the average fluid velocity defined by (27). However, we now have one further fluid variable, the spin vector, and accordingly we let $\mathbf{S} = \langle \mathbf{s}_\alpha \rangle$. From this we can define the microscopic microscopic spin density $\boldsymbol{\mathcal{S}}_\alpha = \mathbf{s}_\alpha - \mathbf{S}$, such that $\langle \boldsymbol{\mathcal{S}}_\alpha \rangle = 0$.

Taking the ensemble average of Eqs. (13) we obtain the continuity equation (29), while we the the ensemble average applied to (14) yield

$$m_e n \left(\frac{\partial}{\partial t} + \mathbf{v} \cdot \nabla \right) \mathbf{v} = -en\left(\mathbf{E} + \mathbf{v} \times \mathbf{B}\right) - \nabla p + \frac{\hbar^2 n}{2m_e} \nabla \left(\frac{\nabla^2 \sqrt{n}}{\sqrt{n}} \right) + \mathbf{F}_{\rm spin} \tag{33}$$

and the average of Eq. (19) gives

$$n \left(\frac{\partial}{\partial t} + \mathbf{v} \cdot \nabla \right) \mathbf{S} = \frac{2\mu_B n}{\hbar} \mathbf{B} \times \mathbf{S} - \nabla \cdot \mathsf{K} + \boldsymbol{\Omega}_{\rm spin} \tag{34}$$

respectively. Here the force density due to the electron spin is

$$\begin{aligned} \mathbf{F}_{\rm spin} &= -\frac{2\mu_B n}{\hbar}(\nabla \otimes \mathbf{B}) \cdot \mathbf{S} - \frac{1}{m_e}\nabla \cdot \left[n(\boldsymbol{\Sigma} + \widetilde{\boldsymbol{\Sigma}}) \right] \\ &\quad - \frac{1}{m_e}\nabla \cdot \left[n(\nabla S_a) \otimes \langle \nabla \mathcal{S}^a_\alpha \rangle + n \langle \nabla \mathcal{S}^a_\alpha \rangle \otimes (\nabla S_a) \right], \end{aligned} \tag{35}$$

consistent with the results in Ref. 39, while the asymmetric thermal-spin coupling is

$$\mathsf{K} = n \langle \mathbf{w}_\alpha \otimes \boldsymbol{\mathcal{S}}_\alpha \rangle \tag{36}$$

and the nonlinear spin fluid correction is

$$\begin{aligned} \boldsymbol{\Omega}_{\rm spin} &= \frac{1}{m_e}\mathbf{S} \times [\partial_a (n \partial^a \mathbf{S})] + \frac{1}{m_e}\mathbf{S} \times [\partial_a (n \langle \partial^a \boldsymbol{\mathcal{S}}_\alpha \rangle)] \\ &\quad + \frac{n}{m_e} \left\langle \frac{\boldsymbol{\mathcal{S}}_\alpha}{n_\alpha} \times \{ \partial_a [n_\alpha \partial^a (\mathbf{S} + \boldsymbol{\mathcal{S}}_\alpha)] \} \right\rangle \end{aligned} \tag{37}$$

where $\boldsymbol{\Sigma} = (\nabla S_a) \otimes (\nabla S^a)$ is the nonlinear spin correction to the classical momentum equation, $\widetilde{\boldsymbol{\Sigma}} = \langle (\nabla \mathcal{S}_{(\alpha)a}) \otimes (\nabla \mathcal{S}^a_{(\alpha)}) \rangle$ is a pressure like spin term (which may be decomposed into trace-free part and trace), and $[(\nabla \otimes \mathbf{B}) \cdot \mathbf{S}]^a = (\partial^a B_b) S^b$. Here the indices $a, b, \ldots = 1, 2, 3$ denotes the Cartesian components of the corresponding tensor. We note that, apart from the additional spin density evolution equation (34), the momentum conservation equation (33) is considerably more complicated compared to the Schrödinger case represented by (30). Moreover, Eqs. (33) and (34) still contains the explicit sum over the N states, and has to be approximated using insights from quantum kinetic theory or some effective theory.

The coupling between the quantum plasma species is mediated by the electromagnetic field. By definition, we let $\mathbf{H} = \mathbf{B}/\mu_0 - \mathbf{M}$ where $\mathbf{M} =$

$-2n\mu_B\mathbf{S}/\hbar$ is the magnetization due to the spin sources. Ampère's law $\nabla \times \mathbf{H} = \mathbf{j} + \epsilon_0\partial_t\mathbf{E}$ takes the form

$$\nabla \times \mathbf{B} = \mu_0(\mathbf{j} + \nabla \times \mathbf{M}) + \frac{1}{c^2}\frac{\partial \mathbf{E}}{\partial t}, \tag{38}$$

where $\mathbf{j}$ is the free current contribution The system is closed by Faraday's law

$$\nabla \times \mathbf{E} = -\frac{\partial \mathbf{B}}{\partial t}. \tag{39}$$

4. The magnetohydrodynamic limit

The concept of a magnetoplasma was first introduced in the pioneering work 47 by Alfvén, who showed the existence of waves in magnetized plasmas. Since then, magnetohydrodynamics (MHD) has found applications in a vast range of fields, from solar physics and astrophysical dynamos, to fusion plasmas and dusty laboratory plasmas.

Magnetic fields, an essential component in the MHD description of plasmas, also couples directly to the spin of the electron. Thus, the presence of spin alters the single electron dynamics, introducing a correction to the Lorentz force term. Indeed, from the experimental perspective, a certain interest has been directed towards the relation of spin properties to the classical theory of motion (see, e.g., Refs. 48–60). In particular, the effects of strong fields on single particles with spin has attracted experimental interest in the laser community.[50–55] However, the main objective of these studies was single particle dynamics, relevant for dilute laboratory systems, whereas our focus will be on collective effects.

We will now include if the ion species, which are assumed to be described by the classical equations and have charge Ze, we may derive a set of one-fluid equations.[20] The ion equations read

$$\frac{\partial n_i}{\partial t} + \nabla \cdot (n_i\mathbf{v}_i) = 0, \tag{40}$$

and

$$m_i n_i \left(\frac{\partial}{\partial t} + \mathbf{v}_i \cdot \nabla\right)\mathbf{v}_i = Zen_i\left(\mathbf{E} + \mathbf{v}_i \times \mathbf{B}\right) - -\nabla p_i. \tag{41}$$

Next we define the total mass density $\rho \equiv (m_e n + m_i n_i)$, the centre-of-mass fluid flow velocity $\mathbf{V} \equiv (m_e n\mathbf{v}_e + m_i n_i\mathbf{v}_i)/\rho$, and the current density $\mathbf{j} = -en\mathbf{v}_e + Zen_i\mathbf{v}_i$. Using these denfinitions, we immediately obtain

$$\frac{\partial \rho}{\partial t} + \nabla \cdot (\rho\mathbf{V}) = 0, \tag{42}$$

from Eqs. (29) and (40). Assuming quasi-neutrality, i.e. $n \approx Zn_i$, the momentum conservation equations (33) and (41) give

$$\rho\left(\frac{\partial}{\partial t}+\mathbf{V}\cdot\nabla\right)\mathbf{V}=\mathbf{j}\times\mathbf{B}-\nabla\cdot\mathbf{\Pi}-\nabla p+\frac{Z\hbar^2\rho}{2m_e m_i}\nabla\left(\frac{\nabla^2\sqrt{\rho}}{\sqrt{\rho}}\right)+\mathbf{F}_{\text{spin}}, \quad (43)$$

where Π is the tracefree pressure tensor in the centre-of-mass frame, and P is the scalar pressure in the centre-of-mass frame. We also note that due to quasi-neutrality, we have $n_e \approx Z\rho/m_i$ and $\mathbf{v} = \mathbf{V} - m_i\mathbf{j}/Ze\rho$, and we can thus express the quantum terms in terms of the total mass density ρ, the centre-of-mass fluid velocity $\mathbf{V}$, and the current $\mathbf{j}$. With this, the spin transport equation (34) reads

$$\rho\left(\frac{\partial}{\partial t}+\mathbf{V}\cdot\nabla\right)\mathbf{S}=\frac{m_e}{Ze}\mathbf{j}\cdot\nabla\mathbf{S}+\frac{2\mu_B\rho}{\hbar}\mathbf{B}\times\mathbf{S}-\frac{m_i}{Z}\nabla\cdot\mathbf{K}+\frac{m_i}{Z}\mathbf{\Omega}_{\text{spin}}. \quad (44)$$

In the momentum equation (43), neglecting the pressure and the Bohm–de Broglie potential for the sake of clarity, we have the force density $\mathbf{j}\times\mathbf{B}+\mathbf{F}_{\text{spin}}$. In general, for a magnetized medium with magnetization density $\mathbf{M}$, Ampère's law gives the free current in a finite volume V according to

$$\mathbf{j}=\frac{1}{\mu_0}\nabla\times\mathbf{B}-\nabla\times\mathbf{M}, \quad (45)$$

where we have neglected the displacement current. The surface current is an important part of the total current when we are interested in the forces on a finite volume, as was demonstrated in Ref. 20 and will be shown below.

It it worth noting that the expression of the force density in the momentum conservation equation can, to lowest order in the spin, be derived on general macroscopic grounds. Formally, the total force density on a volume element V is defined as $\mathbf{F} = \lim_{V\to 0}(\sum_\alpha \mathbf{f}_\alpha/V)$, where $\mathbf{f}_\alpha$ are the different forces acting on the volume element, and might include surface forces as well. For magnetized matter, the total force on an element of volume V is then

$$\mathbf{f}_{\text{tot}}=\int_V \mathbf{j}_{\text{tot}}\times\mathbf{B}\,\mathrm{d}V+\oint_{\partial V}(\mathbf{M}\times\hat{\mathbf{n}})\times\mathbf{B}\,\mathrm{d}S \quad (46)$$

where (neglecting the displacement current) $\mathbf{j}_{\text{tot}} = \mathbf{j}+\nabla\times\mathbf{M}$. Inserting the expression for the total current into the volume integral and using the divergence theorem on the surface integral, we obtain the force density

$$\mathbf{F}_{\text{tot}}=\mathbf{j}\times\mathbf{B}+M_k\nabla B^k, \quad (47)$$

identical to the lowest order description from the Pauli equation (see Eq. (43)). Inserting the free current expression (45), due to Ampère's law, we can write the total force density according to

$$F^i = -\partial^i \left(\frac{B^2}{2\mu_0} - \mathbf{M} \cdot \mathbf{B} \right) + \partial_k (H^i B^k). \tag{48}$$

The first gradient term in Eq. (48) can be interpreted as the force due to a potential (the energy of the magnetic field and the magnetization vector in that field), while the second divergence term is the anisotropic magnetic pressure effect. Noting that the spatial part of the stress tensor takes the form[39]

$$T^{ik} = -H^i B^k + (B^2/2\mu_0 - \mathbf{M} \cdot \mathbf{B})\delta^{ik}, \tag{49}$$

we see that the total force density on the magnetized fluid element can be written $F^i = -\partial_k T^{ik}$, as expected. Thus, the Pauli theory results in the same type of conservation laws as the macroscopic theory. The momentum conservation equation (43) then reads

$$\rho \left(\frac{\partial}{\partial t} + \mathbf{V} \cdot \nabla \right) \mathbf{V} = -\nabla \left(\frac{B^2}{2\mu_0} - \mathbf{M} \cdot \mathbf{B} \right) + B^k \partial_k \mathbf{H} - \nabla p, \tag{50}$$

where for the sake of clarity we have assumed an isotropic pressure, dropped the displacement current term in accordance with the nonrelativistic assumption, and neglected the Bohm potential (these terms can of course simply be added to (50)). This concludes the discussion of the spin-MHD plasma case. Next, we will look at some applications of the derived equations. However, it should be noted that in many cases the spins are close to thermodynamic equilibrium, and we can thus write the paramagnetic electron response in terms of the magnetization[20]

$$\mathbf{M} = \frac{\mu_B \rho}{m_i} \tanh \left(\frac{\mu_B B}{k_B T} \right) \widehat{\mathbf{B}}, \tag{51}$$

instead of using the full spin dynamics. Here B denotes the magnitude of the magnetic field and $\widehat{\mathbf{B}}$ is a unit vector in the direction of the magnetic field, k_B is Boltzmann's constant, and T is the electron temperature.

5. Examples and applications

The above equations are quite complicated, but as such also extremely rich. Suitable and physically relevant approximations, such as the magnetization given by the expression (51), will however lead to considerable simplifications. Below we consider two specific examples where such simplifying assumptions lead to interesting spin effects.

5.1. *Spin solitons*

In Ref. 21, it was shown that the electron spin can introduce novel nonlinear structures in plasmas, with no limiting classical counterpart. In particular, the MHD limit for a *electron-positron pair plasma* is considered.

Neglecting dissipative effect, the governing equations for the system of interest read

$$\frac{\partial \rho}{\partial t} + \nabla \cdot (\rho \mathbf{v}) = 0, \tag{52}$$

$$\rho \left(\frac{\partial}{\partial t} + \mathbf{V} \cdot \nabla \right) \mathbf{V} = \mathbf{j} \times \mathbf{B} - \nabla \cdot \mathbf{\Pi} + \frac{\hbar^2 \rho}{2m_e^2} \nabla \left(\frac{\nabla^2 \sqrt{\rho}}{\sqrt{\rho}} \right) + \mathbf{F}_{\text{spin}}, \tag{53}$$

where $\mathbf{\Pi} = [(T_e + T_p)/2m_e]\mathbf{I} + (m^2/\rho)\mathbf{j} \otimes \mathbf{j}$ is the total pressure tensor in the centre-of-mass frame and

$$\mathbf{F}_{\text{spin}} = 2 \tanh \left(\frac{\mu_B B}{k_B T} \right) \frac{\mu_B}{m_e} \nabla B, \tag{54}$$

where we have assumed equal temperature T of the electrons and positrons. Moreover, we have

$$\frac{\partial \mathbf{B}}{\partial t} = \nabla \times (\mathbf{V} \times \mathbf{B}), \tag{55}$$

while the current is given by

$$\mathbf{j} = \frac{1}{\mu_0} \nabla \times \mathbf{B} - \epsilon_0 \frac{\partial}{\partial t} (\mathbf{V} \times \mathbf{B}) - \frac{\mu_B}{2m} \nabla \times \left[\rho \tanh \left(\frac{\mu_B B}{k_B T} \right) \hat{\mathbf{B}} \right] \tag{56}$$

For one-dimensional Alfvén waves, the above system can be reduces to the modified Korteweg-de Vries equation[21]

$$\left[\frac{\partial}{\partial t} + c_{A,\text{sp}} \cos\theta \frac{\partial}{\partial \xi} + \frac{c_A^3 \cos^3\theta}{2\omega_c^2 \sin^2\theta} \frac{\partial^3}{\partial \xi^3} \right] V_y = -\frac{(\mu_B B_0)^2}{2m_e k_B T} \frac{V_y^2}{c_A^3} \frac{\partial V_y}{\partial \xi} \tag{57}$$

where $c_A = \left[c^2 B_0^2/(c^2 \mu_0 \rho_0 + B_0^2) \right]^{1/2}$ is the Alfvén speed, $c_{A,\text{sp}} = c_A^2/(1 + \delta_{\text{sp}})$ is the spin-modified Alfvén speed, $\omega_c = eB_0/m_e$ is the cyclotron frequency, B_0 is the magnitude of the unperturbed magnetic field, ρ_0 is the unperturbed density, and we have the spatial coordinate $\xi = x \sin\theta + z \cos\theta$. We see that neglecting the spin contribution leads to a purely dispersive equations. Thus, the spin enables the formation of solitons with no limiting classical solution.

5.2. *Ferromagnetic plasma behaviour*

For an ion-electron plasma, we have the governing equations[25]

$$\frac{\partial \rho}{\partial t} + \nabla \cdot (\rho \boldsymbol{V}) = 0, \tag{58}$$

the momentum equation

$$\rho\left(\frac{\partial}{\partial t} + \boldsymbol{V}\cdot\nabla\right)\boldsymbol{V} = -\nabla\left(\frac{B^2}{2\mu_0} - \boldsymbol{M}\cdot\boldsymbol{B}\right) - \nabla p + \boldsymbol{B}\cdot\nabla\left(\frac{1}{\mu_0}\boldsymbol{B} - \boldsymbol{M}\right) + \frac{\hbar^2\rho}{2m_e m_i}\nabla\left(\frac{\nabla^2\sqrt{\rho}}{\sqrt{\rho}}\right), \tag{59}$$

and the idealized Ohm's law

$$\frac{\partial \boldsymbol{B}}{\partial t} = \nabla \times (\boldsymbol{V}\times\boldsymbol{B}), \tag{60}$$

where the variables are defined as above. Using the magnetization (51), we obtain a closed set of equations.

In what follows, we will study the linear modes of this system, with a particular focus on the stability properties. With $\rho = \rho_0 + \rho_1$, $\boldsymbol{B} = \boldsymbol{B}_0 + \boldsymbol{B}_1$, $\boldsymbol{M} = \boldsymbol{M}_0 + \boldsymbol{M}_1$, and $\boldsymbol{v} = \boldsymbol{v}_1$, such that $\rho_1 \ll \rho_0$, $|\boldsymbol{B}_1| \ll |\boldsymbol{B}_0|$, $|\boldsymbol{M}_1| \ll |\boldsymbol{M}_0|$, and $\boldsymbol{B}_0 = B_0\hat{\boldsymbol{z}}$, we linearize our equations in the perturbed variables. Assuming that the background quantities are constants, the general dispersion relation can, after a Fourier decomposition, be written

$$\left(\omega^2 - k_z^2\widetilde{C}_A^2\right)\left[\left(\omega^2 - k^2\widetilde{C}_A^2 - k_x^2\widetilde{V}_A^2(k)\right)\left(\omega^2 - k_z^2 V_A^2(k)\right) + k_x^2 k_z^2 \widetilde{V}_A^4(k)\right] = 0 \tag{61}$$

where $\widetilde{C}_A$ is the spin-modified Alfvén velocity given by

$$\widetilde{C}_A = \frac{C_A}{\left[1 + (\hbar\omega_{pe}^2/2m_e c^2\omega_{ce}^{(0)})\tanh(\mu_B B_0/k_B T)\right]^{1/2}}, \tag{62}$$

C_A is the standard Alfvén velocity $C_A = (B_0^2/\mu_0\rho_0)^{1/2}$,

$$\widetilde{V}_A^2(k) = V_A^2(k) - \frac{\hbar\omega_{ce}}{m_i}\tanh\left(\frac{\mu_B B_0}{k_B T}\right), \tag{63}$$

and

$$V_A^2(k) = c_s^2 + \frac{\hbar^2 k^2}{4m_i m_e} \tag{64}$$

Here $\omega_{pe} = (\rho_0 e^2/\varepsilon_0 m_e m_i)^{1/2}$ is the plasma frequency, $\omega_{ce}^{(0)}$ is the electron cyclotron frequency associated with the external magnetic field (i.e. with the contribution to B_0 from the spin sources excluded). The relation between the full electron cyclotron frequency $\omega_{ce} = eB_0/m_e$ and $\omega_{ce}^{(0)}$ is given by $\omega_{ce} = \omega_{ce}^{(0)} + (\hbar\omega_{pe}^2/m_e c^2)\tanh(\mu_B B_0/k_B T)$. We stress that $\widetilde{V}_A$, which to some extent can be considered as an effective acoustic velocity, may be imaginary for a strongly magnetized plasma due to the spin contribution, a fact which will be explored in some detail below.

We consider propagation perpendicular to the external magnetic field, which is the geometry which leads to instability most easily. For the case $\boldsymbol{k} = k\hat{\boldsymbol{x}}$, Eq. (61) reduces to

$$\omega = k\left[\frac{C_A^2}{1 + (\hbar\omega_{pe}^2/2m_e c^2\omega_{ce}^{(0)})\tanh(\mu_B B_0/k_B T)} + c_s^2 + \frac{\hbar^2 k^2}{4m_i m_e} - \frac{\hbar\omega_{ce}}{m_i}\tanh\left(\frac{\mu_B B_0}{k_B T}\right)\right]^{1/2}. \tag{65}$$

The necessary and sufficient instability condition can thus be written as

$$P_{\text{sp}} + P_{\text{m}} + P + \frac{\rho_0\hbar^2 k^2}{4m_e m_i} < 0, \tag{66}$$

where the total pressure $P_{\text{tot}} = P_{\text{sp}} + P_{\text{m}} + P$ consists of the effective spin pressure $P_{\text{sp}} = -(\rho_0\hbar\omega_{ce}/m_i)\tanh(\mu_B B_0/k_B T)$, which is the only negative pressure term and therefore the source of the instability, the magnetic pressure $P_{\text{m}} = \rho_0 C_A^2/[1 + (\hbar\omega_{pe}^2/2m_e c^2\omega_{ce}^{(0)})\tanh(\mu_B B_0/k_B T)]$, and the particle pressure $P = n_0 m_i c_s^2$, containing both the thermal and Fermi pressure part. Thus, a plasma can contain a magnetization instability, much like the gravitationally induced Jeans instability.[25]

6. Conclusions and future possibilities

We have seen that quantum effects, and in particular the electron spin, can introduce new and interesting aspects in plasma theory and experiments. It is expected that the rapid development of quantum plasma theory will be fueled by recent experiments (see, *e.g*, Ref. 31), and that new regimes of interest will enter the arena as the next-generation laser systems gets online.

Some particular developments that would be of interest is to further look at spin effects from a kinetic perspective.[61,62] Such treatments would be similar to the density matrix approach, and could, by using analogies from classical kinetic theory, spur the experimental interest in collective quantum effects and the transition from quantum to classical behaviour. Moreover, the fluid equations presented here has not been analyzed to their full extent. For example, there are terms which have been neglected, that could produce interesting nonlinear effects in quantum plasmas, such as spin self-interaction. Such topics will be approached in future research.

References

1. D. Pines, J. Nucl. Energy C: Plasma Phys. **2**, 5 (1961).
2. D. Pines, *Elementary Excitations in Solids* (Westview Press, 1999)
3. D. Kremp, M. Schlanges, and W.-D. Kraeft, *Quantum Statistics of Nonideal Plasmas* (Springer, 2005).
4. B. Bezzerides and D. F. DuBois, Ann. Phys. (N.Y.) **70**, 10 (1972).
5. R. Hakim and J. Heyvaerts, Phys. Rev. A **18**, 1250 (1978).
6. G. Manfredi, Fields Inst. Comm **46**, 263 (2005)
7. F. Haas, G. Manfredi, and M. R. Feix, Phys. Rev. E **62**, 2763 (2000).
8. F. Haas, Phys. Plasmas **12**, 062117 (2005).
9. L. G. Garcia, F. Haas, L. P. L. de Oliveira, and J. Goedert, Phys. Plasmas **12**, 012302 (2005).
10. P. K. Shukla, Phys. Lett. A **352**, 242 (2006).
11. F. Haas, L. G. Garcia, J. Goedert, and G. Manfredi, Phys. Plasmas **10**, 3858 (2003).
12. L. G. Garcia, F. Haas, L. P. L. de Oliviera, and J. Goedert, Phys. Plasmas **12**, 012302 (2005).
13. P. K. Shukla and L. Stenflo, Phys. Lett. A **355**, 378 (2006).
14. P. K. Shukla, Phys. Lett. A **357**, 229 (2006).
15. P. K. Shukla, L. Stenflo, and R. Bingham, Phys. Lett. A **359**, 218 (2006).
16. P. K. Shukla and B. Eliasson, Phys. Rev. Lett. **96**, 245001 (2006).
17. P. K. Shukla, S. Ali, L. Stenflo, and M. Marklund, Phys. Plasmas **13**, 112111 (2006).
18. F. Haas, Europhys. Lett. **44**, 45004 (2007).
19. M. Marklund and G. Brodin, Phys. Rev. Lett. **98**, 025001 (2007).
20. G. Brodin and M. Marklund, New J. Phys. **9**, 277 (2007).
21. G. Brodin and M. Marklund, Phys. Plasmas **14**, 112107 (2007).
22. P. K. Shukla and B. Eliasson, Phys. Rev. Lett. **96**, 245001 (2006).
23. P. K. Shukla and B. Eliasson, Phys. Rev. Lett. **99**, 096401 (2007).
24. D. Shaikh and P. K. Shukla, Phys. Rev. Lett. **99**, 125002 (2007).
25. G. Brodin and M. Marklund, Phys. Rev. E, in press (2007). (arXiv:0709.3575)
26. H. G. Craighead, Science **290**, 1532 (2000).
27. G. Manfredi and P.-A. Hervieux, Appl. Phys. Lett. **91**, 061108 (2007).
28. W. Li, P. J. Tanner, and T. F. Gallagher, Phys. Rev. Lett. **94**, 173001 (2005).
29. R. S. Fletcher, X. L. Zhang, and S. L. Rolston, Phys. Rev. Lett. 96, 105003 (2006).
30. A. K. Harding and D. Lai, Rep. Prog. Phys. **69**, 2631 (2006).
31. S. H. Glenzer et al., Phys. Rev. Lett. **98**, 065002 (2007).
32. M. Marklund and P. K. Shukla, Rev. Mod. Phys. **78**, 591 (2006).
33. E. Lundström et al., Phys. Rev. Lett. **96**, 083602 (2006).

34. G. Brodin, M. Marklund, B. Eliasson and P. K. Shukla, Phys. Rev. Lett. **98**, 125001 (2007)
35. J. Lundin, J. Zamanian, M. Marklund and G. Brodin, Phys. Plasmas, **14**, 062112 (2007).
36. P. R. Holland, *The Quantum Theory of Motion* (Cambridge University Press, Cambridge, 1993).
37. P. A. M. Dirac, *Principles of Quantum Mechanics* (Oxford University Press, Oxford, 1981).
38. A. O. Barut and W. D. Thacker, Phys. Rev. D **31**, 2076 (1985).
39. S. R. de Groot and L. G Suttorp, *Foundations of Electrodyanmics* (North-Holland, 1972).
40. J. Dawson, Phys. Fluids **4**, 869 (1961).
41. D. Anderson, B. Hall, M. Lisak, and M. Marklund, Phys. Rev. E **65**, 046417 (2002).
42. E. P. Wigner, Phys. Rev. **40**, 749 (1932).
43. J. E. Moyal, Proc. Cambridge Philos. Soc. **45**, 99 (1949).
44. J. T. Mendonça, *Theory of Photon Acceleration* (IOP Publishing, 2001).
45. W. P. Schleich, *Quantum Optics in Phase Space* (Wiley, 2001).
46. M. Marklund, Phys. Plasmas **12**, 082110 (2005).
47. H. Alfvén, Nature **150**, 405 (1942).
48. B. I. Halperin and P. C. Hohenberg, Phys. Rev. **188**, 898 (1969).
49. A. V. Balatsky, Phys. Rev. B **42**, 8103 (1990).
50. U. W. Rathe, C. H. Keitel, M. Protopapas, and P. L. Knight, J. Phys. B: At. Mol. Opt. Phys. **30**, L531 (1997).
51. S. X. Hu and C. H. Keitel, Phys. Rev. Lett. **83**, 4709 (1999).
52. R. Arvieu, P. Rozmej, and M. Turek, Phys. Rev. A **62**, 022514 (2000).
53. J. R. Vázquez de Aldana and L. Roso, J. Phys. B: At. Mol. Opt. Phys. **33**, 3701 (2000).
54. M. W. Walser and C. H. Keitel, J. Phys. B: At. Mol. Opt. Phys. **33**, L221 (2000).
55. M. W. Walser, D. J. Urbach, K. Z. Hatsagortsyan, S. X. Hu, and C. H. Keitel, Phys. Rev. A **65**, 043410 (2002).
56. Z. Qian and G. Vignale, Phys. Rev. Lett. **88**, 056404 (2002).
57. J. S. Roman, L. Roso, and L. Plaja, J. Phys. B: At. Mol. Opt. Phys. **37**, 435 (2004).
58. R. L. Liboff, Europhys. Lett. **68**, 577 (2004).
59. J. N. Fuchs, D. M. Gangardt, T. Keilman, and G. V. Shlyapnikov, Phys. Rev. Lett. **95**, 150402 (2005).
60. K. Kirsebom *et al.*, Phys. Rev. Lett. **87**, 054801 (2001).
61. S. C. Cowley, R. M. Kulsrud, and E. Valeo, Phys. Fluids **29**, 430 (1986).
62. R. M. Kulsrud, E. J. Valeo, and S. C. Cowley, Nucl. Fusion **26**, 1443 (1986).

REVISED QUANTUM ELECTRODYNAMICS WITH FUNDAMENTAL APPLICATIONS

B. LEHNERT

Alfvén Laboratory, Royal Institute of Technology,
S-10044 Stockholm, Sweden
E-mail: Bo.Lehnert@ee.kth.se

There are important areas within which conventional electromagnetic theory and its combination with quantum mechanics does not provide fully adequate descriptions of physical reality. These difficulties are not removed by and are not directly associated with quantum mechanics. Instead electromagnetic field theory is a far from completed area of research, and modified forms of it have been elaborated by several investigators during the recent decades. The investigation to be described here has the form of a Lorentz and gauge invariant theory which is based on a nonzero electric field divergence in the vacuum state. It aims beyond Maxwell's equations and leads to new solutions of a number of fundamental problems. The applications include a model of the electron with its point-charge-like nature, the associated self-energy problem, the radial force balance, and a quantized minimum of the elementary electronic charge. There are further applications on the individual photon and on light beams, in respect to the angular momentum (spin), the spatially limited geometry, the associated needle radiation, and the particle-wave nature, such as in the photoelectric effect and in two-slit experiments at low light intensities.

1. Introduction

Maxwell's equations in the vacuum state have served as guideline and basis in the development of quantum electrodynamics (QED) which has been successful in many applications and has sometimes manifested itself in an extremely good agreement with experiments. However, as pointed out by Feynman,[1] there nevertheless exist areas within which conventional electromagnetic theory and its combination with quantum mechanics does not provide fully adequate descriptions of physical reality. These difficulties are not removed by and are not directly associated with quantum mechanics. Instead electromagnetic field theory is a far from completed area of research, and QED will therefore also become subject to the topical shortcomings of such a theory in its conventional form.

As a consequence, modified theories leading beyond Maxwell's equations have been elaborated by several investigators during the recent decades. This advancement of research has been described in books, reviews, and conference proceedings all of which cannot be mentioned here, but where some of the more recent one has been listed in a survey by the author.[2]

Among these new approaches there is one theory to be treated in this paper which attaches main importance to conceptual features and leaves out part of the detailed formal deductions which are reported elsewhere.[2–9] The theory will be shown to have a number of fundamental applications, such as deduced models of the electron and photon.

2. Some Unsolved Problems in Conventional Theory

There are a number of important physical features which have so far not been fully explained in terms of conventional theory. The first to be mentioned here is the point-charge-like behaviour of the electron which appears to have an extremely small radius. Second, the question arises why the electron does not "explode" under the action of its self-charge. It has been assumed that its internal force balance is due to some unknown nonelectromagnetic cause.[10] Third, the point-charge-like character seems to end up with an infinite self-energy. This problem has been solved in the renormalization procedure by adding extra counter-terms to the Lagrangian, to obtain a finite result from the difference between two infinities. Such a procedure is not quite satisfactory from the physical point of view.[11] Fourth, there is no explanation why the free electronic charge has a quantized minimum value "e", as first shown in the experiments by Millikan.

When further considering the individual photon, it is first noticed that conventional theory does not explain that it has an angular momentum (spin), and this is also the case of a light beam with limited cross-section.[12] Second, the question arises how a propagating photon can behave as an object with limited spatial extensions, because conventional theory results in divergent solutions and infinite integrals of the field energy when being extended all over space. Third, it still has to be understood how the photon can behave both as a particle and as a wave.

3. Basis of Present Revised Field Equations

To revise the conventional theory we now turn to recent knowledge of and aspects on the vacuum state which is not merely an empty space. Due to quantum mechanics, there is a nonzero level of its ground state, the zero-point energy. Related electromagnetic vacuum fluctuations appear which give rise to the Casimir effect[13] due to which two metal plates attract each other when being brought sufficiently close together. The observed electron-positron pair formation from an energetic photon further demonstrates that electric charges can be created out of an electrically neutral state.

To translate these vacuum properties into a quantitative form, we start with the Lorentz invariant Proca-type field equation in four-space

$$\Box A_\mu \equiv \left(\frac{1}{c^2}\frac{\partial^2}{\partial t^2} - \nabla^2\right) A_\mu = \mu_0 J_\mu \qquad \mu = 1,2,3,4\,. \tag{1}$$

Here

$$A_\mu = (\mathbf{A}, i\phi/c) \qquad\qquad J_\mu = (\mathbf{j}, ic\bar{\rho}) \tag{2}$$

with $\mathbf{A}$ and ϕ standing for the magnetic vector potential and the electrostatic potential, and $\mathbf{j}$ and $\bar{\rho}$ representing the three-space current and charge density parts of a general four-current density J_μ constituting the sources of the electromagnetic field. Maxwell's equations in the vacuum are recovered when the four-current density vanishes.

Equations (1) and (2) are now given a new interpretation, where $\bar{\rho}$ is a nonzero charge density which can arise in the vacuum, and $\mathbf{j}$ then stands for an associated current density. Physical experience supports the field equations to remain Lorentz invariant also with this interpretation. It implies that J_μ still has to behave as a four-vector, its square thereby being invariant when transformed from one inertial frame K to another such frame, K'. Thus

$$\mathbf{j}^{\,2} - c^2\bar{\rho}^{\,2} = \mathbf{j}'^{\,2} - c^2\bar{\rho}'^{\,2} = 0 \tag{3}$$

if we further require that $\mathbf{j}$ should exist only when there is also a charge density $\bar{\rho}$. This merely becomes analogous to a choice of origin for the space and time coordinates. The final form of the four-current thus becomes

$$J_\mu = \bar{\rho}(\mathbf{C}, ic) \qquad\qquad \mathbf{C}^{\,2} = c^2 \tag{4}$$

where $\mathbf{C}$ is a velocity vector having a modulus equal to the velocity constant c of light. In analogy with the direction to be specified for the current density in conventional theory, the unit vector $\mathbf{C}/c$ depends on the specific geometry to be considered. There is a connection between the current density (4) and the electron theory by Dirac.[5,8]

The three-dimensional representation of the present revised and extended field equations *in the vacuum* now becomes

$$\text{curl } \mathbf{B}/\mu_0 = \varepsilon_0(\text{div } \mathbf{E})\mathbf{C} + \varepsilon_0 \partial \mathbf{E}/\partial t \tag{5}$$

$$\text{curl } \mathbf{E} = -\partial \mathbf{B}/\partial t \tag{6}$$

$$\mathbf{B} = \text{curl } \mathbf{A} \qquad \text{div } \mathbf{B} = 0 \tag{7}$$

$$\mathbf{E} = -\nabla \phi - \partial \mathbf{A}/\partial t \qquad \text{div } \mathbf{E} = \bar{\rho}/\varepsilon_0 \tag{8}$$

for the electric and magnetic fields $\mathbf{E}$ and $\mathbf{B}$. Here we also include the relation $\text{div } \mathbf{C} = 0$. In these equations there is a dielectric constant ε_0 and a magnetic permeability μ_0 of the conventional vacuum, because they apply to a state which does not include electrically polarized and magnetized atoms or molecules. The new feature of Eqs. (5)–(8) is the appearance of the electric field divergence terms. In principle, a nonzero electric field divergence in the vacuum should not be less conceivable than the nonzero curl of the magnetic field in the vacuum of Maxwell's theory. A nonzero magnetic field divergence is on the other hand not introduced here, but is with Dirac[14] left as an open question as well as that of magnetic monopoles.

Equations (5) and (6) include the field strengths $\mathbf{E}$ and $\mathbf{B}$ only, and are therefore invariant to a gauge transformation. This does not always become the case for Eqs. (1) and (2) when there are other forms of the four-current J_μ.

Using well-known vector identities, Eqs. (5)–(8) result in the local momentum equation

$$\text{div } {}^2\mathbf{S} = \mathbf{f} + \frac{\partial}{\partial t}\mathbf{g} \tag{9}$$

with

$$\mathbf{f} = \bar{\rho}(\mathbf{E} + \mathbf{C} \times \mathbf{B}) \tag{10}$$

$$\mathbf{g} = \varepsilon_0 \mathbf{E} \times \mathbf{B} = \frac{1}{c^2}\mathbf{S}\,. \tag{11}$$

Here ${}^2\mathbf{S}$ is the electromagnetic stress tensor, $\mathbf{f}$ is the volume force density, $\mathbf{g}$ can be interpreted as an electromagnetic momentum density, and $\mathbf{S}$ is the Poynting vector. Likewise a local energy equation

$$-\text{div } \mathbf{S} = \bar{\rho}\, \mathbf{E} \cdot \mathbf{C} + \frac{\partial}{\partial t} w_f \tag{12}$$

is obtained where

$$w_f = \frac{1}{2}\left(\varepsilon_0 \mathbf{E}^2 + \mathbf{B}^2/\mu_0\right) \tag{13}$$

represents the electromagnetic field energy density. An electromagnetic source energy density

$$w_s = \frac{1}{2}\bar{\rho}\,(\phi + \mathbf{C} \cdot \mathbf{A}) \tag{14}$$

can as well be deduced by recasting the form (13). The volume integrals of w_f and w_s become equal for certain steady configurations which are limited in space.[2,5] In the cases where the volume force (10) can be neglected, the angular momentum density is finally given by

$$\mathbf{s} = \mathbf{r} \times \mathbf{S}/c^2 \tag{15}$$

where $\mathbf{r}$ is the radius vector pointing in the direction from the origin.

It has to be remembered that relations (9) and (12) have merely been obtained from a rearrangement of the basic equations. They therefore have forms by which equivalent expressions are obtained for the momentum and energy from the stress tensor.

This section is ended in summarizing the characteristic features of the present revised field equations:

- The theory is based on the pure radiation field in the vacuum state.
- The theory is both Lorentz and gauge invariant.
- The nonzero electric field divergence introduces an additional degree of freedom which changes the character of the field equations substantially, and leads to new physical phenomena.
- The velocity of light is no longer a scalar c but a vector $\mathbf{C}$ which has the modulus c.

4. The Present Form of QED

As shown by Heitler[15] quantization of the electromagnetic field equations, also with included source terms, ends up with the same equations in which the electromagnetic potentials are replaced by their expectation values. In the present approach, which is based on the pure radiation field, a rather good approximation should therefore be obtained in a simplified procedure where relevant quantum conditions are applied afterwards on the general solutions of the field equations. This could be conceived as a way to the most probable state of the quantized solutions.

In the original and current presentation of conventional QED, Maxwell's equations for a vanishing electric field divergence form the basis of the theory.[15,16] This becomes a further justification for the present theory to use its field equations as a basis for a revised quantum electrodynamic theory.

A special but important question concerns the momentum of the pure radiation field. In conventional QED this momentum is derived from a plane-wave representation and the Poynting vector.[15,16] In the present theory the Poynting vector and the momentum density (11) have an analogous rôle, in cases where the volume force (10) vanishes or can be neglected.

Here it has also to be noticed that there are nonrelativistic forms in conventional quantum mechanics, such as the Schrödinger equation, in which the quantized momentum has been successfully represented by the operator

$$\mathbf{p} = -i\hbar\nabla \tag{16}$$

for a massive particle. However, this concept leads to some inconsistencies when being applied to a photon model, as discussed later in this context and elsewhere.[6]

Thus there are additional points of view which also characterize the present theory:

- Being based on the pure radiation field, the theory includes no ad hoc assumption of particle mass at its outset.
- A possibly arising mass and particle behaviour comes out of "bound" states which result from a type of vortex-like "self-confinement" of the radiation.
- The wave nature results from the "free" states of propagating wave phenomena.
- These "bound" and "free" states can become integrating parts of the same system.

5. Steady Axisymmetric States

Steady electromagnetic field configurations are of special interest, in particular with respect to particles such as the leptons. In contrast to conventional theory, the basic equations (5)–(8) provide a class of steady solutions in the vacuum state. These equations combine to

$$c^2 \mathrm{curl}\,\mathrm{curl}\,\mathbf{A} = -\mathbf{C}\left(\nabla^2\phi\right) = (-\bar{\rho}/\varepsilon_0)\,\mathrm{C}\,. \tag{17}$$

Here we limit ourselves to particle-shaped states in which the configuration becomes bounded in all spatial directions. In a further restriction to axisymmetry, a spherical frame (r, θ, φ) is introduced in which all quantities are independent of φ. With $\mathbf{C} = (0, 0, C)$, $C = \pm c$, $\mathbf{j} = (0, 0, C\bar{\rho})$, and

$\mathbf{A}=(0,0,A)$, Eqs. (17) reduce to

$$\frac{(r_0\rho)^2\bar{\rho}}{\varepsilon_0} = D\phi = \left[D + (\sin\theta)^{-2}\right](CA) \tag{18}$$

where $\rho=r/r_0$ is a dimensionless radial coordinate, r_0 a characteristic radial dimension, $D=D_\rho + D_\theta$ and

$$D_\rho = -\frac{\partial}{\partial\rho}\left(\rho^2\frac{\partial}{\partial\rho}\right) \qquad\qquad D_\theta = -\frac{\partial^2}{\partial\theta^2} - \frac{\cos\theta}{\sin\theta}\frac{\partial}{\partial\theta}\,. \tag{19}$$

5.1. *The Generating Function*

A solution of Eq. (18) can be obtained in terms of a *generating function*

$$F(r,\theta) = CA - \phi = G_0 \cdot G(\rho,\theta) \tag{20}$$

where G_0 stands for its characteristic amplitude and G for a normalized dimensionless part. This yields the general solution

$$CA = -(\sin\theta)^2 DF \qquad\qquad \phi = -\left[1 + (\sin\theta)^2 D\right] F \tag{21}$$

$$\bar{\rho} = -\left(\tfrac{\varepsilon_0}{r_0^2\rho^2}\right) D\left[1 + (\sin^2\theta)D\right] F\,. \tag{22}$$

With the definitions

$$f(\rho,\theta) \;=\; -(\sin\theta)D\left[1 + (\sin\theta)^2 D\right] G \tag{23}$$

$$g(\rho,\theta) \;=\; -\left[1 + 2(\sin\theta)^2 D\right] G \tag{24}$$

the integrated (total) electric charge q_0, magnetic moment M_0, mass m_0, and angular momentum (spin) s_0 then become

$$q_0 = 2\pi\varepsilon_0 r_0 G_0 J_q \qquad\qquad I_q = f \tag{25}$$

$$M_0 = \pi\varepsilon_0 C r_0^2 G_0 J_M \qquad\qquad I_M = \rho\,(\sin\theta) f \tag{26}$$

$$m_0 = \pi(\varepsilon_0/c^2) r_0 G_0^2 J_m \qquad\qquad I_m = fg \tag{27}$$

$$s_0 = \pi(\varepsilon_0 C/c^2) r_0^2 G_0^2 J_s \qquad\qquad I_s = \rho\,(\sin\theta) fg \tag{28}$$

where

$$J_k = \int_{\rho_k}^{\infty}\int_0^{\pi} I_k \, d\rho d\theta \qquad\qquad k = q, M, m, s\,. \tag{29}$$

In Eq. (29) $\rho_k \neq 0$ are small radii of circles being centered around the origin $\rho=0$ when G is divergent there, and $\rho_k=0$ when G is convergent at $\rho=0$. The implication of $\rho_k \neq 0$ will be given later.

A further step is taken by imposing the restriction of a separable generating function

$$G(\rho, \theta) = R(\rho) \cdot T(\theta)\,. \tag{30}$$

This restriction is useful when treating configurations where the sources $\bar{\rho}$ and $\mathbf{j}$ and the corresponding energy density are mainly localized near the origin $\rho = 0$, such as for a particle of limited spatial extent.[2] The integrands of the form (29) then become

$$I_q = \tau_0 R + \tau_1 (D_\rho R) + \tau_2 D_\rho (D_\rho R) \tag{31}$$

$$I_M = \rho\,(\sin\theta) I_q \tag{32}$$

$$I_m = \tau_0 \tau_3 R^2 + (\tau_0 \tau_4 + \tau_1 \tau_3) R (D_\rho R) + \tau_1 \tau_4 (D_\rho R)^2 + \\ + \tau_2 \tau_3 R D_\rho (D_\rho R) + \tau_2 \tau_4 (D_\rho R)\,[D_\rho (D_\rho R)] \tag{33}$$

$$I_s = \rho\,(\sin\theta) I_m \tag{34}$$

where

$$\tau_0 = -(\sin\theta)(D_\theta T) - (\sin\theta) D_\theta \left[(\sin^2\theta)(D_\theta T)\right] \tag{35}$$

$$\tau_1 = -(\sin\theta) T - (\sin\theta) D_\theta \left[(\sin^2\theta) T\right] - \sin^3\theta (D_\theta T) \tag{36}$$

$$\tau_2 = -(\sin^3\theta) T \tag{37}$$

$$\tau_3 = -T - 2(\sin^2\theta)(D_\theta T) \tag{38}$$

$$\tau_4 = -2(\sin^2\theta) T\,. \tag{39}$$

5.2. *Features of the Generating Function*

Among the possible forms to be adopted of the generating function, we will here consider radial parts R which can become convergent or divergent at $\rho = 0$, but which always decrease strongly towards zero when $\rho \to \infty$. The polar part T and its derivatives are always finite. It can be symmetric or antisymmetric with respect to the "equatorial plane" (midplane) defined by $\theta = \pi/2$.

For a radial part R which is convergent at the origin and where $\rho_k = 0$, partial integration yields $J_q = J_{q\rho} J_{q\theta}$ and $J_M = J_{M\rho} J_{M\theta}$ for the integrals (29) where

$$J_{q\rho} = \int_0^\infty R\, d\rho \qquad\qquad J_{M\rho} = \int_0^\infty \rho R\, d\rho \tag{40}$$

$$J_{q\theta} = \left\{ (\sin\theta) \frac{d}{d\theta} \left[(\sin\theta)^2 (D_\theta T)\right] + (\sin\theta) \frac{dT}{d\theta} \right\}_0^\pi = 0 \tag{41}$$

$$J_{M\theta} = \left\{ (\sin\theta)^3 \frac{d}{d\theta} \left[(\sin\theta)(D_\theta T - 2T)\right] \right\}_0^\pi = 0\,. \tag{42}$$

Thus both q_0 and M_0 vanish in this case.

Concerning the polar part T, the integrals (29) with respect to θ become nonzero for symmetric integrands I_k but vanish for antisymmetric ones. The symmetry or antisymmetry of T further leads to a corresponding symmetry or antisymmetry of $D_\theta T$, $D_\theta\left[(\sin\theta)^2 T\right]$, and $D_\theta\left[(\sin\theta)^2(D_\theta T)\right]$. As a result the integrated mass m_0 and angular momentum s_0 always become nonzero, whereas the charge q_0 and magnetic moment M_0 have the following features:

- A neutral state of vanishing q_0 and M_0 is obtained for a radial part R which is convergent at $\rho = 0$, and regardless of the symmetry properties of the polar function T. This leads to models of the neutrino, not to be treated here in detail but being described elsewhere.[2,6,8]
- An electrically charged state of nonzero q_0 and M_0 requires a radial part R which is *divergent* at $\rho = 0$, and a polar part T being symmetric with respect to the equatorial plane. At a first glance this appears to lead to divergent final solutions which become physically unacceptable. However, the analysis of the electron model in the following Section 6 demonstrates a way out of this difficulty. It is based on arbitrarily small limits $\rho_k \neq 0$ in Eq. (29).

5.3. *Quantum Conditions of Particle-shaped States*

The nonzero electric field divergence provides the field equations with a certain degree of freedom, here manifesting itself in the partly arbitrary form of the generating function. To close the system, relevant quantum conditions have to be imposed, as well as conditions on the force balance which is treated later.

The angular momentum (spin) condition on models of the electron as a fermion, or of the neutrino, is combined with expression (28) to result in

$$s_0 = \pi\left(\varepsilon_0 C/c^2\right) r_0^2 G_0^2 J_s = \pm h/4\pi\,. \tag{43}$$

This condition is compatible with the two signs of $C = \pm c$ due to relation (4). In particular for a charged particle, Eqs. (25) and (43) combine to a dimensionless charge

$$q^* \equiv |q_0/e| = \left(f_0 J_q^2/2J_s\right)^{1/2} \qquad f_0 = 2\varepsilon_0 ch/e^2 \tag{44}$$

being normalized with respect to the experimentally determined elementary charge "e", and where $f_0 \cong 137.036$ is the inverted value of the fine-structure constant.

According to Dirac,[17] Schwinger[18] and Feynman[19] the quantum condition on the magnetic moment of a charged particle such as the electron is determined by

$$M_0 m_0 / q_0 s_0 = 1 + \delta_M \qquad \delta_M = 1/2\pi f_0 \tag{45}$$

which shows excellent agreement with experiments.

Conditions (43) and (45) can also be made plausible by elementary physical arguments based on the present picture of a particle-shaped state of "self-confined" radiation.[2,8]

In a charged particle-shaped state with nonzero magnetic moment, the electric current distribution generates a total magnetic flux Γ_{tot}. The quantized value of the angular momentum s_0 further depends on the type of configuration to be considered. It becomes $h/4\pi$ for a fermion, but $h/2\pi$ for a boson. Here the electron is conceived to be a system also having a quantized charge q_0. The magnetic flux should then become quantized as well, and be given by the two concepts s_0 and q_0, in a relation having the dimension of magnetic flux. This leads to the flux quantum condition[5,9]

$$\Gamma_{tot} = |s_0/q_0| \, . \tag{46}$$

6. A Model of the Electron

In this section a model of the electron will be elaborated which in principle also applies to the muon, tauon, and to the corresponding antiparticles.

6.1. *The Form of the Generating Function*

In accordance with the discussion of Section 5.2, a generating function is now chosen having the parts

$$R = \rho^{-\gamma} e^{-\rho} \qquad \gamma > 0 \tag{47}$$

$$T = 1 + \sum_{\nu=1}^{n} \{a_{2\nu-1} \sin[(2\nu - 1)\theta] + a_{2\nu} \cos(2\nu\theta)\} \, . \tag{48}$$

Here the radial part R diverges at $\rho = 0$ as required, but the form (47) may at first glance appear to be somewhat special and artificial. Generally one could thus have introduced a negative power series of ρ instead of the single term $\rho^{-\gamma}$. However, due to the analysis which follows, such a series has to be replaced by one single term only, with a locked special value. The exponential factor in expression (47) has further been included to secure convergence of any moment with R at large distances from the origin.

The polar part T represents a general form of geometry having top-bottom symmetry with respect to the equatorial plane.

6.2. *Integrated Field Quantities at a Shrinking Characteristic Radius*

Since the radial part R is divergent at the origin, its divergence must be outbalanced. This can be done by introducing the concept of a shrinking characteristic radius r_0 to obtain finite integrated field quantities. We therefore define

$$r_0 = c_0 \varepsilon \qquad 0 < \varepsilon \ll 1 \tag{49}$$

where c_0 is a positive constant of the dimension length and ε is a dimensionless smallness parameter. Insertion of the forms (47) and (48) into Eqs. (25)–(39) yields after some deductions the result

$$\begin{aligned} q_0 &= 2\pi\varepsilon_0 c_0 G_0 \left[J_{q\theta}/(\gamma-1)\right](\varepsilon/\rho_q^{\gamma-1}) && (50)\\ M_0 m_0 &= \pi^2(\varepsilon_0^2 C/c^2)c_0^3 G_0^3 \left[J_{M\theta}J_{m\theta}/(\gamma-2)(2\gamma-1)\right]\cdot(\varepsilon^3/\rho_M^{\gamma-2}\rho_m^{2\gamma-1}) && (51)\\ s_0 &= \pi(\varepsilon_0 C/c^2)c_0^2 G_0^2 \left[J_{s\theta}/2(\gamma-1)\right](\varepsilon/\rho_s^{\gamma-1})^2 && (52) \end{aligned}$$

where

$$J_{k\theta} = \int_0^\pi I_{k\theta}\, d\theta \qquad k = q, M, m, s \tag{53}$$

is determined from the quantities (35)–(39) and will later be given in its final form. The reason for introducing the compound quantity $M_0 m_0$ in expression (51) is that this quantity appears as a single entity in all finally obtained results of the present theory. A separation of M_0 and m_0 is in itself an important problem which has so far not been considered.

The integrated quantities (50)–(52) are now required to become finite for all values of the parameter ε and of the radii ρ_k, and to scale in such a way that the field geometry becomes independent of ε and ρ_k in the range of small ε. Such a uniform scaling becomes possible when

$$\rho_q = \rho_M = \rho_m = \rho_s = \varepsilon \tag{54}$$

and when the radial parameter γ approaches the value 2 from above, as given by

$$\gamma(\gamma-1) = 2+\delta \qquad 0 \le \delta \ll 1 \qquad \gamma \approx 2+\delta/3\,. \tag{55}$$

From the earlier results (41) and (42) and with relation (55), the integrands $I_{k\theta}$ of Eq. (53) then reduce to

$$I_{q\theta} = -2\tau_1 + 4\tau_2 \tag{56}$$

$$I_{M\theta}/\delta = (\sin\theta)(-\tau_1 + 4\tau_2) \tag{57}$$

$$I_{m\theta} = \tau_0\tau_3 - 2(\tau_0\tau_4 + \tau_1\tau_3) + 4(\tau_1\tau_4 + \tau_2\tau_3) - 8\tau_2\tau_4 \tag{58}$$

$$I_{s\theta} = (\sin\theta) I_{m\theta}\,. \tag{59}$$

This leads to the finite integrated field quantities

$$q_0 = 2\pi\varepsilon_0 c_0 G_0 A_q \tag{60}$$

$$M_0 m_0 = \pi^2(\varepsilon_0^2 C/c^2)c_0^3 G_0^3 A_M A_m \tag{61}$$

$$s_0 = (1/2)\pi(\varepsilon_0 C/c^2)c_0^2 G_0^2 A_s \tag{62}$$

where

$$A_q \equiv J_{q\theta} \quad A_M \equiv J_{M\theta}/\delta \quad A_m \equiv J_{m\theta} \quad A_s \equiv J_{s\theta} \tag{63}$$

as obtained from Eqs. (53), (55) and (56)–(59).

6.3. *The Magnetic Flux*

Using Eqs. (21), (19), (47), (49) and (55) the magnetic flux function becomes

$$\Gamma = 2\pi\,(c_0 G_0/C)\sin^3\theta\left[(2 + 2\rho + \rho^2)T - D_\theta T\right](\varepsilon/\rho)\,. \tag{64}$$

It increases strongly as ρ decreases towards small values, such as for a point-charge-like behaviour. To obtain a nonzero and finite flux, one has to choose a corresponding dimensionless lower radius limit $\rho = \rho_\Gamma = \varepsilon$, in analogy with relations (49) and (54). There is a main part of the flux the magnetic field lines of which intersect the equatorial plane. This flux is counted from the sphere $\rho_\Gamma = \varepsilon$ and outwards, and is given by

$$\Gamma_0 = -\Gamma(\rho = \varepsilon, \theta = \pi/2) = 2\pi(c_0 G_0/C)A_\Gamma \tag{65}$$

where

$$A_\Gamma = [D_\theta T - 2T]_{\theta=\pi/2}\,. \tag{66}$$

The flux (65) ca be regarded as to be generated by a set of thin current loops which are localized to a spherical surface of radius $\rho = \varepsilon$.

It has to be observed that the flux (65) is not necessarily the total flux generated by the current system as a whole. In the present case it is found that one magnetic island is formed above and one below the equatorial plane, and where each island possesses an isolated flux which does not intersect the same plane.[2,9] The total flux Γ_{tot} thus consists of the main flux $-\Gamma_0$ plus an extra island flux Γ_i which can be deduced from the function (64).

We now introduce the normalized flux function

$$\Psi \equiv \Gamma(\rho = \varepsilon, \theta)/2\pi(c_0 G_0/C) = \sin^3\theta(D_\theta T - 2T) \qquad (67)$$

in the upper half-plane if the sphere $\rho = \varepsilon$. The radial magnetic field component vanishes at the angles $\theta = \theta_1$ and $\theta = \theta_2$ in the range $0 \leq \theta \leq \pi/2$. When θ increases from $\theta = 0$ at the axis of symmetry, the flux Ψ first increases to a maximum at the angle $\theta = \theta_1$. Then there follows an interval $\theta_1 \leq \theta \leq \theta_2$ of decreasing flux, down to a minimum at $\theta = \theta_2$. Finally, in the range $\theta_2 \leq \theta \leq \pi/2$ there is again an increasing flux, up to the total main flux value

$$\Psi_0 = \Psi(\pi/2) = A_\Gamma\,. \qquad (68)$$

This behaviour is due to a magnetic island having dipole-like field geometry with current centra at the angles θ_1 and θ_2 on the spherical surface. We also define the resulting outward island flux

$$\Psi_i = \Psi(\theta_1) - \Psi(\theta_2) \qquad (69)$$

of one magnetic island. The total flux which includes the main flux (68) and that from two magnetic islands then becomes

$$\Psi_{tot} = f_{\Gamma f}\Psi_0 \qquad f_{\Gamma f} = 1 + 2(\Psi_i/\Psi_0) > 1 \qquad (70)$$

where $f_{\Gamma f}$ is a *resulting* flux factor due to the magnetic flied geometry and its magnetic islands.

6.4. *The Quantum Conditions*

Relevant quantum conditions can now be imposed on the system. For the angular momentum (43) the associated normalized charge (44) becomes

$$q^* = (f_0 A_q^2/A_s)^{1/2}\,. \qquad (71)$$

The magnetic moment relation (45) further reduces to

$$A_M A_m/A_q A_s = 1 + \delta_M\,. \qquad (72)$$

Finally the magnetic flux quantization due to condition (46) and expressions (68), (60), (62) and (63) obtains the form

$$8\pi f_{\Gamma q} A_\Gamma A_q = A_s \qquad (73)$$

where $f_{\Gamma q}$ is the flux factor being *required* by the quantization. Only when one arrives at a self-consistent solution will the flux factors of Eqs. (70) and (73) become equal to a common factor

$$f_{\Gamma} = f_{\Gamma f} = f_{\Gamma q} \,. \tag{74}$$

6.5. *Variational Analysis on the Integrated Charge*

The elementary electronic charge has so far been considered as an independent and fundamental physical constant of nature, determined through measurements only. Since it appears to represent the smallest quantum of free electronic charge, however, the question can be raised whether there is a more profound reason for such a minimum charge to exist.

In the present theoretical approach, standard variational analysis was first applied to the normalized charge (71), including Lagrange multipliers when treating relations (72) and (73) as subsidiary conditions, and having the amplitudes $a_{2\nu-1}$ and $a_{2\nu}$ of the expansion (48) as independent variables. This attempt failed, because there was no well-defined extremum point in amplitude space but instead a clearly expressed plateau behaviour. The analysis then proceeded by successively including an increasing number of amplitudes which are "swept" (scanned) across their entire range of variation.[9] The results were as follows:

- In the case of four amplitudes (a_1, a_2, a_3, a_4) the normalized charge q^* was found to behave as shown in Fig. 1. Here conditions (72) and (73) have been imposed with a_3 and a_4 being left as variables for the scanning. There is a steep barrier of large q^* for values of a_3 and a_4 near the origin, and a very flat "plateau" close to the experimental value $q^* = 1$ in the ranges of large $|a_3|$ and $a_4 > 0$. This plateau is slightly "warped", having values which vary along its perimeter from $q^* = 0.969$ with $f_\Gamma = 1.81$ to $q^* = 1.03$ with $f_\Gamma = 1.69$.
- At an increasing number of amplitudes beyond four, there is a similar but slightly changed and somewhat higher plateau. This can be understood in the way that the contributions in the expansion (48) from higher order multipoles should have a limited effect on the integrated profiles of the polar function T. Moreover, an increase of the minimum level of q^* at the inclusion of an extra variable is not in conflict with the variational principle, because any function can have minima when some of its included variables vanish. Thus the four-amplitude case appears to be that which ends up with the lowest value of q^*.

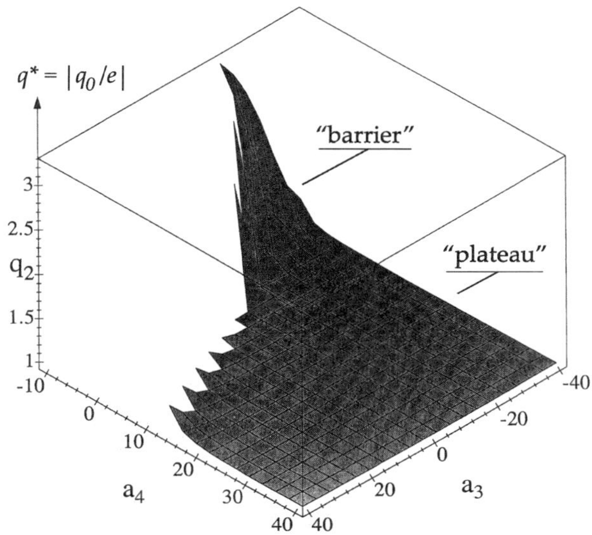

Fig. 1. The normalized electron charge $q^* \equiv |q_0/e|$ as a function of the two amplitudes a_3 and a_4, for solutions satisfying the subsidiary quantum conditions for a fixed flux factor $f_\Gamma = f_{\Gamma q} = 1.82$, and being based on a polar function T having four amplitudes (a_1, a_2, a_3, a_4). The deviations of this profile from that obtained for the self-consistent solutions which obey condition (74) are hardly visible on the scale of the figure.

6.6. *The Radial Force Balance*

The outcome of the variational analysis with its plateau behaviour still includes some remaining degree of freedom to be investigated. In a steady state Eq. (9) shows that the total acting forces can be represented by the volume force density (10). The latter only consists of an electrostatic force due to the electron charge in conventional theory, and this tends to "explode" the electron.[10,20] In the present theory, however, there is an extra magnetic force which under certain conditions can outbalance the electrostatic one, at least when being integrated over the entire volume. A local balance defined by $\mathbf{f} = 0$ does on the other hand not seem to be possible, because this leads to an overdetermined system of equations.

In a straight circular geometry of constant charge density, limited radius and with an axial velocity vector, the radial force (10) vanishes.[2] A local balance can on the other hand not be fully realized in the present geometry, but the integrated radial force can in any case be made to vanish. Thus, with the results obtained from Eqs. (18)–(22) and (30), an integrated radial force

$$F_r = -2\pi\varepsilon_0 G_0^2 \iint \left[DG + D\left(s^2 DG\right)\right] \cdot \left[\frac{\partial G}{\partial \rho} - \frac{1}{\rho} s^2 DG\right] \rho^2 s \, d\rho d\theta \quad (75)$$

is obtained where $s = \sin\theta$ and

$$\rho^2 DG = D_\theta T - 2T \quad (76)$$

in the present point-charge-like model. The force balance (75) has the form

$$F_r = I_+ - I_- \quad (77)$$

where I_+ is a positive radially outward directed contribution due to the electrostatic part of the volume force, and I_- is a negative negative inward directed (confining) contribution due to its magnetic part. Thus $I_+/I_- = 1$ defines an integrated radial force balance. When applied to the four-amplitude case for Fig. 1, the values of the normalized charge q^* and the related values of the ratio I_+/I_- are found to vary along the perimeter of the plateau. The integrated force ratio decreases from $I_+/I_- = 1.27$ at $q^* = 0.98$ to $I_+/I_- = 0.37$ at $q^* = 1.01$, thus passing an equilibrium point $I_+/I_- = 1$ at $q^* \cong 0.988$. The remaining degrees of freedom of this case have then been used up.

To sum up, a combination of a lowest possible normalized charge q^* with the requirement of an integrated radial force balance results in a value $q^* \cong 0.988$ which deviates only by one percent from the experimental value of the elementary free charge. The reason for this small deviation is not clear at this stage. One possible explanation could be due to a necessary small quantum mechanical correction of the magnetic flux, in analogy with that of the magnetic moment in Eq. (45). Another possibility may be due to a small error resulting from the large number of steps to be performed in numerical analysis which includes matching of the quantum conditions (71)–(73), of the flux factors (74), and of the contributions (77) to the radial force balance.

7. Models of the Photon

In a model of the individual photon as a propagating boson, a wave or wave packet with preserved and limited geometrical shape as well as with undamped motion in a defined direction, has to be taken as a starting point. This leads to cylindrical waves in a frame (r, φ, z) , with z in the direction of propagation. As in conventional theory, an initial arbitrary disturbance can in principle be represented by a spectrum of plane waves with normals in different directions, but would then become disintegrated at later times.[20] In this revised analysis we further introduce a velocity vector

$$\mathbf{C} = c(0, \cos\alpha, \sin\alpha) \tag{78}$$

of helical geometry where the angle α is constant and

$$0 < \cos\alpha \ll 1 \tag{79}$$

for reasons to be clarified later. In fact, $\cos\alpha$ and $\sin\alpha$ can have either sign, as determined by the two directions of spin and propagation, but are here restricted to positive values for the sake of simplicity.

The basic Eqs. (5)–(8) can be combined to the wave equation

$$\left(\frac{\partial^2}{\partial t^2} - c^2\nabla^2\right)\mathbf{E} + \left(c^2\nabla + \mathbf{C}\frac{\partial}{\partial t}\right)(\operatorname{div}\mathbf{E}) = 0\,. \tag{80}$$

of the electric field, and further in a cylindrical frame to

$$\left(D_1 - \frac{1}{r^2} + \frac{1}{r^2}\frac{\partial^2}{\partial\varphi^2}\right)E_r - \frac{2}{r^2}\frac{\partial}{\partial\varphi}E_\varphi = \frac{\partial}{\partial r}(\operatorname{div}\mathbf{E}) \tag{81}$$

$$\left(D_1 - \frac{1}{r^2} + \frac{1}{r^2}\frac{\partial^2}{\partial\varphi^2}\right)E_\varphi + \frac{2}{r^2}\frac{\partial}{\partial\varphi}E_r = \left[\frac{1}{r}\frac{\partial}{\partial\varphi} + \frac{1}{c}(\cos\alpha)\frac{\partial}{\partial t}\right](\operatorname{div}\mathbf{E}) \tag{82}$$

$$\left(D_1 + \frac{1}{r^2}\frac{\partial^2}{\partial\varphi^2}\right)E_z = \left[\frac{\partial}{\partial z} + \frac{1}{c}(\sin\alpha)\frac{\partial}{\partial t}\right](\operatorname{div}\mathbf{E}) \tag{83}$$

where

$$D_1 = \frac{\partial^2}{\partial r^2} + \frac{1}{r}\frac{\partial}{\partial r} + \frac{\partial^2}{\partial z^2} - \frac{1}{c^2}\frac{\partial^2}{\partial t^2}\,. \tag{84}$$

A divergence operation on Eq. (5) yields

$$\left[\frac{\partial}{\partial t} + c(\cos\alpha)\frac{1}{r}\frac{\partial}{\partial\varphi} + c(\sin\alpha)\frac{\partial}{\partial z}\right](\operatorname{div}\mathbf{E}) = 0 \tag{85}$$

when $\operatorname{div}\mathbf{C} = 0$. Equation (85) sometimes becomes useful, but it does not introduce more information than that already contained in Eq. (80). In the present theory where $\operatorname{div}\mathbf{E} \neq 0$ the symmetry between the fields $\mathbf{E}$ and $\mathbf{B}$ has been broken. The magnetic field has instead to be given by the electric field through the induction law (6).

In a normal mode analysis every field quantity Q is here represented by the form

$$Q = Q(\rho)\exp\left[i(-\omega t + \bar{m}\varphi + kz)\right] \equiv Q(\rho)\exp(i\theta_m) \tag{86}$$

where $\rho = r/r_0$, ω is the angular frequency, and k the axial wave number. In connection with the operator (84) and the form (86) we further define

$$K_0^2 = (\omega/c)^2 - k^2 . \tag{87}$$

7.1. *Conventional Wave Modes*

In conventional theory div $\mathbf{E}$ drops out of Eqs. (5)–(8) and (81)–(83). The condition of a vanishing electric field divergence can be taken implicitly into account by introducing the Herz vector.[20] For $K_0^2 > 0$ the phase velocity ω/k becomes larger and the group velocity $\partial\omega/\partial k$ smaller than c, as obtained from relation (87). The general solution then has field components in terms of Bessel functions $Z_{\bar{m}}(K_0 r)$ of the first and second kind, where the r-dependence of every component is of the form $Z_{\bar{m}}/r$ or $Z_{\bar{m}+1}$.[20] These solutions can be applied to wave guides with boundary conditions given by surrounding metal walls. Application of the same solutions, as well as of those for any value of K_0^2 to a model of an individual photon with angular momentum (spin) leads on the other hand to physically irrelevant results:

- Already the special purely axisymmetric case $\bar{m} = 0$ results in $s_z = 0$ due to Eq. (15), and thus in zero spin.
- The photon model cannot be bounded by walls but has to concern the entire surrounding vacuum space. But then the total integrated field energy becomes divergent. This also applies to an attempt to form a wave packet for each of the field components.

Consequently, conventional theory based on Maxwell's equations does not lead to a physically acceptable photon model.

7.2. *Axisymmetric Space-charge Wave Modes*

As a next step Eqs. (81)–(85) are applied to purely axisymmetric space-charge waves where $\partial/\partial\varphi = 0$ but div $\mathbf{E} \neq 0$. Equation (85) then results in the dispersion relation

$$\omega = kv \qquad v = c(\sin\alpha) \tag{88}$$

which has phase and group velocities both being equal to v for a constant value of α. Combination of Eqs. (81), (88) and (82) then yields

$$ik(\cos\alpha)^2 E_r = \frac{\partial E_z}{\partial r} \tag{89}$$

and

$$\left(D - \frac{1}{\rho^2}\right) E_\varphi = -(\mathrm{tg}\alpha) D E_z \tag{90}$$

where

$$D = \bar{D}_\rho - \bar{\theta}^2 (\cos\alpha)^2 \,, \qquad \bar{D}_\rho = \frac{\partial^2}{\partial\rho^2} + \frac{1}{\rho}\frac{\partial}{\partial\rho} \,, \qquad \bar{\theta} = k r_0 \,. \tag{91}$$

Here a *generating function*

$$G_0 \cdot G = E_z + (\cot\alpha) E_\varphi \qquad G = R(\rho) e^{i(-\omega t + kz)} \tag{92}$$

can be found which combines with Eqs. (89)–(91) and (6) to the field components

$$E_r = -\mathrm{i} G_0 \left[\theta (\cos\alpha)^2\right]^{-1} \frac{\partial}{\partial\rho} \left[\left(1 - \rho^2 D\right) G\right] \tag{93}$$

$$E_\varphi = G_0 \left(\mathrm{tg}\,\alpha\right) \rho^2 D G \tag{94}$$

$$E_z = G_0 \left(1 - \rho^2 D\right) G \tag{95}$$

and

$$B_r = -G_0 \left[c(\cos\alpha)\right]^{-1} \rho^2 D G \tag{96}$$

$$B_\varphi = -i G_0 (\sin\alpha) \left[\theta c (\cos\alpha)^2\right]^{-1} \frac{\partial}{\partial\rho} \left[\left(1 - \rho^2 D\right) G\right] \tag{97}$$

$$B_z = -i G_0 \left[\theta c (\cos\alpha)\right]^{-1} \left(\frac{\partial}{\partial\rho} + \frac{1}{\rho}\right) \left(\rho^2 D G\right) . \tag{98}$$

These solutions give rise to a nonzero spin. By a proper choice of the generating function the integrated field energy also becomes finite.

With the dispersion relation (88) it is seen that condition (79) has to be satisfied for the group velocity not to get in conflict with experiments of the Michelson-Morley type. For $\cos\alpha \leq 10^{-4}$ the deviation of this velocity from c would thus become less than a change in the eight decimal of c.

We are free to rewrite the amplitude factor of the generating function (92) as

$$G_0 = g_0 (\cos\alpha)^2 \,. \tag{99}$$

With this notation and the solutions (93)–(98), the components E_r and B_φ are of zero order in the smallness parameter $\cos\alpha$, E_φ, B_r and B_z of first order, and E_z of second order. There is thus essentially a radially polarized cylindrical wave.

A wave packet can be formed from the normal mode solutions, having a narrow line width, as required from experiments and observations, and with the spectral amplitude distribution

$$A_k = \left(\frac{k}{k_0^2}\right) \exp\left[-z_0^2 (k - k_0)^2\right] \tag{100}$$

where k_0 is the main wave number, $2z_0$ represents the axial packet length, and $k_0 z_0 \gg 1$ in the narrow line limit. Integration over the spectrum is performed with the notation $\bar{z} = z - vt$ and

$$\bar{E}_0 = E_0(\bar{z}) = \left(\frac{g_0}{k_0 r_0}\right)\left(\frac{\sqrt{\pi}}{k_0 z_0}\right) \exp\left[-\left(\frac{\bar{z}}{2z_0}\right)^2 + \mathrm{i}k_0\bar{z}\right]. \tag{101}$$

It results in the average packet field components

$$\bar{E}_r = -\mathrm{i}E_0\left[R_5 + (\theta_0')^2 R_1\right] \tag{102}$$

$$\bar{E}_\varphi = E_0\theta_0(\sin\alpha)(\cos\alpha)\left[R_3 - (\theta_0')^2 R_1\right] \tag{103}$$

$$\bar{E}_z = E_0\theta_0(\cos^2\alpha)\left[R_4 + (\theta_0')^2 R_1\right] \tag{104}$$

$$\bar{B}_r = -\left(\frac{1}{c}\right)(\sin\alpha)^{-1}E_\varphi \tag{105}$$

$$\bar{B}_\varphi = \left(\frac{1}{c}\right)(\sin\alpha)E_r \tag{106}$$

$$\bar{B}_z = -i\left(\frac{1}{c}\right)E_0(\cos\alpha)[R_8 - (\theta_0')^2 R_7] \tag{107}$$

where

$$R_1 = \rho^2 R \qquad R_2 = dR_1/d\varphi \qquad R_3 = \rho^2 \bar{D}_\rho R \tag{108}$$

$$R_4 = R - R_3 \qquad R_5 = dR_4/d\rho \qquad R_6 = \bar{D}_\rho R_4 \tag{109}$$

$$R_7 = \left(\frac{d}{d\rho} + \frac{1}{\rho}\right) R_1 \qquad R_8 = \left(\frac{d}{d\rho} + \frac{1}{\rho}\right) R_3\,. \tag{110}$$

Since expressions (105)–(107) have been obtained in the narrow-line approximation, the condition $\operatorname{div}\mathbf{B} = 0$ is only satisfied approximately, whereas it holds exactly for the normal mode solutions (96)–(98).

In the following analysis a generating function is chosen which is symmetric with respect to the axial centre $\bar{z} = 0$ of the moving wave packet. Then

$$G = R(\rho)\cos k\bar{z} \tag{111}$$

when the real parts of (92) and (101) are adopted. Here G and (E_φ, E_z, B_r) are symmetric and (E_r, B_φ, B_z) are antisymmetric functions of z with respect to $\bar{z} = 0$.

The analysis now proceeds in forming the spatially integrated average field quantities which represent an electric charge q, magnetic moment M, total mass m, and spin s. The limits of z are $\pm\infty$, and those of ρ will later be specified. The integrated charge becomes

$$q = 2\pi\varepsilon_0 \int \left\{ \frac{\partial}{\partial r}(r^2 \int_{-\infty}^{+\infty} \bar{E}_r d\bar{z}) + r[\bar{E}_z]_{-\infty}^{+\infty} \right\} dr = 0 \tag{112}$$

and the integrated magnetic moment

$$M = \pi\varepsilon_0 c(\cos\alpha) \int \left\{ r\frac{\partial}{\partial r}(r \int_{-\infty}^{+\infty} \bar{E}_r d\bar{z}) + r^2[\bar{E}_z]_{-\infty}^{+\infty} \right\} dr = 0 \tag{113}$$

due to the symmetry properties of the field components. It should be observed that, even if q and M vanish, the local charge density and magnetic field strength are nonzero.

For the total mass the Einstein relation yields

$$m = \left(\frac{1}{c^2}\right) \int w_f \, dV \tag{114}$$

with dV as a volume element and the field energy density given by Eq. (13). Using Eqs. (102)–(107) and the energy relation by Planck, the narrow-line limit then gives the result

$$m \cong 2\pi \left(\varepsilon_0/c^2\right) \int_{-\infty}^{+\infty} \int r|\bar{E}_r^2| \, dr d\bar{z} = a_0 W_m = h\nu_0/c^2 \,. \tag{115}$$

Here $\nu_0 = c/\lambda_0$ is the average frequency related to the average wave length $\lambda_0 = 2\pi/k_0$ of the packet,

$$a_0 = \varepsilon_0 \pi^{5/2} \sqrt{2} z_0 \left(g_0/ck_0 z_0\right)^2 \equiv 2a_0^* g_0^2 \tag{116}$$

and

$$W_m = \int \rho R_5^2 \, d\rho \,. \tag{117}$$

The slightly reduced phase and group velocity of Eq. (89) becomes associated with a very small nonzero rest mass

$$m_0 = m\left[1 - (v/c)^2\right]^{1/2} = m(\cos\alpha) \,. \tag{118}$$

This can be further verified,[2] by comparing the total energy of the wave packet in the laboratory frame K with that in a frame K' following the packet motion at the velocity $c(\sin\alpha) < c$ of Eq. (88).

Turning to the integrated angular momentum, we first notice that the volume force (10) contains the vector $\mathbf{E} + \mathbf{C} \times \mathbf{B}$. From Eqs. (78) and (102)–(107) is readily seen that the volume force has a vanishing component $\bar{f}_\varphi$

and that its components $\bar{f}_r$ and $\bar{f}_z$ are of second order in the smallness parameter $\cos\alpha$. Consequently, and somewhat in analogy with the force balance of the electron in Section 6.6, there is here a local transverse force balance, as provided by a confining magnetic force contribution.

The density of angular momentum (15) in the axial direction now becomes

$$\bar{s}_z = \varepsilon_0 r(\bar{E}_z\bar{B}_r - \bar{E}_r\bar{B}_z) \cong -\varepsilon_0 r\bar{E}_r\bar{B}_z \tag{119}$$

which is on the other hand of first order in $\cos\alpha$. Consequently, the volume force can be neglected and the contribution from the momentum density (11) due to the Poynting vector dominates the right hand member of the momentum Eq. (9). The total spin becomes

$$s = \int \bar{s}_z\, dV = a_0 r_0 c(\cos\alpha)W_s = h/2\pi \tag{120}$$

for the photon as a boson, and where

$$W_s = -\int \rho^2 R_5 R_8\, \mathrm{d}\rho\,. \tag{121}$$

The results (120) and (118) show that a nonzero spin s requires a nonzero rest mass m_0 to exist. These two concepts are associated with the component C_φ of the velocity vector. This component circulates around the axis of symmetry and has two opposite possible spin directions.

To proceed further the radial function $R(\rho)$ has to be specified. A form

$$R(\rho) = \rho^\gamma \mathrm{e}^{-\rho} \qquad\qquad \gamma > 0 \tag{122}$$

being convergent at $\rho = 0$ is first considered. It has a maximum at the radius $\hat{r} = \gamma r_0$ and drops rapidly towards zero at large ρ. When evaluating the integrals (117) and (121), the Euler integral

$$J_{2\gamma-2} = \int_0^\infty \rho^{2\gamma-2} e^{-2\rho}\, d\rho = 2^{-(2\gamma+3)}\Gamma(2\gamma+1) \tag{123}$$

appears in terms of the gamma function. For $\gamma \gg 1$ only the dominant terms prevail, $R_8 \simeq -R_5$, and the result becomes

$$W_m = W_s/\gamma\,. \tag{124}$$

Combination of relations (115), (120) and (124) finally leads to an effective photon diameter

$$2\hat{r} = \frac{\lambda_0}{\pi(\cos\alpha)} \tag{125}$$

being independent of γ and of the exponential factor in Eq. (122), The diameter (125) is limited but large as compared to atomic dimensions when the wave length λ_0 is in the visible range.

We next turn to a radial function

$$R(\rho) = \rho^{-\gamma} e^{-\rho} \qquad \gamma > 0 \tag{126}$$

which *diverges* at the axis. Here $\hat{r} = r_0$ can be taken as an effective radius. This situation becomes similar to that of the electron model in Section 6.1 and a discussion of its radial form will not be repeated. To obtain finite integrated values of the total mass m and spin s, small lower radial limits ρ_m and ρ_s are introduced in the integrals (117) and (121). We further make the Ansatz

$$r_0 = c_r \cdot \varepsilon \qquad g_0 = c_g \cdot \varepsilon^\beta \qquad 0 < \varepsilon \ll 1 \tag{127}$$

of shrinking values for both the characteristic radius and the amplitude of the generating function, and where c_r, c_g and β are positive constants. Equations (115) and (120) combine to

$$m = a_0^* \gamma^5 c_g^2 \left(\varepsilon^{2\beta}/\rho_m^{2\gamma}\right) \cong h/\lambda_0 c \tag{128}$$

$$s = a_0^* \gamma^5 c_g^2 c_r c(\cos\alpha) \left(\varepsilon^{2\beta+1}/\rho_m^{2\gamma-1}\right) = h/2\pi\,. \tag{129}$$

To obtain finite m and s it is then necessary that

$$\rho_m = \varepsilon^{\beta/\gamma} \qquad \rho_s = \varepsilon^{(2\beta+1)(2\gamma-1)}\,. \tag{130}$$

We are here free to choose $\beta = \gamma \gg 1$ by which $\rho_s \cong \rho_m = \varepsilon$. This leads to a similar set of geometrical configurations in the range of small ε. Combination of Eqs. (128) and (129) yields an effective photon diameter

$$2r_0 = \frac{\varepsilon\lambda_0}{\pi(\cos\alpha)} \tag{131}$$

being independent of γ and β. Here ε and $\varepsilon/(\cos\alpha)$ can be made small enough to result in "needle radiation" at a diameter (131) which becomes comparable to atomic dimensions.

The obtained results (125) and (131) of an axisymmetric photon model can be illustrated by a simple example where $\cos\alpha = 10^{-4}$. For a wave length $\lambda_0 = 3 \times 10^{-7}$ m Eq. (125) yields a photon diameter of about 10^{-3} m, and Eq. (131) results in a diameter smaller than 10^{-7} m when $\varepsilon < \cos\alpha$ for needle-like radiation.

The individual photon models resulting from the present theory appear to be relevant in respect to the particle-wave dualism. A subdivision into a "bound" particle part associated with the component C_φ and a "free" pilot wave part associated with the component C_z is imaginable but not necessary. This is because the rest mass here merely constitutes an integrating part of the total field and its energy. In other words, the wave packet behaves as an entirety, having particle and wave properties at the same time. Such a joint particle-wave nature of the single individual photon reveals itself in the comparatively small effective radius, especially in the case of a needle-like shape. This is reconcilable with the photoelectric effect where a photon knocks out an electron from an atom, and also with the dot-shaped marks which form an interference pattern on a screen in two-slit experiments at low light intensity,[21] as well as with recent such experiments under different boundary conditions.[22] Thereby the interference patterns should also arise in the case of cylindrical waves. The nonzero rest mass may further make it possible for the photon to perform spontaneous transitions between different wave modes, by means of proposed "photon oscillations",[2,4,6] in analogy with the neutrino oscillations.

7.3. *Screw-shaped Space-charge Wave Modes*

In a review by Battersby[23] twisted light is described where the energy travels along cork-screw-shaped paths. This discovery is expected to become important in communication and microbiology.

Corresponding modes should exist in the present theory for a nonzero $\bar{m}$ in Eqs. (81)–(85). As compared to the purely axisymmetric normal modes, these screw-shaped modes lead to a more complex analysis, partly on account of the second term in Eq. (5). In a first iteration we attempt to neglect this term due to its small factor $\cos\alpha$, and then end up again with the dispersion relation (88). From Eq. (83) the component E_z is seen to be of the order $(\cos\alpha)^2$ as compared to E_r and E_φ. Thereby Eq. (81) would take the form

$$E_r \cong -\left(\frac{r}{\bar{m}}\right)\left[1 - k^2(\cos\alpha)^2\left(\frac{r}{\bar{m}}\right)^2\right]\left(\frac{\partial}{\partial r} + \frac{1}{r}\right)(iE_\varphi)\,. \tag{132}$$

When inserting this relation into Eq. (82), the latter becomes identically satisfied up to first order in $\cos\alpha$. Consequently the component iE_φ can be used in this approximation as a generating function

$$iE_\varphi = F = G_0 G\,, \qquad G = R(\rho)\exp(i\theta_m)\,. \tag{133}$$

In analogy with the deductions in Section 7.2, wave packet solutions can be formed, q and M be found to vanish, the volume force (10) to be neglected in Eq. (9), and expressions for m and s to be obtained.[4] With the same convergent radial function as in Eq. (122), there is a nearly radially polarized wave in which

$$|\bar{E}_r/i\bar{E}_\varphi| = |\gamma + 1 - \rho| \gg 1 \tag{134}$$

for $\rho \ll \gamma$ and $\gamma \gg 1$. Moreover, insertion of E_r from expression (132) into Eq. (83) shows that E_φ and E_z are of the same order in the parameter γ. Thus Eq. (134) shows that also $|E_r/iE_z| \gg 1$. The effective photon diameter would then become

$$2\hat{r} = \frac{\lambda_0 \bar{m}^{3/2}}{\pi(\cos\alpha)} \qquad \bar{m} \neq 0\,. \tag{135}$$

However, this result is not fully consistent with the basic equations. Relation (88) and insertion of $r = \hat{r}$ from Eq. (135) namely shows that the second term in Eq. (85) is of second order in $\cos\alpha$ as well as the sum of its other two terms. To remove this difficulty, the analysis has to be restricted to configurations where R is peaked at a nearly constant value of r, to form a ring-shaped radial distribution. This is in fact what happens with the form (122) which becomes strongly peaked at $\hat{r} = \gamma r_0$ in the limit of large γ.

As a next "iteration" we therefore replace the variable r in the second term of Eq. (85) by a constant value

$$\bar{r} = \bar{m}^{3/2}/c_1 k(\cos\alpha) \tag{136}$$

where c_1 is a positive constant. This implies that the dispersion relation is modified to

$$k^2 - (\omega/c)^2 \cong k^2(\cos\alpha)^2 C_0 \qquad C_0 = 1 - 2c_1/\sqrt{\bar{m}}\,. \tag{137}$$

As a result, the effective photon diameter becomes

$$2\hat{r} = \frac{\lambda_0 \bar{m}^{3/2}}{\pi(\cos\alpha)\sqrt{C_0}} \qquad \bar{m} \neq 0\,. \tag{138}$$

This result is consistent with expression (136) for $r = \hat{r}$, provided that

$$C_0 = 1 + (2/\bar{m}) - (2/\bar{m})\sqrt{(1/\bar{m}) + 1} \tag{139}$$

where one solution has been discarded because c_1 and C_0 have to be positive. The value of $\sqrt{C_0}$ ranges from 0.414 for $\bar{m} = 1$ to 1 for large values of $\bar{m}$.

The result (138) as well as those of Eqs. (125) and (131) are applicable both to individual photons and to dense light beams of N photons per unit length, because the corresponding integrated mass and angular momentum both become proportional to N. In the beam case, the effective diameters then stand for those of the corresponding beam models. This also applies to the present screw-shaped mode with a convergent and ring-shaped radial part of the generating function which seems to be consistent with experimental observations.[23] Here screw-shaped normal modes of radii (138) with slightly different values of $\cos\alpha$ can be superimposed to form a ring-shaped beam profile of a certain width.

Attempts to analyze screw-shaped modes having a divergent radial part R and aiming at a needle-shaped behaviour are faced with the difficulties of Eq. (85) when the configuration extends all the way to the axis.

8. The Linearly Polarized Photon Beam

The photon and beam models studied here have so far essentially been radially polarized. We now turn to the case of a linearly polarized light beam of circular cross section. Elliptically or circularly polarized beams are obtained from the superposition of linearly polarized modes being ninety degrees out of phase. For linear polarization a rectangular frame of reference would become suitable, whereas a cylindrical one becomes preferable for a circular cross-section.

Without changing the essential features the analysis is simplified by the restriction to a homogeneous core with plane wave geometry, limited radially by a narrow boundary region in which the light intensity drops to zero. The radius of the beam is large as compared to the characteristic wave length, and the boundary conditions can in a first approximation be applied separately to every small local segment of the boundary. A localized analysis is then performed in which the electric field vector of the core wave forms a certain angle with the boundary, and where local rectangular coordinates can be introduced. In its turn, the core wave is then subdivided into two waves of the same frequency and wavelength, but having electric field vectors being perpendicular and parallel to the local segment of the boundary region.

8.1. *Flat-shaped Beam Geometry*

In the analysis on a segment of the boundary region, a local frame (x, y, z) is now chosen with z in the axial direction of propagation, and with the normal of the boundary in the x-direction. There is no y-dependence. The velocity vector is given by

$$\mathbf{C} = c(0, \cos\alpha, \sin\alpha) \qquad 0 < \cos\alpha \ll 1 \tag{140}$$

thus having C_y along the boundary and C_z in the direction of propagation. From a divergence operation on Eq. (5), a dispersion relation of the form (88) is again obtained. The wave Eq. (80) for normal modes now reduces to the three relations

$$E_x = -\left[\frac{i}{k(\cos\alpha)^2}\right]\frac{\partial E_x}{\partial x} \tag{141}$$

$$\left[k^2(\cos\alpha)^2 - \frac{\partial^2}{\partial x^2}\right]E_y - ik(\cos\alpha)(\sin\alpha)\left(\frac{\partial E_x}{\partial x} + ikE_z\right) = 0 \tag{142}$$

$$\frac{\partial E_x}{\partial x} = -\left[\frac{i}{k(\cos\alpha)^2}\right]\frac{\partial^2 E_z}{\partial x^2} \tag{143}$$

where relation (143) is merely obtained from derivation of (141). Combination of Eqs. (141) and (142) gives

$$\left[k^2(\cos\alpha)^2 - \frac{\partial^2}{\partial x^2}\right]\left(E_y + \frac{\sin\alpha}{\cos\alpha}E_z\right) = 0\,. \tag{144}$$

Equations (141) and (144) thus show that the component E_z can serve as a generating function for the transverse components E_x and E_y. At least one solution of Eq. (144) is readily found when E_y and E_z have the same spatial profiles and

$$E_y = -\frac{\sin\alpha}{\cos\alpha}E_z \tag{145}$$

The ordering of the electric field components with respect to the smallness parameter $\cos\alpha$ is thus $E_x = O(1)$, $E_y = O(\cos\alpha)$ and $E_z = O(\cos^2\alpha)$.

The magnetic field components are finally given by

$$\omega(B_x, B_y, B_z) = \left(-kE_y, kE_x + i\frac{\partial E_z}{\partial x}, -i\frac{\partial E_y}{\partial x}\right) \tag{146}$$

From relation (88) we have

$$B_x = -E_y/c(\sin\alpha) \tag{147}$$

$$B_y = E_x/c(\sin\alpha) + \frac{i}{kc(\sin\alpha)}\frac{\partial E_z}{\partial x} = (\sin\alpha)E_x/c \tag{148}$$

$$B_z = -\frac{i}{kc(\sin\alpha)}\frac{\partial E_y}{\partial x} = -(\cos\alpha)E_x/c \tag{149}$$

where $B_x = O(\cos\alpha)$, $B_y = O(1)$ and $B_z = O(\cos\alpha)$.

Application of relations (88) and (145) on the result (147)–(149) finally yields

$$\mathbf{E} + \mathbf{C}\times\mathbf{B} = 0\,. \tag{150}$$

This implies that the volume force (10) vanishes identically in rectangular geometry, and the Poynting vector is again the source of the electromagnetic momentum.

8.2. *Two Special Flat-shaped Cases*

Two special plane cases are now studied of a beam which has a core region defined by $-a < x < a$ and two narrow boundary regions $-b < x < -a$ and $a < x < b$ of the small thickness $d = b - a$. There is symmetry in respect to $x = 0$ and we discuss only the region at $x = a$ henceforth. With the chosen frame there is first the case in which E_x is the main component. Within the core region a homogeneous linearly polarized plane wave is assumed to exist, having the constant components E_{x0} and B_{y0}.

Inside the boundary region an axial field E_z is assumed to increase linearly with x from a small value near $x = a$, and in such a way that E_x in Eq. (141) becomes matched to E_{x0} at $x = a$. Within the same region E_z further passes a maximum, after which it drops to zero at the edge $x = b$. There is then a reversed field $E_x$in the outer part of the boundary layer, and the maximum strength of E_x is of the same order as E_{x0} according to Eq. (141). From Eq. (145) the spatial profile of E_y further becomes the same as that for E_z . The component B_y of Eq. (148) is matched to B_{y0} at the edge of the core region and to lowest order in $\cos\alpha$. Then the Poynting vector components become $S_x = 0$ and

$$S_y \cong -E_x B_z/\mu_0 = c(\cos\alpha)\varepsilon_0 E_x^2 \tag{151}$$

$$S_z \cong E_x B_y/\mu_0 = c(\sin\alpha)\varepsilon_0 E_x^2 \,. \tag{152}$$

Thus there is a primary flow S_z of momentum in the direction of propagation, a secondary flow S_y along the boundary, and no flux S_x across it. The field energy density (13) becomes

$$w_f \cong \varepsilon_0 E_x^2 \tag{153}$$

to lowest order in $\cos\alpha$.

Turning then to the second case where E_y is the main field component, being parallel with the boundary, there is a plane wave in the core region with the components E_{y0} and B_{x0}.

Within the boundary region, in a small range of x near $x = a$, the axial field E_z is now assumed to be constant, and E_x thus vanishes due to Eq. (141). Then relation (145) makes it possible to match E_y to E_{y0} at $x = a$. The field E_z is further chosen to decrease towards zero when approaching the outer edge $x = b$. According to Eqs. (141) and (145) this results in a perpendicular field component

$$E_x = \left[\frac{i}{k(\cos\alpha)(\sin\alpha)}\right]\frac{\partial E_y}{\partial x} \tag{154}$$

which first reaches a maximum and then drops to zero at $x = b$. For a characteristic length L_{cy} of the derivative of E_y the ratio

$$|E_x/E_y| = \lambda/2\pi L_{cy}(\cos\alpha) \tag{155}$$

can then become smaller than unity. Thus with $\lambda/L_{cy} = 10^{-4}$ and $\cos\alpha = 10^{-4}$ this ratio is about 0.16. The magnetic field components (147)–(149) now have the ordering $|B_x| = O(1) > |B_y|$ and $B_z = O(\cos\alpha)$ when E_x is smaller than E_y. Here B_x can be matched to B_{x0} at $x = a$, since both B_y and E_x vanish at $x = a$ due to Eq. (148). The Poynting vector components finally become $S_x = 0$ and

$$S_y = c(\cos\alpha)\varepsilon_0 E_y^2\left[1 + (\sin\alpha)^2(E_x/E_y)^2\right]/(\sin\alpha)^2 \tag{156}$$

$$S_z = c\varepsilon_0 E_y^2\left[1 + (\sin\alpha)^2(E_x/E_y)^2\right]/(\sin\alpha)^2\,. \tag{157}$$

Also here the flow of momentum is essentially of the same character as in the first case. The field energy density becomes

$$w_f \cong \varepsilon_0 E_y^2 \tag{158}$$

as long as $E_x^2 \ll E_y^2$.

8.3. *The Plane Core Wave*

There is an additional problem with the matching of the solutions at the edge of the beam core. This is due to the fact that the phase and group velocity of expression (88) for the present electromagnetic space-charge (EMS) wave in the boundary region is slightly smaller than that of a conventional plane electromagnetic (EM) wave in the core. However, this problem can be solved by introducing a plane EMS wave in the core which becomes hardly distinguishable from a plane EM core wave.

We start with the basic Eqs. (5)–(8) for a plane wave where all quantities vary as

$$Q = Q_0 \exp\left[i(-\omega t + \mathbf{k}\cdot\mathbf{r})\right] \tag{159}$$

and Q_0 is a constant. For the velocity vector we now use the form

$$\mathbf{C} = c\left[(\cos\beta)(\cos\alpha), (\sin\beta)(\cos\alpha), \sin\alpha\right]\,. \tag{160}$$

From the last of Eqs. (8) the dispersion relation (88) is then recovered, and a matching of the phase velocity becomes possible. The basic equations further result in

$$c^2k(-B_y, B_x) = [(kE_z)C_x - \omega E_x, (kE_z)C_y - \omega E_y] \quad (161)$$

$$c(\sin\alpha)(B_x, B_y) = (-E_y, E_x) \quad (162)$$

and $B_z = 0$. Combination of Eqs. (161) and (162) yields

$$E_z(C_x, C_y) = -(\cos\alpha)^2 c(E_x, E_y) \quad (163)$$

and

$$E_x/E_y = C_x/C_y = (\cos\beta)/(\sin\beta)\,. \quad (164)$$

Here E_z is small due to Eq. (163). The first flat-shaped case corresponds to the choice of a small $\sin\beta$, whereas the second one is represented by a small $\cos\beta$. In this way a plane EMS wave is obtained which differs very little from a plane EM wave, with the exception of the phase velocity which now can be matched to that of the wave in the boundary region. A matching of the field components can be made as in Section 8.2.

8.4. *Simplified Analysis on the Spin of a Circular Beam*

The results of the analysis on the small segments of the circular perimeter are now put together to form a first simplified approach to a circular beam in a frame (x, y, z), thus consisting of a homogeneous field $\mathbf{E}_0 = (E_0, 0, 0)$ and $\mathbf{B}_0 = (0, B_0, 0)$ in its core, and with a narrow boundary layer with large derivatives. Introducing the angle φ between the y-direction and the radial direction counted from the axis, the electric field components of the core are expressed by

$$E_{0\perp} = E_0 \sin\varphi \qquad E_{0\|} = E_0 \cos\varphi \quad (165)$$

in the perpendicular and parallel directions of the boundary.

In the boundary region $\mathbf{E}^2 = E_0^2$ at the edge of the core. With the restriction $E_x^2 \ll E_y^2$ in expressions (156)–(157), the contributions (151) and (156) to the Poynting vector add up in the transverse direction along the boundary to

$$S_t = c(\cos\alpha)\varepsilon_0\mathbf{E}^2 \quad (166)$$

being independent of the angle φ. The energy density of the beam core can further be written as

$$w_{fc} = \varepsilon E_0^2 = n_p hc/\lambda \quad (167)$$

where n_p is the number of equivalent photons per unit volume. With a parallel spin $h/2\pi$ of each photon, the core would possess a total equivalent (imagined) angular momentum per unit length

$$s_c = r_0^2 n_p h/2 = \varepsilon_0 E_0^2 \lambda r_0^2/2c \tag{168}$$

as obtained from combination with Eq. (167), and with r_0 standing for the radius of the core. Due to Eqs. (166) and (15) the real angular momentum generated per unit axial length in the boundary layer becomes on the other hand

$$s_b = 2\pi(\cos\alpha)\varepsilon E_0^2 f_E r_0^2 d/c \tag{169}$$

where $d \ll r_0$ is the thickness of the layer and $f_E < 1$ is a profile factor of order unity obtained from integration of $\mathbf{E}^2$ over the same layer. This yields the ratio

$$\frac{s_b}{s_c} = \frac{4\pi(\cos\alpha) f_E d}{\lambda}\,. \tag{170}$$

For the equivalent angular momentum s_c of the core to be replaced by a real angular momentum s_b being generated in the boundary layer, the result thus becomes $s_b = s_c$ or

$$d = \frac{\lambda}{4\pi f_E(\cos\alpha)} \tag{171}$$

which is analogous to the effective photon diameter relation (125) in the case of a convergent radial part of the generating function (92). As an example, $\lambda = 3\times 10^{-7}$ m, $f_E = 0.2$ and $\cos\alpha = 10^{-4}$ leads to a physically relevant value of $d \cong 10^{-3}$ m for a beam with a radius $r_0 = 10^{-2}$ m, say.

It should be observed that Eq. (168) applies to a case where all imagined core photons have spin in the same axial direction. There is also an imagined situation where two unequal fractions of the photons could have opposite spin directions. This still results in a net beam angular momentum as observed in experiments,[12,15] but requiring a smaller layer thickness (171) at given values of the remaining parameters. In this consideration it should also be remembered that a plane wave does not give rise to a spin.

We finally notice that the obtained results should also hold for a corresponding field configuration in the limit of a single photon model. Such a photon could thus become linearly polarized in its "core", and be limited in the transverse direction by an outer "mantle" of radially decreasing field intensity within which an angular momentum is being generated.

9. Comments on the Momentum of the Radiation Field

The pure radiation field has a momentum density **g** being based on Eqs. (9)–(11) and the Poynting vector. The latter has also been used as a basis for the conventional and original QED theory described by Schiff[16] and Heitler[15] among others.

In the case of a massive particle, the quantized momentum has on the other hand been represented successfully by the operator **p** of expression (16) in the nonrelativistic Schrödinger equation.[16] For the normal wave modes (86) treated in this context the corresponding axial component becomes

$$p_z = \hbar k = h/\lambda = h\nu/c. \tag{172}$$

In conventional theory this component is related to a photon of energy $h\nu$, moving along z at the velocity c of light.

A comparison between the concepts of **g** and **p** leads, however, to a number of not quite clear questions concerning **p**. These are as follows:

- In the present theory based on the concept of **g**, individual photons as well as light beams are limited spatially in the directions being perpendicular to the axial direction of propagation and have vanishing or negligible transverse losses of momentum. This is not the case when applying the concept **p** which has a radial component and leads to a corresponding transverse loss, thus becoming questionable from the physical point of view.
- In a pure axisymmetric case the concept **g** results in a momentum directed around the axis of symmetry. The same momentum vanishes when applying the concept **p**.
- In the present Lorentz invariant photon model the momentum **g** has a component around the axis which provides a spin at the expense of the axial velocity of propagation which becomes slightly reduced below c. With the concept **p** the result (172) is in conventional theory indicating that the photon moves at the full velocity c in the axial direction. But for the same photon to possess a nonzero spin, there should also exist an additional transverse momentum p_φ corresponding to an additional velocity v_φ which circulates around the z axis. However, this would lead to a superlumial total velocity within the photon configuration.

10. Conclusions

In the present revised quantum electrodynamic theory, the nonzero electric field divergence introduces an additional degree of freedom and modifies the basic field equations to a considerable extent, thereby also giving rise to new results and interpretations in respect to a number of fundamental applications.

Considering the resulting models of leptons such as the electron, the field equations contain additional electric field divergence terms which constitute large contributions already at the outset, and which give rise to a number of new features:

- The point-charge behaviour comes out as a necessity from the theory, with the requirement of a nonzero net electric charge.
- The infinite self-energy problem of the point charge is eliminated in the present theory in which a divergent behaviour of the generating function is outbalanced by a shrinking characteristic radius. This provides a physically more acceptable and realistic alternative to the renormalization process in which extra ad hoc counter terms are merely added to the Lagrangian, to outbalance one infinity by another.
- The integrated electrostatic force of the electron configuration can be outbalanced by an integrated magnetic force. This prevents the electron from "exploding" under the action of its self-charge. This so far not understood balance can be conceived as a kind of electromagnetic confinement.
- Variational analysis, in combination with the requirement of an integrated electromagnetic force equilibrium, leads to a deduced and quantized elementary charge which deviates by only one percent from its experimental value. If this small deviation could be understood and removed, the electronic charge would no longer remain an independent constant of nature, but become deduced from the velocity of light, Planck's constant, and the dielectric constant.

In the applications to photon and light beam physics, the nonzero electric field divergence appears at a first sight as a small quantity, but it still comes out to have essential effects on the end results:

- The theory leads to a spin of the individual photon, not being obtained from conventional theory in a physically relevant field configuration. The present cylindrical field geometry is helical.

- It is explained how a propagating photon can behave as an object of limited spatial extensions, and even in the form of needle radiation, whereas conventional theory results in solutions which are extended over space and lead to a divergent integrated field energy.
- The individual photon wave packet solutions have simultaneous particle and wave properties. They become reconcilable both with the photoelectric effect and with the dot-shaped marks and their interference patterns in two-slit experiments at low light intensities.
- There is electromagnetic confinement of the local transverse forces in a propagating photon wave packet, somewhat in analogy with that of the electron.
- Also a photon beam of limited cross-section has a spin which can be explained by the present theory, and not by conventional analysis.
- The observed ring-shaped intensity profile of screw-shaped light beams is consistent with the present theory.

References

1. R. P. Feynman, *Lectures in Physics: Mainly Electromagnetism and Matter* (Addison-Wesley, Reading, Massachusetts, 1964).
2. B. Lehnert, *International Review of Electrical Engineering* (I.R.E.E.), **1**, n.4 452 (2006); Report TRITA-ALF-2004-02 (Royal Institute of Technology, Stockholm, 2004).
3. B. Lehnert, *Speculations in Science and Technology* **9**, 117 (1986).
4. B. Lehnert, *Physica Scripta* **T82**, 89 (1999); **66**, 105 (2002); **T113**, 41 (2004); **72**, 359 (2005); **74**, 139 (2006).
5. B. Lehnert, *Modern Nonlinear Optics* (M. W.Evans, I. Prigogine, S. A. Rice, Editors), Vol. **119**, PartII,1 (John Wiley and Sons, Inc., New York, 2001).
6. B. Lehnert, *Progress in Physics* **2**, 78 (2006); **3**, 43 (2006); **1**, 1 (2007).
7. B. Lehnert, *Dirac Equation, Neutrinos and Beyond* (V. V. Dvoeglazov, Editor), Ukrainian Journal: Electromagnetic Phenomena **T.3**, No1(9) 35 (2003).
8. B. Lehnert and S. Roy, *Extended Electromagnetic Theory* (World Scientific Publishers, Singapore, 1998).
9. B. Lehnert and J. Scheffel, *Physica Scripta* **65**, 200 (2002).
10. J. D. Jackson, *Classical Electrodynamics* Ch. 17.4, p. 589 (John Wiley and Sons, Inc., New York, 1962).
11. L. H. Ryder, *Quantum Field Theory* Ch. 9.3, p. 321 (Cambridge Univ. Press, 1996; see also edition of 1987).
12. R. W. Ditchburn, *Light* Sec. 17.24 (Third Edition, Academic Press, London, 1976).
13. H. B. G. Casimir, *Proc. K. Ned. Akad. Wet.* **51**, 793 (1948).
14. P. A. M. Dirac, *Directions in Physics* (Wiley-Interscience, New York, 1978).
15. W. Heitler, *The Quantum Theory of Radiation* p. 409, Ch. II (Third Edition, Clarendon Press, Oxford, 1954).

16. L. I. Schiff, *Quantum Mechanics* Ch. XIV, Ch. II (McGraw-Hill Book Comp.,Inc., New York, 1949).
17. P. A. M. Dirac, *Proc. Roy. Soc.* **117**, 610 (1928) and **118**, 351 (1928).
18. J. Schwinger, *Phys. Rev.* **76**, 790 (1949).
19. R. P. Feynman, *QED: The strange theory of light and matter* (Penguin, London, 1990).
20. J. A. Stratton, *Electromagnetic Theory* Ch. II, Sec. 2.5; Ch. VI, Sec. 6.7 (McGraw-Hill Book Comp., New York and London, 1941).
21. T. Tsuchiya, E. Inuzuka, T. Kurono, and M. Hosada, *Advances in Electronics and Electron Physics* **64A** 21 (1985).
22. S. S. Afshar, E. Flores, K. F. McDonald, and E. Knoesel, *Foundations of Physics* **37**, 295 (2007).
23. S. Battersby, *New Scientist* p. 37, 12 June 2004.

QUANTUM METHODOLOGIES IN BEAM, FLUID AND PLASMA PHYSICS

R. Fedele

Dipartimento di Scienze Fisiche, Università Federico II and INFN Sezione di Napoli, Complesso Universitario di M.S. Angelo, via Cintia, I-80126 Napoli, Italy
E-mail: renato.fedele@na.infn.it

The quantum methodologies useful for describing in a unified way several problems of nonlinear and collective dynamics of fluids, plasmas and beams are presented. In particular, the pictures given by the Madelung fluid and the Moyal-Ville-Wigner phase-space quasidistribution, including the related quantum tools such as marginal distributions for the tomographic representations, are described. Some relevant applications to soliton and modulational instability theories are presented.

Keywords: Quantum methodologies; quantumlike models; Madelung's fluids; nonlinear phenomena; nonlinear Schrödinger equation; Korteweg-de Vries equation; modulational instability

1. Introduction

1.1. *General role of the quantum methodologies*

The quantum methodologies, such as the tools provided by Schrödinger-like equations, Madelung fluid picture,[1] Moyal-Ville-Wigner transform[2–4] or von Neumann-Weyl formalism,[5,6] quantum tomography[7] and n-waves parametric processes,[8] are widely used in almost all branches of nonlinear physics. In particular, they are frequently encountered in dispersive media such as laboratory, space and astrophysical plasmas, fluids, Kerr media, optical fibers, electrical transmission lines and many other physical systems including cosmological and biological systems; in optical beam physics and charged particle beam physics they play a very relevant role.

Since the recently past years, the quantum methodologies are intensively applied in all the above branches as result of international collaborations belonging to the frontiers of the physics researches and they are one of the main topics of several important interdisciplinary scientific international conferences.[9] In fact, each of the above physical systems exhibit behavior that can be described with a quantum formalism. Typically, their evolution in space and time is governed by suitable linear or nonlinear Schrödinger-like equations for complex functions that are coupled, through an effective potential, with a set of equations describing the interaction of the system with the surroundings. For instance, in plasmas, the nonlinearity arises from the harmonic generation and the ponderomotive force,[10] while in nonlinear optics its origin is due to a Kerr nonlinear refractive index.[11] The nonlinear collective charged-particle beam dynamics in accelerating machines is due to the interaction between beam and the surroundings by means of both image charges and image currents that are created on the walls of the accelerator vacuum chamber. This interaction is conveniently described in terms of the so-called "coupling impedance", whose imaginary part accounts for both the space charge blow up and the magnetic self attraction, and whose real part accounts for the resistive effects occurring on the walls. In the physics of the surface gravity waves the nonlinearity is introduced by the high values of the wave elevation.[12]

The system dynamics governed by linear or nonlinear Schrödinger equations can be described in an equivalent way by means of the Madelung fluid equations.

Alternatively, one may use the Moyal-Ville-Wigner transform (quasidistribution) that allows to transit from the configuration space description to the phase space one, providing a kinetic approach.

Additionally, there is a tomographic map which provides a description in terms of a marginal distribution (classical probability function), starting from the quasidistribution.

In this scenario, the study of the quantum methodologies have been recognized as very important for a synergetic developments of the above branches of physics with very powerful multidisciplinary as well as interdisciplinary approaches. For instance, the intense study on nonlinear and collective effects in the several physical systems have stimulated a number of interdisciplinary approaches and transfer of know how from one discipline to another, obtaining, in turn, a big growing of importance of the methodologies used to investigate very different physical phenomena governed by formally identical equations. Two advantages of this interdisciplinary strategy are absolutely fruitful. One is that communities of physicists from different areas are stimulated to collaborate more an more exchanging their own experiences and make available their own expertise; the other one is the subsequent very rapid improvement of both methodologies to be used

and goals to be reached in each specific discipline. These aspects are connected with the efforts done during the last decades in transferring know how and methodologies from one discipline to another trying to predict new effects as well as to give answers to scientific and technological problems of international expectation. For instance, the applications of the quantum methodologies:
(i) to gravity ocean waves, touche the very important and hot problem of the environmental risk due to natural catastrophes, as the one that recently took place in the South-Est of Asia;
(ii) to beam physics, open up the possibility to develop an emerging area of physics, called Quantum Beam Physics, which in the limit of very low temperature should provide the realization of non-classical (but collective and nonlinear) states of charged particle beams fully similar to the ones obtained for the light (optical beams) and for Bose-Einstein condensation;
(iii) to nonlinear optics (f.i., optical fibers) and electric transmission lines, deal with important and modern aspects of telecommunications;
(iv) to discrete systems, are relevant for the very recent development of nanotechnologies.

1.2. *Some important aspects of the phenomenological platform investigated by means of the quantum methodologies*

The modulational instability (MI), also known as Benjamin-Feir instability, is a general phenomenon encountered when a quasi-monochromatic wave is propagating in a weak nonlinear medium. It has been predicted and experimentally observed in almost all field of physics where these conditions are present. We mention especially the wave propagation in deep waters (ocean gravity waves), in plasma physics (electrostatic and electromagnetic plasma waves), in particle accelerators (high-energy charged-particle beam dynamics), in nonlinear optics (Kerr media, optical fibers), in electrical transmission lines, in matter wave physics (Bose-Einstein condensates), in lattice vibrations physics (molecular crystals) and in the physics of antiferromagnetism (dynamics of the spin waves). For ocean gravity waves, the modulational instability has been discovered independently by Benjamin and Feir and by Zakharov in the Sixties; the instability predicts that in deep water a monochromatic wave is unstable under suitable small perturbations. This phenomenon is well described by the nonlinear Schrödinger equation (NLSE). In this framework, it has been established that the MI can be responsible for the formation of freak waves. In plasmas, Langmuir waves of finite amplitude can be created when some free energy sources, such as electron and laser beams, are available in the system as a result of a nonlinear coupling between high-frequency Langmuir and low-frequency ion-acoustic waves. Under suitable physical conditions, the dynamics can be described by a NLSE and the MI can be analyzed directly with this equation. In nonlinear optics, the propagation of large amplitude electromagnetic waves produces a modification of the refractive index which, in turn, affects the propagation itself and makes possible the formation of wave envelopes. In the slowly-varying amplitude approximation, this propagation is governed again by suitable NLSE and the MI plays a very important role. In electrical transmission lines, the propagation of modulated non-linear waves is governed by discrete equations of the LC circuit which, in turn, can be reduced to single or two coupled NLSEs.

The large amplitude wave propagation is common of many environmental and technological processes. In nature, wave interactions exhibit a random character. Therefore, in order to predict the wave behavior, taking into account the statistical properties of the medium, an adequate statistical description is needed. This kind of approach started in physics of fluids at least forty years ago to study the random behavior of the large amplitude surface gravity waves. The time evolution of the small amplitude surface gravity waves is characterized by a Gaussian statistics for most of the time. But, nevertheless, in still quite unknown conditions, the behavior of these waves may become highly non-Gaussian and the wave field is characterized by the appearance of very large amplitude waves whose effects can be devastating for fixed off-shore structures, for boats and also for coastal regions. Figure 1 shows one of the most spectacular recordings of a freak wave measured in the North Sea the first of January 1995 from Draupner platform (Statoil operated platform, Norway). The time series of the surface elevation shows a single wave whose height from crest to through is 26 meters (a 9th floor building) in a 10 meters height sea state. In deep water those large amplitudes have been recently attributed to MI.[12]

On the other hand, MI is of fundamental importance for the formation of robust nonlinear excitations of the medium. In fact, the asymptotic behaviour of MI may be characterized by the formation of very stable localized solutions, such as envelope solitons, cavitons, holes, etc., which, in turn, are involved in a long timescale dynamics that have been shown to be of a great importance in all nonlinear systems. Furthermore, as a peculiarity of any nonlinear system, MI can be understood as a four wave parametric process.

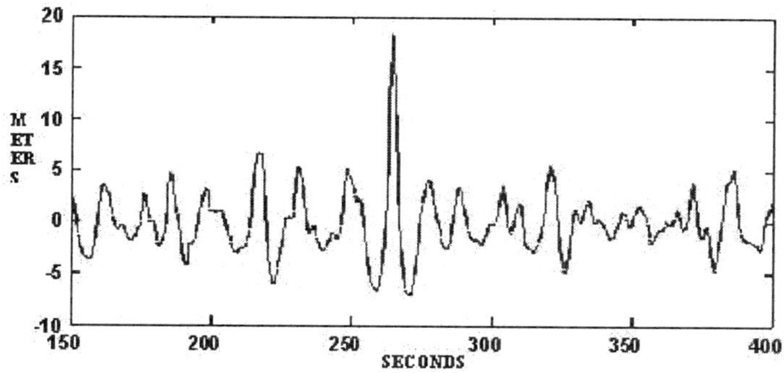

Fig. 1. Freak wave measured in the North Sea the first of January 1995 from Draupner platform (Statoil operated platform, Norway).[12]

1.3. *Impacts of the quantum methodologies in nonlinear physics*

In this paper, we confine our attention on three subjects only, namely, physics of fluids, plasma physics and beam physics (optical and charged particle beams). In order to present in the next section some relevant applications of the quantum methodologies involved in these three subjects, we discuss here the impact that they have produced in nonlinear physics with special emphasis for the transfer of the methodologies from one discipline to another.

- One of the most relevant example of using quantum methodologies in nonlinear physics is surely given by the inverse scattering method[13] which allows to find soliton solutions of the standard Korteweg-de Vries equation (KdVE) by constructing a correspondence between the latter and a linear Schrödinger equation (LSE) whose potential term coincides with the solution of the KdVE. This way, the problem of solving the KdVE is reduced to a quantumlike problem, i.e., to an inverse eigenvalue problem of the LSE. Very important theorems have been found for this method[14] and it has been successfully extended to NLSE.[15] The capability and the richness of similar methods currently applied to nonlinear partial differential equations for solving a number of physical problems have produced an autonomous research activity in mathematical physics called "inverse problems".
- More recently, outside the framework of the inverse scattering method, a

correspondence between solitonlike and envelope solitonlike solutions, in the form of travelling waves, of wide families of generalized Korteweg-de Vries equation (gKdVE) and generalized nonlinear Schrödinger equation (gNLSE), respectively, has been constructed within the framework of the Madelung fluid.[16,17] Under suitable constraints, this correspondence can be made invertible. Starting from the gNLSE, the travelling wave solutions of the associated gKdVE are found taking the square modulus of the travelling envelope wave solutions of the gNLSE. Viceversa, starting from the gKdVE, an arbitrary non-negative definite travelling wave solutions can be used to construct a map which, apart from an arbitrary linear phase, gives a travelling envelope wave solutions of the associated gNLSE. Remarkably, this correspondence has been used to find envelope solitonlike solutions of a wide family of gNLSE. In particular: (i) soliton solutions, in the form of bright and/or gray/dark envelope solitons of the cubic NLSE[17] and the modified one with the nonlinearity of the type $|\Psi|^{2\beta}\Psi$ (β being an arbitrary positive real number) have been recovered or found, respectively;[16,18] (ii) bright and gray/dark envelope solitons of "cubic-quintic" NLSE (i.e., the nonlinear term is of the form $c_1|\Psi|^2\Psi + c_2|\Psi|^4\Psi$) have been found, as well;[16] the method, extended to the cubic-quintic NLSE with and additional "anti-cubic" nonlinearity (i.e., the nonlinear term is of the form $|\Psi|^{-4}\Psi$), has allowed to find new soliton-like solutions[19] .

- Quantum methodologies have been recently employed to describe the charged-particle beam optics and dynamics in terms of a Schrödinger-like equation for a complex function whose square modulus is proportional to the beam density. The related model is called Thermal Wave Model (TWM)[20–22] . In general, the potential term of this equation accounts for both external and self-consistent field forces. The self-consistent fields are due to collective effects (reactive and resistive interaction schematized by the coupling impedance). The inclusion of the collective effects leads to a sort of NLSE.[23] For purely reactive coupling impedance,[24] the NLSE becomes formally identical to the one governing the propagation of optical beams with a cubic nonlinearity.[11] So that, a transfer of know how from nonlinear optics to accelerator physics has allowed to predict new results, such as soliton density structures associated with the longitudinal dynamics of a charged-particle bunch in a circular high–energy accelerating machine that the conventional approach, based on Vlasov equation, was not yet capable to predict.[24] Later on, by including the resistive part of the coupling impedance, the resulting integro-differential

NLSE was capable to describe the nonlocal and distortion effects, non dissipative shock waves and wave breaking on an initially given soliton-like particle beam density profile.[25]

- A further methodological transfer from nonlinear optics to accelerator physics was done with the analysis of modulational instability of macroscopic matter waves as described by the TWM. The results of these transfers may be summarized as follows. (i) The well known coherent instability (for instance, positive or negative mass instability), described by the Vlasov theory, is nothing but a sort of MI predicted by TWM for macroscopic matter waves with the above integro-differential NLSE.[23] (ii) The phenomenon of Landau damping[26] and its stabilizing role against the coherent instability was recovered,[27] and then extended in a more general framework.[28–30] This was done by adding a new element to the nonlinear macroscopic matter wave description: the extension to phase space of the MI analysis in TWM context by using the Moyal-Ville-Wigner quasidistribution, whose evolution is governed by a quantumlike kinetic equation (von Neumann equation). Actually, this approach was first introduced by Klimontovich and Silin in plasma theory[31] and, later on, in nonlinear fluid theory to develop the modulational instability theory of the surface gravity waves (nonlinear dynamics of deep waters waves[32]). However, it should be pointed out that the notion of "density matrix", also referred to as "statistical operator", very useful to give the definition of the quasidistribution, has been introduced for the first time by Landau[33] in 1927 and later, in 1932, it was mathematically treated by von Neumann.[5]
- Until few years ago, the modulational instability description in nonlinear optics was not yet capable to include the stabilizing effects. On the contrary, in accelerator physics, the stabilizing effects of the Landau damping was, already included in the coherent instability description. In particular, the classical version was given by the Vlasov-Maxwell system, whilst the quantumlike version was predicted by the TWM. The results given by the TWM were then transferred back to nonlinear optics to extend the standard modulational instability theory of optical beams and bunches to the context of ensemble of partially incoherent waves whose dynamics includes the statistical properties of the medium. At the present time, as result of the quantum methodology advances described above, we can say that two distinct ways to treat MI are possible. The first and the most used one, corresponding to the standard one, is a *deterministic approach*, where the linear stability analysis around a carrying wave is considered. This corresponds to consider the stability/instability of monochromatic

wave trains (system of coherent waves). The second one, is a *statistical approach* and its main goal is to introduce the statistical properties of the medium (whether continuum or discrete). In these physical conditions, the stability analysis cannot be carried out as in the monochromatic case. An ensemble of partially incoherent waves has, in fact, to be taken into account. Thus, MI reveals to be strongly dependent on the parameters characterizing the initial conditions (initial wavenumber distribution or initial momentum distribution of waves). This second approach stimulated very recently a new branch of investigation devoted to MI of ensembles of partially incoherent waves with both theoretical and experimental aims.[34,35] It was rapidly applied to Kerr media[36,37] and soon extended to plasma physics (ensemble of partially incoherent Langmuir wave envelopes)[38] and physics of lattice vibrations[39,40] where the discrete self-trapping equation represents an useful model for several properties of one-dimensional nonlinear molecular crystals. New improvements were also registered in the statistical formulation of MI for large amplitude surface gravity waves.[41]

- In the very recent years, thanks to a methodological transfer from nonlinear optics and plasma physics to matter wave physics, the deterministic approach to modulational instability has been applied to Bose Einstein condensates, as well. It has been both predicted and experimentally confirmed. For instance, MI conditions for the phonon spectrum takes place for an array of traps containing Bose-Einstein condensates (BEC) with each trap linked to adjacent traps by tunneling;[42] additionally, MI of matter waves of BECs periodically modulated by a laser beam takes place in a number of physical situations, as well.[43] The 3D dynamics of BECs is, in fact, governed by the well known Gross-Pitaevskii equation[44] which is a sort of NLSE. In the same effort of methodology and know how transfer, several valuable predictions and experimental confirmations concerning the formation of soliton-like structures in Bose Einstein condensates should be mentioned.[45–47] Remarkably, a very valuable scientific and technological feedback of this transfer was a production of dark-bright BEC solitons within the framework of the nanotechnologies.
- Another useful quantumlike tool is the method of filtering and controlling soliton states of the NLSE. It has been recently proposed to find analytically controlled 3D localized solutions of the Gross-Pitaevskii equation[48] and seems to be suitable to find the experimental conditions to control a soliton state of a Bose-Einstein condensate. According to this method, under suitable controlling conditions, the 3D Gross-Pitaevskii equation can

be decomposed into two coupled equation. One is a 2D linear Schrödinger equation (governing the transverse BEC dynamics) and the other one is a 1D controlled NLSE (i.e., 1D controlled Gross-Pitaevskii equation governing the longitudinal BEC dynamics). While controlling the system with appropriate external potential well, the BEC exhibits a transverse quantum dynamics (for instance, quantum interference) and a classical longitudinal nonlinear dynamics.

2. The Madelung fluid picture

2.1. *Hydrodynamical description of quantum mechanics*

During the period 1924-1925, L. de Broglie elaborated his theory of "pilot waves",[49] introducing the very fruitful idea of wave-particle dualism, funding the theory of matter waves. However, until 1926 a wave equation for particles, thought as waves, was not yet proposed. In that year, Schrödinger proposed a wave equation that today has his name (the Schrödinger equation), funding the wave mechanics.[50] On the pilot waves de Broglie published a series of articles during 1927[51] but they did not produce great excitation within the scientific community. During October of the same year, in fact, de Broglie presented a simplified version of his recent studies on pilot waves at the Fifth Physical Conference of the Solvay Institute in Brussels. The criticism received pushed him to abandon this theory to start to study the complementary principle. He came back to the pilot waves during the period 1955-1956, proposing a more organic theory.[52] Nevertheless, a very valuable seminal contribution to quantum mechanics was given by de Broglie while developing the pilot wave theory with the concept of "quantum potential", but a systematic presentation of this idea came only several years later.[53] At the beginning of Fifties, Bohm also have considered the concept of quantum potential.[54] However, the concept was naturally appearing in a hydrodynamical description proposed in 1926 by Madelung[1] (first proposal of a hydrodynamical model of quantum mechanics), followed by the proposal of Korn in 1927.[55] The Madelung fluid description of quantum mechanics revealed to be very fruitful in a number of applications: from the pilot waves theory to the hidden variables theory, from stochastic mechanics to quantum cosmology.

In the Madelung fluid description, the wave function, say Ψ, being a complex quantity, is represented in terms of modulus and phase which, substituted in the Schrödinger equation, allow to obtain a pair of nonlinear fluid equations for the "density" $\rho = |\Psi|^2$ and the "current velocity" $\mathbf{V} =$

$\nabla \text{Arg}(\Psi)$: one is the continuity equation (which accounts for the probability conservation) and the other one is a Navier-Stokes-like motion equation, which contains a force term proportional to the gradient of the quantum potential, i.e., $(\nabla^2|\Psi|)/|\Psi| = (\nabla^2\rho^{1/2})/\rho^{1/2}$. The nonlinear character of this system of fluid equations naturally allows to extend the Madelung description to systems whose dynamics is governed by one ore more NLSEs. Remarkably, during the last four decades, this quantum methodology was imported practically into all the nonlinear sciences, especially in nonlinear optics[34,56,57] and plasma physics[58,59] and it revealed to be very powerful in solving a number of problems. Let us consider the following (1+1)D nonlinear Schrödinger-like equation (NLSE):

$$i\alpha\frac{\partial\Psi}{\partial s} = -\frac{\alpha^2}{2}\frac{\partial^2\Psi}{\partial x^2} + U\left[|\Psi|^2\right]\Psi \quad , \tag{1}$$

where $U\left[|\Psi|^2\right]$ is, in general, a functional of $|\Psi|^2$, the constant α accounts for the dispersive effects, and s and x are the timelike and the configurational coordinates, respectively. Let us assume

$$\Psi = \sqrt{\rho(x,s)}\exp\left[\frac{i}{\alpha}\Theta(x,s)\right] \quad , \tag{2}$$

then substitute (2) in (1). After separating the real from the imaginary parts, we get the following Madelung fluid representation of (1) in terms of pair of coupled fluid equations:

$$\frac{\partial\rho}{\partial s} + \frac{\partial}{\partial x}(\rho V) = 0\,, \tag{3}$$

(continuity)

$$\left(\frac{\partial}{\partial s} + V\frac{\partial}{\partial x}\right)V = -\frac{\partial U}{\partial x} + \frac{\alpha^2}{2}\frac{\partial}{\partial x}\left[\frac{1}{\rho^{1/2}}\frac{\partial^2\rho^{1/2}}{\partial x^2}\right], \tag{4}$$

(motion) where the current velocity V is given by

$$V(x,s) = \frac{\partial\Theta(x,s)}{\partial x}\,. \tag{5}$$

In the next subsections, we present solme relevant applications of Madelung fluid methodology to the soliton and MI theories.

2.2. *Applications to soliton theory*

In order to apply the Madelung description to the soliton theory, let us manipulate the system of equations (3) and (4), in such a way to transform the motion equation into a third-order partial differential equaion for ρ.

By multiplying Eq. (3) by V, the following equation can be obtained:

$$\rho\left(\frac{\partial}{\partial s}+V\frac{\partial}{\partial x}\right)V \;=\; -V\frac{\partial\rho}{\partial s} \;-\; V^2\frac{\partial\rho}{\partial x} \;+\; \rho\frac{\partial V}{\partial s} \quad . \tag{6}$$

Note that:

$$\frac{\partial}{\partial x}\left(\frac{1}{\rho^{1/2}}\frac{\partial^2\rho^{1/2}}{\partial x^2}\right) \;=\; \frac{1}{\rho}\left(\frac{1}{2}\frac{\partial^3\rho}{\partial x^3} \;-\; 4\,\frac{\partial\rho^{1/2}}{\partial x}\,\frac{\partial^2\rho^{1/2}}{\partial x^2}\right) \quad . \tag{7}$$

Furthermore, multiplying Eq. (4) by ρ and combining the result with (6) and (7) one obtains

$$\rho\left(\frac{\partial}{\partial s}+V\frac{\partial}{\partial x}\right)V = -\frac{\partial U}{\partial x}\rho + \frac{\alpha^2}{4}\frac{\partial^3\rho}{\partial x^3} \;-\; 2\alpha^2\,\frac{\partial\rho^{1/2}}{\partial x}\,\frac{\partial^2\rho^{1/2}}{\partial x^2} \quad , \tag{8}$$

which combined again with Eq. (6) gives:

$$-V\frac{\partial\rho}{\partial s} \;-\; V^2\frac{\partial\rho}{\partial x} \;+\; \rho\frac{\partial V}{\partial s} = -\frac{\partial U}{\partial x}\rho + \frac{\alpha^2}{4}\frac{\partial^3\rho}{\partial x^3} \;-\; 2\alpha^2\,\frac{\partial\rho^{1/2}}{\partial x}\,\frac{\partial^2\rho^{1/2}}{\partial x^2} \quad . \tag{9}$$

On the other hand, by integrating Eq. (4) with respect to x and multiplying the resulting equation by $\rho^{1/2}\;\left(\partial\rho^{1/2}/\partial x\right)$, we have:

$$-2\alpha^2\,\frac{\partial\rho^{1/2}}{\partial x}\,\frac{\partial^2\rho^{1/2}}{\partial x^2} = -2\frac{\partial\rho}{\partial x}\int\left(\frac{\partial V}{\partial s}\right)dx - V^2\frac{\partial\rho}{\partial x} - 2\,U\frac{\partial\rho}{\partial x} + 2\,c_0(s)\,\frac{\partial\rho}{\partial x} \;, \tag{10}$$

where $c_0(s)$ is an arbitrary function of s. By combining (9) and (10) the following equation is finally obtained:

$$-\frac{\partial V}{\partial s}\rho + V\frac{\partial\rho}{\partial s} + 2\left[c_0 - \int\left(\frac{\partial V}{\partial s}\right)dx\right]\frac{\partial\rho}{\partial x} - \left(\frac{\partial U}{\partial x}\rho + 2\,U\frac{\partial\rho}{\partial x}\right) \;+\; \frac{\alpha^2}{4}\frac{\partial^3\rho}{\partial x^3} = 0 \;. \tag{11}$$

Note that now Eq. (1), or the pair of equations (3) and (4), is reduced to the pair of equations (3) and (11).

Let us denote with $\mathcal{E} = \{\Psi\}$ the set of all the envelope solutions of the generalized nonlinear Schrödinger equation (gNLSE), Eq. (1), in the form of travelling wave envelope, i.e. $\Psi(x,s) = \sqrt{\rho(\xi)}\exp\left\{\frac{i}{\alpha}\;\Theta(x,s)\right\}$, where $\xi = x - u_0 s$ (u_0 being a real constant).

Let us also denote with $\mathcal{S} = \{u(\xi) \geq 0\}$ the set of all non-negative stationary-profile solutions (travelling waves) of the following generalized Korteweg-de Vries equation (gKdVE)

$$a\,\frac{\partial u}{\partial s} \;-\; G\left[u\right]\,\frac{\partial u}{\partial x} \;+\; \frac{\nu^2}{4}\,\frac{\partial^3 u}{\partial x^3} \;=\; 0 \quad , \tag{12}$$

where $G[u]$ is a functional of u. In particular, when is assumed that $G[u] \propto u$, Eq. (12) reduces to the standard KdVE. In order to construct a

correspondence between $\mathcal{E}$ and $\mathcal{S}$, we observe that if $\Psi \in \mathcal{E}$, thus ρ and V have the form $\rho = \rho(\xi)$ and $V = V(\xi)$, respectively.
Under the above hypothesis, it is easy to see that: (a). $c_0(s)$ becomes constant (so that, let us put $c_0(s) \equiv c_0$); (b). continuity equation (3) becomes:

$$u_0 \frac{d\rho}{d\xi} = \frac{d}{d\xi}(\rho V) \quad , \tag{13}$$

which integrated gives:

$$V(\xi) = u_0 + \frac{A_0}{\rho(\xi)} \quad , \tag{14}$$

where A_0 is an arbitrary constant. By combining (11) and (13), we easily have:

$$\left(u_0^2 + 2c_0\right) \frac{d\rho}{d\xi} \; - \; \mathcal{I}[\rho] \frac{d\rho}{d\xi} \; + \; \frac{\alpha^2}{4} \frac{d^3\rho}{d\xi^3} \; = \; 0 \quad , \tag{15}$$

where the functional $\mathcal{I}[\rho]$ is defined as:

$$\mathcal{I}[\rho] \; = \; \rho \frac{dU[\rho]}{d\rho} \; + \; 2U[\rho] \quad . \tag{16}$$

On the other hand, for stationary-profile solution $u = u(\xi)$, Eq. (12) becomes:

$$-u_0 a \, \frac{du}{d\xi} \; - \; G[u] \frac{du}{d\xi} \; + \; \frac{\nu^2}{4} \frac{d^3u}{d\xi^3} \; = \; 0 \quad . \tag{17}$$

Consequently, (15) and (17) have the same solutions, if the same boundary conditions are taken for them and provided that their coefficients are respectively proportional. In particular, it follows that $u(\xi)$ is a non-negative travelling wave solution of the following gKdVE ($u_0 \neq 0$):

$$-\frac{u_0^2 + 2c_0}{u_0} \frac{\partial\rho}{\partial s} \; - \; \mathcal{I}[\rho] \frac{\partial\rho}{\partial x} \; + \; \frac{\alpha^2}{4} \frac{\partial^3\rho}{\partial x^3} \; = \; 0 \quad , \tag{18}$$

where the following identies have been used:

$$a \equiv -\left(u_0^2 + 2c_0\right)/u_0 \;, \qquad G[u] \equiv \mathcal{I}[u] \;, \qquad \nu \equiv \alpha \quad . \tag{19}$$

Thus, it results that, starting from Eq. (1), we have constructed the following correspondence:

$$\begin{aligned} \mathcal{F} \; &: \; \Psi \in \mathcal{E} \rightarrow u \in \mathcal{S} \quad , \\ u \; &= \; \mathcal{F}[\Psi] \; = \; |\Psi|^2 \; = \; \rho \quad . \end{aligned} \tag{20}$$

$\mathcal{F}$ associates a travelling wave envelope solution of (1) to a travelling wave solution of the associated gKdVE (12). In particular, it may associate an envelope solitonlike solution of (1) with a solitonlike solution of (11). It can be proven that:[16]

$$\Theta(x,s) \;=\; \phi_0 \;-\; \left(c_0 + u_0^2\right) s \;+\; u_0 x \;+\; A_0 \int \frac{d\xi}{\rho(\xi)} \quad , \tag{21}$$

where ϕ_0 is an arbitrary real constant.
Now, let $u(\xi)$ be a positive stationary-profile solution of (12). Thus, u satisfies an equation similar to (17) and, provided that (15) and (17) have still proportional coefficients, in correspondence to the same boundary conditions, u is also solution of (15). Thus, by defining the phase $\Theta(x,s)$ given by (21), one can go back defining $V = \partial\Theta(x,s)/\partial x$. It follows that $\rho(\xi) = u(\xi)$ and V are solutions of the following system of coupled equations:

$$\frac{\partial u}{\partial s} + \frac{\partial}{\partial x}(uV) = 0\,, \tag{22}$$

$$-\frac{\partial V}{\partial s}u + V\frac{\partial u}{\partial s} + 2\left[c_0(s) - \int \left(\frac{\partial V}{\partial s}\right) dx\right]\frac{\partial u}{\partial x} - \left(\frac{\partial \mathcal{U}}{\partial x}u + 2\mathcal{U}\frac{\partial u}{\partial x}\right) + \frac{\nu^2}{4}\frac{\partial^3 u}{\partial x^3} = 0\,, \tag{23}$$

where the functional $\mathcal{U}$ is solution of the following differential equation:

$$u\,\frac{d\mathcal{U}}{du} \;+\; 2\mathcal{U} \;=\; G\left[u\right]\,, \tag{24}$$

namely

$$\mathcal{U}\left[u\right] \;=\; \frac{1}{u^2}\left[K_0 \;+\; \int G\left[u\right]\; u\; du\right]\,, \tag{25}$$

where K_0 is an arbitrary real integration constant. Consequently, the complex function

$$\Psi = \sqrt{u(\xi)}\;\exp\left[\frac{i}{\alpha}\Theta(x,s)\right] \tag{26}$$

is a travelling envelope wave solution of the following gNLSE:

$$i\nu\frac{\partial\Psi}{\partial s} \;+\; \frac{\nu^2}{2}\frac{\partial^2\Psi}{\partial x^2} \;-\; \left[\frac{K_0 \;+\; \int G\left[|\Psi|^2\right]\; |\Psi|^2\; d|\Psi|^2}{|\Psi|^4}\right]\Psi \;=\; 0 \quad . \tag{27}$$

The substitution of (26) in (27) (after separating real and imaginary parts) gives the following equation for u:

$$-\frac{\nu^2}{2}\frac{d^2u^{1/2}}{d\xi^2} + \frac{K_0}{u^{3/2}} + \frac{1}{u^{3/2}}\int G\left[u\right]\; u\; du \;=\; \left(c_0 \;+\; \frac{u_0^2}{2}\right)u^{1/2} - \frac{A_0^2}{2u^{3/2}} \quad . \tag{28}$$

It results that for each given $u \in \mathcal{S}$ and for each given set of constants ϕ_0, c_0, and A_0, the modulus and the phase of Ψ are uniquely determined and, consequently, the solution of (27) is uniquely determined.
In conclusion, starting from the gKdVE (12), we have constructed the following correspondence:

$$\mathcal{H} \; : \; u \in \mathcal{S} \rightarrow \Psi \in \mathcal{E} \quad ,$$

$$\Psi = \mathcal{H}\left[u\right] = \sqrt{u(\xi)}\,\exp\left\{\frac{i}{\nu}\left[\phi_0 - \left(c_0 + u_0^2\right)s + u_0 x + A_0 \int \frac{d\xi}{u(\xi)}\right]\right\} \,, \quad (29)$$

which, for each given set of real constants ϕ_0, c_0, and A_0, associates a positive stationary-profile solution $u(\xi)$ of (12) to a stationary-profile envelope solution $\Psi(x, s)$ of (27) which is of the type (1). It is clear that, as the above parameters vary over all their accessible ranges of values, $\mathcal{H}\left[u\right]$ describes the subset of stationary-profile envelope solutions of (27) whose squared modulus equals $u(\xi)$. In particular, if $u(\xi)$ is a localized solution of (12), thus $\mathcal{H}\left[u\right]$ describes the subset of envelope localized solutions of the associated equation (27), where ϕ_0 is still arbitrary and the values allowed for c_0 and A_0 are determined by the specific boundary conditions required for such a kind of localized solution. It has been proven that for bright and dark soliton solutions, satisfying the standard boundary conditions, i.e., $\lim_{\xi\rightarrow\pm\infty} u(\xi) = 0$, $K_0 = A_0 = 0$ and therefore the phase of Ψ is linear (see Eq. (29)).[16] If $K_0 = 0$, the case $A_0 \neq 0$ is compatible with gray and up-shifted solitons,[16] while the case of $K_0 \neq 0$ and $A_0 \neq 0$ exhibits some new insights.[19]

In the next subsections we present some examples where the above mapping is applied to find solitonlike solution taking some special forms of the nonlinear potential $U[|\Psi|^2]$ and the ones of the corresponding nonlinear term $G[u]$. We divide the analysis into two cases: the case of $K_0 = 0$ and the one of $K_0 \neq 0$.

2.2.1. *Case of* $K_0 = 0$.

(a). Let us take the following nonlinear term of the gNLSE (1): $U\left[|\Psi|^2\right] = q_0|\Psi|^{2\beta}$, where q_0 and β are real and positive real numbers, respectively. Thus, according to Eq.s (16) and (19), the corresponding nonlinear term of the gKdVE (12) is $G[u] = p_0 u^\beta$, where $p_0 = (\beta+2)q_0$. For $au_0 > 0$ and $p_0 < 0$, Eq. (12) admits the following positive solution in the form of bright travelling solitons (for any positive real β):

$$u(\xi) = \left[\frac{au_0\,(\beta+2)\,(\beta+1)}{2|p_0|}\right]^{1/\beta} \operatorname{sech}^{2/\beta}\left[\frac{\beta\sqrt{au_0}}{|\nu|}\xi\right] \quad (30)$$

(note that $\beta = 1$ recovers the bright KdV soliton[17]).

Consequently, by virtu of (29), we can write the following travelling bright envelope soliton solutions of (1) for the case under discussion (note that here $A_0 = 0$):

$$\Psi(x,s) = \left[\frac{|E_0|\,(\beta+1)}{|q_0|}\right]^{1/2\beta} \operatorname{sech}^{1/\beta}\left[\frac{\beta\sqrt{2|E_0|}}{|\alpha|}\xi\right] \times \exp\left\{\frac{i}{\alpha}\left[\phi_0 - \left(E_0 + u_0^2/2\right)s + u_0 x\right]\right\} \tag{31}$$

where $E_0 < 0$ and $q_0 < 0$. In Table 1 the above correspondence between modified Korteweg-de Vries equations (mKdVEs) and modified nonlinear Schrödinger equations (mNLSEs) is shown for several values of β. In Table 2 the bright solitary wave solutions of mKdVE and mNLSE, respectively, are given for the same values of β displayed in Table 1.

Table 1. Correspondence between mNLSE and mKdVE for several values of β

β	**mKdVE**	**mNLSE**
1/10	$a\frac{\partial u}{\partial s} - \frac{21}{10}q_0 u^{1/10}\frac{\partial u}{\partial x} + \frac{\alpha^2}{4}\frac{\partial^3 u}{\partial x^3} = 0$	$i\alpha\frac{\partial \Psi}{\partial s} = -\frac{\alpha^2}{2}\frac{\partial^2 \Psi}{\partial x^2} - q_0\|\Psi\|^{1/5}\Psi$
1/8	$a\frac{\partial u}{\partial s} - \frac{17}{8}q_0 u^{1/8}\frac{\partial u}{\partial x} + \frac{\alpha^2}{4}\frac{\partial^3 u}{\partial x^3} = 0$	$i\alpha\frac{\partial \Psi}{\partial s} = -\frac{\alpha^2}{2}\frac{\partial^2 \Psi}{\partial x^2} - q_0\|\Psi\|^{1/4}\Psi$
1/3	$a\frac{\partial u}{\partial s} - \frac{7}{3}q_0 u^{1/3}\frac{\partial u}{\partial x} + \frac{\alpha^2}{4}\frac{\partial^3 u}{\partial x^3} = 0$	$i\alpha\frac{\partial \Psi}{\partial s} = -\frac{\alpha^2}{2}\frac{\partial^2 \Psi}{\partial x^2} - q_0\|\Psi\|^{2/3}\Psi$
1/2	$a\frac{\partial u}{\partial s} - \frac{5}{2}q_0 u^{1/2}\frac{\partial u}{\partial x} + \frac{\alpha^2}{4}\frac{\partial^3 u}{\partial x^3} = 0$	$i\alpha\frac{\partial \Psi}{\partial s} = -\frac{\alpha^2}{2}\frac{\partial^2 \Psi}{\partial x^2} - q_0\|\Psi\|\Psi$
1	$a\frac{\partial u}{\partial s} - 3q_0 u\frac{\partial u}{\partial x} + \frac{\alpha^2}{4}\frac{\partial^3 u}{\partial x^3} = 0$	$i\alpha\frac{\partial \Psi}{\partial s} = -\frac{\alpha^2}{2}\frac{\partial^2 \Psi}{\partial x^2} - q_0\|\Psi\|^{2}\Psi$
3/2	$a\frac{\partial u}{\partial s} - \frac{7}{2}q_0 u^{3/2}\frac{\partial u}{\partial x} + \frac{\alpha^2}{4}\frac{\partial^3 u}{\partial x^3} = 0$	$i\alpha\frac{\partial \Psi}{\partial s} = -\frac{\alpha^2}{2}\frac{\partial^2 \Psi}{\partial x^2} - q_0\|\Psi\|^{3}\Psi$
2	$a\frac{\partial u}{\partial s} - 4q_0 u^{2}\frac{\partial u}{\partial x} + \frac{\alpha^2}{4}\frac{\partial^3 u}{\partial x^3} = 0$	$i\alpha\frac{\partial \Psi}{\partial s} = -\frac{\alpha^2}{2}\frac{\partial^2 \Psi}{\partial x^2} - q_0\|\Psi\|^{4}\Psi$
5/2	$a\frac{\partial u}{\partial s} - \frac{9}{2}q_0 u^{5/2}\frac{\partial u}{\partial x} + \frac{\alpha^2}{4}\frac{\partial^3 u}{\partial x^3} = 0$	$i\alpha\frac{\partial \Psi}{\partial s} = -\frac{\alpha^2}{2}\frac{\partial^2 \Psi}{\partial x^2} - q_0\|\Psi\|^{5}\Psi$
3	$a\frac{\partial u}{\partial s} - 5q_0 u^{3}\frac{\partial u}{\partial x} + \frac{\alpha^2}{4}\frac{\partial^3 u}{\partial x^3} = 0$	$i\alpha\frac{\partial \Psi}{\partial s} = -\frac{\alpha^2}{2}\frac{\partial^2 \Psi}{\partial x^2} - q_0\|\Psi\|^{6}\Psi$
10	$a\frac{\partial u}{\partial s} - 12q_0 u^{10}\frac{\partial u}{\partial x} + \frac{\alpha^2}{4}\frac{\partial^3 u}{\partial x^3} = 0$	$i\alpha\frac{\partial \Psi}{\partial s} = -\frac{\alpha^2}{2}\frac{\partial^2 \Psi}{\partial x^2} - q_0\|\Psi\|^{20}\Psi$

Dark (i.e., black) and grey solitons have been recovered by using the

Table 2. Bright solitary wave solutions related to the correspondence of Table 1

β	**mKdVE solution**	**modulus of mNLSE solution**
1/10	$\left(\frac{11\|E_0\|}{10\|q_0\|}\right)^{10}\mathrm{sech}^{20}\left(\frac{\sqrt{2\|E_0\|}}{10\|\alpha\|}\xi\right)$	$\left(\frac{11\|E_0\|}{10\|q_0\|}\right)^{5}\mathrm{sech}^{10}\left(\frac{\sqrt{2\|E_0\|}}{10\|\alpha\|}\xi\right)$
1/8	$\left(\frac{9\|E_0\|}{8\|q_0\|}\right)^{8}\mathrm{sech}^{16}\left(\frac{\sqrt{2\|E_0\|}}{8\|\alpha\|}\xi\right)$	$\left(\frac{9\|E_0\|}{8\|q_0\|}\right)^{4}\mathrm{sech}^{8}\left(\frac{\sqrt{2\|E_0\|}}{8\|\alpha\|}\xi\right)$
1/3	$\left(\frac{4\|E_0\|}{3\|q_0\|}\right)^{3}\mathrm{sech}^{6}\left(\frac{\sqrt{2\|E_0\|}}{3\|\alpha\|}\xi\right)$	$\left(\frac{4\|E_0\|}{3\|q_0\|}\right)^{3/2}\mathrm{sech}^{3}\left(\frac{\sqrt{2\|E_0\|}}{3\|\alpha\|}\xi\right)$
1/2	$\left(\frac{3\|E_0\|}{2\|q_0\|}\right)^{2}\mathrm{sech}^{4}\left(\frac{\sqrt{2\|E_0\|}}{2\|\alpha\|}\xi\right)$	$\left(\frac{3\|E_0\|}{2\|q_0\|}\right)\mathrm{sech}^{2}\left(\frac{\sqrt{2\|E_0\|}}{2\|\alpha\|}\xi\right)$
1	$\left(\frac{2\|E_0\|}{\|q_0\|}\right)\mathrm{sech}^{2}\left(\frac{\sqrt{2\|E_0\|}}{\|\alpha\|}\xi\right)$	$\left(\frac{2\|E_0\|}{\|q_0\|}\right)^{1/2}\mathrm{sech}\left(\frac{\sqrt{2\|E_0\|}}{\|\alpha\|}\xi\right)$
3/2	$\left(\frac{5\|E_0\|}{2\|q_0\|}\right)^{2/3}\mathrm{sech}^{4/3}\left(\frac{3\sqrt{2\|E_0\|}}{2\|\alpha\|}\xi\right)$	$\left(\frac{5\|E_0\|}{2\|q_0\|}\right)^{1/3}\mathrm{sech}^{2/3}\left(\frac{3\sqrt{2\|E_0\|}}{2\|\alpha\|}\xi\right)$
2	$\left(\frac{3\|E_0\|}{\|q_0\|}\right)^{1/2}\mathrm{sech}\left(\frac{2\sqrt{2\|E_0\|}}{\|\alpha\|}\xi\right)$	$\left(\frac{3\|E_0\|}{\|q_0\|}\right)^{1/4}\mathrm{sech}^{1/2}\left(\frac{2\sqrt{2\|E_0\|}}{\|\alpha\|}\xi\right)$
5/2	$\left(\frac{7\|E_0\|}{2\|q_0\|}\right)^{2/5}\mathrm{sech}^{4/5}\left(\frac{5\sqrt{2\|E_0\|}}{2\|\alpha\|}\xi\right)$	$\left(\frac{7\|E_0\|}{2\|q_0\|}\right)^{1/5}\mathrm{sech}^{2/5}\left(\frac{5\sqrt{2\|E_0\|}}{2\|\alpha\|}\xi\right)$
3	$\left(\frac{4\|E_0\|}{\|q_0\|}\right)^{1/3}\mathrm{sech}^{2/3}\left(\frac{3\sqrt{2\|E_0\|}}{\|\alpha\|}\xi\right)$	$\left(\frac{4\|E_0\|}{\|q_0\|}\right)^{1/6}\mathrm{sech}^{1/3}\left(\frac{3\sqrt{2\|E_0\|}}{\|\alpha\|}\xi\right)$
10	$\left(\frac{11\|E_0\|}{\|q_0\|}\right)^{1/10}\mathrm{sech}^{1/5}\left(\frac{\sqrt{2\|E_0\|}}{10\|\alpha\|}\xi\right)$	$\left(\frac{11\|E_0\|}{\|q_0\|}\right)^{1/20}\mathrm{sech}^{1/10}\left(\frac{10\sqrt{2\|E_0\|}}{\|\alpha\|}\xi\right)$

above Madelung's fluid method.[16–19] In particular, Figure 2 displays the modulus of the wave function Ψ and the one of the current velocity V, respectively. They show the solution of a cubic NLSE ($\beta = 1$) in the form of a grey soliton.[17]

(b). As second example, let us find envelope solitonlike solutions of the Eq. (1) with the following nonlinear potential: $U = U\left[|\Psi|^2\right] = q_1|\Psi|^2 + q_2|\Psi|^4$ (cubic-quintic NLSE). Thus the corresponding nonlinear terms in Eq. (12) is: $G[u] = p_1 u + p_2 u^2$, where $p_1 = 3q_1$ and $p_2 = 5q_2$. This term can be also cast in the form: $G[u] = p_0(u - \overline{u})^2 + \delta_0$, where $p_0 = 4q_2$, $\overline{u} = -3q_1/(8q_2)$, and $\delta_0 = -9q_1^2/16q_2$. By using the results of the previous example for $\beta = 2$, travelling solitons of the modified KdVE under discussion are easily

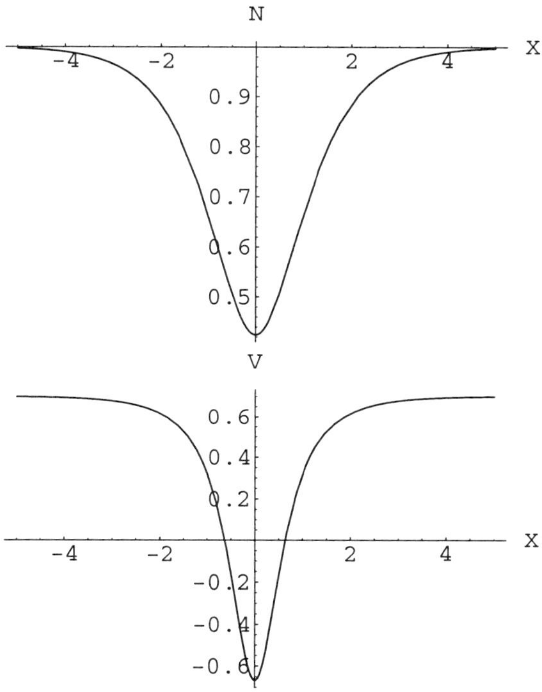

Fig. 2. Grey soliton for cubic NLSE ($\beta = 1$). $N \equiv |\Psi|/\sqrt{\rho_0} = \sqrt{\rho/\rho_0}$ versus $X \equiv \xi/|\alpha|$ (picture at the left) and V versus $X \equiv \xi/|\alpha|$ (picture at the right), where $\xi = x - u_0 s$.[17] The parameters are: $\rho_0 q_0 = 0.5$, $u_0 = 1$, and $V_0 = 0.7$.

found:[16]

$$u(\xi) = \overline{u} \; [1 \; + \; \epsilon \; \text{sech} \, (\xi/\Delta)] \quad , \tag{32}$$

where

$$\epsilon = \pm\sqrt{1 - 32|q_2| \, (u_0 - V_0)^2 \, / \, (3q_1^2)} \quad ,$$

$$\Delta = |\alpha| / \left(2\sqrt{2|E_0'|} \right) \quad ,$$

and

$$E_0' = -3q_1^2 / \, (64|q_2|) \; + \; (u_0 - V_0)^2 \, /2 \quad ,$$

provided that $E_0' < 0$ and $q_2 < 0$ and

$$-\sqrt{\frac{3q_1^2}{32|q_2|}} + V_0 < u_0 < \sqrt{\frac{3q_1^2}{32|q_2|}} + V_0 \quad .$$

Equation (32) shows that we can distinguish the following four cases.

(i). $0 < \epsilon < 1$ $(u_0 - V_0 \neq 0)$:

$$u(\xi = 0) = \overline{u}(1+\epsilon)\ , \quad \text{and} \quad \lim_{\xi\to\pm\infty} u(\xi) = \overline{u}$$

which corresponds to a bright soliton of maximum amplitude $(1+\epsilon)\overline{u}$ and up-shifted by the quantity $\overline{u}$. We could call it *up-shifted bright soliton.*

(ii).$- < \epsilon < 0$ $(u_0 - V_0 \neq 0)$:

$$u(\xi = 0) = \overline{u}(1-\epsilon)\ , \quad \text{and} \quad \lim_{\xi\to\pm\infty} u(\xi) = \overline{u}$$

which is a dark soliton with minimum amplitude $(1-\epsilon)\overline{u}$ and reaching asimptotically the upper limit $\overline{u}$. It corresponds to a standard *gray soliton.*

(iii). $\epsilon = 1$ $(u_0 - V_0 = 0)$:

$$u(\xi = 0) = 2\overline{u}\ , \quad \text{and} \quad \lim_{\xi\to\pm\infty} u(\xi) = \overline{u}$$

which corresponds to a bright soliton of maximum amplitude $2\overline{u}$ and up-shifted by the maximum quantity $\overline{u}$. We could call it *upper-shifted bright soliton.*

(iv).$\epsilon = -1$ $(u_0 - V_0 = 0)$:

$$u(\xi = 0) = 0\ , \quad \text{and} \quad \lim_{\xi\to\pm\infty} u(\xi) = \overline{u}$$

which is a dark soliton (zero minimum amplitude), reaching asimptotically the upper limit $\overline{u}$. It correspond to a standard *dark soliton.*

Consequenlty, by using mapping (29), we can conclude that the cubic-quintic NLSE under discussion has the following travelling envelope solitons:

$$\Psi(x,s) = \sqrt{\overline{u}\ [1\ +\ \epsilon\ \mathrm{sech}\,(\xi/\Delta)]}\exp\left\{\frac{i}{\alpha}\left[\phi_0 - As + u_0 x\right]\right\}$$
$$\times\exp\left\{\frac{iB}{\alpha}\left[\frac{\xi}{\Delta} + \frac{2\epsilon}{\sqrt{1-\epsilon^2}}\arctan\left(\frac{(\epsilon-1)\tanh\,(\xi/2\Delta)}{\sqrt{1-\epsilon^2}}\right)\right]\right\} \quad (33)$$

where ϕ_0 still plays the role of arbitrary constant,

$$A = \frac{15q_1^2}{64|q_2|} + \frac{(u_0 - V_0)^2}{2} + \frac{u_0^2}{2}\ , \quad (34)$$

and

$$B = -\frac{|\alpha|\,(u_0 - V_0)}{2\sqrt{2|E_0'|}}\ . \quad (35)$$

2.2.2. *Case of* $K_0 \neq 0$

It is clear from (28) that, for $K_0 \neq 0$, a family of solitary wave solutions of (27) can be obtained by imposing the following condition

$$A_0 = \pm\sqrt{-2K_0}, \tag{36}$$

which implies that such a kind of family of solutions exists for negative values of K_0. In fact, condition (36) select the suitable values of the arbitrary constant A_0 for finding solitonlike solutions, by providing the cancellation of the term $K_0/u^{3/2}$, located at left hand side of (28), with $-A_0^2/u^{3/2}$, located a the right hand side. Consequently, equation (28) becomes the following NLSE for stationary states

$$-\frac{\alpha^2}{2}\frac{d^2u^{1/2}}{d\xi^2} + \mathcal{U}[u]\,u^{1/2} = E_0 u^{1/2} \quad , \tag{37}$$

where $E_0 = c_0 + u_0^2/2$. Note that, by virtue Eq. (24) the functional G[u] remains unchanged switching from the case of $K_0 = 0$ to the one of $K_0 \neq 0$. Consequently, the corresponding gKdVE equations are undistinguishable. This implies that the travelling envelope soliton solutions of (27) with $K_0 \neq 0$ can be constructed by using the ones known in the case of $K_0 = 0$. Recently, this implication has allowed to find travelling envelope solitons of (1) with a nonlinear potential of the form: $q_1|\Psi|^2 + q_2|\Psi|^4 + Q_0|\Psi|^{-4}$ (cubic-quintic-anticubic NLSE).[19] It has been shown that for any $Q_0 < 0$, under condition (36) ($Q_0 \equiv K_0$), the cubic-quintic NLSE under discussion, with the additional anti-cubic nonlinear term, has the following envelope soliton solutions:

$$\Psi_{\pm}(x,s) = \sqrt{\overline{u}\left[1 + \epsilon\,\mathrm{sech}\left(\frac{\xi}{\Delta}\right)\right]} \tag{38}$$
$$\times \exp\left\{\frac{i}{\alpha}\left[\phi_0 - \left(E_0 - \frac{u_0^2}{2}\right)s + u_0\xi \pm \sqrt{2|Q_0|}\int\frac{d\xi}{\sqrt{\overline{u}\,[1+\epsilon\mathrm{sech}\,(\xi/\Delta)]}}\right]\right\}$$

where here, in principle, ϵ should be taken in the following range[19]

$$-1 < \epsilon \leq 1 \quad , \tag{39}$$

which excludes the standard "dark" solitary waves ($\epsilon = -1$), namely the condition for which the modulus of Ψ vanishes at $\xi = 0$.
Actually, the direct substitution of $u = |\Psi|^2$ given by (39) into the eigenvalue equation (37) allows us to find

$$\epsilon = 1 \quad , \tag{40}$$

$$\overline{u} = -\frac{3q_1}{8q_2} \quad , \tag{41}$$

$$q_1 > 0 \quad , \qquad q_2 < 0 \quad , \tag{42}$$

$$E_0 = -\frac{15q_1^2}{64q_2} \quad , \tag{43}$$

$$\Delta = \frac{2|\alpha|}{q_1}\sqrt{\frac{2|q_2|}{3}} \quad . \tag{44}$$

Consequently, solution (39) can be cast as

$$\Psi_\pm(x,s) = \sqrt{\frac{3q_1}{8|q_2|}\left[1+\mathrm{sech}\left(\frac{\xi}{\Delta}\right)\right]}\exp\left\{\frac{i}{\alpha}\left[\phi_0 - \left(\frac{15q_1^2}{64|q_2|}-\frac{u_0^2}{2}\right)s + u_0\xi\right]\right\}$$
$$\times\exp\left\{\frac{i}{\alpha}\left[\pm\sqrt{2|Q_0|}\,\frac{16|\alpha||q_2|}{3q_1^2}\sqrt{\frac{2|q_2|}{3}}\left(\frac{\xi}{\Delta}-\tanh\left(\frac{\xi}{2\Delta}\right)\right)\right]\right\} \tag{45}$$

where u_0 is a fully arbitrary soliton velocity. Equation (45) represents an upper-shifted bright envelope solitonlike solution, provided that the coefficients Q_0, q_1 and q_2 satisfy the conditions $Q_0 < 0$ and (42), respectively.[16] It is clear that equation (40), which does not contradict condition (39), implies that also grey solitary solutions do not exist in the solution form (39).

2.3. *Application to the modulational instability theory in the presence of nonlocal effects*

In this subsection, we present an example of using the Madelung fluid description to analyze the MI of NLSE in the presence of nonlocal effects.[60,61] We consider the case of longitudinal charged-particle beam dynamics in high-energy accelerating machines.

2.3.1. *The NLSE governing the longitudinal dynamics of a charged particle beam including nonlocal effects in the framework of TWM.*

Within the TWM framework, the longitudinal dynamics of particle bunches is described in terms of a complex wave function $\Psi(x,s)$, where s is the distance of propagation (time-like coodinate) and x is the longitudinal instantaneous location of a generic particle, measured in the moving frame of reference. The particle density, $\lambda(x,s)$, is related to the wave function according to the assumption[20] $\lambda(x,s) = |\Psi(x,s)|^2$. The collective longitudinal evolution of the beam in a circular high-energy accelerating machine is governed by the Schrödinger-like equation

$$i\epsilon\frac{\partial\Psi}{\partial s}+\frac{\epsilon^2\eta}{2}\frac{\partial^2\Psi}{\partial x^2}-U(x,s)\Psi=0\ , \qquad (46)$$

where ϵ is the longitudinal beam emittance and η is the slip factor,[62] defined as $\eta=\gamma_T^{-2}-\gamma^{-2}$ (γ_T being the transition energy, defined as the inverse of the momentum compaction,[62] and γ being the relativistic factor); $U(x,s)$ is the effective dimensionless (with respect to the nominal particle energy, $E_0=m\gamma c^2$) potential energy given by the interaction between the bunch and the surroundings. Note that η can be positive (above transition energy) or negative (below transition energy). Above transition energy, in analogy with quantum mechanics, $1/\eta$ plays the role of an effective mass associated with the beam as a whole. Below transition energy, $1/\eta$ plays the role of a "negative mass".

Equation (46) has to be coupled with an equation for U. If no external sources of electromagnetic fields are present and the effects of charged-particle radiation damping is negligible, the self-interaction of the beam with the surroundings, due to the image charges and the image currents originated on the walls of the vacuum chamber, makes U a functional of the beam density. It can be proven that, in a torus-shaped accelerating machine, characterized by a toroidal radius R_0 and a poloidal radius a, for a coasting beam of radius $b << a$ travelling at velocity βc ($\beta\leq 1$ and c being the speed of light), the self-interaction potential energy is given by[25] (however, a more general expression can be provided[63]):

$$U[\lambda_1(x,s)]=\frac{q^2\beta c}{E_0}\left(R_0Z_I'\lambda_1(x,s)+Z_R'\int_0^x\lambda_1(x',s)\,dx'\right), \qquad (47)$$

where $\lambda_1(x,s)$ is an (arbitrarily large) line beam density perturbation, q is the charge of the particles, ϵ_0 is the vacuum dielectric constant, Z_R' and Z_I' are the resistive and the total reactive parts, respectively, of the longitudinal coupling impedance per unit length of the machine. Thus, the coupling impedance per unit length can be defined as the complex quantity $Z'=Z_R'+iZ_I'$. In our simple model of a circular machine, it is easy to see that:[62,63]

$$Z_I'=\frac{1}{2\pi R_0}\left(\frac{g_0Z_0}{2\beta\gamma^2}-\omega_0\mathcal{L}\right)\equiv\frac{Z_I}{2\pi R_0}, \qquad (48)$$

where Z_0 is the vacuum impedance, $\omega_0=\beta c/R_0$ is the nominal orbital angular frequency of the particles and $\mathcal{L}$ is the total inductance. This way, Z_I represents the total reactance as the difference between the total space charge capacitive reactance, $g_0 Z_0/(2\beta\gamma^2)$, and the total inductive reactance, $\omega_0\mathcal{L}$. Consequently, in the limit of negligible resistance, Eq. (47) reduces to

$$U[\lambda_1] = \frac{q^2\beta c}{2\pi E_0}\left(\frac{g_0 Z_0}{2\beta\gamma^2} - \omega_0\mathcal{L}\right)\lambda_1. \tag{49}$$

By definition, an unperturbed coasting beam has the particles uniformly distributed along the longitudinal coordinate x. Denoting by $\rho(x,s)$ the line density and by $\rho(x,0)$ the unperturbed one, in the TWM framework we have the following identifications: $\rho(x,s) = |\Psi(x,s)|^2$, $\rho_0 = |\Psi(x,0)|^2 \equiv |\Psi_0|^2$, where Ψ_0 is a complex function and, consequently, $\lambda_1(x,s) = |\Psi(x,s)|^2 - |\Psi_0|^2$. Thus, the combination of Eq. (46) and Eq. (47) gives the following evolution equation for the beam

$$i\frac{\partial\Psi}{\partial s} + \frac{\alpha}{2}\frac{\partial^2\Psi}{\partial x^2} + \mathcal{X}\left[|\Psi|^2 - |\Psi_0|^2\right]\Psi + \mathcal{R}\,\Psi\int_0^x \left[|\Psi(x',s)|^2 - |\Psi_0|^2\right]dx' = 0\,, \tag{50}$$

where

$$\alpha = \epsilon\eta\,, \tag{51}$$

$$\mathcal{X} = \frac{q^2\beta c R_0}{\epsilon E_0} Z_I', \tag{52}$$

$$\mathcal{R} = \frac{q^2\beta c}{\epsilon E_0} Z_R'. \tag{53}$$

Equation (50) is a sort of NLSE governing the propagation and dynamics of wave packets in the presence of nonlocal effects. The modulational instability of such an integro-differential equation has been investigated for the first time in literature by Anderson *et al.*[23] Some nonlocal effects associated with the collective particle beam dynamics have been recently described with this equation.[25] Note that Eq. (50) can be cast in the form of Eq. (1), provided that (51)-(53) are taken and the following expression for the nonlinear potential is assumed, i.e.,

$$U[|\Psi|^2] = -\alpha\left\{\mathcal{X}\left[|\Psi|^2 - |\Psi_0|^2\right] + \mathcal{R}\int_0^x \left[|\Psi(x',s)|^2 - |\Psi_0|^2\right]dx'\right\}. \tag{54}$$

2.3.2. *MI analysis of a monochromatic coasting beam*

Under the conditions assumed above, let us consider a monochromatic coasting beam travelling in a circular high-energy machine with the unperturbed velocity V_0 and the unperturbed density $\rho_0 = |\Psi_0|^2$ (equilibrium state). In these conditions, all the particles of the beam have the same velocity and their collective interaction with the surroundings is absent. In the Madelung fluid representation, the beam can be thought as a fluid with both current velocity and density uniform and constant. In this state, the Madelung fluid equations (3) and (4) vanish identically. Let us now introduce small perturbations in $V(x, s)$ and $\rho(x, s)$, i.e.,

$$V = V_0 + V_1\,, \quad |V_1| << |V_0|\,, \tag{55}$$

$$\rho = \rho_0 + \rho_1\,, \quad |\rho_1| << \rho_0\,. \tag{56}$$

By introducing (55) and (56) in the pair of equations (3) and (4), after linearizing, we get the following system of equations (for details see Fedele *et al.*[61]):

$$\frac{\partial \rho_1}{\partial s} + V_0 \frac{\partial \rho_1}{\partial x} + \rho_0 \frac{\partial V_1}{\partial x} = 0\,, \tag{57}$$

$$\frac{\partial V_1}{\partial s} + V_0 \frac{\partial V_1}{\partial x} = \alpha \mathcal{R} \rho_1 + \alpha \mathcal{X} \frac{\partial \rho_1}{\partial x} + \frac{\alpha^2}{4\rho_0} \frac{\partial^3 \rho_1}{\partial x^3}\,. \tag{58}$$

In order to find the linear dispersion relation, we take the Fourier transform of the system of equations (57) and (58), i.e. we express the quantities $\rho_1(x, s)$ and $V_1(x, s)$ in terms of their Fourier transforms $\tilde{\rho_1}(k, \omega)$, respectively,

$$\rho_1(x, s) = \int dk\, d\omega\, \tilde{\rho_1}(k, \omega) e^{ikx - i\omega s}\,, \tag{59}$$

$$V_1(x, s) = \int dk\, d\omega\, \tilde{V_1}(k, \omega) e^{ikx - i\omega s}\,, \tag{60}$$

and, after substituting in (57) and (58), we get the following system of algebraic equations:

$$-\rho_0 k \tilde{V_1} = (kV_0 - \omega)\, \tilde{\rho_1}\,, \tag{61}$$

$$i\,(kV_0 - \omega)\, \tilde{V_1} = \left(\alpha \mathcal{R} + i\alpha k \mathcal{X} - i\frac{\alpha^2}{4\rho_0} k^3\right) \tilde{\rho_1}\,. \tag{62}$$

By combining (61) and (62) we finally get the dispersion relation

$$\left(\frac{\omega}{k} - V_0\right)^2 = i\alpha\rho_0 \left(\frac{\mathcal{Z}}{k}\right) + \frac{\alpha^2 k^2}{4}\,, \tag{63}$$

where we have introduced the complex quantity $\mathcal{Z} = \mathcal{R} + ik\mathcal{X} \equiv \mathcal{Z}_R + i\mathcal{Z}_I$, proportional to the longitudinal coupling impedance per unity length of the beam. In general, in Eq. (63), ω is a complex quantity, i.e., $\omega \equiv \omega_R + i\omega_I$. If $\omega_I \neq 0$, the modulational instability takes plase in the system. Thus, by substituting the complex form of ω in Eq. (63), separating the real from the imaginary parts and using (51), we finally get:

$$\mathcal{Z}_I = -\eta \frac{\epsilon k \rho_0}{4\omega_I^2} \mathcal{Z}_R^2 + \frac{1}{\eta} \frac{\omega_I^2}{\epsilon k \rho_0} + \eta \frac{\epsilon k^3}{4\rho_0} \,. \tag{64}$$

This equation fixes, for any values of the wavenumber k and any values of the growth rate ω_I a relationship between real and imaginary parts of the longitudinal coupling impedance. For each $\omega_I \neq 0$, running the values of the slip factor η, it describes two families of parabolas in the complex plane $(\mathcal{Z}_R - \mathcal{Z}_I)$. Each pair $(\mathcal{Z}_R, \mathcal{Z}_I)$ in this plane represents a working point of the accelerating machine. Consequently, each parabola is the locus of the working points associated with a fixed growth rate of the MI. According to Figure 3, below the transition energy $(\gamma < \gamma_T)$, η is negative and therefore the instability parabolas have a positive concavity, whilst above the transition energy $(\gamma > \gamma_T)$, since η is positive the instability parabolas have a negative concavity (negative mass instability). It is clear from Eq. (64) that, approaching $\omega_I = 0$ from positive (negative) values of η, the two families of parabolas reduce asymptotically to a straight line lower (upper) unlimited located on the imaginary axis. The straight line represent the only possible region above (below) the transition energy where the system is modulationally stable against small perturbations in both density and velocity of the beam, with respect to their unperturbed values ρ_0 and V_0, respectively (note that density and velocity are directly connected with amplitude and phase, respectively, of the wave function Ψ). Any other point of the complex plane belongs to an instability parabola $(\omega_I \neq 0)$.

In the limit of weak dispersion, i.e., $\epsilon k << 1$, the second term of the right hand side of Eq. (63) can be neglected and Eq. (64) reduces to

$$\mathcal{Z}_I \approx -\eta \frac{\epsilon k \rho_0}{4\omega_I^2} \mathcal{Z}_R^2 + \frac{1}{\eta} \frac{\omega_I^2}{\epsilon k \rho_0} \,. \tag{65}$$

Furthermore, for purely reactive impedances $(\mathcal{Z}_R \equiv 0)$, Eq. (50) reduces to the cubic NLSE and the corresponding dispersion relation gives (note that in this case $\omega_R = V_0 k$)

$$\frac{\omega_I^2}{k^2} = -\epsilon \eta \rho_0 \left(\frac{\mathcal{Z}_I}{k} \right) + \frac{\alpha^2 k^2}{4} \,, \tag{66}$$

from which it is easily seen that the system is modulationally unstable $(\omega_I^2 > 0)$ under the following conditions

$$\eta \, \mathcal{Z}_I > 0 \tag{67}$$

$$\rho_o > \frac{\epsilon \eta k^2}{4\mathcal{X}_I} \,. \tag{68}$$

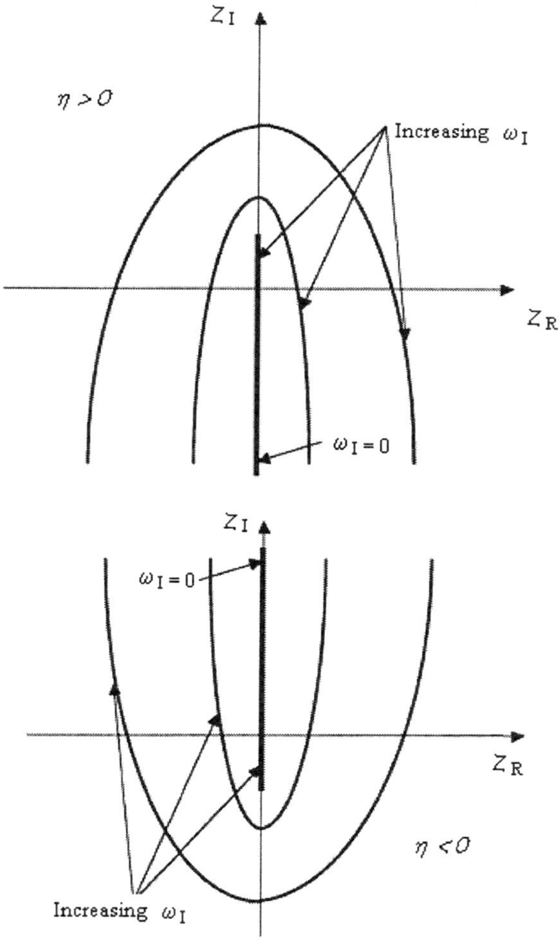

Fig. 3. Qualitative plots of the modulational instability curves in the plane $(\mathcal{Z}_R - \mathcal{Z}_I)$ of a coasting beam below the transition energy $(\eta < 0)$ and above the transition energy $(\eta > 0)$, respectively.[23] *The bold face vertical straight lines represent the stability region* $(\omega_I = 0)$.

Condition (67) is a well known coherent instability condition for purely reactive impedances which coincides with the well known "Lighthill criterion"[64] associated with the cubic NLSE. This aspect has been pointed out for the first time by Fedele *et al.*[24,65] According to Table 3, this condition implies that the system is modulationally stable below (above) transition energy and for capacitive (inductive) impedances and unstable in the other different possible circumstances. Condition (68) implies that the instability

Table 3. Coherent instability scheme of a monochromatic coasting beam in the case of a purely reactive impedance.

η / $\mathcal{Z}_I$	$\mathcal{Z}_I > 0$ (capacitive)	$\mathcal{Z}_I < 0$ (inductive)
$\eta < 0$ (below transition energy)	**stable**	**unstable**
$\eta > 0$ (above transition energy)	**unstable**	**stable**

threshold is given by the nonzero minimum intensity $\rho_{0m} = \epsilon\eta k^2/4\mathcal{X}_I$.

2.3.3. *MI analysis of a non-monochromatic coasting beam*

The dispersion relation (63) allows to write an expression for the admittance of a monochromatic coasting beam $\mathcal{Y} \equiv 1/\mathcal{Z}$:

$$k\mathcal{Y} = \frac{i\alpha\rho_0}{(\omega/k - V_0)^2 - \alpha^2 k^2/4}. \tag{69}$$

Let us now consider a non-monochromatic coasting beam. Such a system may be thought as an ensemble of coasting beams with different unperturbed velocities. Let us call $f_0(V)$ the distribution function of the velocity at the equilibrium. The subsystem corresponding to a coasting beam collecting the particles having velocities between V and $V + dV$ has an elementary admittance $d\mathcal{Y}$. Provided, in Eq. (69), to replace ρ_0 with $f_0(V)dV$, the expression for the elementary admittance is easily given:

$$kd\mathcal{Y} = \frac{i\alpha\, f_0(V)\, dV}{(V - \omega/k)^2 - \alpha^2 k^2/4}. \tag{70}$$

All the elementary coasting beams in which we have divided the system suffer the same electric voltage per unity length along the longitudinal direction. This means that the total admittance of the system is the sum of the all elementary admittances, as it happens for a system of electric wires connected in parallel. Therefore,

$$k\mathcal{Y} = i\alpha \int \frac{f_0(V)\,dV}{(V-\omega/k)^2 - \alpha^2 k^2/4}\,. \tag{71}$$

Of course, this dispersion relation can be cast also in the following way:

$$1 = i\alpha \left(\frac{\mathcal{Z}}{k}\right) \int \frac{f_0(V)\,dV}{(V-\omega/k)^2 - \alpha^2 k^2/4}\,, \tag{72}$$

where we have introduced the total impedance of the system which is the inverse of the total admittance, i.e., $\mathcal{Z} = 1/\mathcal{Y}$.

An interesting equivalent form of Eq. (72) can be obtained. To this end, we first observe that the following identity holds:

$$\frac{1}{(V-\omega/k)^2 - \alpha^2 k^2/4} = \frac{1}{\alpha k}\left[\frac{1}{(V-\alpha k/2) - \omega/k} - \frac{1}{(V+\alpha k/2) - \omega/k}\right].$$

Then, using this identity in Eq. (72) it can be easily shown that:

$$1 = i\left(\frac{\mathcal{Z}}{k}\right)\frac{1}{k}\left[\int \frac{f_0(V)\,dV}{(V-\alpha k/2) - \omega/k} - \int \frac{f_0(V)\,dV}{(V+\alpha k/2) - \omega/k}\right], \tag{73}$$

which, after defining the variables $p_1 = V - \alpha k/2$ and $p_2 = V + \alpha k/2$, can be cast in the form:

$$1 = i\left(\frac{\mathcal{Z}}{k}\right)\frac{1}{k}\left[\int \frac{f_0(p_1 + \alpha k/2)\,dp_1}{p_1 - \omega/k} - \int \frac{f_0(p_2 - \alpha k/2)\,dp_2}{p_2 - \omega/k}\right], \tag{74}$$

and finally in the following form:

$$1 = i\alpha\left(\frac{\mathcal{Z}}{k}\right)\int \frac{f_0(p+\alpha k/2) - f_0(p-\alpha k/2)}{\alpha k}\,\frac{dp}{p-\omega/k}\,. \tag{75}$$

We soon observe that, assuming that $f_0(V)$ is proportional to $\delta(V - V_0)$, from Eq. (75) we easily recover the dispersion relation for the case of a monochromatic coasting beam (see Eq. (63)). In general, Eq. (75) takes into account the velocity (or energy) spread of the beam particles, but it has not obtained with a kinetic treatment. We have only assumed the existence of an equilibrium state associated with an equilibrium velocity distribution, without taking into account any phase-space evolution of a kinetic distribution function. Our result has been basically obtained within the framework of Madelung fluid description, extending the standard MI analysis for monochromatic wave train to a non-monochromatic wave packets.

Nevertheless, Eq. (75) can be also obtained within the kinetical description provided by the Moyal-Ville-Wigner description, as it has been done for the first time in the context of the TWM.[27] In the next section, we present the quantum methodologies provided by the quantum kinetic formalism where Eq. (75) is recovered. In this perspective, the main features of the Eq. (75) will be also presented through specific examples. In particular, we will show that for the dynamics governed by the NLSE:

- The wavepacket propagation can be also suitably described in phase space with a kinetic-like equation
- MI of a monochromatic wavetrain concides with coherent instability of a coasting beam
- A Landau–type damping for a non-monochromatic wavepacket is predicted

3. Wave Kinetics and Moyal-Ville-Wigner picture

As we have pointed out in the Introduction, two distinct ways to treat MI are possible. The first, and the most used one, is a deterministic approach, whilst the second one is a statistical approach. In the latter, the basic idea is to transit from the configuration space description, where the NLS equation governs the particular wave-envelope propagation, to the phase space, where an appropriate kinetic equation is able to show a random version of the MI. This has been accomplished by using the mathematical tool provided by the *quasidistribution*[2–4] (Fourier transform of the density matrix[33]) that is widely used for quantum systems. For any nonlinear system, whose dynamics is governed by a nonlinear Schrödinger equation, one can introduce a two-point correlation function which plays the role similar to the one played by the density matrix of a quantum system. Consequently, the governing kinetic equation is nothing but a sort of nonlinear von Neumann-Weyl equation.[5,6] In the statistical approach to modulational instability, a linear stability analysis of the von Neumann-Weyl equation leads to a phenomenon fully similar to the well known Landau damping, predicted by L.D. Landau in 1946 for plasma waves.[26] This approach has been carried out in several contexts, such as quantum plasmas,[31,66,67] surface gravity waves,[32] charged-particle beam dynamics in high-energy accelerators,[27–30] optical beam propagation in Kerr media[29,36,37] and Langmuir envelope wave propagation in classical[38] and in quantum[68] plasmas. All these investigations have predicted a stabilizing effect provided by a Landau-type damping against the occurrence of the MI.

3.1. *Wigner picture associated with NLSE*

Let us assume a system whose dynamics is governed by the gNLSE (1). This equation provides a description in configuration space. In order to transit to phase space and give a kinetic description, let us introduce the conjugate momentum associated with x, i.e., $p = dx/ds$. Thus, according to the formalism of quantum mechanics, we introduce the Moyal-Ville-Wigner function for a pure state $\Psi(x,s)$, i.e.,

$$w(x,p,s) = \frac{1}{2\pi\alpha}\int_{-\infty}^{\infty} \Psi^*\left(x+\frac{y}{2},s\right)\,\Psi\left(x-\frac{y}{2},s\right)\,\exp\left(i\frac{py}{\alpha}\right)\,dy \quad . \tag{76}$$

In phase space $w(x,p,s)$ plays the role of a distribution function. In fact, the integrals $\int w(x,p,s)\,dp$ and $\int w(x,p,s)\,dx$ are proportional to the probability density in configuration space and momentum space, respectively. However, due to the uncertainty principle, w is not positive definite and for this reason it is usually referred to as "quasidistribution". In particular, (76) implies that

$$\rho(x,s) = |\Psi(x,s)|^2 = \int_{-\infty}^{\infty} w(x,p,s)\,dp \quad . \tag{77}$$

Consequently, $U[|\Psi|^2]$ appearing in Eq. (1) can be expressed here as a functional of w:

$$U = U\left[\int_{-\infty}^{\infty} w\,dp\right] . \tag{78}$$

We observe that, if Ψ satisfies Eq. (1), then w satisfies the following nonlinear von Neumann-Weyl equation :

$$\frac{\partial w}{\partial s}+p\frac{\partial w}{\partial x}-\sum_{n=0}^{\infty}\frac{(-1)^n}{(2n+1)!}\left(\frac{\alpha}{2}\right)^{2n}\frac{\partial^{2n+1}}{\partial x^{2n+1}}\left(U\left[\int_{-\infty}^{\infty} w\,dp\right]\right)\frac{\partial^{2n+1}w}{\partial p^{2n+1}} = 0 \quad . \tag{79}$$

3.2. *Landau-type damping in charged-particle beam dynamics*

In order to specialize our kinetic description to the longitudinal dynamics of charged-particle beams in a circular high-energy accelerating machine, let us take the governing equation (46), where the functional form for $U[|\Psi|^2]$ is given by Eq. (54). Within the present kinetic picture, the latter reads as:

$$U[w] = -\alpha\mathcal{X}\left(\int_{-\infty}^{\infty} w\,dp - |\Psi_0|^2\right) - \alpha\mathcal{R}\int_0^x\left(\int_{-\infty}^{\infty} w(x',p,s)\,dp - |\Psi_0|^2\right)dx'. \tag{80}$$

Thus, (79) and (80) constitute a set of coupled equations governing the nonlinear longitudinal dynamics of a charged-particle coasting beam in the TWM context. We want to find the linear dispersion relation of this system. The procedure is similar to the one used in plasma physics for Vlasov-Poisson system.[26,69] We perturb the system around an initial state. To this end, we start from the equilibrium state: $w = w_0(p)$, where $\rho_0 = |\Psi_0|^2 = \int_{-\infty}^{\infty} w_0(p)\, dp$ and $U = U_0 \equiv U[w_0] = 0$. Then, we introduce the following small perturbations in w and U, respectively:

$$w(x,p,s) = w_0(p) + w_1(x,p,s) \quad , \tag{81}$$

$$U(x,s) \;=\; U_0 \;+\; U_1(x,s) \;=\; U[w_1(x,s)] \quad , \tag{82}$$

where $w_1(x,p,s)$ and $U_1(x,s)$ are first-order quantities. Consequently, the resulting linearized set of equation is:

$$\frac{\partial w_1}{\partial s} + p\frac{\partial w_1}{\partial x} - \sum_{n=0}^{\infty} \frac{(-1)^n}{(2n+1)!}\left(\frac{\alpha}{2}\right)^{2n} \frac{\partial^{2n+1} U_1}{\partial x^{2n+1}}\, w_0^{(2n+1)} = 0 \tag{83}$$

$$U_1 = -\alpha \mathcal{X} \int_{-\infty}^{\infty} w_1\, dp \;-\alpha \mathcal{R} \int_0^x \int_{-\infty}^{\infty} w_1(x',p,s)\, dp\, dx' \, . \tag{84}$$

where $w_0^{(2n+1)} \equiv d^{2n+1} w_0 / dp^{2n+1}$.

In order to perform the Fourier analysis on the set of equations (83) and (84), let us introduce the Fourier transform of $U_1(x,s)$ and $w_1(x,p,s)$, i.e.:

$$U_1(x,s) = \int_{-\infty}^{\infty} dk \int_{-\infty}^{\infty} d\omega\; \widetilde{U_1}(k,\omega)\; \exp(ikx - i\omega s) \quad , \tag{85}$$

$$w_1(x,p,s) = \int_{-\infty}^{\infty} dk \int_{-\infty}^{\infty} d\omega\; \widetilde{w_1}(k,p,\omega)\; \exp(ikx - i\omega s) \quad . \tag{86}$$

After substituting (85) and (86) in the pair of equation (83) and (84), the following dispersion relation is easily obtained:

$$1 \;=\; i\,\alpha \left(\frac{\mathcal{Z}}{k}\right) \int_{-\infty}^{\infty} \frac{w_0\,(p + \alpha k/2) \;-\; w_0\,(p - \alpha k/2)}{\alpha k} \; \frac{dp}{p - \omega/k} \quad , \tag{87}$$

which is formally identical to Eq. (75) obtained in the Madelung fluid description. Since the monochromatic case, i.e., $w_0(p) \propto \delta(p - p_0)$, has been already discussed in the Madelung fluid context, we confine here our attention to the non-monochromatic case which corresponds to the assumption that $w_0(p)$ has a finite spread in the p-space.

3.2.1. *Case of $\alpha k << 1$*

Since $\alpha k << 1$, we have:

$$\frac{w_0\,(p+\alpha k/2)\ -\ \rho_0\,(p-\alpha k/2)}{\alpha k} \approx dw_0/dp\ \equiv w_0^{'} \quad . \tag{88}$$

Consequently, Eq. (87) becomes:

$$1\ =\ i\alpha\left(\frac{\mathcal{Z}}{k}\right)\int_{-\infty}^{\infty}\frac{w_0^{'}}{p-\omega/k}\,dp \quad . \tag{89}$$

We note that the limit of small wavenumbers here considered allows us to predict a sort of weak Landau damping, as described in the Vlasov theory of charged-particle beam physics.[62]

A. If we assume that $\mathcal{Z}_R = 0$, thus $\mathcal{Z} = i\mathcal{Z}_I$, (89) becomes:

$$1\ =\ -\,\alpha\left(\frac{\mathcal{Z}_I}{k}\right)\int_{-\infty}^{\infty}\frac{w_0^{'}}{p-\omega/k}\,dp \quad . \tag{90}$$

Provided that $\alpha\ Z_I < 0$, by replacing $-\alpha\ Z_I$ with the ratio ω_p^2/k^2 (ω_p being the electron plasma frequency), Eq. (90) becomes formally identical to the linear dispersion relation for a warm unmagnetized electron plasma, which predicts the existence of Landau damping. Consequently, following the well known Landau method,[26,69] and using the small wavenumber approximation, we easily get the following dispersion relation:

$$1\ =\ -\,\alpha\ k\ \mathcal{X}\left[D(\omega,k)\ +\ i\frac{\pi}{k}\ w_0^{'}(\omega/k)\right] \quad , \tag{91}$$

where

$$D(\omega,k)\ \equiv\ \int_{PV}\frac{w_0^{'}}{kp-\omega}\ dp \tag{92}$$

is the principal value of the integral in (90).

B. In the case $\mathcal{Z}_R \neq 0$, the analysis can be carried out as follows. We note that Eq. (90) can be cast in the following way:

$$V_R + i\,V_I\ \equiv\ \alpha\left(\frac{\mathcal{Z}_R}{k}\ +\ i\ \frac{\mathcal{Z}_I}{k}\right)\ =\ -\left[i\int_{PV}\frac{w_0^{'}}{p-\beta_{ph}}\ dp\ +\ \pi\ \rho_0^{'}(\beta_{ph})\right]^{-1} , \tag{93}$$

where $V_R = \alpha\mathcal{Z}_R/k$, $V_I = \alpha\mathcal{Z}_I/k$, and $\beta_{ph} = \omega/k$. This equation determines a relationship between V_R, V_I, and β_{ph}. In principle, β_{ph} is a complex quantity. Thus, we put: $\beta_{ph}\ \equiv\ \gamma_R + i\,\gamma_I$. Consequently, we can plot curves in the

V_R-V_I plane for a given equilibrium distribution funtion $\rho_0(p)$ and for different growth rates γ_I. For instance, we assume $\rho_0(p) = \left(1 - p^2/4\right)^2 / 2.133$ which is plotted in Figure 4.

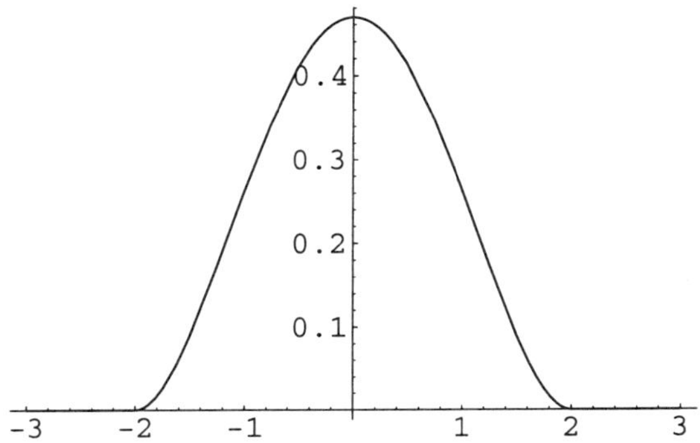

Fig. 4. Plot of the equilibrium distribution $\rho_0(p) = \left(1 - p^2/4\right)^2 / 2.133$, defined in the intervall of p $(-2, 2)$.

Figure 5 shows these curves, for $\alpha k = .01$ and for the smooth distribution plotted in Figure 4. This picture describes the coherent instability (for instance, the negative-mass instability) in circular accelerating machines in which coherent instabilities competes with the stabilizing effect of the Landau damping. We would like to stress that Figure 5 represents a sort of universal stability chart of the MI predicted by NLSE (50). Any impedance $\mathcal{Z}$ leading to a (V_R, V_I) pair belonging the area surrounded by the curve with $\gamma_I = 0$ corresponds to a stable operation.

3.2.2. *Case of arbitrary αk*

When the αk assumes arbitrary values, approximation (88) is no longer valid. In this case, the instability analysis must be carried out directly with Eq. (87). To perform the integration, the residui theory can be applied as in the previous case. Figure 6 shows the instability curves for $\alpha k = 0.4$ for the equilibrium distribution $w_0(p)$ shown in Figure 4, as well. It is evident that the stabilizing effect of the Landau damping is enhanced in comparison to the case shown in Figure 5. In fact, the stability region is sensitively enlarged.

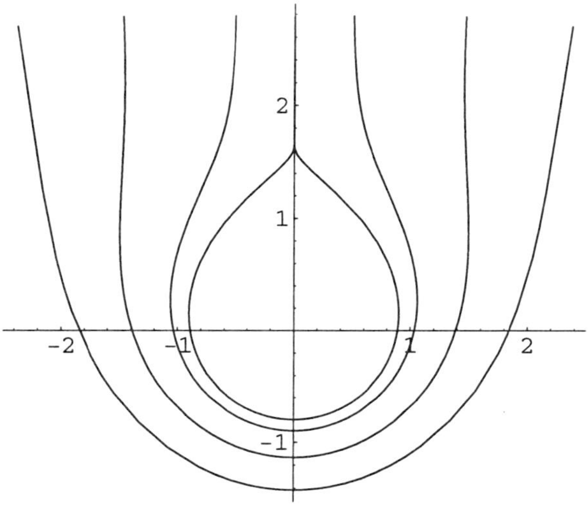

Fig. 5. Instability chart for $\alpha k = .01$ and for ρ_0 plotted in Figure 4.[28] The area inside the curve with $\gamma_I = 0$ represents the stability region. For increasing values of γ_I ($\gamma_I = 0, .1, .3, .5$), the curves plotted cover in the instability region. They are a sort of "deformed parabolas". However, as γ_I increases more and more, their shapes become more and more similar to the parabolas described by Eq. (64) as given in the monochromatic case. We note that in the present case the stability region is larger. This effect, together with the deformation of the above parabolas, is due to the stabilizing effect of the quantum-like Landau damping of the wave packet which exists for a p-distribution with non–negligible spread. In fact, in this case, the stabilizing effect is in competition with the modulational instability.

3.3. *Landau-type damping of partially-incoherent Langmuir wave envelopes*

We now describe the MI of partially incoherent Langmuir wave envelopes in a unmagnetized, collisionless, warm plasma with unperturbed number density n_0. To this end, we start from the pair of equations governing the nonlinear propagation of a Langmuir wavepackets (the so-called "Zacharov system of equations"), i.e.,

$$i\frac{\partial E}{\partial t} + \frac{3v_{te}^2}{2\omega_{pe}}\frac{\partial^2 E}{\partial x^2} - \omega_{pe}\frac{\delta n}{2n_0}E = 0, \tag{94}$$

$$\left(\frac{\partial^2}{\partial t^2} - c_s^2\frac{\partial^2}{\partial x^2}\right)\frac{\delta n}{2n_0} = -\frac{c_s^2}{2}\frac{\partial^2}{\partial x^2}\left(\frac{|E|^2 - |E_0|^2}{4\pi n_0 T_e}\right), \tag{95}$$

where E is the slowly varying complex electric field amplitude, δn is the density fluctuation, v_{te} is the electron thermal velocity, ω_{pe} is the electron

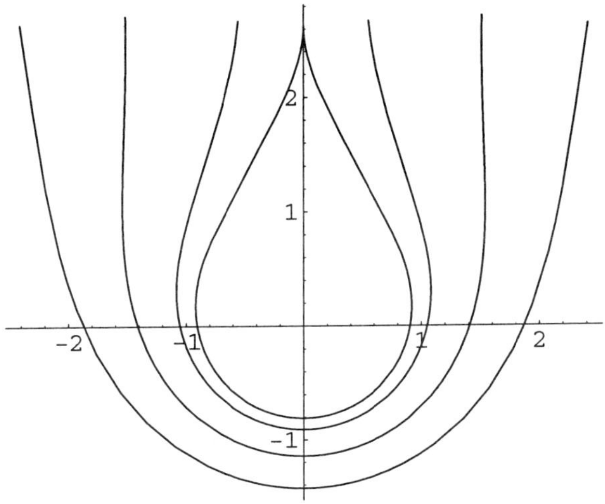

Fig. 6. Instability chart for $\alpha k = .4$, for ρ_0 plotted in Figure 4 and for increasing values of γ_I ($\gamma_I = 0, .1, .3, .5$) .[28] In this case, the area inside of the stability region ($\gamma_I = 0$) is larger than the one of $\alpha k << 1$ (compare with Figure 5). The "deformed parabolas" do not coincide with the case of Figure 4. However, as γ_I is increasing, also in this case the deformation becomes more and more negligible and the parabolas shown is Figure 3 are recovered for very large values of this parameter.

plasma frequency, T_e is the electron temperature, c_s is the sound speed, and E_0 is the unperturbed electric field amplitude. Moreover, x and t play the role of configurational space coordinate (longitudinal coordinate with respect to the wave packet centre) and time, respectively. The above equations have been obtained by assuming that the electric field of the wave has the envelope form

$$\mathrm{E} = E(x,t)\exp\left(-i\Omega t\right) \quad ,$$

where $\Omega \leq \omega_{pe}$ and $\left| \ \Omega^{-1}\partial E(x,t)/\partial t \ \right| \ll 1$. The above system of equations (94) and (95) can be cast in the form

$$i\alpha\frac{\partial\Psi}{\partial s} + \frac{\alpha^2}{2}\frac{\partial^2\Psi}{\partial x^2} - U\,\Psi = 0 \quad , \tag{96}$$

$$\left(\frac{\partial^2}{\partial s^2} - \mu^2\frac{\partial^2}{\partial x^2}\right)U = -\frac{\partial^2}{\partial x^2}\left(\langle|\Psi|^2 - |\Psi_0|^2\rangle\right) \quad , \tag{97}$$

where the angle brackets account for the statistical ensemble average due to the partial incoherence of the waves, $s = \sqrt{3}\lambda_{De}\omega_{pe}t$ ($\lambda_{De} \equiv v_{te}/\omega_{pe}$ is the electron Debye length), $\alpha = \sqrt{3}\lambda_{De}$, $U = \delta n/2n_0$, $\mu = \sqrt{m_e/3m_i}$ (m_e and m_i are the electron and the ion masses, respectively), and

$$\Psi = \frac{\mu E}{\sqrt{8\pi n_0 T_e}} \quad , \tag{98}$$

$$\Psi_0 = \frac{\mu E_0}{\sqrt{8\pi n_0 T_e}} \quad . \tag{99}$$

Note that (99) can be written as

$$\rho_0 \equiv |\Psi_0|^2 = \mu^2 \frac{\epsilon}{\epsilon_T} \quad , \tag{100}$$

where $\epsilon \equiv E_0^2/8\pi$ and $\epsilon_T \equiv n_0 T_e$ are the electric energy density and the electron thermal energy density, respectively. According to the theory modelled by the Zakharov system (96) and (97), $\epsilon/\epsilon_T \gg 1$ (i.e, the ponderomotive effect is larger than the thermal pressure effect). Note also that Eq. (97) implies that U is a functional of $|\Psi|^2$, namely $U = U\left[|\Psi|^2\right]$. Consequently, once (97) is combined with (96), the latter becomes a nonlinear Schrödinger equation (NLSE) with the nonlinear term $U\left[|\Psi|^2\right]$. One can introduce the correlation function (whose corresponding meaning in Quantum Mechanics is nothing but the density matrix for mixed states)

$$\rho(x, x', s) = \langle \Psi^*(x, s)\Psi(x', s)\rangle \quad , \tag{101}$$

where the statistical ensemble average takes into account the incoherency of the waves. The technique was successfully introduced in the field of statistical quantum mechanics to describe the dynamics of a system in the classical space language.[32,70,71] Thus, one can define the following Moyal-Ville-Wigner transform

$$w(x, p, s) = \frac{1}{2\pi\alpha} \int_{-\infty}^{\infty} \rho\left(x + \frac{y}{2},\ x - \frac{y}{2},\ s\right) \exp\left(i\frac{py}{\alpha}\right) dy \quad , \tag{102}$$

which extends to the "mixed states" the definition of w given in the previous subsection for "pure states". Consequently, it is easy to show that the Zacharov system is transformed into the following pair of equatons:

$$\frac{\partial w}{\partial s} + p\frac{\partial w}{\partial x} - \sum_{n=0}^{\infty} \frac{(-1)^n}{(2n+1)!} \left(\frac{\alpha}{2}\right)^{2n} \frac{\partial^{2n+1} U}{\partial x^{2n+1}} \frac{\partial^{2n+1} w}{\partial p^{2n+1}} = 0 \quad . \tag{103}$$

and

$$\left(\frac{\partial^2}{\partial s^2} - \mu^2 \frac{\partial^2}{\partial x^2}\right) U = -\frac{\partial^2}{\partial x^2}\left(\int_{-\infty}^{\infty} w\, dp - \int_{-\infty}^{\infty} w_0\, dp\right) \quad , \tag{104}$$

where w_0 is the Moyal-Ville-Wigner transform of Ψ_0 ($\rho_0 = |\Psi_0|^2 = \int_{-\infty}^{\infty} w_0\, dp$).

Note that Eq. (102) implies that

$$\langle|\Psi|^2\rangle = \int_{-\infty}^{\infty} w(x,p,s)\, dp \quad . \tag{105}$$

The above scheme allows us to carry out a stability analysis of the stationary Langmuir wave by considering the linear dispersion relation of small amplitude perturbations in a way similar to the one presented in the previous subsection. Then we find the following dispersion relation:

$$\omega^2 - \mu^2 k^2 = k^2 \int_{-\infty}^{\infty} \frac{w_0\,(p+\alpha k/2) - w_0\,(p-\alpha k/2)}{\alpha k} \frac{dp}{p-\omega/k} \quad . \tag{106}$$

Since the assumption $w_0(p) \propto \delta(p-p_0)$ leads to the standard results of MI of monochromatic Langmuir wave packets, we focus here our attention to non-monochromatic wave packets to predict the Landau-type damping.

We consider a Lorenzian spectrum of the form

$$w_0(p) = \frac{\rho_0}{\pi} \frac{\Delta}{\Delta^2 + p^2}, \tag{107}$$

where Δ is the width of the spectrum. The dispersion relation (106) then becomes

$$\left(\omega^2 - \mu^2 k^2\right)\left(\omega^2 - \alpha^2 k^4/4 + 2ik\Delta\omega - k^2\Delta^2\right) = \rho_0 k^4 \quad , \tag{108}$$

(Note that the limit of $\Delta \to 0$ corresponds to a coherent monochromatic wave-train and the dispersion relation reduces to $\left(\omega^2 - \mu^2 k^2\right)\left(\omega^2 - \alpha^2 k^4/4\right) = \rho_0 k^4$.

In the following, we concentrate on the limits $\alpha k \ll 1$ and $\omega/k \gg 1$. We will show that during its nonlinear evolution the Langmuir wave packet can exhibit a phenomenon, similar to the Landau damping,[72] between parametrically-driven non-resonant density ripples and Langmuir quasi-particles. It is easy to verify that the condition $\alpha k \ll 1$ (i.e., $k\lambda_{De} \ll 1$, λ_{De} being the electron Debye length) implies that

$$\omega^2 = k^2 \int \frac{w_0'}{p-\omega/k}\, dp \quad , \tag{109}$$

where, using the assumption $\omega/k \gg 1$, we can neglect the term involving μ^2 because $\mu \ll 1$. Consequently, the dispersion relation (109) allows us to predict a sort of Landau damping,[26] as described in plasma physics for linear plasma waves. We can investigate this phenomenon also for large-amplitude Langmuir waves, just using the standard procedure of Landau damping description. The result is

$$\omega^2 = k^2 \int_{PV} \frac{w_0'}{p - \omega/k} \, dp + i\pi k^2 w_0'(\beta) \quad , \tag{110}$$

For a symmetric sufficiently smooth stationary background distribution, $w_0(p)$, we have

$$\omega^2 = \left(1 + \frac{3\Delta^2}{\rho_0}\right) k^2 + i\pi k^2 w_0'(\beta) \quad , \tag{111}$$

where $\Delta \equiv \left(\langle p^2 \rangle\right)^{1/2}$ denotes the r.m.s. width of $w_0(p)$, and $\beta \equiv \rho_0^{1/4}\left(1 + 3\Delta^2/4\rho_0^2\right)$. Note that, according to the above hypothesis, we have $\beta \gg 1$ and $\Delta^2/\rho_0 \ll 1$. This implies that $\rho_0 = |\Psi_0|^2 \gg 1$, and from (100) it follows that $\epsilon/\epsilon_T \gg 1/\mu^2$. Eq.

(111) clearly shows that the damping rate is proportional to the derivative of the Wigner distribution w_0. This is formally similar to the expression for the Landau damping of a linear plasma wave in a warm unmagnetized plasma. In fact, writing ω in the complex form $\omega = \omega_R + i\omega_I$, substituting it in (110) and then separating the real and imaginary parts, we obtain

$$\omega_R^2 - \omega_I^2 = k^2 \int_{PV} \frac{w_0'}{p - \omega/k} \, dp \quad , \tag{112}$$

and

$$\omega_I = \frac{\pi}{2} \frac{k^2}{\omega_R} w_0'(\beta) \quad . \tag{113}$$

The Landau damping rate of the Langmuir wave in the appropriate units is $\gamma = \sqrt{3}\omega_{pe}\lambda_{De}\omega_I$. For example, for a Gaussian wave packet spectrum, i.e. $w_0(p) = \left(\rho_0/\sqrt{2\pi\Delta^2}\right)\exp\left(-p^2/2\Delta^2\right)$, one obtains

$$\gamma = -\left(\frac{3\pi}{8}\right)^{1/2} \frac{\mu^{5/2}}{\Delta^3} \left(\frac{\epsilon}{\epsilon_T}\right)^{5/4} \omega_{pe}\lambda_{De}k \, \exp\left[-\frac{\mu}{2\Delta^2}\left(\frac{\epsilon}{\epsilon_T}\right)^{1/2}\right] \quad , \tag{114}$$

where higher-order terms have been neglected.

The present damping mechanism differs from the standard Landau damping in that here $w_0(p)$ does not represent the equilibrium velocity distribution of the plasma electrons, but it can be considered as the "kinetic" distribution of all Fourier components of the partially incoherent large-amplitude Langmuir wave in the warm plasma. In terms of the plasmons, we realize that $w_0(p)$ represents the distribution of the partially incoherent plasmons that are distributed in p-space (i.e. in k-space) with a finite "temperature". Consequently, the Landau damping described here is due to the partial incoherence of the wave which corresponds to a finite-width Wigner distribution of the plasmons (ensemble of partially incoherent plasmons) which acts in competition with the modulational instability.

4. Related tools: marginal distributions for tomographic representations

The picture presented in the previous section provides a phase space description in terms of the quasidistribution w. However, as we have already pointed out, it can be negative and it does not match with usual classical picture that is usually given by the Boltzmann-Vlasov description. Actually, there is a possibility to transit from the quasidistribution to a positive definite function, called "marginal distribution", that has the features of a classical probability distribution.[7] It is widely employed for a number of tomographic applications. In particular, in quantum optics and quantum mechanics it is involved in both optical [7,73] and symplectic tomographic[74] methods and it has been suggested for measuring quantum states. The marginal distribution application reveals to be important in spin tomography, as well.[75,76] Its definitions establishes an invertible map with the quasidistribution. For instance, in the symplectic tomographic methods, the marginal distribution is defined as the Fourier transform, say $F_w\left(X,\mu,\nu,s\right)$, of the quasidistribution $w(x,p,s)$:

$$F_w\left(X,\mu,\nu,s\right) = \int \exp\left[-ik(X-\mu x-\nu p)\right] w(x,p,s) \frac{dk\, dx\, dp}{(2\pi)^2}. \tag{115}$$

The function $F_w\left(X,\mu,\nu,s,\right)$ is a real function of the random variable X with the properties: $F_w\left(X,\mu,\nu,s\right) \geq 0$, and $\int F_w\left(X,\mu,\nu,s\right) dX = 1$. If the marginal distribution is known, the quasidistribution can be obtained by the inverse Fourier transform of (115). It has been shown in[74] that, for a suitable choice of the parameters μ and ν, the expression for the quasidistribution in terms of the marginal distribution can be reduced to the Radon transform [77] used in optical tomography.

Furthermore, the marginal distribution $F_w\left(X,\mu,\nu,s\right)$ obeys to a sort of Fokker–Planck-like equation[78] that can be also nonlinear.[79,80] In fact, the use of marginal distributions seems to be important in providing a tomographic representation of several nonlinear processes. In particular, it has been very recently applied to the envelope soliton propagation governed by some kinds of NLSEs[79,80] and used to find both a novel approach to the wave function reconstruction based on Fresnel representation of tomograms[81] and a new uncertainty relations for tomographic entropy.[82]

In the next subsections, we briefly introduce the concept of tomogram and discuss its properties and give some relevant examples of tomogram of envelope solitons.

4.0.1. *Tomographic map*

One of the main reasons to use the tomogram technique is justified by the natural possibility of measuring the states usually described by the complex wave function Ψ, in principle solution of Eq. (1). In fact, it is possible to prove that, by using the mapping between F_w and w and the one between w and Ψ, the following direct connection between F_w and Ψ holds (expression of the tomogram in terms of the wave function):[79,80]

$$F_w(X,\mu,\nu) = \frac{1}{2\pi|\nu|}\left|\int \psi(y)\exp\left(\frac{i\mu}{2\nu}\,y^2 - \frac{iX}{\nu}\,y\right)\,dy\right|^2 . \tag{116}$$

One can prove that the tomogram has the following homogeneity property, very useful for the optical tomography:[79]

$$F_w(\lambda X,\lambda\mu,\lambda\nu,s) = \frac{1}{|\lambda|}F_w(X,\mu,\nu,s). \tag{117}$$

Actually, a relation among the parameters can be in principle assumed. In particular, one can take $\mu = \cos\theta$ and $\nu = \sin\theta$ and the optical tomogram becomes:

$$F_w(X,\theta) = \frac{1}{2\pi|\sin\theta|}\left|\int \psi(y)\exp\left(\frac{i\,\cot\theta}{2}\,y^2 - \frac{iX}{\sin\theta}\,y\right)\,dy\right|^2 . \tag{118}$$

In the next subsection, we use this formula to provide tomograms of envelope solitons.

4.0.2. *Soliton envelopes in tomographic representation*

For the case of a modified NLSE, with $U[|\psi|^2] = q_0|\psi|^{2\beta}$, for $q_0 < 0$ (β being an arbitrary real positive value) one has the following bright envelope soliton solutions (see section II):

$$\Psi(x,s) = \left[\frac{|E|\,(1+\beta)}{|q_0|}\right]^{1/2\beta}\operatorname{sech}^{1/\beta}\left[\beta\sqrt{2\,|E|}\xi\right]\exp\left\{i\left[V_0x - \left(E + \frac{V_0^2}{2}\right)s\right]\right\}. \tag{119}$$

Thus, by virtue of Eq. (118), the corresponding tomograms can be computed, respectively. 3D plots and density plots of tomograms of bright solitons for $\beta = 0.5$, 1.0, 2.0, and 2.5 are given in Figure 7. The free parameters

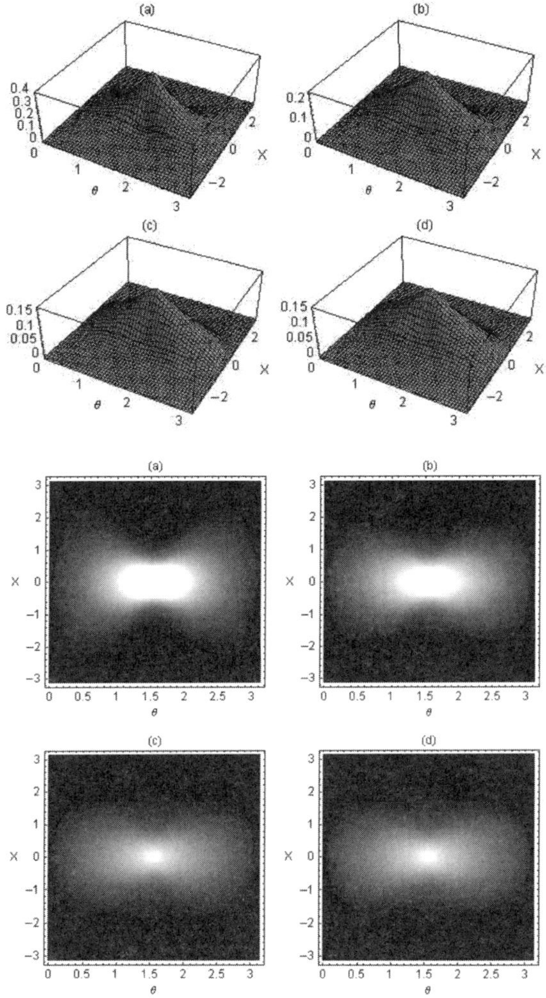

Fig. 7. Tomogram of the soliton for various β: 3D plots (at the left hand side) and density plots (at the right hand side).[79] For both sides: (a) $\beta = 0.5$, (b) $\beta = 1.0$, (c) $\beta = 2$, (d) $\beta = 2.5$.

have been fixed as follows: $V_0 = 0$, $E = -1$ and $q_0 = -1$. The left picture in Figure 7 represents 3D plots of the tomogram of solitons for different values of β. The corresponding density plots are displayed in the right picture of Figure 7. Additionally, the probability description of collective states as-

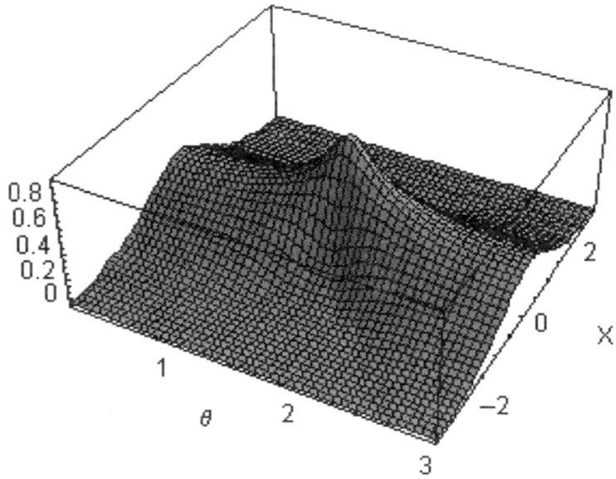

Fig. 8. Tomogram of the bright soliton-like solution as function of X and θ. According to the BEC experimental conditions (Khayakovich *et al.*[47]), $\gamma = L/\ell_z \approx 0.82$ ($L = 1.4$ μm, $\ell_z = 1.7$ μm).

Fig. 9. Density plot in the (X,θ) plane .[80] According to the BEC experimental conditions (Khayakovich *et al.*[47]), $\gamma = L/\ell_z \approx 0.82$ ($L = 1.4$ μm, $\ell_z = 1.7$ μm).

sociated with Bose-Einstein condensates has been recently given in terms of tomographic map.[79] A tomogram of a quasi-1D bright soliton, which is solution of the Gross-Pitaevskii equation, is displayed in Figures 8 and 9.[80]

5. Conclusions

In this paper, we have presented the main quantum methodologies as tools useful for describing a number of problems in nonlinear physics. We have first discussed the role played by the quantum methodologies in the development of the physical theories. In particular, we have considered the very-recently progress registered in the modulational instability and soliton theories involving quantum tools given by the Madelung fluid description, the Moyal-Ville-Wigner kinetic approaches and the tomographic techniques. Valuable methodological transfer among physics of fluids, plasma physics, nonlinear optics and particle accelerator physics have been discussed in terms of recently done applications.

References

1. E. Madelung, *Zeitschrift für Physik* **40**, 332 (1926).
2. E. Wigner, *Phys. Rev.*, **40** 749 (1932).
3. J.E. Moyal, *Proc. Cambidge Phil. Soc.*, **45**, 99 (1949).
4. J. Ville, *Cables et Transmission* **2**, 61 (1948).
5. von Neumann J., *Mathematische Grundlagen der Quantenmechanik* (Springer, Berlin, 1932); *Collected Works* (Oxford, Pergamon, 1963).
6. H. Weyl, *Gruppentheorie und Quantenmechanik* (1931); engl. transl.: *The Theory of Groups and Quantum Mechanics* (Dover, Publ., 1931).
7. K. Vogel and H. Risken, *Phys. Rev. A* **40**, 2847 (1989).
8. B. Kadomtsev, *Phénomenes Collectifs dans les Plasmas* (Mir, Moscow 1979) and references therein.
9. See for instance, R. Fedele and P.K. Shukla (editors), *Quantum-like Models and Coherent Effects* (World Scientific, Singapore, 1995), Proc. of the 27th Workshop of the INFN Eloisatron Project, Erice, Italy 13-20 June 1994; S. Marticno, S. De Nicola, S. De Siena, R. Fedele and G. Miele (editors), *New Perspectives in the Physics of Mesoscopic Systems* (World Scientifc, Singapore, 1997), Proc. of the Workshop "New Perspectives in the Physics of Mesoscopic Systems: Quantum-like Description and Macroscopic Coherence Phenomena", Caserta, Italy, 18-20 April 1996; P. Chen (editor), *Quantum Aspects of Beam Physics* (World Scientific, Singapore, 1999), Proc. of the Advanced ICFA Beam Dynamics Workshop on "Quantum Aspects of Beam Physics", Monterey, California (USA), 4-9 January 1998; P. Chen (editor), *Quantum Aspects of Beam Physics* (World Scientific, Singapore, 2002), Proc. of the 18th Advanced ICFA Beam Dynamics Workshop on "Quantum Aspects of Beam Physics", Capri, Italy, 15-20 October 2000; P. Chen and K. Reil (editors), *Quantum Aspects of Beam Physics* (World Scientific, Singapore, 2004), Proc. of the Joint 28th ICFA Advanced Beam Dynamics and Advanced and Novel Accelerator Workshop on "Quantum Aspects of Beam Physics", Hiroshima, Japan, 7-11 January 2003; P.K. Shukla and L. Stenflo (editors), *New Frontiers in Nonlinear Sciences*, Proc. of the Inter. Topical

Conf. on Plasma Physics, Univ. do Algarve, Faro, Portugal, 6-10 September, 1999, published in *Physica Scripta* **T84** (2000); P.K. Shukla and L. Stenflo (editors), *New Plasma Horizons*, Proc. of the Inter. Topical Conf. on Plasma Physics, Univ. do Algarve, Faro, Portugal, 3-7 September, 2001, published in *Physica Scripta* **T98** (2002); P. K. Shukla and L. Stenflo (editors), *Modern Plasma Sciences*, Proc. of the Int. Workshop on Theoretical Plasma Physics, Adbus Salam ICTP, Trieste, Italy, 5-16 July 2004, published in *Physica Scripta* **T116** (2005)

10. V. I. Karpman, *Nonlinear Waves in Dispersive Media* (Pergamon Press, Oxford, 1975).
11. P. Sulem and C. Sulem, *Nonlinear Schrödinger Equation* (Springer-Verlag, Berlin, 1999).
12. K. Trulsen and K.B. Dysthe, *Freak Waves - A three-dimensional wave simulation*, Pro. of the 21st Symposium on Naval Hydrodynamics (National Academy Press, Washington, DC, 1997), 55-560, 1997; M. Onorato, A.R. Osborne, M. Serio and S. Bertone, *Phys. Rev. Lett.*, **86**, 5831 (2001); K.B. Dysthe and K. Trulsen, *Physica Scripta* **T82**, 48 (1999); A.R. Osborne, M. Onorato and M. Serio, *Phys. Lett. A* **275**, 386 (2000).
13. C.S. Gardner et al., *Phys. Rev. Lett.* **19**, 1095 (1967).
14. G.B. Whitham *Linear and Nonlinear Waves* (John Wiley and Sons, New York 1974).
15. V.E. Zakharov and A.B. Shabat, Sov. Phys. JETP **34**, 62 (1972) .
16. R. Fedele, Physica Scripta **65**, 502 (2002).
17. R. Fedele and H. Schamel, Eur. Phys. J. B **27**, 313 (2002).
18. R. Fedele, H. Schamel and P.K. Shukla, *Physica Scripta* **T98**, 18 (2002).
19. R. Fedele, H. Schamel, V.I. Karpman and P.K. Shukla, *J. Phys. A: Math. and Gen.* **36**, 1169 (2003).
20. R.Fedele and G. Miele, *Il Nuovo Cimento* **D13**, 1527 (1991).
21. R.Fedele and G. Miele, *Phys. Rev. A* **46**, 6634 (1992).
22. R.Fedele and P.K. Shukla, *Phys. Rev. A* **45**, 4045 (1992).
23. D. Anderson, R Fedele, V. Vaccaro, M. Lisak, A. Berntson, S. Johanson, *Phys. Lett. A* **258**, 244 (1999).
24. R. Fedele, G. Miele, L. Palumbo and V.G. Vaccaro, *Phys. Lett. A* **179**, 407 (1993).
25. P. Johannisson, D. Anderson, M. Lisak, M. Marklund, R. Fedele and A. Kim, *Phys. Rev. E.* **69**, 066501 (2004).
26. L.D. Landau, *J. Phys. USSR*, **10**, 25 (1946).
27. D. Anderson, R. Fedele, V.G. Vaccaro, M. Lisak, A. Berntson, S. Johansson, *Quantum-like Description of Modulational and Instability and Landau Damping in the Longitudinal Dynamics of High-Energy Charged-Particle Beams*, Proc. of 1998 ICFA Workshop on "Nonlinear Collective Phenomena in Beam Physics". Arcidosso, Italy, September 1-5, 1998, S. Chattopadhyay, M. Cornacchia, and C. Pellegrini (Ed.s), (AIP Press, New York, 1999) p.197; D. Anderson, R. Fedele, V.G. Vaccaro, M. Lisak, A. Berntson, S. Johansson, *Modulational Instabilities and Landau damping within the Thermal Wave Model Description of High-Energy Charged-Particle Beam Dynamics*,

INFN/TC-98/34, 24 November (1998); R. Fedele, D. Anderson, and M. Lisak, *Role of Landau damping in the Quantum-Like Theory of Charged-Particle Beam Dynamics*, Proc. of Seventh European Particle Accelerator Conference (EPAC2000), Vienna, Austria, 26-30 June, 2000, p.1489.

28. R. Fedele, S. De Nicola, V.G. Vaccaro, D. Anderson and M. Lisak, *Landau Damping in Nonlinear Schrödinger Equations*, Proc. of the 18th Advanced ICFA Beam Dynamics Workshop on Quantum Aspect of Beam Physics, Capri, 15-20 October 2000, P. Chen ed. (World Scientific, Singapore, 2002), p.483.
29. R. Fedele and D. Anderson, *J. Opt. B: Quantum Semiclass. Opt.*, **2**, 207 (2000).
30. R. Fedele, D. Anderson and M. Lisak, *Physica Scripta* **T84**, 27 (2000).
31. Y. Klimontovich and V. Silin, *Sov. Phys. Usp.* **3**, 84 (1960).
32. I.E. Alber, *Proc. R. Soc. London*, Ser. A **636**, 525 (1978).
33. L.D. Landau, *Zeitschrift für Physik* **45**, 430 (1927).
34. G. P. Agrawal, *Nonlinear Fiber Optics* (Academic, 2001)
35. See, for instance: A. Sauter, S. Pitois, G. Millot and A. Picozzi, *Opt. Lett.* **30**, 2143 (2005); K.G. Makris, H. Sarkissian, D.N. Christodoulides and G. Assanto, *J. Opt. Soc. Am. B* **22**, 1371 (2005); M. Soljacic, M. Segev, T. Coskun, D. N. Christodoulides and A. Vishwanath, *Phys. Rev. Lett.* **84**, 467 (2000); J. P. Torres, C. Anastassiou, M. Segev, M. Soljacic, and D. N. Christodoulides, *Phys. Rev. E* **65**, 015601(R) (2001); H. Buljan, A. Siber, M. Soljacic, and M. Segev, *Phys. Rev. E* **66**, 035601(R) (2002); D. Kip, M. Soljacic, M. Segev, E. Eugenieva, and D. N. Christodoulides, *Science* **290**, 495 (2000); T. Coskun, D. N. Christodoulides, Y. Kim, Z. Chen, M. Soljacic, and M. Segev, *Phys. Rev. Lett.* **84**, 2374 (2000); J. Klinger, H. Martin, and Z. Chen, *Opt. Lett.* **26**, 271 (2001); D. Kip, M. Soljacic, M. Segev, S. M. Sears, and D. N. Christodoulides, *J. Opt. Soc. Am. B* **19**, 502 (2002); T. Schwartz, T. Carmon, H. Buljan, and M. Segev, *Phys. Rev. Lett.* **93**, 223901 (2004).
36. B. Hall, M. Lisak, D. Anderson, R. Fedele, and V.E. Semenov, *Phys. Rev. E*, **65**, 035602(R) (2002).
37. L. Helczynski, D. Anderson, R. Fedele, B. Hall, and M. Lisak, *IEEE J. of Sel. Topics in Q. El.*, **8**, 408 (2002)
38. R. Fedele, P.K. Shukla, M. Onorato, D. Anderson, and M. Lisak, *Phys. Lett. A* **303**, 61 (2002).
39. A. Visinescu and D. Grecu, *Eur. Phys. J. B* **34**, 225 (2003); A. Visinescu, D. Grecu AIP Conf. Proc. Vol. 729, p. 389 (2004).
40. D. Grecu and A. Visinescu, *Rom. J. Phys.* **50**, nr.1-2 (2005).
41. M. Onorato, A. Osborne, R. Fedele, and M. Serio, *Phys. Rev. E* **67**, 046305 (2003).
42. J. Javanainen, *Phys. Rev. A* **60**, 4902 (1999).
43. V. V. Konotop and M. Salerno, *Phys. Rev. E* **62**, 021602(R) (2002).
44. E. P. Gross, *Il Nuovo Cimento* **20**, 454 (1961); L. P. Pitaevskii, *Zh. Eksp Teor. Fiz.* **40**, 646 (1961) [*Sov. Phys. JETP* **13**, 451 (1961)].
45. F. Dalfovo et al., *Rev. Mod. Phys.* **71**, 463 (1999).

46. S. Burger et al., *Phys. Rev. Lett.* **83**, 15198 (1999); E. A. Donley et al., *Nature* **412**, 295 (2001); B. P. Anderson et al., *Phys. Rev. Lett.* **86**, 2926 (2001); B. Eiermann et al., *Phys. Rev. Lett.* **92**, 230401 (2004).
47. L. Khayakovich, et al., *Science* **296**, 1290 (2002).
48. R. Fedele, P. K. Shukla, S. De Nicola, M. A. Man'ko, V. I. Man'ko and F. S. Cataliotti, *JETP Letters*, **80**, 535 (2004) [Pis'ma Zh. Eksp. Teor. Fiz., 80, 609 (2004)]; *Physica Scripta* **T116**, 10 (2005); S. De Nicola, R. Fedele, D. Jovanovic, B. Malomed, M.A. Man'ko, V.I. Man'ko and P.K. Shukla, *Eur. Phys. J. B* **54**, 113 (2006); D. Jovanović and R. Fedele, Stability of two-dimensional, controlled, Bose-Einstein coherent states, submitted to *Eur. Phys. J. B* (2007).
49. L. de Broglie, *Comptes Rendus l'Academie des Sciences* **183**, 447 (1926)
50. E. Schrödinger, *Annalen der Physik* **79**, 361 (1926); **79**, 489 (1926).
51. L. de Broglie, *Comptes Rendus l'Academie des Sciences* **184**, 273 (1927); **185**, 380 (1927); *Journal de Physique* **8**, 255 (1927).
52. L. de Broglie, *Il Nuovo Cimento* **1**, 37 (1955).
53. L. de Broglie, *Une tentative d'Interpretation Causale et Non-linere de la Meccanique Ondulatoire* (Gauthier-Villars, Paris, 1956).
54. D. Bohm, *Phys. Rev.* **85**, 166 (1952).
55. A. Korn, *Zeitschrift für Physik* **44**, 745 (1927).
56. S.A. Akhmanov, A.P. Sukhuorukov, and R.V. Khokhlov, *Sov. Phys. Usp.* **93**, 609 (1968).
57. Y.R. Shen, *The Principles of Nonlinear Optics* (Wiley-Interscience Publication, New York, 1984).
58. P.K. Shukla, N.N. Rao, M.Y. Yu and N.L. Tasintsadze, *Phys. Rep.* **138**, 1 (1986).
59. P.K. Shukla and L. Stenflo (editors), *Modern Plasma Science*, Proc. of the Int. Workshop on Theoretical Plasma Physics, Abdus Salam ICTP, Trieste, Italy, July 5-16, 2004, in *Physica Scripta* **T116** (2005).
60. R. Fedele, D. Anderson and M. Lisak, *Coherent instabilities of intense high-energy "white" charged-particle beams in the presence of nonlocal effects within the context of the Madelung fluid description* [arXiv:Physics/0509175 v1 21 september 2005]
61. R. Fedele, D. Anderson and M. Lisak, *Eur. Phys. J. B* **49**, 275 (2006).
62. J. Lawson, *The Physics of Charged Particle Bea ms* (Clarendon, Oxford, 1988), 2nd ed.
63. H. Schamel and R. Fedele, *Phys. Plasmas* **7**, 3421 (2000).
64. M.J. Lighthill, *J. Inst. Math. Appl.* **1**, 269 (1965); Proc. Roy. Soc. **229**, 28 (1967).
65. R. Fedele, L. Palumbo and V.G. Vaccaro, *A Novel Approach to the Nonlinear Longitudinal Dyanamics in Particle Accelerators*, Proc. of the Third European Particle Accelerator Conference (EPAC 92), Berlin, 24-28 March, 1992 edited by H. Henke, H. Homeyer and Ch. Petit-Jean-Genaz (Edition Frontieres, Singapore, 1992), p. 762.
66. F. Haas, L. G. Garcia, J. Goedert, and G. Manfredi, *Phys. Plasmas* **10**, 3858 (2003); L. G. Garcia, F. Haas, L. P. L. de Oliveira, and J. Goedert,

Physics of Plasmas **12**, 2302 (2005).
67. A. Luque, H. Schamel, and R. Fedele, *Physics Letters A* **324**, 185 (2004); A. Luque and H. Schamel, *Phys. Rep.* **415**, 261 (2005) and reference therein.
68. D. Jovanović and R. Fedele, *Phys. Lett. A* **364**, 304 (2007) and references therein.
69. P. A. Sturrock, *Plasma Physics*, (Cambridge University Press, Cambridge, 1994).
70. D.R. Crawford, P.G. Saffman, and H.C. Yuen, *Wave Motion* **2**, 1 (1980).
71. P.A.E.M. Janssen, *J. Fluid Mech.* **133**, 113 (1983); P.A.E.M. Janssen, in *The Ocean Surface*, edited by Y. Toba and H. Mitsuyasu (Reidel, Dordrecht, The Netherlands, 1985), p.39.
72. J.M. Dawson, *Phys. Rev.* **118**, 381 (1960).
73. J. Bertrand and P. Bertrand, *Found. Phys.* **17**, 397 (1987).
74. S. Mancini, V.I. Man'ko and P. Tombesi, *Phys. Lett.* **A213**, 1 (1996); *Found. Phys.* **27**, 801 (1997).
75. V.V. Dodonov and V.I. Man'ko, *Phys. Lett.* A **229**, 335 (1997).
76. G.M. D'Ariano, L. Maccone and M. Paini, *J. Opt. B: Quantum and Sem. Opt.* **5**, 77 (2003).
77. J. Radon, *Ber. Sachs. Akad. Wiss. Leipzig*, **69**, 262 (1917).
78. Man'ko V I and Mendes R V 2000 *Physica* D **145** 330.
79. S. De Nicola, R. Fedele, M.A. Man'ko and V.I. Man'ko, *J. Opt. B: Quantum and Sem. Opt.* **5**, 95 (2003).
80. S. De Nicola, R. Fedele, M.A. Man'ko and V.I. Man'ko, *Eur. Phys. J. B* **36**, 385 (2003).
81. S. De Nicola, R. Fedele, M.A. Man'ko and V.I. Man'ko, *Theor. Math. Phys.* **144**(2), 1206 (2005).
82. S. De Nicola, R. Fedele, M.A. Man'ko, V.I. Man'ko, *J. Phys.: Conf. Series* **70** 012007 (2007); *Theor. Math. Phys.* **152**(2), 1081 (2007); *Eur. Phys. J. B*, **52**, 191 (2006).

PLASMA EFFECTS IN COLD ATOM PHYSICS

J.T. MENDONÇA[1,2], J. LOUREIRO[1], H. TERÇAS[2], and R. KAISER[3]

CFP[1] and CFIF[2], Instituto Superior Técnico, Av. Rovisco Pais 1, 1049-001 Lisboa, Portugal

[3]Institut Non Linéaire de Nice, UMR 6618, 1361 Route des Lucioles, F-06560 Valbonne, France

We discuss collective effects that can be relevant in cold atom physics. Similarities with plasma physics are emphasized. Both neutral and ionized atomic clouds are considered We establish the basic frequencies and wave modes of a cloud of ultra-cold neutral atoms confined in a magneto-optical trap. The existence of a hybrid mode, Tonks-Dattner resonances and Mie oscillations are studied. Landau damping and resonant neutral atom-density wave interactions are also considered. Finally, free expansion and ambipolar diffusion regimes for a cold ionized cloud of atoms are discussed.

1. Introduction

Recently, an increasing interest has been given to the physics of ultra-cold atoms. This interest was driven by the study of Bose Einstein condensates,[1,2] but attention is now turning to the understanding of collective oscillations of non-condensed atomic molasses in the magneto-optical traps.[3–5] Collective behavior similar to that observed in plasma physics was discovered, leading to the discovery of an effective electric charge for the neutral atoms, and of electrostatic type of interactions between nearby atoms,[6] and to Coulomb like explosions of the atomic cloud when the magnetic confinement is switched off.[7]The theory of such collective processes is still not well understood, but it becomes clear that strong similarities can be found between the atomic molasses and a plasma. The main purpose of this paper is to call the attention for such interesting and unexpected similarities.

On the other hand, the cold atomic molasses can be ionized, leading to the creation of very cold plasmas, with electron temperatures below 1 Kelvin. This strongly contrasts with the usual concept of a plasma medium as a very hot gas, which electron temperatures of the order or higher than the energy of ionization, which is typically of a few electron-Volt. This extends the domain of application of plasma physics into a quite new direction, that of a very cold ionized gas. This cold plasma state can be achieved by photo-ionizing the ultra-cold gas,[8] but it can also be achieved by spontaneous evolution on a gas of neutral atoms, excited in Rydberg quantum states, into a plasma.[9] Such a spontaneous ionization process can eventually be achieved by a cascading process on the distribution level of the Rydberg atoms, from higher to lower energies. The physics of this new area of cold plasmas has been recently reviewed,[11] but here we are primarily interested in the similarities between neutral and ionized gas, which have been partly disregarded in this review.

In this paper we discuss the two kinds of collective processes associated with the cold atom physics, those of a neutral gas, and those in ionized state. We first consider the collective behavior of ultra-cold atom gas, in order to identify the basic mechanisms of oscillation, and to derive dispersion relations for the basic propagation modes. We use both fluid and kinetic descriptions where the main forces associated with laser cooling are retained. We also consider oscillations at various lengths scales with respect to the dimensions of the atomic cloud. We then describe the expansion of the cold plasma, resulting from Rydberg ionization, or from photoionization. This expansion[10] is treated here in two different regimes. First we consider the free electron expansion, where collisional effects are neglected. We then compare it with the ambipolar diffusion regime, which is more general than the approaches used in the recently published literature.

2. Laser cooling forces

The process of cooling and confining a gas of ultra-cold atoms, in the mili and micro-Kelvin domain, became possible due to the use of nearly resonant lasers beams and static magnetic fields in magneto-optical traps.[12,13] The laser frequency is tuned to a given atomic transition, between two quantum states $|a>$ and $|b>$, with energy eigenvalues E_a and E_b, in such a way that laser frequency ω_L is slightly lower than the atomic frequency, $\omega_L < \omega_{ba} \equiv (E_b - E_a)/\hbar$, where $\hbar$ is the Planck constant divide by 2π. In such conditions, absorption of a laser photon by an atom with velocity $\vec{v}$ can only be possible, by Doppler shift of the incident radiation, in such a way that $\omega_{ba} = \omega_L + \vec{k}_L \cdot \vec{v}$, where $\vec{k}_L$ is the photon wavevector. This means that, a cycle of photon absorption with frequency ω_L, followed by spontaneous emission at a frequency ω_{ba}, can be performed at the expenses of the kinetic energy of the atom, as illustrated in Figure 1. After a series of such cycles, the kinetic energy of the atoms will be significantly reduced, thus leading to atom cooling. On the other hand, the existence of a static magnetic field with a minimum at the centre of the confining device, as produced by a system of Helmoltz coils, will created a potential minimum for the atoms, which will then add to the laser forces and complement the trapping configuration.

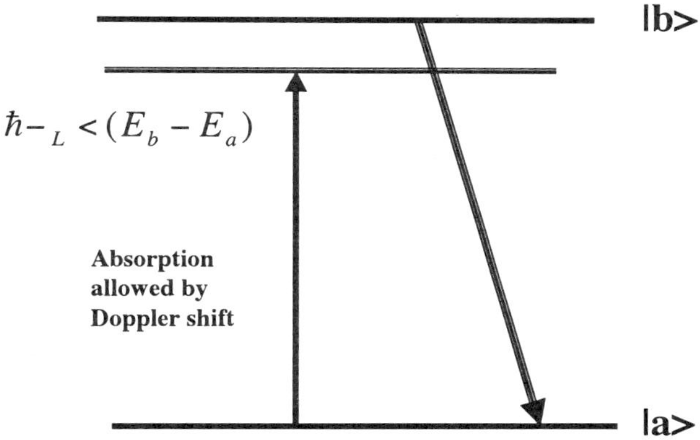

Fig. 1. *Laser cooling scheme.*

The laser action on the gas of cold atoms confined in a magneto-optical trap can be described in terms of three different of forces. One is the induced light pressure force, $\vec{F}_D$, first considered by Ashkin[16] and responsible for the cooling effect. It leads to a dissipation of the atomic kinetic energy and can be written, to the lowest order of the atom velocity $\vec{v}$, as

$$\vec{F}_D = F_0 \frac{\vec{k}_L}{k_L} - M\alpha\vec{v} \quad , \quad F_0 = \hbar k_L \frac{I_0 \Gamma^3}{\Gamma^2 + \Delta^2} \quad , \quad \beta = \frac{2\hbar k_L^2 I_0}{[1 + (\Delta/\Gamma)^2]^2} \frac{\Delta}{\Gamma} \qquad (1)$$

where I_0 is the laser intensity and $\vec{k}_L$ its wavevectors, M is the mass of the atom, Γ is the natural line width of the transition used in the cooling process, Δ the frequency detuning between the laser frequency $\omega_L = k_L c$ and the atomic transition frequency.

The second force is absorption force, $\vec{F}_A$, and was first discussed by Dalibar.[17] This is associated with the gradient of the incident laser intensity due to laser absorption by the atomic cloud. This is an attractive force, also called the shadow force. Finally, the third force, $\vec{F}_R$, can be called repulsive force or radiation trapping force, and was first considered by Sesko et al.[18] It describes atomic repulsion, due to the radiation pressure of scattered photons on nearby atoms. Both the shadow and the repulsive force can be determined by a Poisson type of equation, of the form

$$\nabla \cdot \vec{F}_A = -\frac{\sigma_L^2 I_0}{c} n(\vec{r}) \quad , \quad \nabla \cdot \vec{F}_R = \frac{\sigma_R \sigma_L I_0}{c} n(\vec{r}) \tag{2}$$

where $n(\vec{r})$ is the atom density, σ_L the laser absorption cross section, and σ_R is the atom scattering cross section. A detailed discussion of these forces and explicit expressions for the cross sections σ_R and σ_L can be found for instance in.[7] These expressions for the three forces acting on the atomic clouds and due to the laser cooling beams, correspond to the simplest possible description of the laser cloud interaction, and can be used as a first approximation to model the collective processes in the ultra-cold gas. We can now define the force $\vec{F} = \vec{F}_A + \vec{F}_R$, which is determined by the Poisson equation resulting from equations (2),

$$\nabla \cdot \vec{F} = Qn \quad , \quad Q = (\sigma_R - \sigma_L)\sigma_L I_0 / c \tag{3}$$

In typical experimental conditions the repulsive forces largely dominate over the shadow effect, and the quantity Q is positive[3,6,7] . The physical implications of this quantity will be discussed later.

3. Wave kinetic description

The analysis of collective oscillations in a cold gas can be done by using a wave kinetic description, based on the Wigner quasi-distribution. This new description will allow us to include resonant kinetic processes, which enhance the energy exchanges between part of the atomic population and the collective oscillations that can be excited in the medium. In order to establish the basis of such a wave kinetic description, we start with the Schroedinger equation describing the space time evolution of the atomic state vector $|\psi(\vec{\rho}, \vec{r}, t)>$, where $\vec{\rho}$ corresponds to the electronic position and $\vec{r}$ to the centre of mass position of the atom. This equation can be written as

$$i\hbar\frac{\partial}{\partial t}|\psi>= H(\vec{\rho}, \vec{r}, t)|\psi> \tag{4}$$

where the Hamiltonian operator is $H(\vec{\rho}, \vec{r}, t) = [\vec{p} + e\vec{A}(\vec{r}, t)]/2m + V(\rho)$, where $V(\rho)$ represents the Coulomb potential of the atomic nucleus, and the vector potential can be written in the dipole approximation as

$$\vec{A}(\vec{r}, t) = \sum_j \vec{A}_0 \exp(i\vec{k}_j \cdot \vec{R} - i\omega_L t) + \vec{A}_s(\vec{R}, t) + \vec{A}_c(R) \tag{5}$$

The first term in this expression represents the laser cooling laser beams, the second term describes the field scattered by the nearby atoms, and the last one the confining static field. We can then represent the atomic state vector as the product of two independent state vectors, $|\psi(\vec{\rho}, \vec{r}, t)>=|\vec{\rho}>|\vec{r}>$, which correspond to the internal atomic structure and to the atom kinetic or translational state. In the spirit of perturbation theory, we can then write:

$$|\vec{\rho}>=\sum_n C_n(t)|n> \exp(-iE_n t/\hbar) \tag{6}$$

where E_n are the energy eigenvalues associated with the quantum states $|n>$, and the coefficients $C_n(t)$ obey the normalization condition $\sum_n |C_n(t)|^2 = 1$. We can then define kinetic state vectors $\psi_n(\vec{r}, t)>= C_n(t)|\vec{r}>$, and consider the Wigner matrix, as defined by

$$W_{nk}(\vec{r}, \vec{q}, t) = \int \psi_n^*(\vec{r} + \vec{s}/2, t)\, \psi_k(\vec{r} - \vec{s}/2, t)\, \exp(-i\vec{q} \cdot \vec{s})\, d\vec{s} \tag{7}$$

It is well known that, starting from the wave equation for the atom in the magneto-optical trap, it is possible to derive an evolution equation for the trace of the Wigner matrix $W(\vec{r}, \vec{p}, t)$, in the the form of a Fokker-Planck equation[22]

$$\left(\frac{\partial}{\partial t} + \frac{\hbar\vec{q}}{M} \cdot \nabla\right) W = -\frac{\partial}{\partial \vec{q}}[\vec{F}_{tot} W] + \sum_{ij} D_{ij} \frac{\partial^2 W}{\partial q_i \partial q_J} \tag{8}$$

where the total force $\vec{F}_{tot}$ includes the radiative fore and the damping force, and the diffusive tensor D_{ij} is due to the fluctuations of the radiative force and spontaneous emission, as discussed by several authors.[22,23] Here, for simplicity, and because we want to focus on the oscillating modes, we neglect the diffusion term. But diffusion effects will not completely be ignored, as we will see later. We are then led to a kinetic equation of the Vlasov type, with a damping correction, as given by

$$\left(\frac{\partial}{\partial t} + \vec{v} \cdot \nabla + \frac{\vec{F}}{M} \cdot \frac{\partial}{\partial \vec{v}}\right) W = -\alpha(\vec{v}) W \tag{9}$$

where $\vec{v} = \hbar \vec{q}/M$ is the atom velocity, and the collective (shadow minus repulsive) force $\vec{F}$ is determined by the Poisson equation

$$\nabla \cdot \vec{F} = Q \int W(\vec{r}, \vec{p}, t) d\vec{v} \tag{10}$$

This equation is obvious identical to (3), because the integral is nothing but the density $n(\vec{r}, t)$. In order to focus our attention on the purely kinetic processes, we assume that $\alpha = 0$ but, at the end of this section, we will discuss the influence of this parameter. We now consider some equilibrium state $W_0(\vec{v})$ and assume a sinusoidal perturbation, such that $\delta \vec{F}$ and $\tilde{W}$ evolve in space and time as $\exp(i\vec{k} \cdot \vec{r} - i\omega t)$. After linearization, the two previous equations reduce to

$$\tilde{W} = -\frac{i}{M} \frac{\delta \vec{F} \cdot \partial W_0 / \partial \vec{v}}{(\omega - \vec{k} \cdot \vec{v})} \quad , \quad i\vec{k} \cdot \delta \vec{F} = Q \int \tilde{W}(\vec{v}) \, d\vec{v} \tag{11}$$

From here we get the dispersion relation for collective cod atom oscillations with frequency ω and wavevector $\vec{k}$

$$1 + \frac{Q}{Mk^2} \int \frac{\vec{k} \cdot \partial W_0 / \partial \vec{v}}{(\omega - \vec{k} \cdot \vec{v})} = 0 \tag{12}$$

This is similar to that of electrostatic waves in unmagnetized plasmas, and can be rewritten as $1 + \chi(\omega, \vec{k}) = 0$, where the quantity $\chi(\omega, \vec{k})$ is the susceptibility. In order to understand the physical implications of such a dispersion relation, let us consider first a simple mono-kinetic atomic distribution, of the form $W_0(\vec{v}) = n_0 \delta(\vec{v} - \vec{v}_0)$, corresponding to a beam of atoms with density n_0 and velocity $\vec{v}_0$. In this case, equation (11) reduces to

$$1 - \frac{Qn_0}{M(\omega - \vec{k} \cdot \vec{v}_0)^2} = 0 \tag{13}$$

This is nothing but the Doppler shifted plasma oscillations. For $\vec{v}_0 = 0$, this reduces to $\omega = \omega_P$, where ω_P can be defined as the effective plasma frequency for the cold neutral gas

$$\omega_P^2 = \frac{Qn_0}{M} \tag{14}$$

Comparing the first of these expressions with the usual definition of the electron plasma frequency ω_{pe} in an ionized medium, we conclude that neutral atoms behave as if they had an equivalent electric charge, as first noticed by,[6] with the value $q_{eff} = \sqrt{\epsilon_0 Q}$, where ϵ_0 is the vacuum electric permitivity. The experimental value observed for this effective atomic charge is $q_{eff} \sim 10^{-4}$ to 10^{-6} times the electron charge. It is clear that plasma like oscillations are only possible for $Q > 0$. Therefore, they cannot occur when the shadow force dominates over the repulsive force.

4. Hybrid modes

We consider the basic oscillations of the gas associated with such forces by considering the fluid equations for the ultra-cold gas, which can then be written as

$$\frac{\partial n}{\partial t} + \nabla \cdot (n\vec{v}) = 0 \tag{15}$$

$$\frac{\partial \vec{v}}{\partial t} + \vec{v} \cdot \nabla \vec{v} = -\frac{\nabla P}{Mn} + \frac{\vec{F}}{M} - \alpha \vec{v} \tag{16}$$

where n and $\vec{v}$ are the mean density and mean velocity of the gas, and P is the gas pressure. Equations (15)-(16) are identical to those used previously to describe the non-condensate cold background coupled with a Bose Einstein condensate phase,[19,20] but with the force $\vec{F}$ determined by equation (3).

We first assume oscillations that can be excited in the cold gas with a wavelength much smaller that its radius. The medium can therefore be assumed as infinite. We then assume that the equilibrium state of the gas is perturbed by oscillations with frequency ω and wavevector $\vec{k}$. Linearizing the above fluid and Poisson equations with respect to the perturbations

$\tilde{n}, \delta\vec{F}$ and $\vec{v}$, and using an equation of state of the form $P \sim n^\gamma$, with the adiabatic constant γ, we can easily obtain

$$\left[\frac{\partial}{\partial t}\left(\alpha + \frac{\partial}{\partial t}\right) + \omega_P^2 - u_S^2\nabla^2\right]\tilde{n} = \left(\frac{u_S^2\nabla\tilde{n}}{n_0} - \frac{\delta\vec{F}}{M}\right)\cdot\nabla n_0 \qquad (17)$$

where n_0 is the equilibrium density, and the quantity u_S can be identified with the sound speed, as defined by

$$u_S^2 = \gamma\frac{P_0}{Mn_0} \qquad (18)$$

where P_0 is the equilibrium gas pressure. We will now assume that the atomic density is uniform, therefore neglecting the right hand side of equation (17). This assumption will be used in order to identify the basic wave modes in the cold gas. The influence of boundary conditions and inhomogeneities on the collective oscillations of the gas will be discussed later. Assuming a space-time dependence of the perturbations $\tilde{n}$ and $\delta\vec{F}$ of the form $\exp(i\vec{k}\cdot\vec{r} - i\omega t)$, with a complex frequency $\omega = \omega_r + i\omega_i$, we obtain for the dispersion relation and for the corresponding damping rate, the values

$$\omega_r^2 = \omega_P^2 + k^2u_S^2 + \frac{3}{4}\alpha \quad , \quad \omega_i = \frac{\alpha}{2} \qquad (19)$$

In the limit of very small viscosity $\alpha \ll \omega_P$, this tends to the following dispersion $\omega^2 = \omega_P^2 + k^2u_S^2$ which his formally identical to that of electron plasma waves in ionized media, but where the electron thermal velocity $v_{the} = \sqrt{T_e/m_e}$, (T_e and m_e are the electron temperature and mass) is replaced by the sound velocity derived by a numerical factor, $u_S/\sqrt{3}$. This shows that the wave mode described by equations (19) contains elements of both electron plasma waves and acoustic waves. It possesses a lower cut-off, given by $\omega_r = \omega_P + 3\alpha/4$, which is typical of electron plasma waves. But its phase velocity tends to the sound velocity u_S and becomes weakly dispersive as for an acoustic wave. Its corresponding quasi-particles can therefore be seen as hybrid entities, somewhere between plasmons and phonons.[24]

The dispersion relation (19) is only valid for an infinite homogeneous medium. In physical terms, it can only be applied to waves that propagate locally, with wavelength scales much smaller than the cloud diameter. Let us consider now oscillations with a wavelength that is comparable with the size of the atomic cloud. In this case we can no longer neglect the boundary conditions. Going back to equation (17), we assume that the density perturbations oscillate at a frequency ω as previously, but the corresponding spatial structure will be determined by the expressions

$$\left[\nabla^2 + k^2(\vec{r})\right] \tilde{n} = \frac{\delta \vec{F}}{M u_S^2} \cdot \nabla n_0 + \nabla \ln n_0 \cdot \nabla \tilde{n} \quad , \quad \nabla \cdot \delta \vec{F} = Q \tilde{n} \tag{20}$$

where the space dependent wavenumber $k(\vec{r})$ is defined by $k^2(\vec{r}) = [\omega^2 - \omega_P^2(\vec{r})]/u_S^2$. Let us first consider the simple one-dimensional problem.[21] In the case of a uniform slab of cold gas, we have $\nabla n_0 = 0$, except at the boundaries $x = 0$ and $x = L$. Equations (20) reduce to a simple one-dimensional equation

$$\frac{d^2 \tilde{n}}{dx^2} + \frac{1}{u_S^2} [\omega^2 - \omega_P^2(x)] \tilde{n} = 0 \tag{21}$$

Taking the boundary conditions $\tilde{n}(0) = \tilde{n}(L) = 0$, we obtain the frequency eigenvalues

$$\omega_n^2 = \omega_P^2 \left\{ 1 + \left[2\pi(2n+1) \frac{\lambda_D}{L} \right]^2 \right\} \tag{22}$$

where the quantum number n can take the values $0, 1, 2, 3, ...$, and the quantity $\lambda_D = u_S/\omega_P$ is the Debye length for a cold neutral gas, in analogy with the plasma definition (where however the sound speed u_S is replaced by $\sqrt{3} v_{the}$). This shows that the finite dimensions of the cloud imply the existence of a series of resonant modes with an integer number of half-wavelengths. These are the well known Tonks-Dattner resonances. The cylindrical geometry was considered, for the plasma case, in a famous paper by Parker, Nickel and Gould in 1964,[14] but it is more natural here to consider a spherical geometry for the ultra-cold gas. Analytical solutions can be found for a spherical cloud with radius a in the homogeneous case, where $\nabla n_0(r) = 0$, for $0 \geq r < a$. In this case we obtain again an infinite series of resonances, similar to equation (22) but given now by[24]

$$\omega_{n,l}^2 = \omega_P^2 + z_{n,l}^2 \frac{u_S^2}{a^2} \tag{23}$$

where $z_{n,l}$ represents the n-th zero of the Bessel function of order $(l + 1/2)$. Comparing with the rectangular case of equation (22) we see that the allowed eigen-frequencies for a spherical cloud now depend on two quantum numbers n and l. But, in contrast with the similar quantum mechanical solutions for hydrogen like atoms, we have no hierarchical relation between these quantum numbers. The normalized radial profiles for the lowest order solutions are illustrated in Figure 2.

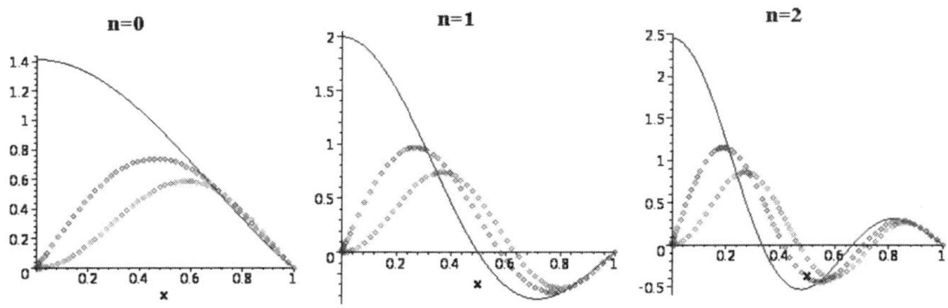

Fig. 2. *Profile of Tonks-Dattner modes, for* $n = 0, 1, 2$ *and* $l = 0, 1, 2$.

We consider next a different kind of oscillation, where the atomic cloud in a magneto-optical trap can oscillate as a rigid body. In contrast with the Tonks-Dattner resonances, there will be no density perturbations and the collective atomic velocity will be uniform and independent of position. In order to consider such a kind of oscillation we have to define the centre of mass position of the cloud, as

$$\vec{R}(t) = \frac{1}{N} \int_V \vec{r}\, n(\vec{r}, t) d\vec{r} \tag{24}$$

where N is the total number of atoms in the cloud. Using the fluid equations, (15)-(16), and the Poisson-like equation (3), and neglecting the nonlinear terms, we obtain the following harmonic oscillator equation

$$\frac{d^2\vec{R}}{dt^2} + \omega_M^2 \vec{R} = 0 \tag{25}$$

where, for a spherically symmetric cloud, the oscillating frequency is determined by the expression

$$\omega_M^2 = \frac{Q}{MR^3} \int_0^a n(r) r^2 dr \tag{26}$$

In the case of a uniform density profile $n(r) = n_0$, this reduces to

$$\omega_M = \frac{\omega_P}{\sqrt{3}} \equiv \sqrt{\frac{Qn_0}{3M}} \tag{27}$$

This is the well known Mie frequency. It characterizes the oscillation of a spherical gas with respect to its equilibrium position. In the plasma case, the restoring force was due to the background ions. Here, for the cold neutral atom gas, this is replaced by the magneto-optical trapping force. Due to the geometric factor of $\sqrt{3}$, this oscillation cannot be confused with the hybrid and Tonks-Dattner modes. On the order hand, due to the intrinsic nonlinearity of the fluid equations describing the collective behavior of the medium, it is possible to couple these centre of mass oscillations with the hybrid modes.[24]

5. Atomic Landau damping

Finally, we consider a generic equilibrium quasi-distribution $W_0(\vec{v})$. The atomic susceptibility can be split into its real and imaginary parts, $\chi = \chi_r + i\chi_i$. Assuming that most of the atoms have velocity smaller than the phase velocity ω/k, which is a very plausible assumption for a gas of ultra-cold atoms, we obtain

$$\chi_r(\omega, \vec{k}) = -\frac{1}{\omega^2}(\omega_P^2 + k^2 u_S^2) \quad , \quad \chi_i(\omega, \vec{k}) = i\pi \frac{Q}{Mk^2}\left(\frac{\partial G_0}{\partial v}\right)_{\omega/k} \tag{28}$$

where v is here the parallel component of the atom velocity, and we have identified the sound speed with the integral

$$u_S^2 = \frac{Q}{M\omega^2}\int G_0(v) v^2 dv \quad , \quad G_0(v) = \int W_0(\vec{v}_\perp, v) d\vec{v}_\perp \tag{29}$$

The quantity $G_0(v)$ introduced here is the average of the Wigner distribution over the perpendicular velocities. From this we can easily obtain the dispersion relation, by using $1 + \chi_r(\omega_r, \vec{k}) = 0$, which coincides with equation (19), and the wave Landau damping defined by the expression

$$\omega_i = -\frac{\chi_i(\omega_r, \vec{k})}{(\partial\chi_r/\partial\omega)_{\omega=\omega_r}} = \frac{\pi}{\omega}\frac{Q}{Mk^2}\left(\frac{\partial G_0}{\partial v}\right)_{\omega/k} \tag{30}$$

Here it describes the resonant interactions between the wave and the atomic population with a parallel velocity nearly equal to the wave phase velocity.

Usually, for a thermal equilibrium distribution $W_0(\vec{v})$ this quantity is negative, and corresponds to wave damping. But the sign of ω_i can change for a non-thermal distribution, eventually leading to wave instability.

We can take a step further in the kinetic description of the collective oscillations in the cold atom cloud, and consider a broad spectrum of fluctuations, described by the total wave intensity

$$I(t) = \int I(\vec{k}, t) \frac{d\vec{k}}{(2\pi)^3} \tag{31}$$

where the spectral intensity is defined by $I(\vec{k}, t) = \tilde{W}^*(\vec{k}, t)\tilde{W}(\vec{k}, t)$. Following the usual steps of the plasma quasi-linear theory, and adapting it to the present context, we can say that, each spectral component behaves in accordance with the above description, and evolves in time according to $dI(\vec{k}, t)/dt = 2\omega_i(\vec{k}, t)I(\vec{k}, t)$, and the damping rate varies slowly with time, due to the slow time evolution of the quasi-equilibrium distribution $W_0(\vec{v}, t)$, which can only be considered constant in a short time scale. The temporal evolution of $W_0(\vec{v}, t)$ under the influence of the fluctuation spectrum is determined by a diffusion equation of the form

$$\left(\frac{\partial}{\partial t} + \vec{v} \cdot \nabla + \frac{\partial}{\partial \vec{v}} \cdot \bar{\bar{D}} \cdot \frac{\partial}{\partial \vec{v}} \right) W_0(\vec{v}, t) = 0 \tag{32}$$

where the new diffusion tensor $\bar{\bar{D}}$, associated with the collective oscillations, is determined by

$$\bar{\bar{D}}(\vec{v}, t) = \frac{\omega_P^2}{n_0^2} \int I(\vec{k}, t) \frac{\vec{k}\vec{k}}{(\omega - \vec{k} \cdot \vec{k})} \frac{d\vec{k}}{(2\pi)^3} \tag{33}$$

Comparing this with our previous kinetic equation (8) it can be seen that the existence of a collective spectrum of oscillations introduces an additional diffusion effect in atomic velocity space, which tends to prevent the atomic cooling process. As we have noticed, these results are only valid in the limit of a negligible viscosity parameter, $\alpha \to 0$. A finite value of α it will have two distinct consequences. First, it will lead to the damping coefficient already stated by equation (19), adding to the purely kinetic Landau damping. Second, it will broaden the Landau resonance, therefore reducing the efficiency of the resonant interactions.

6. Ambipolar expansion of a Rydberg plasma

Ultracold neutral plasmas have been produced by photoionizing a small cloud of laser-cooled atoms confined in a magneto-optical trap, and the subsequent expansion into the surrounding vacuum was studied. Two different situations are usually considered in the literature: *i)* the ultracold atoms are firstly laser excited into high Rydberg states and then the Rydberg gas spontaneously evolves into a plasma; *ii)* the ultracold atoms are directly ionized by the laser and Rydberg atoms are formed by electron-ion recombination as the plasma expands.

In the case *ii)*, a spherical cloud of cold atoms is photoionized using a laser pulse. At first, some electrons promptly leave the plasma until sufficient space charge builds up to trap low energy electrons. Since the electrons are generated with a narrow band laser, they start with a nonthermal energy distribution, but they are rapidly thermalized by electron-electron collisions. While the electrons thermalize, some high energy electrons boil way, which reduces the electron number and temperature. After the electrons thermalize, the plasma as a whole expands.[10] The plasma expansion results from the electron thermal pressure, $p_e = n_e k_B T_e$, and it gives to the ions a radially directed velocity. The velocity of the expansion is faster than that it would be expected at low temperatures. Since the electron-ion recombination depends on ${T_e}^{-9/2}$, the faster observed expansion results from the enhancement of electron-ion recombination at low temperatures.

The simplest model for expansion of an ultracold neutral plasma is a fluid model, in which the ions are treated as a zero temperature fluid, since the ion thermal energy remains always small as compared to the radial kinetic energy associated with the plasma expansion. We notice that the ions gain very little thermal energy from the electrons due to the large mass ratio. We start by neglecting collision processes in the cold plasma. The momentum conservation equation for electrons an ions can be written as

$$m_\alpha n_\alpha \frac{\partial \vec{u}_\alpha}{\partial t} + m_\alpha n_\alpha \left(\vec{u}_\alpha \cdot \nabla\right) \vec{u}_\alpha + \nabla \left(n_\alpha \, k_B T_\alpha\right) - q_\alpha n_\alpha \vec{E} = 0 \tag{34}$$

with n_α and $\vec{u}_\alpha$ denoting the particle density and the particle mean velocity and $\vec{E} = -(\partial\phi/\partial\vec{r}_\alpha)$ the space-charge field. Here, q_α and m_α are the charge and the mass of species α, and ϕ is the mean-field potential. For electrons, we can neglect inertia, because of the smallness of its mass, and assume an homogeneous temperature in the kinetic pressure term. Moreover, we can assume that the electrons are instantaneously in equilibrium with the field, leading to a time-independent equation, $k_B T_e \nabla n_e \simeq - e n_e \vec{E}$. Since $\mathbf{E} = - \partial\phi/\partial\vec{r}_\alpha$, we obtain in spherical geometry $k_B T_e(\partial n_e \; \partial r) \simeq e n_e(\partial\phi/\partial r)$, from which a Boltzmann-like distribution can be derived

$$n_e(r,t) = n_{e0} \exp\left(\frac{e\,\phi}{k_B T_e}\right) , \tag{35}$$

with $\phi(0)$=0 and $\phi(r)$ <0 for $r > 0$. With respect now to equation (34) for the ions, we can no more neglect the inertial term due to an heavier ion mass. Further, assuming no ion pressure (T_i=0), we may write

$$\frac{\partial \vec{u}_i}{\partial t} + (\vec{u}_i \cdot \nabla)\,\vec{u}_i \simeq \frac{e\vec{E}}{m_i} \tag{36}$$

In spherical geometry we simply use $\vec{u}_i \cdot \nabla = u_i(\partial u_i/\partial r)$ On the other hand, the ions can initially be assumed to obey a gaussian-type distribution

$$n_i(r,t) = N_i \left[\frac{\beta(t)}{\pi}\right]^{3/2} \exp\left[-\,\beta(t)\,r^2\right] \tag{37}$$

where N_i is the total number of ions in the atomic cloud. For a spherically symmetric cloud we have

$$N_i = \int_0^\infty n_i(r,t)\; 4\pi r^2\; dr \tag{38}$$

It is worth noting here that equation (37) is the solution of the free-diffusion equation for the ions (*i.e.* the continuity equation, assuming $n_i\vec{u}_i = - D_i \nabla n_i$) in spherical geometry, which can be written as

$$\frac{\partial n_i}{\partial t} - D_i\,\frac{1}{r^2}\,\frac{\partial}{\partial r}\left(r^2\,\frac{\partial n_i}{\partial r}\right) = 0 \tag{39}$$

whose solution takes the form

$$n_i(r,t) = N_i\,\frac{1}{(4\pi\,D_i\,t)^{3/2}}\,\exp\left(-\,\frac{r^2}{4\,D_i\,t}\right) , \tag{40}$$

Comparing with equation (37), we conclude that $\beta(t) = 1/4D_i t$. Expressing now the potential $\phi(r,t)$ with the help of equation (35), and assuming the quasi-neutrality condition $n_e(r,t) = n_i(r,t)$, since $(n_i - n_e) \ll n_i$, we can write,

$$\phi(r,t) = -\,\frac{k_B T_e(t)\beta(t)}{e}\,r^2 + C(t) \tag{41}$$

Since the space charge field is $E(r,t) = -\partial\phi/\partial r$, we obtain for the ion acceleration

$$a_i(r,t) = \frac{2k_B T_e(t)\beta(t)}{m_i}\, r \tag{42}$$

The ion aceleration is hence linear with respect to the distance r, and the same linear dependence can be assumed for the ion velocity. We can therefore define a function $\gamma(t)$, such that $u_i(r,t) = \gamma(t)\, r$. Using the above equations, we can then derive an evolution for $\gamma(t)$, which can be written as

$$\frac{d\gamma}{dt} + \gamma^2 = \frac{2}{m_i} k_B T_e(t)\beta(t) \tag{43}$$

A relation between $\gamma(t)$ and $\beta(t)$ can also be derived from the continuity equation for the ions, as long as collisions can be neglected

$$\frac{\partial n_i}{\partial t} + \frac{1}{r^2}\frac{\partial}{\partial r}\left(r^2\, n_i\, u_i\right) = 0 \tag{44}$$

Replacing $n_i(r,t)$ and $u_i(r,t)$ in equation (37), we obtain

$$\frac{d\beta}{dt} = -\, 2\, \gamma(t)\, \beta(t)\ , \tag{45}$$

whose solution is

$$\beta(t) = \beta(0)\, \exp\left(-2\int_0^t \gamma(t')\, dt'\right)\ . \tag{46}$$

In some recent work on ultra-coul plasmas,[11] the gaussian distribution for the ions is written with the form

$$n_i(r,t) \propto \exp\left[-\frac{r^2}{2\,\sigma^2(t)}\right] \tag{47}$$

which allows us to obtain, instead of equation (45), the following alternative expression

$$\frac{d\sigma^2}{dt} = 2\,\gamma(t)\,\sigma^2(t) \tag{48}$$

Finally, we have to complement equations (43, 46) with the total energy conservation equation, where the total energy is given by the thermal electron energy plus the directed kinetic energy of ions

$$\frac{3}{2}\,k_B T_e(0) = \frac{3}{2}\,k_B T_e(t) + \frac{1}{2}\,m_i < u_i^2 > \tag{49}$$

where <> denotes the radial average. Taking into account the result for the mean square distance

$$< r^2 > (t) = \frac{1}{N_i}\int_0^\infty r^2\, n_i(r,t)\, 4\pi r^2\, dr \tag{50}$$

equal to $< r^2 > = 6D_i t$ or $= 3/(2\beta)$, whether we use equation (40) or (37), respectively. We obtain

$$\gamma(t)^2 = 2k_B[T_e(0) - T_e(t)]\frac{\beta(t)}{m_i} \tag{51}$$

Equations (43, 46, 51) can be solved, with initial conditions $\gamma(0) = 0$, *i.e.* with $u_i = 0$, and the solutions for $\gamma(t)$, $\beta(t)$ and $T_e(t)$ derived. Notice that equation (51) allows us to obtain the characteristic expansion velocity $u_{exp} \propto \sqrt{k_B T_e/m_i}$ of the ionized cold gas. This result is typical of a plasma expansion into vacuum. Such a free expansion regime is only valid for $< r^2 >\gg \lambda_D^2$, or for long times such that $t \gg \epsilon_0 k_B T_e/6e^2 n_e D_i$, where the ion diffusion coefficient D_i is defined below. In the opposite case, we have to retain collisions.

Let us now see how these equations should be modified for a collisional regime. In this regime we have to include an additional term in the momentum equation, which now appears on the right hand side of equation (34), with the usual form $-m_\alpha n_\alpha \nu_{c\alpha}\vec{u}_\alpha$. Here $\nu_{c\alpha}$ is the collision frequency for the particle species α. Using a similar procedure, we derive for electrons

$$n_e\mathbf{u_e} \simeq -\, D_e\,\nabla n_e -\, n_e\,\mu_e\,\mathbf{E} \tag{52}$$

where the electron free-diffusion coefficient and the electron mobility are $D_e = k_B T_e/(m_e\nu_{ce})$ and $\mu_e = e/(m_e\nu_{ce})$. In the limit of ambipolar loss for the electrons, we have $n_e\mathbf{u_e} \simeq 0$. In what concerns now the ion momentum conservation equation, it is usually assumed that the space-charge field is high enough to neglect inertial term, leading to a similar equation

$$n_i\vec{u}_i \simeq -\, D_i\,\nabla n_i +\, n_i\,\mu_i\,\mathbf{E} \tag{53}$$

with similar expressions for D_i and μ_i. On the other hand, since the right-hand side member of the continuity equation for electrons and ions is the sam, we have the continuity equations for both electrons and ions as

$$\frac{\partial n_\alpha}{\partial t} + \nabla \cdot (n_\alpha \, \mathbf{u}_\alpha) = n_\alpha \, \nu_{ion} \tag{54}$$

where ν_{ion} denotes the ionization frequency: $e + X_0 \longrightarrow \mathrm{e} + X^+ + e$, we obtain in the case of a slowly time-dependent variation of $n_e(t)$ and $n_i(t)$, as it is proper under ambipolar loss condition $\nabla \cdot (n_e \, \vec{u}_e) = \nabla \cdot (n_i \, \vec{u}_i)$. From this we can assume the congruence condition $n_e \, \mathbf{u_e} = n_i \, \vec{u}_i$. In the limit of ambipolar diffusion, we can still assume the coupled transport of electrons and ions, with $n_e = n_i$ and $vecu_e = \vec{u}_i$, so that we can write from equation (53) the following expression for the space-charge field , making use of the two inequalities $D_e \gg D_i$ and $\mu_e \gg \mu_i$

$$\vec{E} \; \simeq \; - \; \frac{k_B T_e}{e} \, \frac{\nabla n_e}{n_e} \; . \tag{55}$$

This equation is the same in the collisionless approximation. Since in the ambipolar loss limit, we have $n_e \vec{u}_e = \; - \, D_a \, \nabla n_e$, with an identical expression for the ions, where

$$D_a = \; \mu_i \; \frac{k_B T_e}{e} \;\; = \;\; \frac{k_B T_e}{m_i \nu_{ci}} \tag{56}$$

denotes the ambipolar diffusion coefficient, in which the inequalities $\mu_e \gg \mu_i$ and $T_e \gg T_i$ have been used, we obtain from the continuity equation for electrons and ions in spherical geometry, *i.e.* from an equivalent equation (39) with D_i replaced by D_a, the usual gaussian solution

$$n_e(r,t) = n_i(r,t) = N_0 \; \frac{1}{(4\pi \; D_a \; t)^{3/2}} \; \exp\left(- \; \frac{r^2}{4 \; D_a \; t} \right) \; . \tag{57}$$

Comparing with equation (55), we see that the electron density has to be also proportional to $\exp(e\phi/k_B T_e)$, in the same way as in equations (35), and as they also obey to the gaussian law (57), the potential $\phi(r)$ has to vary as r^2, and the ion acceleration is proportional to r, as in equations (42). An acceleration linear in r, implies a velocity which is also linear in r, therefore preserving the Gaussian spatial profile along the expansion process. Finally, the energy conservation equation can be still written in the form (51) , and the velocity for ambipolar loss expansion will preserve the characteristic trend $u_{exp} \propto \sqrt{k_B T_e / m_i}$. These results extend the previously predicted properties of an expanding ultra-cold plasma, and are particularly adequate for experimental verification.

7. Conclusions

In this work we have discussed two different aspects of the physics of an ultra-cold gas. First, the plasma like aspects of the neutral gas, due to the repulsive forces associated with the exchange of photons between nearby atoms. Second, the expansion of an ultra-cold plasma resulting from spontaneous or induced ionization of the neutral gas.

In order to understand the collective processes in the neutral cold gas, we have used a very simple but realistic model for the laser forces acting on the atoms confined in a magneto-optical trap, which has been verified by recent experiments. The most relevant aspect of this approach is that the laser forces can be described by a Poisson equation, similar to that describing electrostatic interactions.[3,7,18] The existence of such forces leads to the possible excitation of collective modes, which can be called hybrid modes, in the sense that they have properties common to both plasmons and phonons. The corresponding dispersion relation reveals a cutt-off at the atomic plasma frequency, but the phase velocity for high frequencies is nearly equal to to the sound velocity.

We have also shown that internal resonances of these hybrid oscillations can be excited in a finite cloud of gas, similar to the Tonks-Dattner resonances, well known in plasma physics, but here related to the oscillations of a neutral gas.[24] We have considered a spherical geometry, which is relevant to the atomic cloud produced in the magneto-optic traps. We have also discussed the centre of mass oscillations of the atomic cloud around its equilibrium position, which corresponds to the well known Mie oscillations.

Our analysis of the hybrid modes was extended to the kinetic regime. We were able to derive a kinetic dispersion relations for the oscillations in the cold gas, where Landau damping was included. Finally, we have discussed a quasi-linear kinetic equation, which shows that a broad spectrum of collective oscillations in the gas produces diffusion in atomic velocity space and acts as an addition process that reduces the efficiency of laser cooling.[24]

In the present work we have explored the similarities of the could atom cloud with a plasma, which can be associated with the existence of an effective electric charge for the neutral atoms. The resulting wave modes however are not identical to the plasma wave modes, but show an hybrid character. We have also discussed the behavior of ultra-cold plasmas, which can be produced after spontaneous or induced ionization processes. Such a cold plasma expands after formation, in free electron expansion regime if collisions are negligible, and in accordance with the laws of ambipolar diffusion in the collisional regime. In both cases the electrons produce an outer negatively charged cloud, progressively dragging the central core of ultra-cold ions. This work shows that plasma physics is now extending to new and quite unexpected domains.

References

1. A. J. Legget, Rev. Mod. Phys. **73**, 307 (2001).
2. F. Dalfovo *et al.*, Rev. Mod. Phys. **71**, 463 (1999).
3. G. Labeyrie, F. Michaud and R. Kaiser, *Phys. Rev. Lett.*, **64**, 023003 (2006).
4. T. Pohl, G. Labeyrie and R. Kaiser, *Phys. Rev. A*, **74**, 023409 (2006).
5. A. di Stefano et al., *Eur. Phys. J. D*, **30**
6. T. Walker, D. Sesko and C. Wieman, *Phys. Rev. Lett.*, **64**, 408 (1990).
7. L. Pruvost et al., *Phys. Rev. A*, **61**, 053408 (2000).
8. T.C. Killian et al. *Phys. Rev. Lett.*, **83**, 4776 (1999).
9. M.P. Robinson et al., *Phys. Rev. Lett.*, **85**, 4466 (2000).
10. F. Robicheaux and J.D. Hanson, *Phys. Rev. Lett.*, **88**, 055002 (2002).
11. T.C. Killian, T. Pattard, T. Pohl and J.M. Rost, *Phys. Rep.*, **449**, 77 (2007).
12. W.D. Phillips, *Rev. Mod. Phys.*, **70**, 721 (1998).
13. C.N. Cohen-Tannoudji, *Rev. Mod. Phys.*, **70**, 707 (1998).
14. J.V. Parker, J.C. Nickel and R.W. Gould, *Phys. Fluids*, **7**, 1489 (1964).
15. R. Guerra and J.T. Mendonça, *Phys. Rev. E*, **62**, 1190 (2000).
16. A. Ashkin, *Phys. Rev. Lett.*, **25**, 1321 (1970).
17. J. Dalibar, *Opt. Commun.*, **68**, 203 (1988).
18. D.W. Sesko, T.G. Walker and C.E. Wieman, *J. Opt. Soc. Am. B*, **8**, 946 (1991).
19. Nikuni and Griffin, *Phys. Rev. A*, **69**, 023604 (2004).
20. E. Zaremba, A. Griffin and T. Nikuni, *Phys. Rev. A*, **57**, 4695 (1998).
21. T.J.M. Boyd and J.J. Sanderson, *The Physics of Plasmas*, Cambridge University Press (2003).
22. J. Dalibard and C. Cohen-Tannoudji, *J. Phys. B: At. Mol. Phys.*, **18**, 1661 (1985).
23. S. Stenholm, *Rev. Mod. Phys.*, **58**, 699 (1986).
24. J.T. Mendonça, R. Kaiser, H. Terças and J. Loureiro, in preparation (2007).

GENERAL PROPERTIES OF THE RAYLEIGH-TAYLOR INSTABILITY IN DIFFERENT PLASMA CONFIGURATIONS: THE PLASMA FOIL MODEL

F. PEGORARO

Physics Dept., University of Pisa, Pisa, Italy
**E-mail: pegoraro@df.unipi.it*

S.V. BULANOV

Kansai Photon Science Institute, Japan Atomic Energy Agency, Japan
**E-mail: bulanov.sergei@jaea.go.jp*

The nonlinear development of the nonrelativistic Rayleigh-Taylor instability of a fluid (plasma) system is investigated within the thin foil model (long wavelength approximation). Explicit solutions are obtained for two- and three-dimensional configurations with the help of various mathematical techniques that include Lagrange variables, complex variable representations, Lie Symmetry transformations and the Hodograph transformation.

1. Introduction

In fluid dynamics the Rayleigh-Taylor instability represents a basic phenomenon that occurs when a lighter fluid is accelerated into a heavier fluid. For an inhomogeneous fluid system in a gravitational field this instability was first discovered by Lord Rayleigh[1] in the 1880's. This instability is responsible for the fact that, if surface tension effects are neglected, it is not possible to keep water inside an inverted container (i.e., open at its bottom) balanced by atmospheric pressure. Later the same concept was applied to accelerated fluids by Sir Geoffrey Taylor[2] .

Understanding the rate of mixing caused by the Rayleigh-Taylor instability is important for different applications ranging from inertial confinement fusion,[3] nuclear weapons explosions,[4] supernova explosions[5] and supernova remnants[6] , to oceanography[7] , and atmospheric physics[8] , to laboratory[9] and space plasmas[10] etc..

A family of different instabilities can be grouped under the general name of Rayleigh-Taylor instabilities, with e.g., an inhomogeneous pressure playing the role of the inhomogeneous density, or electromagnetic radiation pressure taking the role of the lighter fluid or even, in the case of a magnetized confined plasma, magnetic field pressure[a] and magnetic field line curvature playing the role of the lighter fluid and of gravity, respectively.

In the simple case of two immiscible incompressible fluids with densities ρ_1 and $\rho_2 < \rho_1$ respectively, where the denser fluid 1 is initially on top of fluid 2, the instability linear growth rate γ can be written as

$$\gamma^2 = |k|g\,\frac{\rho_1 - \rho_2}{\rho_1 + \rho_2}, \qquad \gamma^2 \sim |k|g, \quad \text{for } \rho_2 \ll \rho_1, \tag{1}$$

with k the wavenumber in the interface plane and g the gravity acceleration. Eq.(1) can be understood by a simple energy argument. If the instability displaces two volume elements of mass σh (where σ is the fluid density per unit surface) vertically by a height h, one up and one down so as to keep the fluid volume constant, the fluid potential energy changes by a quantity of order $\sigma g h^2$. The kinetic energy involved in this displacement is not limited to the two volume elements but involves a fluid column of height $\sim k^{-1}$, and thus of mass σk^{-1}, as can be easily understood from the fact that the spatial dependence of the instability in the direction perpendicular to the interface is exponentially decreasing, as characteristic of surface waves. Thus the kinetic energy change is of order $\sigma k^{-1}\gamma^2 h^2$. By equating the potential and the kinetic energy changes, we recover Eq.(1) in the limit $\rho_2 \ll \rho_1$.

The Rayleigh-Taylor instability is of special importance in the case of inertial fusion where a fuel pellet is compressed by the reaction force exerted by the surface layers of the pellet ablated by the energy deposited by a high intensity laser pulse[11] . The apparently unbound increase of the growth rate γ in Eq.(1) at small wavelengths is interrupted by effects, such as surface tension for a real fluid system, that are not included in Eq.(1). At small wavelengths these effects first reduce the mode growth rate and finally stabilize the mode. Different stabilizing mechanisms have been shown to arise for different forms of Rayleigh-Taylor instabilities, ranging from matter ablation in the pellet compression in inertial fusion, to field line tension in the case of magnetically confined plasmas.

---

[a]Note however that the tensor nature of the magnetic pressure, the Maxwell stress tensor, makes a magnetic Rayleigh-Taylor instability evolve differently from a fluid Rayleigh-Taylor instability.

2. The foil geometry

Let us now consider the Rayleigh-Taylor instability of a thin accelerated material foil. This configuration is of both of practical interest in a number of physical conditions and is amenable to analytical solutions. Indeed, the Rayleigh-Taylor instability of a thin plasma slab[12] provides one of the best examples of the basic nonlinear behavior of a fluid when its equilibrium configuration is unstable against infinitesimal perturbations. In addition, in some simplified limits exact mathematical solutions can be found that make it possible to study the formation and the properties of singularities produced in the nonlinear evolution of the instability.

More specifically we will consider an initially planar plasma slab (in the y-z plane)[b] of width d (along x). We suppose that the characteristic transverse size of the slab in the y-z plane is much larger than d and thus take for the sake of simplicity the slab to be infinitely extended in this plane. We suppose that the foil is acted upon by a spatially uniform pressure difference that is constant in time and that acts at all times along the normal to the foil surface.

If we restrict ourselves to motions and deformations of the plasma slab that do not involve spatial scales smaller that its width d we can describe the slab as an infinitely thin foil with an assigned surface density given by the slab density (assumed to be uniform) multiplied times d. This description may appear as a restrictive approximation in view of the fact that Eq.(1) would predict that the growth rate of the Rayleigh-Taylor instability increases with its wavenumber k, while d^{-1} can be obviously taken as an upper bound on k for the foil model to be valid. However, as mentioned before, physical effects non included in Eq.(1) may stabilize short wavelength perturbations and thus the thin foil approximation can be taken as a valid model as long as the slab dynamics is correctly described by "long wavelength" perturbations defined by the condition $kd \ll 1$.

Within this approximation, the basic equations that describe the motion of a foil surface element are its surface mass conservation equation and the momentum equation. If we include the presence of a friction force[c], these equations take the form

$$\frac{d}{dt}\left(\sigma\ \mathbf{d\Sigma}\right) = 0, \tag{2}$$

---

[b]Different geometries, such as e.g., spherical or cylindrical configurations are also easily treated.

[c]A friction force is of interest for example in the case of a plasma foil accelerated through a through a diffuse medium.

and

$$\sigma \left[\frac{d}{dt} + \nu^{(in)} \right] \mathbf{v} = \mathcal{P}\, \mathbf{n}. \tag{3}$$

where $\mathbf{v}$ is the velocity of the foil, σ its surface mass density, $\mathcal{P}$ is the pressure jump through the plasma slab with respect to the normal vector $\mathbf{n}$, $\nu^{(in)}$ is an effective friction frequency, d/dt is the Lagrangian time derivative and $\mathbf{d\Sigma}$ is the (oriented) surface element on the shell.

First we consider a general 3-D case where the foil position depends on all three spatial coordinates, x, y, z and on time t. We assume that the foil is initially located on a smooth surface that we parametrize as $x = \mathcal{X}(y, z)$. In order to obtain the equations for the foil evolution, we introduce the Lagrange variables, α, and β, related to the Euler coordinates by

$$x = x(\alpha, \beta, t), \quad y = y(\alpha, \beta, t), \quad \text{and} \quad z = z(\alpha, \beta, t), \tag{4}$$

where α and β are a set of variables marking the foil elements. A convenient choice of α and β is given e.g., by a set of (local) orthogonal coordinates on the surface where the shell is located at $t = 0$. In the simple case where the shell is initially planar we can choose $\mathcal{X} \equiv 0$ and $y = \alpha$, $z = \beta$ at $t = 0$. Then the surface density conservation gives $\sigma_0\, d\Sigma_0 = \sigma d\Sigma$, with $d\Sigma_0 = d\alpha \wedge d\beta$, from which we obtain

$$\sigma_0 \left[\frac{d}{dt} + \nu^{(in)} \right] \mathbf{v} = \mathcal{P}\; \mathbf{d\Sigma}/d\Sigma_0. \tag{5}$$

3. Two dimensional solutions: linear equations in Lagrange variables

An important simplification occurs in the case of two dimensional (2D) foil evolutions defined e.g. by the condition $\partial/\partial z = 0$ in which case $\mathbf{v} \equiv (dx/dt)\,\mathbf{e}_x + (dy/dt)\,\mathbf{e}_y$, $\mathbf{d\Sigma}/d\Sigma_0 \equiv (\partial y/\partial\alpha)\,\mathbf{e}_x - (\partial x/\partial\alpha)\,\mathbf{e}_y$ with α the element initial position along y, $\mathbf{e}_{x,y}$ the unit vectors in the x-y plane. In this 2-D case, the evolution equations (5) expressed in Lagrangian variables are linear. Note that the evolution equations are not linear if expressed in Eulerian variables.

3.1. *Frictionless solutions*

Let us first set $\nu^{(in)} = 0$. The 1-D solution $x_0 = x_0(t)$ (i.e., $\partial y_0/\partial\alpha = 1$, $\partial x_0/\partial\alpha = 0$) for the accelerated foil reads

$$d^2 x_0/dt^2 = \mathcal{P}/\sigma_0, \tag{6}$$

while the perturbations $\tilde{x}(\alpha, t)$, $\tilde{y}(\alpha, t)$ obey the equations

$$d^2\tilde{x}/dt^2 = (\mathcal{P}/\sigma_0)(\partial\tilde{y}/\partial\alpha), \qquad d^2\tilde{y}/dt^2 = -(\mathcal{P}/\sigma_0)(\partial\tilde{x}/\partial\alpha), \tag{7}$$

where no linearization has been performed. Choosing solutions of the form $\exp(ik\alpha + \gamma t)$, with k the Lagrangian wavenumber, we recover the square root scaling of the growth rate with wave number

$$\gamma^4 = k^2(\mathcal{P}/\sigma_0)^2. \tag{8}$$

3.1.1. *Complex function representation*

It is interesting to observe that more general solutions can be obtained by introducing[12,13] the complex function $w(\alpha, \tau) = x + iy$ which then obeys the equation (in terms of a properly normalized time variable τ)

$$\partial_{\tau\tau} w = -i\partial_\alpha w, \tag{9}$$

while for the complex conjugate function $w^*(\alpha, \tau) = x - iy$ we have

$$\partial_{\tau\tau} w^* = i\partial_\alpha w^*, \tag{10}$$

where w and w^* are considered as independent functions. For the sake of illustration we recall that the solution

$$w = i\alpha + \tau^2/2 \tag{11}$$

corresponds to a uniformly accelerated plane foil, while the solution

$$w(\alpha, \tau) = x + iy = i\alpha^3 - i\alpha\tau^4/4 - \tau^6/120 + 3\alpha^2\tau^2/2 \tag{12}$$

describes the local structure[d] of wave breaking (x is no longer a single valued function of y along the foil surface, i.e. the foil is folding), see Fig.(1). For $w(\alpha, \tau) \propto \exp(i\alpha q)$ from Eq.(9) we recover exponentially growing and decaying modes for $q > 0$ and oscillatory modes with real frequency for $q < 0$, the intervals in q are interchanged in Eq.(10) for $w^*(\alpha, \tau) \propto \exp(-i\alpha q)$.

3.1.2. *Lie symmetries*

Eq.(9) admits 7 symmetry transformations[e] represented by the operators:

$$X_\infty = w_1(\alpha, \tau)\partial_w, \tag{13}$$

[d] In this solution a non homogeneous initial surface density is assumed corresponding to $dy(t = 0) = \sigma_0(\alpha)d\alpha$.

[e] A discussion of the theory of the Lie group analysis of differential equations is presented e.g., in Ref. ([14]).

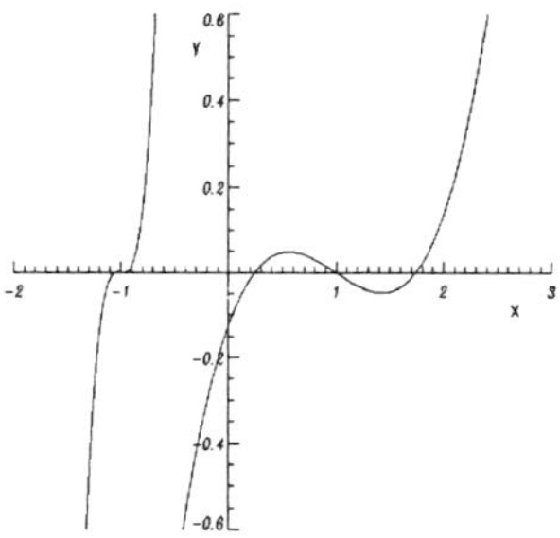

Fig. 1. Breaking solution from Eq.(12) as a function of α at $\tau = 0.5$, left curve, and at $\tau = 1$, right curve. The two curves have been shifted along x for graphical clarity.

$$X_1 = i\partial_\alpha, \tag{14}$$

$$X_2 = \partial_\tau, \tag{15}$$

$$X_3 = 2\alpha\partial_\alpha + \tau\partial_\tau, \tag{16}$$

$$X_4 = 2i\alpha\partial_\tau - \tau w\partial_w, \tag{17}$$

$$X_5 = i\tau\alpha\partial_\tau - i\alpha^2\partial_\alpha - (\tau^2 + 2i\alpha)(w/4)w\partial_w, \tag{18}$$

$$X_6 = w\partial_w. \tag{19}$$

The operator X_∞ stems from the fact that Eq.(9) is linear with respect to $w(\alpha,\tau)$ so that, to any solution $w(\alpha,\tau)$, one can add any other solution $w_1(\alpha,\tau)$. The operators X_1 to X_4 and X_6 correspond to time and space translations and to the invariance with respect to stretching of the variables respectively. The operator X_5 represents the transformation

$$(\bar{\alpha},\bar{\tau}) = (\alpha.\tau)/(1-a\alpha), \qquad \bar{w} = (1-a\alpha)^{1/2}\exp\left[ia\tau^2/(4-4a\alpha)\right]w. \tag{20}$$

If we choose $w = i/(4\pi)^{1/2}$ and $a = -1/h$, and superpose $w_0 = i\alpha + \tau^2/2$, we obtain a solution of the form

$$\bar{w}(\alpha,\tau) = i\alpha + \frac{\tau^2}{2} + \frac{i(h)^{1/2}}{[4\pi(\alpha+h)]^{1/2}}\exp\left[i\frac{\tau^2}{4(\alpha+h)}\right], \tag{21}$$

where $i(h)^{1/2}$ is the initial perturbation amplitude and h is a complex parameter. The foil is initially a planar with perturbations localized in a region with size of order $|h|$. If h is imaginary and positive, $h = i|h|$, this solution describes perturbations that grow faster than exponential: $\propto \exp(\tau^2/4|h|)$.

3.2. *Frictional solutions*

If friction dominates over inertia we obtain[13]

$$\partial_\tau x = \partial_\alpha y, \qquad \partial_\tau y = -\partial_\alpha x, \tag{22}$$

where we have suitably normalized the time variable τ. The above equations are the Cauchy-Riemann conditions for the real and imaginary parts of an analytical function $W(\zeta) = x + iy$ of a complex variable $\zeta = \alpha + i\tau$. Their solutions are thus given by the conformal mapping from the complex plane $\alpha + i\tau$ to the plane $x + iy$. Choosing as an example the analytical function

$$W(\zeta) = -i\zeta + \kappa \frac{1}{1+\zeta^2} \tag{23}$$

where $\kappa = \kappa_R + i\kappa_I$ is a complex constant, we find

$$x(\alpha,\tau) = \tau + \kappa_R \frac{1-\tau^2+\alpha^2}{(1-\tau^2+\alpha^2)^2+4\alpha^2\tau^2} + \kappa_I \frac{2\alpha\tau}{(1-\tau^2+\alpha^2)^2+4\alpha^2\tau^2}, \tag{24}$$

$$y(\alpha,\tau) = -\alpha + \kappa_I \frac{1+\tau^2+\alpha^2}{(1-\tau^2+\alpha^2)^2+4\alpha^2\tau^2} - \kappa_R \frac{2\alpha\tau}{(1-\tau^2+\alpha^2)^2+4\alpha^2\tau^2}. \tag{25}$$

These expressions describe the growth of perturbations that are faster than exponential. It easy to see that at the finite time $\tau = 1$ the Jacobian $|W'|$ of the transformation becomes infinite at the point $\alpha = 0$.
A more extended description of the use of conformal mapping in order to solve Eq.(22) for the foil motion can be found in Ref.([13])

4. Three dimensional frictional solutions; equations remain nonlinear in Lagrange variables

For three dimensional friction dominated foil evolutions, we obtain the equations of motion[15]

$$\frac{\partial x}{\partial \tau} = \{y, z\}_{\alpha,\beta}\,, \quad \frac{\partial y}{\partial \tau} = \{z, x\}_{\alpha,\beta}\,, \quad \frac{\partial z}{\partial \tau} = \{x, y\}_{\alpha,\beta}\,, \tag{26}$$

where the Poisson brackets with respect to the Lagrange variables α and β are defined by

$$\{\ ,\ \}_{\alpha,\beta} \equiv \frac{\partial}{\partial\alpha}\frac{\partial}{\partial\beta} - \frac{\partial}{\partial\beta}\frac{\partial}{\partial\alpha}. \tag{27}$$

These equations are nonlinear[f] but reduce to the linear 2-D equations if x and y are independent of β (with $\partial z/\partial\beta = 1$).

[f]This also would apply to the corresponding 3-D frictionless equations.

Eqs.(26) can be written in the notation of differential forms as the equality of three 3-forms involving exterior products of the 1-forms obtained by differentiating the independent and the dependent variables

$$dx \wedge d\alpha \wedge d\beta = d\tau \wedge dy \wedge dz, \tag{28}$$

$$dy \wedge d\alpha \wedge d\beta = d\tau \wedge dz \wedge dx,$$

$$dz \wedge d\alpha \wedge d\beta = d\tau \wedge dx \wedge dy,$$

where now d denotes the exterior derivative and $\wedge$ the exterior product[g]. In order to recover Eqs.(26) we express the 1-form dx in terms of the three 1-forms $d\alpha$, $d\beta$, $d\tau$ as

$$dx = \frac{\partial x}{\partial \alpha} d\alpha + \frac{\partial x}{\partial \beta} d\beta + \frac{\partial x}{\partial \tau} d\tau, \tag{29}$$

and analogously for dy and dz and use the antisymmetric properties of the exterior product. If we interchange dependent and independent variables (hodograph transformation) and write

$$d\alpha = \frac{\partial \alpha}{\partial x} dx + \frac{\partial \alpha}{\partial y} dy + \frac{\partial \alpha}{\partial z} dz, \tag{30}$$

and analogously for $d\beta$ and $d\tau$, we obtain the hodograph transformed equations

$$\frac{\partial \tau}{\partial x} = \frac{\partial \alpha}{\partial y} \frac{\partial \beta}{\partial z} - \frac{\partial \beta}{\partial y} \frac{\partial \alpha}{\partial z}, \tag{31}$$

$$\frac{\partial \tau}{\partial y} = \frac{\partial \alpha}{\partial z} \frac{\partial \beta}{\partial x} - \frac{\partial \beta}{\partial z} \frac{\partial \alpha}{\partial x}, \qquad \frac{\partial \tau}{\partial z} = \frac{\partial \alpha}{\partial x} \frac{\partial \beta}{\partial y} - \frac{\partial \beta}{\partial x} \frac{\partial \alpha}{\partial y}.$$

These can be combined in the vector equation

$$\nabla \tau = \nabla \alpha \times \nabla \beta, \tag{32}$$

where $\nabla \tau$ is the gradient of the function $\tau = \tau(x, y, z)$ in x, y, z space, etc.. In the generalization from 2-D to 3-D the harmonic property

$$\nabla^2 \tau = 0, \tag{33}$$

where the Laplace operator is taken with respect to x, z in 2-D and to x, y, z in 3-D, is preserved. From Eq.(32) we also obtain $\nabla \tau \cdot \nabla \alpha = \nabla \tau \cdot \nabla \beta = 0$ and the τ-independent compatibility equation $\nabla \times (\nabla \alpha \times \nabla \beta) = 0$. Note that, in contrast to the 2-D case where all variables play the same role, in

---

[g] For a detailed definition of these operations see e.g., Ref.([16]).

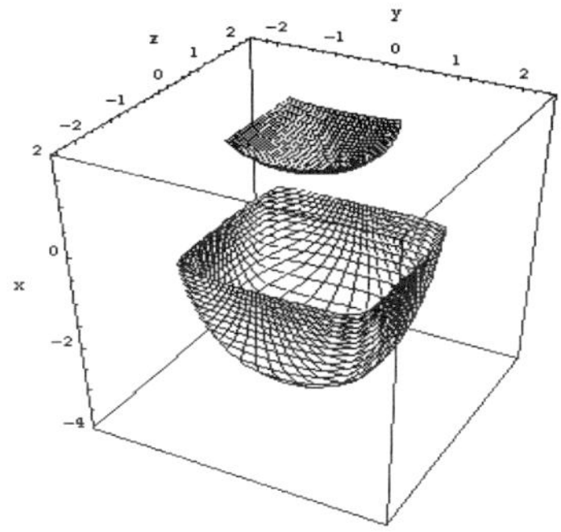

Fig. 2. Time evolution of a small square foil pushed along the negative x according to Eq. (35). The position of the foil is shown at $\tau = -2$, top foil, and at at $\tau = -0.05$, bottom foil. Again a shift along x has been added for graphical clarity.

general $\nabla^2\alpha \neq 0$ and $\nabla^2\beta \neq 0$.
Exponential-type solutions are obtained by taking

$$\alpha = \sqrt{2}\sin(y)\exp(x/\sqrt{2}), \qquad \beta = \sqrt{2}\sin(z)\exp(x/\sqrt{2}),$$

$$\tau = \sqrt{2}\cos(y)\cos(z)\exp(x\sqrt{2}), \tag{34}$$

which can be easily inverted and give

$$x = \frac{1}{\sqrt{2}}\ln\left(\frac{\alpha^2+\beta^2 \pm \sqrt{(\alpha^2-\beta^2)^2+8\tau^2}}{4}\right), \tag{35}$$

$$(y,z) = \arcsin\left[\frac{(\alpha,\beta)}{\left((\alpha^2+\beta^2)/2 \pm \sqrt{(\alpha^2-\beta^2)^2/4+2\tau^2}\right)^{1/2}}\right],$$

where we will consider only the plus sign in front of the square root which is defined to be positive. Physically, for $\tau < 0$, this solution can be taken to represent the evolution of a small square foil that is pushed along x and that breaks at $\alpha = \beta = 0$ at $\tau = 0$, having reached $x = -\infty$ as shown in Fig.(2). Additional solutions of Eqs.(32) can be obtained in terms of the well known systems of orthogonal coordinates that are commonly used in mathematical physics, as shown in Ref.([15]).

5. Conclusions

Although restricted to the long wavelength approximation, the thin foil model makes it possible to gain insight into the nonlinear development of the Rayleigh- Taylor instability in 2-D and in 3-D configurations. In the former case the use of Lagrangian coordinates makes the equations that govern the foil evolution in time linear. Moreover, as shown in detail in Refs.([13] ,[15]), this method allow us to obtain a wide range of explicit solutions and to investigate, within the limitations of the long wavelength approximation, the formation in time of singularities in the foil motion.

When the velocity of the foil approaches the speed of light, a new type of nonlinearity appears in the foil equations through the relationship between the foil momentum and its velocity. This relativistic nonlinearity remains present also when Lagrangian variables are used for 2-D foil configurations and has important effects on the foil dynamics. In an accompanying paper[h] the stability of a relativistic foil pushed by the radiation pressure of an intense laser pulse is investigated.

References

1. Rayleigh, Lord (John William Strutt), *Proceedings of the London Mathematical Society*, **14**, 170 (1883).
2. Taylor, Sir Geoffrey Ingram, *Proceedings of the Royal Society of London, Series A, Mathematical and Physical Sciences*, **201**, 192 (1950).
3. Lindl J.D , in "*Inertial confinement fusion: The quest for ignition and energy gain using indirect drive*, Springer-Verlag, New York, (1998).
4. Fermi, E., 1951, in "*The Collected Papers of Enrico Fermi*", (ed. E. Segre *et al.*), **201**, 816, Chicago, University of Chicago Press, (1961).
5. Blinnikov S., Sorokina E., *Astroph. Space Sci.*, **290**, 13 (2004).
6. Ribeyre X., Hallo L., Tikhonchuk V.T., Bouquet S., Sanz J., *Astroph. Space Sci.*, **307**, 169 (2007).
7. Debnath L., n "*Nonlinear Water Waves*" Academic Press, Inc., (1994).
8. Basu,B., *JGR*, **104**, 6859 (1999).
9. Coppi B., *Phys. Rev. Lett.*, **39** 939 (1977).
10. Isobe H., Miyagoshi T., Shibata K., Yokoyama, T., *Nature*, **434**, 478 (2005).
11. Atzeni S., Meyer-ter-Vehn J., in "*The Physics of Inertial Fusion*", Clarendon Press, Oxford, (2004).
12. Ott E., *Phys. Rev. Lett*, **29**, 1429 (1972).
13. Bulanov S.V., Pegoraro F., and Sakai J.-I., *Phys. Rev.*, **E 59**, 2292 (1999).
14. Ovsiannikov, L.V., *Group Analysis of Differential Equations*, (Academic Press, New York, 1982).
15. Pegoraro F., Bulanov S.V., Sakai J-I., Tomassini G., *Phys. Rev.*, **E 64**, 016415 (2001).
16. Flanders H., "*Differential Forms with Applications to the Physical Sciences*", (Dover Publ.: New York, 1989)

[h]In this same issue: Pegoraro F., Bulanov S.V., *The Rayleigh-Taylor instability of a plasma foil accelerated by the radiation pressure of an ultraintense laser pulse.*

THE RAYLEIGH-TAYLOR INSTABILITY OF A PLASMA FOIL ACCELERATED BY THE RADIATION PRESSURE OF AN ULTRA INTENSE LASER PULSE

F. PEGORARO

Physics Dept., University of Pisa, Pisa, Italy
**E-mail: pegoraro@df.unipi.it*

S.V. BULANOV

Kansai Photon Science Institute, Japan Atomic Energy Agency, Japan
**E-mail: bulanov.sergei@jaea.go.jp*

The effective ion acceleration during the interaction of an ultra short and ultra intense laser pulse with matter is possibly one of most important results in the investigation of the interaction of multi-terawatt and petawatt power laser pulses with plasmas. At high laser intensities a very efficient acceleration regime has been predicted where the radiation pressure of the laser pulse plays a major role. The stability of this regime against the onset of the relativistic Rayleigh-Taylor instability is investigated.

1. Introduction

Radiation pressure can be a very effective mechanism of particle acceleration. The importance of this mechanism was discovered long ago[1,2] . Physical conditions of interest range from stellar structures and radiation generated winds[3–5] , to high accuracy optical experiments[6,7] and optical traps, to the formation of photon bubbles in very hot stars and accretion disks[8–10] and to the investigation of different high energy astrophysical environments[11–13] . Particle acceleration by radiation pressure has also been considered in the laboratory[14–19] in laser plasma interactions[a]. Radiation pressure arises from the coherent interaction of the radiation with the particles in the medium which absorbs or reflects the incoming electromagnetic radiation and, in the process, acquires momentum.

[a]For a review on laser plasma interactions see e.g., Refs.([21,22]) and references therein.

In the case of an electron-ion plasma, which is the case considered here, radiation pressure acts mostly on the lighter particles, the electrons, with a force that is quadratic in the wave field amplitude. Ions on the contrary are accelerated by the charge separation field caused by the electrons pushed by the radiation pressure. As first pointed out some fifty years ago[20] , this collective acceleration mechanism is very efficient when the number of ions inside the electron cloud is much smaller than that of the electrons.

2. Laser plasma acceleration of ions

The electric fields produced by the interaction of ultra-short and ultra-intense laser pulses with a thin target make it possible to obtain multi-MeV, high density, highly collimated proton and ion beams[23–26] of extremely short duration, in the sub-picosecond range. Such laser pulses may also open up the possibility of exploring high energy astrophysical phenomena, such as in particular the formation of photon bubbles in the laboratory.

2.1. *The Radiation Pressure Dominant Acceleration Regime*

Different regimes of plasma ion acceleration have been discussed in the literature[b]. A critical factor for a number of applications is the efficiency of the energy conversion. In the Radiation Pressure Dominant Acceleration (RPDA) the ion acceleration in a plasma is directly due to the radiation pressure of the electromagnetic pulse.[14,15] In this regime, the ions move forward with almost the same velocity as the electrons and thus have a kinetic energy well above that of the electrons. In contrast to the other regimes, this acceleration process is highly efficient and the ion energy per nucleon is proportional to the electromagnetic pulse energy.

This acceleration mechanism can be illustrated by considering a thin, dense plasma foil, made of electrons and protons, pushed by an ultra intense laser pulse in conditions where the radiation cannot propagate through the foil, while the electron and the proton layers move together and can be regarded as forming a (perfectly reflecting) relativistic plasma mirror co-propagating with the laser pulse. The frequency of the reflected electromagnetic wave is reduced by

$$\frac{(1-v/c)}{(1+v/c)} \approx \frac{1}{4\gamma^2}, \qquad \text{with} \quad \gamma = \left(1 - v^2/c^2\right)^{-1/2} \tag{1}$$

[b]See e.g., the following recent conference review papers ([27,28]).

and v the mirror velocity. Thus the plasma mirror is accelerated and acquires from the laser the energy $[1 - 1/(4\gamma^2)]\mathcal{E}$, where $\mathcal{E}$ is the incident laser pulse energy in the laboratory frame. For large values of γ practically all the electromagnetic pulse energy is transferred to the mirror, essentially in the form of proton kinetic energy. This high efficiency of the electromagnetic energy conversion into the fast protons opens up a wide range of applications[29] . For example it can be exploited in the design of proton dump facilities for spallation sources or for the production of large fluxes of neutrinos[30] .

3. Rayleigh-Taylor instabilities

Both in the astrophysical and in the laser plasma contexts, the onset of Rayleigh-Taylor-like instabilities[c] may affect the interaction of the plasma with the radiation pressure. In this case the electromagnetic radiation may eventually dig through the plasma and make it porous to the radiation (and allow e.g., for super-Eddington luminosities) or, in the case of a plasma foil accelerated by a laser pulse, may tear it into clumps[31] and broaden the energy spectrum of the fast ions.

In the present article we will discuss the stability of a plasma foil in the ultra relativistic conditions that are of interest for the RPDA regime along the line of Ref.([32]). We shall show that in the relativistic regime the growth of the instability is slower than in the nonrelativistic regime and that by proper tailoring of the pulse amplitude can allow for stable foil acceleration. As mentioned in the accompanying article (see footnote (c)), the foil model is based on a number of simplifications that are justifiable in the long wavelength approximation but that may turn out to be invalid in the nonlinear phase of the instability. In particular the description of the interaction between the electromagnetic field in the laser pulse and the plasma foil simply in terms of the action of the radiation pressure, which neglects light polarization and diffraction effects, may become inaccurate when spatial features of the order of the radiation wavelength start to form on the foil. Numerical simulations based on the Particle in cell (PIC) method allow us to overcome these restrictions and to follow the nonlinear development of the instability and the breaking of the plasma foil. The results obtained with the help of two-dimensional (2D) PIC simulations, confirm the analytical scaling of the instability growth rate with the laser pulse intensity and ion mass and

[c]See e.g the article in this same issue: Pegoraro F., Bulanov S.V., *General Properties of the Rayleigh-Taylor Instability in different plasma configurations; the plasma foil model.*

in addition show that the nonlinear development of the instability leads to the formation of high-density, high-energy plasma clumps.

3.1. *Radiation Pressure Acceleration of a Thin foil Mirror*

The equation of motion of an element of area $|d\Sigma|$ of a perfectly reflecting mirror can be written in the laboratory frame as (see accompanying article, footnote (c), and Refs.([33–35]))

$$d\mathbf{p}/dt = P d\mathbf{\Sigma}, \tag{2}$$

where $\mathbf{p}$ is the momentum of the mirror element, $d\mathbf{\Sigma}$ is normal to the mirror surface and P is the Lorentz invariant radiation pressure.[36,37] For the sake of geometrical simplicity, we refer to a 2D configuration where the mirror velocity and the mirror normal vector remain coplanar (in the x-y plane, x being the direction of propagation of the electromagnetic pulse), i.e., we assume that the mirror does not move or bend along z.

The radiation pressure P is given in terms of the amplitude of the electric field E_M of the incident electromagnetic pulse and of the pulse incidence angle θ_M in the comoving frame by

$$P = (E_M^2/2\pi)\cos^2\theta_M, \tag{3}$$

$$E_M^2 = (\omega_M^2/\omega_0^2)\, E_0^2, \quad \text{with} \quad \omega_M^2/\omega_0^2 = (1-\beta\cos\phi)^2/(1-\beta^2), \tag{4}$$

where the subscript 0 denotes quantities in the laboratory frame and ϕ the angle the mirror velocity β makes with the x-axis in the laboratory frame. The angle θ_M vanishes when the incidence angle θ_0 in the laboratory frame vanishes and $\phi = 0, \pi$, but is a fast increasing function of γ for $\theta_0 \neq 0$, or $\phi \neq 0, \pi$: indeed kinematic considerations show that the laser pulse can no longer reach the receding mirror when $|\sin\phi| > 1/\gamma$. This inequality constrains the maximum value of γ that can be obtained with a non perfectly collimated beam, see also Ref.([38]).

Then, the equation of motion of a mirror element of unit length along z and uniform density n_0 in the laboratory frame is

$$\frac{\partial p_x}{\partial t} = \frac{P}{n_0 l_0}\frac{\partial y}{\partial s}, \qquad \frac{\partial p_y}{\partial t} = -\frac{P}{n_0 l_0}\frac{\partial x}{\partial s} \tag{5}$$

with

$$\frac{\partial x}{\partial t} = \beta_x c, \quad \frac{\partial y}{\partial t} = \beta_y c \quad \text{and} \quad \frac{p_{x,y}}{m_i c} = \frac{\beta_{x,y}}{(1-\beta^2)^{1/2}}. \tag{6}$$

Here $p_{x,y}$ are the spatial components of the momentum 4-vector of the mirror element, l_0 is the mirror thickness and m_i is the ion mass. Lagrangian coordinates, x_0 and y_0, have been adopted such that $x, y = x, y(x_0, y_0, t)$ and $ds = (dx_0^2 + dy_0^2)^{1/2}$.

In the non relativistic limit and for constant P, Eqs.(5) coincide with Ott's equations[33] for the motion of a thin foil. Note that in the present case "relativistic nonlinearities" appear both in the relationship between the foil momentum and velocity and in the dependence of the radiation pressure on the foil momentum due to the Doppler effect in Eq.(4).

3.1.1. *One dimensional acceleration and the phase variable*

Assuming that the unperturbed mirror moves along the x-axis, i.e. that the initial conditions correspond to a flat mirror along y_0, so that $dx_0 = 0$, $dy_0 = ds$ and $\theta_M = 0$, we write Eqs.(5) as[d]

$$\frac{dp_x^0}{dt} = \frac{E_0^2}{2\pi n_0 l_0} \frac{\gamma_0 - p_x^0/(m_i c)}{\gamma_0 + p_x^0/(m_i c)}, \tag{7}$$

where p_x^0 is the unperturbed x component of momentum and depends on the variable t only and $\gamma_0^2 = 1 + \left[p_x^0/(m_i c)\right]^2$.

In the general case the amplitude of the electric field in the electromagnetic pulse is not constant in time and, at the mirror position $x(t)$, it depends on time through the combination $E_0 = E_0[t - x(t)/c]$ which introduces an additional nonlinearity into the foil equations. It is then convenient to introduce the phase of the wave $\psi = \omega_0[t - x^0(t)/c]$, at the unperturbed mirror position $x^0(t)$ as a new independent variable. Differentiating with respect to time, we obtain

$$\frac{d\psi}{dt} = \omega_0 \frac{\gamma_0 - p_x^0/(m_i c)}{\gamma_0}. \tag{8}$$

Using the variable ψ and the normalized fluence of the electromagnetic pulse

$$w(\psi) = \int_0^{\psi} (R(\psi')/\lambda_0)\, d\psi', \quad \text{with} \quad R(\psi) = E_0^2(\psi)/(m_i n_0 l_0 \omega_0^2), \tag{9}$$

$R(\psi)$ has the dimensions of a length and $\lambda_0 = 2\pi c/\omega_0$, and choosing $p_x^0(0) = 0$ as initial condition, the solution of Eq.(7) is

$$\frac{p_x^0(\psi)}{m_i c} = \frac{w(\psi)[w(\psi)+2]}{2[w(\psi)+1]}, \tag{10}$$

---

[d]Eq.(7) has been analyzed in detail in Refs.([14,15]).

while from Eq.(8) we obtain that t and ψ are related by

$$\psi + \int_0^{\psi} \left[w(\psi') + w^2(\psi')/2\right] d\psi' = \omega_0 t. \tag{11}$$

For a constant amplitude electromagnetic pulse, i.e., for $R = R_0$, Eqs.(9-11) reduce to

$$w(\psi) = (R_0/\lambda_0)\,\psi, \qquad \psi + (R_0/\lambda_0)\,\psi^2/2 + (R_0/\lambda_0)^2\,\psi^3/6 = \omega_0 t, \tag{12}$$

with

$$p_x^0/(m_i c) \approx (R_0/\lambda_0)\,\omega_0 t \tag{13}$$

for $t \ll \omega_0^{-1}\,(\lambda_0/R_0)$ and, for $t \gg \omega_0^{-1}\,(\lambda_0/R_0)$,

$$p_x^0/(m_i c) \approx (3R_0\omega_0 t/2\lambda_0)^{1/3}. \tag{14}$$

3.1.2. *The instability of the accelerated foil in the relativistic regime*

We shall now investigate the linear stability of the accelerated mirror with respect to perturbations $x^1(y_0, \psi)$, $y^1(y_0, \psi)$ that bend the plasma foil. Linearizing Eqs.(5) around the solution given by Eq.(10) we obtain

$$\frac{\partial}{\partial\psi}\left[\frac{p_x^0(\psi)}{m_i c}\frac{\partial x^1}{\partial\psi}\right] = \frac{R(\psi)}{2\pi}\frac{\partial y^1}{\partial y_0}, \tag{15}$$

$$\frac{\partial}{\partial\psi}\left[\frac{m_i c}{p_x^0(\psi)}\frac{\partial y^1}{\partial\psi}\right] = -\frac{R(\psi)}{2\pi}\frac{\partial x^1}{\partial y_0}. \tag{16}$$

Here we retain only leading terms in the ultrarelativistic limit $p_x^0/m_i c \gg 1$ for the foil motion and neglect a term proportional to $(\partial R/\partial\psi)\,x^1/\lambda_0$. To gain insight in the development of the foil instability it is convenient to consider WKB solutions of the form

$$y^1(y_0, \psi) \propto \exp\left[\int_0^{\psi} \Gamma(\psi')d\psi' - iky_0\right], \tag{17}$$

with (dimensionless) growth rate $\Gamma \gg 1$. We find

$$\Gamma(\psi) = \left[kR(\psi)/2\pi\right]^{1/2}, \quad \text{with} \quad x^1 \sim -iy^1(m_i c/p_x^0). \tag{18}$$

Note the relativistic contraction of the displacement along x, as compared to the displacement along y, in the laboratory frame . For a constant amplitude pulse, using Eq.(12), we obtain

$$y^1 \propto \exp\left[(t/\tau_r)^{1/3} - iky_0\right], \tag{19}$$

where

$$\tau_r = \omega_0^{-1}(2\pi)^{3/2}R_0^{1/2}/(6k^{3/2}\lambda_0^2) \tag{20}$$

is the characteristic time of the instability in the ultrarelativistic limit. Note that the dependence of the dimensionless growth rate Γ, which expresses the growth of the instability as a function ot the phase variable ψ and is proportional to the square root of R, is reversed when the growth rate is re-expressed in terms of the time variable t in the laboratory frame. Indeed, Eq.(20) shows that the time scale of the instability is proportional to the square root of the ratio between the radiation pressure and the ion mass. Thus, the larger the ion mass, the faster the perturbation grows while the larger the radiation pressure the slower the perturbation.
In the non-relativistic limit instead the perturbation grows as

$$x^1, y^1 \;\propto\; \exp\left[t/\tau - iky_0\right], \tag{21}$$

where

$$\tau = \omega_0^{-1}(2\pi/kR_0)^{1/2}. \tag{22}$$

We see that there are two major differences between the relativistic and the nonrelativistic regimes: in the ultrarelativistic limit the instability develops more slowly with time than in the non-relativistic case: $t^{1/3}$ instead of t and in addition the ultrarelativistic instability time scale τ is proportional (instead of inversely proportional) to the square root of the radiation pressure.

3.2. *Stabilization with tailored EM pulses*

We can also describe the development of the instability using, instead of either ψ or t, the unperturbed momentum p_x^0 as the independent variable. If we express Eqs.(19,21) in terms of the unperturbed momentum p_x^0, in both limits we find an exponential growth of the form

$$y^1(y_0, p_x^0) \propto \exp\left[\kappa p_x^0/(m_i c) - iky_0\right], \tag{23}$$

where

$$\kappa = (k\lambda_0)^{1/2}/(2\pi R_0/\lambda_0)^{1/2}. \tag{24}$$

This exponential growth of the perturbation with the unperturbed momentum for a constant amplitude pulse can be stopped by tailoring the shape of the electromagnetic pulse. We refer to the ultrarelativistic limit and define

$\Phi(\psi) = \int_0^\psi \Gamma(\psi')d\psi'$. Then we see that the different scalings of p_x^0, which in this limit is linear in $R(\psi)$:

$$p_x^0(\psi) = m_i c \int_0^\psi (R(\psi')/2\lambda_0)\, d\psi', \tag{25}$$

and of Γ, which is proportional to the square root of $R(\psi)$, make, at fixed p_x^0, a short intense pulse comparatively less affected by the instability.
As in a number of other cases where the Rayleigh-Taylor instability cannot be fully avoided but its effects can be effectively limited, in this sense an appropriate stability condition can be formulated as follows: it is possible to choose the dependence of the electromagnetic pressure $R(\psi)$ on the phase ψ such that, for ψ approaching a limiting value ψ_m, the ion momentum $p_x^0(\psi)$ grows, formally to infinity, while $\Phi(\psi)$ remains finite. As an illustration we can take $R(\psi)$ of the form

$$R(\psi) = R_0(1 - \psi/\psi_m)^{-\alpha}\ \chi(\psi_1 - \psi), \tag{26}$$

with $1 < \alpha < 2$, $\quad \chi(x) = 1$ for $x > 0$, and $\chi(x) = 0$ for $x < 0$ and $\psi_m > \psi_1$ so as to keep the pulse fluence finite. In this case the maximum value of the ion momentum

$$p_x^0/(m_i c) \approx R_0\psi_m(1 - \psi_1/\psi_m)^{1-\alpha}/[2\lambda_0(\alpha - 1)], \tag{27}$$

tends to infinity for $\psi_1 \to \psi_m$, while

$$\Phi(\psi_m) = (2kR_0/\pi)^{1/2}\psi_m[1 - (1 - \psi_1/\psi_m)^{1-\alpha/2}]/(2 - \alpha) \tag{28}$$

remains finite. In addition, for $\alpha < 3/2$ the acceleration time is finite.

4. Particle in Cell Simulations

In the last part of this article we shall briefly recall some numerical results presented in Refs.([14,15]) and in particular in Ref.([32]). The aim here is simply to illustrate how Particle in Cell (PIC) simulations can be used to confirm the analytical models adopted, such as e.g., the use of the foil model and the restriction to long wavelength perturbation, and, most importantly, to extend these results beyond the scope of these models.

First we refer to the PIC simulations presented in Figs.(1,2) of Ref.([14]) for an electron-proton plasma foil. Indeed these simulations show a stable phase of the RPDA regime where a portion of the foil, with the size of the pulse focal spot, is pushed forward by a super-intense electromagnetic pulse. The wavelength of the reflected radiation is substantially larger than that of the incident pulse, as consistent with the light reflection from the co-moving relativistic mirror with the electromagnetic energy transformation into the kinetic energy of the plasma foil. At this stage, the initially planar plasma slab is transformed into a "cocoon" that almost confines the laser pulse. The protons at the front of the cocoon are accelerated up to energies in the multi-GeV range at a rate that agrees with the $t^{1/3}$ scaling predicted by the analytical model. The proton energy spectrum, see Fig.(3) of Ref.([14]), has a narrow feature corresponding to a quasi-monoenergetic beam, but part of it extends over a larger energy interval.

In order to investigate the onset and the nonlinear evolution of the instability of the foil in Ref.([32]) a series of 2-D numerical simulations has been performed with heavier ions (aiming at a faster growth rate as follows from the scaling of the instability time in the ultrarelativistic regime given by Eq.(20)). The detailed list of the parameters of the simulations and of the code used can be found in Ref.([32]). Here we recall that a thin plasma slab, of width 20λ and thickness 0.5λ, is considered. The plasma is made of fully ionized aluminum ions with $Z = 13$, the ion to electron mass ratio is 26.98×1836. The electron density is equal to $64n_{cr}$, with n_{cr} the so-called critical density. An s-polarized laser pulse with electric field along the z-axis (i.e., perpendicular to the simulation plane) is initialized at the left-hand side of the plasma slab. The pulse has a "gaussian" envelope given by $a_0 \exp(-x^2/2l_x^2 - y^2/2l_y^2)$, with $l_x = 40\lambda$, $l_y = 20\lambda$. The dimensionless pulse amplitude, $a_0 = eE_0/(m_e c\omega_0) = 320$, corresponds for $\lambda = 1\mu m$ to the intensity $I = 1.37 \times 10^{23}\ W/cm^2$.
The results of these simulations are shown in the following four figures where the ion density distribution in the x-y plane, (Fig.1), the distribution of the electric field E_z in the x-y plane, (Fig.2), the ion phase plane (p_x, x), (Fig.3), and the energy spectrum of the ions (Fig.4), are shown at $t = 75,\ 87.5$.

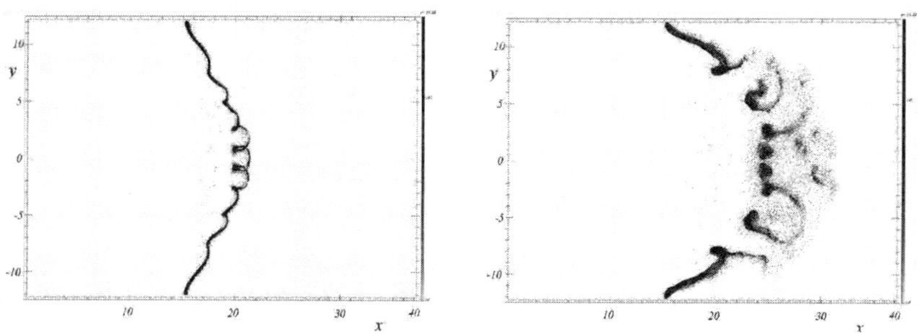

Fig. 1. Aluminum ion density distribution at $t = 75$ (left) and at $t = 87.5$ (right).

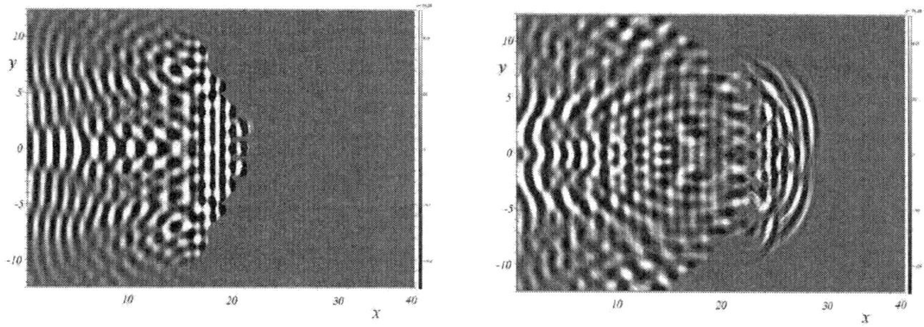

Fig. 2. Distribution of the electric field E_z at $t = 75$ (left) and at $t = 87.5$ (right).

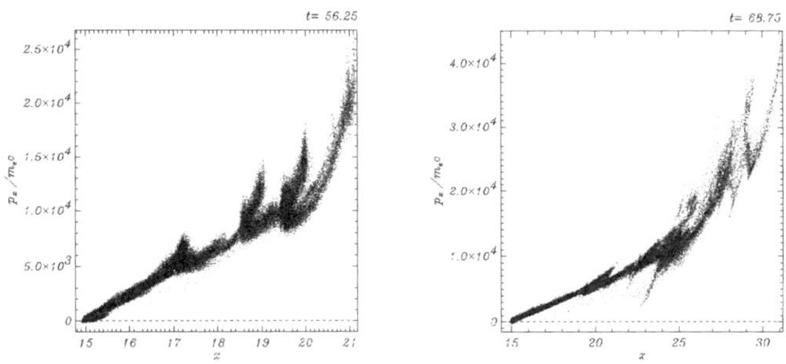

Fig. 3. Aluminum ion phase plane (p_x, x) at $t = 75$ (left) and at $t = 87.5$ (right).

The ion momentum and kinetic energy are given in GeV/c and in GeV, respectively. The wavelength λ of the incident radiation and its period $2\pi/\omega_0$ are chosen as units of length and time.

In the left frames, at $t = 75$, we see the typical initial stage of the Rayleigh-Taylor instability with the formation of cusps and of multiple-bubbles in the plasma density distribution. These are accompanied by a modulation of the electromagnetic pulse at its front. At this time the ions are accelerated forward, as seen in their phase plane. Their energy spectrum is made up of quasi-monoenergetic beamlets which correspond to the cusp regions, and of a relatively high energy tail which is formed by the ions at the front of the bubbles. In the right frames at $t = 87.5$ we see that the fully nonlinear stage of the instability results in the formation of several clumps in the ion density distribution with more diffuse, lower density plasma clouds

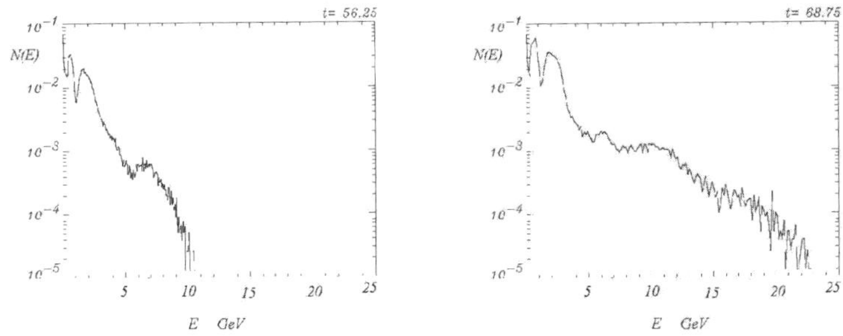

Fig. 4. Aluminum ion energy spectrum at $t = 75$ (left) and at $t = 87.5$ (right).

between them. The electromagnetic wave partially penetrates through, and partially is scattered by, the clump-plasma layer. The high energy tail in the ion spectrum grows much faster than in the stable case. At later times, because of the mass reduction of the diffuse clouds at the front of the pulse, the maximum ion energy scales linearly with time. The local maxima at relatively lower energy correspond to the plasma clumps.

5. Conclusions

In the relativistic regime the Rayleigh-Taylor instability of a plasma foil accelerated by the radiation pressure of the reflected electromagnetic pulse develops much more slowly than in the non-relativistic regime. In the former limit its timescale is inversely proportional to the square root of the ratio between the radiation pressure and the ion mass while in the latter this dependence is reversed. The use of a properly tailored electromagnetic pulse with a steep intensity rise can stabilize the shell acceleration.

Numerical simulations show that the nonlinear development of the instability leads to the formation of high-density, high-energy plasma clumps and to a relatively higher rate of ion acceleration in the regions between the clumps.

References

1. Lebedev, P. N., *Ann. Phys.*. (Leipzig) **6**, 433 (1901);
2. Eddington A.S., *MNRAS* **85**, 408 (1925).
3. Milne E.A., *MNRAS* **86**, 459 (1926).
4. Chandrasekhar S., *MNRAS*, **94**, 522 (1934).
5. Shaviv N.J., *ApJ*, **532**, L137 (2000).
6. Cohadon, P.F., *et al.*, *Phys. Rev. Lett.*, **83**, 3174 (1999).

7. Ashkin A., *Phys. Rev. Lett.*, **24**, 156 (1970).
8. Arons J., *ApJ*, **388**, 561 (1992).
9. Gammie C.F., *MNRAS*, **297**, 929 (1998);
10. Begelman M.C., *ApJ*, **551**, 897 (2001).
11. Goldreich P. , *Phys. Scripta*, **17**, 225 (1978).
12. Piran T., *ApJ*, **257**, L23 (1982).
13. Berezinskii V. S., *et al.*, in "*Astrophysics of Cosmic Rays*", (Elsevier, Amsterdam, 1990).
14. Esirkepov T.Zh., *et al.*, *Phys. Rev. Lett.*, **92**, 175003 (2004).
15. Bulanov S.V., *et al.*, *Plasma Phys. Rep.*, **30**, 196 (2004).
16. Pegoraro F., *et al.*, *Phys. Lett. A*, **347**, 133 (2005).
17. Yu W., *et al.*, *Phys. Rev. E*, **72**, 046401 (2005).
18. Macchi A., *et al.*, *Phys. Rev. Lett.*, **94**, 165003, (2005).
19. Badziak J., *et al.*, *Appl. Phys. Lett.*, **89**, 061504, (2006).
20. V.I. Veksler, in "*Proc. CERN Symposium on High Energy Accelerators and Pion Physics*", Geneva, **1**, 80 (1956).
21. Bulanov S.V. , *et al.*, in *Reviews of Plasma Physics*, ed. V.D. Shafranov, **22**, 227 (Kluwer Acad., N.Y., 2001).
22. Mourou G.A., *et al.*, *Rev. Mod. Phys.* **78**, 309 (2006).
23. Borghesi M., *et al.*, *Fus. Sc. & Techn.*, **49**, 412 (2006).
24. Hegelich B.M., *et al.*, *Nature*, **439**, 441 (2006).
25. Schwoerer H., *et al.*, *Nature* **439**, 445 (2006).
26. Willingale L., *et al.*, *Phys. Rev. Lett.*, **96**, 245002 (2006) and references therein.
27. Bulanov S. V., in "*Super strong electromagnetic fields and their applications*", AIP Conf. Series, **920**, 3 (2007).
28. Mora,P., in "*Super strong electromagnetic fields and their applications*", AIP Conf. Series, **920**, 98 (2007).
29. Borghesi M., *et al.*, *J.Phys. Conf, Ser.*, **58**, 74 (2007).
30. Bulanov S.V, *et al.*, *Nucl. Instr. Meth. A*, **540**, 25 (2005).
31. Kifonidis K., *et al.*, *Astron. Astrophys.*, **408**, 621 (2003).
32. Pegoraro F., Bulanov S.V., *Phys. Rev. Lett.*, **99**, 065002 (2007).
33. Ott E., *Phys. Rev. Lett*, **29**, 1429 (1972).
34. Bulanov S.V., *et al.*, *Phys. Rev.*, **E 59**, 2292 (1999).
35. Pegoraro F., *et al.*, *Phys. Rev.*, **E 64**, 016415 (2001).
36. Landau L.D., Lifshitz E.M., in "*The Classical Theory of Fields*", Pergamon Press, Oxford, (1980).
37. Pauli W., in "*Theory of Relativity* ", Dover, New York, (1981).
38. Phinney E.S., *MNRAS, **198**, 1109 (1982)*..

GENERATION OF GALACTIC SEED MAGNETIC FIELDS

H. Saleem

Theoretical Plasma Physics Division (TPPD),
PINSTECH, P. O. Nilore, Islamabad, Pakistan.

A theoretical model for the generation of 'seed' magnetic field and plasma flow on galactic scales driven by externally given baro-clinic vectors is presented. The incompressible plasma fields can grow from zero values at initial time $t = 0$ from a non-equilibrium state $T_e \neq T_i$ (where $T_e(T_i)$ are electron(ion) temperatures, respectively) due to pressure gradients. An exact analytical solution of the set of two fluid equations is obtained which is valid for both small and large plasma β-values. The magnetic field generated by this mechanism has three dimensional structure. Weaknesses of previous single fluid models for seed magnetic field generation are also pointed out. The estimate of the magnitude of the galactic seed magnetic field turns out to be $10^{-15}G$ and may vary depending upon the scale lengths of the density and temperature gradients. The seed magnetic field may be amplified later by $\alpha\omega$ -dynamo (or by some other mechanism) to the present observed values of $\sim (2 - 10)\mu G$. The theory has been applied to laser-induced plasmas as well and the estimate of the magnetic field's magnitude is in agreement with the experimentally observed values.

1. Introduction

In spite of a great deal of research work in this direction there is still not a convincing theory for the generation of seed magnetic field on cosmological and laboratory scales [1]. Extensive literature has appeared on galactic and intergalactic magnetic field [2–5]. The simplest justification can be that the generation of magnetic fields was a feature of initial conditions of the universe. However a more appealing hypothesis is that they are created by the physical mechanism operating after the Big Bang. The magnetic fields have been observed in almost all astronomical environments like the galaxies, intergalactic space, active galactic nuclei (AGN), galaxy clusters etc.

Most of the studies of magnetic field generation are based on single fluid magnetohydrodynamics (MHD). This model is very useful for the study of large scale magnetic phenomena. Therefore the MHD-based theoretical models and numerical simulations are very helpful to study the astrophysical magnetic fields [6–8]. In principal, the dynamo paradigm is incomplete because it is unable to explain the creation of initial magnetic field, the seed [1]. It is possible that the magnetic field is highly amplified by the $\alpha\omega$ -dynamo effect later, but there must be some seed field already present for this action.

The magnetic fields of magnitudes of the order of 2-10 μG have been observed in many galaxies [9,10]. The galactic magnetic fields have components parallel to the galactic disk planes as well as along the vertical directions [5,11,12]. Biermann battery effect [13] is the most widely studied mechanism for the generation of seed magnetic fields. Several modifications to the basic Biermann model and computer simulations have also been presented [14–17]. More than a decade ago, it was shown that the seed magnetic field on galactic scale can be produced by the electron Biermann-type diffusion processes [18]. Unlike the original Biermann process these mechanisms do not require rotation of the system.

The strong magnetic fields produced in laser-induced plasmas [19,20] strengthen the argument that plasma dynamics can create large magnetic fields. Most of the theoretical models to explain the creation of magnetic fields in laser plasma systems are also based on electron baroclinic term ($\nabla T_e \times \nabla n_e$) (where n_e and T_e are the electron density and temperature, respectively). In these works, the ions are assumed to be stationary and electrons are treated to be inertia-less. The equations of the single fluid electron magnetohydrodynamics (EMHD) are used in these models [21,22]. The weaknesses and contradictions of EMHD theory have also been discussed in a few research papers [23] with reference to magnetic field generation.Therefore we note that the seed magnetic field generation has not been explained in a true sense by MHD and EMHD models. The ideal MHD equations conserve the magnetic flux as "circulation", so that they cannot explain the generation of magnetic field. A theory for the self-excitation of transverse electromagnetic waves due to anisotropy of electron velocity distribution has also been presented many decades ago [24].

The magnetic fields in galaxies have coherent structures of the scales of tens of kilo parsec (kpc) winding around the galaxy which is larger than the seed from any given star or stellar binary system. These regular structures have superimposed on them shorter scale irregular structures as well [25]. These fields are believed to have been growing with the evolution of the universe during times of the order of billions of years$\sim 10^9 yrs$.

On the other hand, magnetic fields have been observed in laser-induced plasma experiments of the order of mega Gauss. In early experiments [19] the growth time of the field was $\tau \sim 10^{-9}S$ and the spatial scale size was of the order of fuel pallet diameter ($\sim \mu m$). In later experiments with intense lasers of short pulse duration $\tau \sim 10^{-12}S$, the growth times were correspondingly much shorter than a nanosecond. The recent experiments use very intense laser beams and produced plasmas are relativistic and degenerate. Such systems are not under consideration here. The classical laser-plasma dynamics are also discussed because they have a similarity with the galactic magnetic field problem, in our opinion. The previous theoretical models based on EMHD assume ions to be static and electrons to be inertial-less. These assumptions need to be analyzed very carefully. First we note that to assume ions to be static and electrons to be inertia-less (to ignore displacement current in Maxwell's equation) one needs the limits (i) $\omega_{pi} \ll |\partial_t| \ll \omega_{pe}$(where $\omega_{pj} = \left(\frac{4\pi n_0 e^2}{m_j}\right)^{\frac{1}{2}}$ are the plasma oscillation frequency of the jth species, and $j = e, i$) and (ii) $|\partial_t| \ll c|\nabla|$. The assumption of inertia-less electrons also helps in using the steady state equation of motion for electrons. Note that for hydrogen plasma $\frac{\omega_{pi}}{\omega_{pe}} \sim \frac{1}{43}$. If we define a smallness parameter $\epsilon \sim O\left(\frac{1}{6.5}\right)$ or $\epsilon \sim \frac{1}{7}$, only then the limit (i) can be applicable and we can work on a time scale where ions are static and electrons are inertia-less. But even for this narrow window, the length of time say τ, for the generation of field may not satisfy the condition (i) easily. Secondly as soon as the magnetic field is created, the nonlinear Lorentz force term $\mathbf{v}_e \times \mathbf{B}$ comes into play and it's role must be taken into account.

In laser induced plasmas, it is assumed that the magnetic field is produced in time τ such that $\omega_{pe}^{-1} << \tau << \omega_{pi}^{-1}$ and hence $m_e \rightarrow 0$ and $m_i \rightarrow \infty$. The electron equation of motion is written as,

$$0 = -en_e\mathbf{E} - \nabla p_e$$

The electron flow is ignored and $B = 0$ at $t = 0$ is assumed. The back reaction of the field on electron motion at times $0 < t$ is assumed to be negligible (which is not justified). The curl of the above equation gives,

$$\nabla \times \mathbf{E} = -\frac{1}{e}\nabla \times \left(\frac{\nabla p_e}{n_e}\right)$$

Then using Faraday's law one obtains,

$$\partial_t \mathbf{B} = \frac{c}{e}\left(\nabla T_e \times \nabla \ln n_e\right)$$

If the vectors ∇T_e and ∇n_e are assumed to be in xy-plane with $\frac{\nabla T_e}{T_e} = \kappa_T \mathbf{y}$ = constant and $\frac{\nabla n_e}{n_e} = \kappa_n \mathbf{x}$ =constant,then $\mathbf{B}$ is along z-axis as,

$$\partial_t \mathbf{B} = \left(\frac{c}{e}\frac{T_e}{L_n L_T}\right)\mathbf{z}$$

where $L_n = \kappa_n^{-1}$ and $L_T = \kappa_T^{-1}$ are density and temperature gradient scale lengths, respectively. Here $\mathbf{x}.\mathbf{y}$ and $\mathbf{z}$ denote unit vectors. Integrating the above equation from 0 to τ, one gets,

$$B = \frac{c}{e}\frac{T_e}{L_n L_T}\tau$$

Let $c_s = \left(\frac{T_e}{m_i}\right)^{\frac{1}{2}}$ be the ion sound speed and assume $\tau = \frac{L_n}{c_s}$. Then for laser plasmas with $T_e \sim 1keV$, $n_0 \sim 10^{20}cm^{-3}$, $c_s \sim 3 \times 10^7 cms^{-1}$ and $L_T \sim 0.005cm$, one obtains [20] $|B| \sim \frac{cT_e}{ec_s L_T} = 0.64 \times 10^6 G$.

There are many contradictions in this model. For example $\mathbf{B}$ grows with time and hence it's back reaction on the electron fluid after initial times $0 < t$ does not remain negligible. It is also important to note that the time $\tau = \frac{L_n}{c_s}$ is the ion time scale, therefore to consider ions to be static is not reasonable. Moreover in laser-induced dense plasmas $\omega_{pi} \sim 10^{12} - 10^{13} radS^{-1}$ and pulse duration (specially in the early experiments) was of the order $\sim 10^{-9}S$. Even in later experiments $\tau \leq 10^{-12}S$ was the case while $\omega_{pi}^{-1} \leq 10^{-12}S^{-1}$.

Therefore ions should not be treated as static [23]. On galactic scales, the time is so long ($\sim 10^9$ years) and density gradient length is so large ($\sim$ tens of pc) that electrons seem to be in equilibrium and their motions on λ_e scale is irrelevant. In this situation only the electron current can not be responsible for magnetic field generation. If $\partial_t \mathbf{B} \neq 0$ due to ($\nabla T_e \times \nabla n_e \neq 0$), then ion dynamics due to ($\nabla T_i \times \nabla n_i$)-term should also be taken into account. Even if $n_e \sim n_i = n$ is assumed, the condition $T_i \neq T_e$ should hold otherwise plasma is in thermal equilibrium and significant currents can not exist. Moreover, the magnetic field can not be time-dependent if plasma is in equilibrium. Hence in equilibrium $\partial_t \mathbf{B} = 0$ as well as $\nabla T_j \times \nabla n = 0, (j = e, i)$ should hold.

The above discussion shows that Biermann battery formalism is very restrictive and can be valid only on a very limited time scale which may not be useful for finding a general mechanism for the creation of seed magnetic fields.

A few years ago, a theory for the generation of magnetic field and plasma flow based on two-fluid equations has been presented [26]. In this work it is assumed that at time $t = 0$, the plasma has $\nabla n \times \nabla T_j \neq 0$ (where $n_e \sim n_i \simeq n$ and T_j denote the temperature of jth species). The plasma evolves on a slow time scale due to given form of baro-clinic vectors ($\nabla \psi \times \nabla T_j$) where $\psi = \ln n$. A particular solution of the two-fluid equations is obtained and it is shown that incompressible field and flow can be generated by the external gradients when the plasma is in a non-equilibrium state with $T_e \neq T_i$. The terms ($\nabla \psi \times \nabla T_j$) are considered to be the functions of only (x, y) coordinates and all the plasma fields are assumed to have the similar form as that of the source terms. The self-consistent fields are separated into growing and ambient (static) parts, and each part has different geometric character- the toroidal magnetic field and poloidal flow grow simultaneously, while the poloidal magnetic field and toroidal flow are static. Then the instabilities of some ambient inhomogeneous plasma (the stationary fields in the model) may create magnetic fields and flows with specific forms due to the baroclinic terms. A particular form of the baroclinic vectors is chosen such that all the nonlinear terms vanish and we succeed to obtain two linear equations which have an exact solution.

However, in this theoretical model there is a weakness that poloidal components of magnetic field and toroidal component of plasma flow are assumed to be time-independent. Therefore some static non-zero magnetic field and flow field have to prevail in the system.

In the present investigation [27], we modify the previous theory [26] to explain the creation of all the components of seed magnetic field and plasma flow from $t = 0$ due to externally given forms of baroclinic vectors ($\nabla n \times \nabla T_j \neq 0$). This theoretical model is applicable to very large plasma β-values as well. Furthermore the spatial and temporal scales of two fluid and one fluid plasma models will be discussed briefly in the context of magnetic field generation and the problems associated with these scales will be pointed out. It seems necessary to mention here that the inclusion of electron pressure term in the ideal MHD set of equations changes the scenario.

It is important to note that the present theory is different from Biermann battery concept and dynamo mechanism in the sense that it takes into account the dynamics of both ions and electrons. This work also points out that the Biermann battery is not necessary for the creation of magnetic field. Rather the field and plasma flow both are simultaneously created by the pressure gradients. However, the pioneering idea of Biermann is used to generate magnetic field and flow. That is the non-collinear density and temperature gradients are assumed to be the source of plasma evolution. The term $\nabla \times \left(\frac{\nabla p_e}{n_e}\right) \neq 0$ is one of the source terms for the generation of magnetic field and it modifies the scope of MHD and Hall magnetohydrodynamics (HMHD).

2. Mathematical model

We show that the set of nonlinear two-fluid partial differential equations (with inertia-less electrons) along with Maxwell's equations can have an exact analytical solution when the pressure gradients ∇p_e and ∇p_i have particular spatial structures. It is interesting to find out an analytical solution of a complicated system of equations to understand the basic physics of plasma evolution along with the creation of magnetic field and flow on the lines of Ref. [26].

The flows $\mathbf{v}_j$ and magnetic field $\mathbf{B}$ vectors will be defined in terms of a few scalar fields (ϕ, u, χ, h) which will be assumed to be correlated. These relationships will cause cancellations of all nonlinear terms in the set of equations. The density will be defined as $\ln n = \psi$, and it will become independent of time. All scalar fields (ϕ, u, χ, h) will be assumed to be functions of ψ. We shall choose particular spatial dependence of ∇n, ∇T_e and ∇T_i. Then all fields $\mathbf{v}_i$ and $\mathbf{B}$ will automatically become functions of externally given baroclinic vectors. We shall see that for such a choice of the form of gradients, the nonlinear equations for electrons and ions reduce to two linear equations and all the assumptions made on the way are satisfied. The flow $\mathbf{v}_i$ and magnetic field $\mathbf{B}$ will appear to grow from $t=0$ due to $(\nabla\psi \times \nabla T_j)$ vectors from a non-equilibrium state with $T_e \neq T_i$ and $n_e \simeq n_i = n$ The equation of motion for inertial-less electrons is,

$$0 \simeq -en_e \left(\mathbf{E} + \frac{1}{c}\mathbf{v}_e \times \mathbf{B}\right) - \nabla p_e \tag{1}$$

The ion momentum conservation gives,

$$m_i n_i \left(\partial_t + \mathbf{v}_i.\nabla\right)\mathbf{v}_i = en_i \left(\mathbf{E} + \frac{1}{c}\mathbf{v}_i \times \mathbf{B}\right) - \nabla p_i \tag{2}$$

The continuity equation for jth species (j = e,i) can be written as,

$$\partial_t n_j + \nabla.\left(n_j \mathbf{v}_j\right) = 0 \tag{3}$$

Ampere's law becomes,

$$\nabla \times \mathbf{B} = \frac{4\pi}{c}\mathbf{J} \tag{4}$$

under the approximation $|\partial_t| \ll \omega_{pe}, |c\nabla|$. We need Faraday's law as well i.e.

$$\nabla \times \mathbf{E} = -\frac{1}{c}\partial_t \mathbf{B} \tag{5}$$

Instead of Poisson equation, the quasi-neutrality ($n_e \sim n_i \sim n$) is used which defines the current as $\mathbf{J} = en(\mathbf{v}_i - \mathbf{v}_e)$ and it yields

$$\mathbf{v}_e = \mathbf{v}_i - \frac{\mathbf{J}}{en} \tag{6}$$

The equations of state are defined as,

$$p_j = n_j T_j \tag{7}$$

Let both electron and ion fluids be incompressible ($\nabla.\mathbf{v}_j = 0$) and densities be time-independent ($\partial_t n_j = 0$). Then equation(3) demands,

$$\nabla\psi.\mathbf{v}_j = 0 \tag{8}$$

Let us assume $\mathbf{E} = -\nabla\Phi - \frac{1}{c}\partial_t \mathbf{A}$. After taking curl of equation (1), we use Eqs. (5) and (6) to obtain,

$$\partial_t \mathbf{B} + \nabla \times \left[\mathbf{B} \times (\mathbf{v}_i - \frac{c}{4\pi e}\frac{\nabla \times \mathbf{B}}{n})\right] = \frac{c}{e}(\nabla T_e \times \nabla\psi) \tag{9}$$

Then using $\mathbf{v}.\nabla\mathbf{v} = \frac{1}{2}\nabla v^2 - \mathbf{v} \times (\nabla \times \mathbf{v})$ and taking curl of Eq.(2), we obtain,

$$\partial_t(a\mathbf{B} + \nabla \times \mathbf{v}_i) - \nabla \times [\mathbf{v}_i \times (a\mathbf{B} + \nabla \times \mathbf{v}_i)] = -\frac{1}{m_i}(\nabla T_i \times \nabla\psi) \tag{10}$$

where $a = \frac{e}{m_i c}$. Now we define the fields as,

$$\mathbf{v}_i = (\nabla\phi \times \mathbf{z} + u\mathbf{z})f(t) = (\partial_y\phi, -\partial_x\phi, u)f \tag{11}$$

$$\mathbf{B} = (\nabla\chi \times \mathbf{z} + h\mathbf{z})f(t) = (\partial_y\chi, -\partial_x\chi, h)f \tag{12}$$

where the scalar fields ϕ, u, χ and h are functions of x and y only. Note that ϕ is a scalar field different from electrostatic potential Φ . Equation (8) demands,

$$\{\phi, \psi\} = 0 \tag{13}$$

where $\{\phi, \psi\} = \partial_y\phi\partial_x\chi - \partial_x\phi\partial_y\chi$. Using definitions (11) and (12) one finds,

$$\nabla \times [\mathbf{v}_i \times (a\mathbf{B} + \nabla \times \mathbf{v}_i)]$$

$$= f^2 \left[-\partial_y\{\phi, (a\chi + u)\}, \partial_x\{\phi, (a\chi + u)\}, \{(ah - \nabla_\perp^2\phi), \phi\} + \{(a\chi + u), u\}\right] \tag{14}$$

If we assume

$$\{\phi, \chi\} = 0, \{\phi, u\} = 0, \{h, \phi\} = 0, \{\nabla^2\phi, \phi\} = 0 \tag{15}$$

then the nonlinear terms of Eq. (10) vanish, and it reduces to

$$\partial_t(a\mathbf{B} + \nabla \times \mathbf{v}_i) = -\frac{1}{m_i}(\nabla T_i \times \nabla\psi) \tag{16}$$

The condition $\{\nabla^2\phi, \phi\} = 0$ may be satisfied if,

$$\nabla_\perp^2\phi = -\lambda\phi \tag{17}$$

where λ is an arbitrary constant. Note that $\nabla \times (\mathbf{v}_e + \mathbf{B}) = -\nabla \times \left[\mathbf{B} \times \left(\mathbf{v}_i - b\frac{\nabla\times\mathbf{B}}{n}\right)\right]$ where $b = \frac{c}{4\pi e}$, and the above term also disappears because due to (15) the following conditions also hold,

$$\left\{\nabla^2\chi, \chi\right\} = 0; \{\psi, h\} = 0; \{\psi, \chi\} = 0 \tag{18}$$

Now we make another important assumption to simplify the problem that the generated plasma vorticity is parallel to **B**.

$$\mathbf{B} = g\,(\nabla \times \mathbf{v}_i) \tag{19}$$

where g is an arbitrary constant. Equation (19) along with (11) and (12) gives,

$$\chi = gu \tag{20}$$

and

$$h = -g\nabla^2\phi \tag{21}$$

Then the equations (9) and (10), respectively, yield

$$\partial_t \mathbf{B} = \frac{c}{e}\left(\nabla T_e \times \nabla\psi\right) \tag{22}$$

and

$$\left(a + g^{-1}\right)\partial_t \mathbf{B} = -\frac{1}{m_i}\left(\nabla T_i \times \nabla\psi\right) \tag{23}$$

where $a + g^{-1} \neq 0$ must hold.

Equations (22) and (23) are similar to Eqs. (17) and (19) of [26] if these are normalized in the same way and if we assume $\partial_t\chi = 0, \partial_t u = 0$. In this case the x and y components of the baroclinic vectors $(\nabla T_j \times \nabla\psi)$ must vanish.

If the Hall term and pressure term both are ignored, then curl of Eq.(1) yields the simplest form of Ohm's law as,

$$\partial_t \mathbf{B} = \nabla \times (\mathbf{v}_i \times \mathbf{B}) \tag{24}$$

The nonlinear term on the right-hand side vanishes due to the forms of $\mathbf{v}_i$ and $\mathbf{B}$ assumed in Eqs. (11) and (12), respectively. This means that the magnetic field generation mechanism is killed when ∇p_e-term is ignored in Ohm's law. If Hall-term is ignored, then ideal MHD equations are modified by the inclusion of ∇p_e-term in Ohm's law. Hence a source of magnetic field generation is added to the MHD equations. Since the nonlinear terms vanish, therefore the present theory is applicable to both MHD and HMHD scales. The HMHD spatial scale is the ion skin depth $\lambda_i = \frac{c}{\omega_{pi}}$ and MHD can be used even for larger scales. It will be shown later that the present model is applicable in the limit of infinitely large plasma β-values as well. The collisions have been ignored here just for simplicity.

It is important to note that we must have $\partial_t(\varphi, \chi, u, h) \neq 0$ so that all the components of magnetic field $\mathbf{B}$ and flow $\mathbf{v}_i$ evolve from $t = 0$. If the nonlinear terms of Eqs.(9) and (10) do not vanish, then an analytical solution of these equations is not easy to find. A numerical simulation is required in this case.

There is another important point to be noticed here. In equations (22) and (23), the scales of $\mathbf{B}$ (and correspondingly $\mathbf{v}_i$ through (19)) should be the same as that of $(\nabla T_j \times \nabla\psi)$ which are the driving terms for plasma evolution. The short scale nonlinear phenomena have disappeared automatically under the conditions (15) and (18). Then $\mathbf{B}$ is directly proportion to $(\nabla T_j \times \nabla\psi)$.

3. Comments on MHD and HMHD scales

A few comments on the scales of applicability of the present model seem to be necessary at this stage. In the previous work [26], the set of Eq.(1-7) was normalized using an arbitrary magnetic field $\mathbf{B}_0 \neq 0$ as was done in the study of double curl Beltrami steady state[28]. A few years ago [29], the analysis of two interesting scales of magneto-plasmas has been presented. In this work the Hall-term is treated as a singular perturbation in the two-fluid equations. A smallness parameter $\epsilon_i = \frac{\lambda_i}{L_0}$ is introduced where $\lambda_i = \frac{c}{\omega_{pi}}$ is the ion skin depth and L_0 is a characteristic length and may be equal to the size of the system. The HMHD has both the macroscopic and microscopic scales superimposed on each other. In the limit $\epsilon_i \to 0$, the system degenerates into standard MHD with a single relevant (the macroscopic) scale. In this work, variables have been normalized as follows, $\mathbf{x} = L_0\tilde{\mathbf{x}}, \mathbf{B} = B_0\tilde{\mathbf{B}}, n = n_0\tilde{n}, t = \left(\frac{L_0}{v_a}\right)\tilde{t}$, $p_j = \left(\frac{B_0^2}{4\pi}\right)\tilde{p}_j, \mathbf{v}_j = v_A\tilde{\mathbf{v}}, \mathbf{E} = \left(\frac{v_A B_0}{c}\right)\tilde{\mathbf{E}}$, where $v_A = \frac{B_0}{\sqrt{4\pi n_0 m_i}}$ is the Alfven speed. Under this normalization, the Eqs.(9) and (10) become, respectively, as

$$\partial_{\tilde{t}}\tilde{\mathbf{B}} + \tilde{\nabla} \times \left[\left(\tilde{\mathbf{B}} \times \tilde{\mathbf{v}}\right) - \epsilon_i \frac{1}{\tilde{n}} \tilde{\mathbf{B}} \times \left(\tilde{\nabla} \times \tilde{\mathbf{B}}\right)\right] = -\epsilon_i \left(\tilde{\nabla}\tilde{\psi} \times \tilde{\nabla}\tilde{T}_e\right) \tag{25}$$

and

$$\partial_{\tilde{t}}\left(\tilde{\mathbf{B}} + \epsilon_i \tilde{\nabla} \times \tilde{\mathbf{v}}_i\right) - \nabla \times \left[\tilde{\mathbf{v}} \times \left(\tilde{\mathbf{B}} + \epsilon_i \nabla \times \tilde{\mathbf{v}}\right)\right] = \epsilon_i \left(\tilde{\nabla}\tilde{\psi} \times \tilde{\nabla}\tilde{T}_i\right) \tag{26}$$

The superscript tilde ($\sim$) denotes the normalized quantities and operators. Note that if $L_0 = \lambda_i$ is assumed, then Eq.(25) and (26) are the same as Eq.(6) and (7) of Ref.[26] because $\frac{v_A}{L_0} = \frac{v_A}{\lambda_i} = \Omega_i$ which normalizes time in this case.

We have shown that if all plasma fields are assumed to be driven by baroclinic vectors and $\nabla\psi$ and ∇T_j have suitable spatial structures, then the Eqs. (25) and (26) will reduce, respectively, to

$$\partial_{\tilde{t}}\tilde{\mathbf{B}} = -\epsilon_i \left(\tilde{\nabla}\tilde{\psi} \times \tilde{\nabla}\tilde{T}_e\right) \tag{27}$$

and

$$\left(1 + \epsilon_i g^{-1}\right) \partial_{\tilde{t}}\tilde{\mathbf{B}} = \epsilon_i \left(\tilde{\nabla}\tilde{\psi} \times \tilde{\nabla}\tilde{T}_i\right) \tag{28}$$

where $\tilde{\psi} = \ln \tilde{n}$ and

$$\tilde{\mathbf{B}} = \tilde{g}\left(\tilde{\nabla} \times \tilde{\mathbf{v}}\right) \tag{29}$$

has been assumed with $\tilde{g}$ to be an arbitrary constant.

The theoretical studies of Refs.[28,29] do not discuss any limit on plasma $\beta = c_s/v_A$. Since they deal with steady state, therefore we think that they are more relevant for the case $\beta < 1$ otherwise plasma will be expanding. The expansion rate may be assumed to be constant. However, we do not discuss the steady state problem here.

For an interesting comparison with the case of evolving plasma with $1 < \beta$, we analyze the two fluid equations using another normalization scheme 2 (say) and let the previous one be called as scheme 1.

Since we are interested in the seed magnetic field generation by plasma dynamics and assume $\mathbf{B} = 0$ at $t = 0$, therefore it is preferable to normalize the pressures with some arbitrary pressure $p_0 = n_0 T_0$. Let us use the following normalization, $\mathbf{x} = L_0\tilde{\mathbf{x}}, \mathbf{B} = B_0\tilde{\mathbf{B}}, n = n_0\tilde{n}, t = \frac{L_0}{c_s}\tilde{t}$, $p_j = \frac{p_j^2}{n_0 T_0}, \mathbf{v}_j = c_s\tilde{\mathbf{v}}_j, \mathbf{E} = \frac{c_s B_0}{c}\tilde{\mathbf{E}}$ Then Eqs.(25) and (26) can be written, respectively, as

$$\partial_{\tilde{t}}\tilde{\mathbf{B}} + \tilde{\nabla} \times \left[\left(\tilde{\mathbf{B}} \times \tilde{\mathbf{v}}\right) - \frac{1}{\tilde{n}}\frac{\rho_s}{L_0}\frac{1}{\beta}\tilde{\mathbf{B}} \times \left(\tilde{\nabla} \times \tilde{\mathbf{B}}\right)\right] = -\frac{\rho_s}{L_0}\left(\tilde{\nabla}\tilde{\psi} \times \tilde{\nabla}\tilde{T}_e\right) \tag{30}$$

and

$$\partial_{\tilde{t}}\left(\tilde{\mathbf{B}} + \frac{\rho_s}{L_0}\tilde{\nabla} \times \tilde{\mathbf{v}}\right) - \tilde{\nabla} \times \left[\left(\tilde{\mathbf{v}} \times \tilde{\mathbf{B}}\right) + \frac{\rho_s}{L_0}\tilde{\mathbf{v}} \times \left(\tilde{\nabla} \times \tilde{\mathbf{v}}\right)\right] = \frac{\rho_s}{L_0}\left(\tilde{\nabla}\tilde{\psi} \times \tilde{\nabla}\tilde{T}_i\right) \tag{31}$$

where$\rho_s = \frac{c_s}{\Omega_i}$ and $\rho_s \longrightarrow \infty$ as $t \longrightarrow 0$ with $\mathbf{B} = 0$.

Note that $\rho_s^2/\lambda_i^2 = \frac{c_s^2}{v_A^2} = \beta$ and hence for $\beta \lesssim 1$, the scheme 1 is suitable and for $1 < \beta$, the scheme 2 is useful, in our opinion.

If $1 < \beta$ is assumed, then we may consider two cases,a) $1 < \beta \lesssim m_i/m_e$ and b) $\frac{m_i}{m_e} < \beta$.

Let $\rho_s \sim L_0$ and $\epsilon = \frac{1}{\beta}$ in case a, then the Hall-term again becomes a singular perturbation defining a microscopic scale. In the limit $1 \ll \beta$, the Hall term becomes negligibly small and the system reduces to MHD case having only the macroscopic scale.

Let us use the common parameters of interstellar medium (ISM) $T_e \sim 1kev$ and $n_0 \sim 1cm^{-3}$ with $\lambda_i \sim 2.3 \times 10^7 cm$ assuming Hydrogen plasma. We shall present an estimate of the seed field to be $10^{-15}G$ in Sec. 5. Previous studies have predicted the galactic seed field generated in $\tau =$

$10^9 yrs$ as $[B] \sim (10^{-15} - 10^{-17})\ Gauss$ [18] and these values are still flexible. So, we choose $[B] \sim 10^{-15}G$ which predicts $\rho_s \sim 1pc$ (parsec) and it is comparable to the characteristic length $(L_0 \sim 1pc \simeq 3 \times 10^{18} cm)$ in this case while $\epsilon_i = \frac{\lambda_i}{L_0} \sim 0.76 \times 10^{-11}$, $\beta = 0.2 \times 10^{20}$ and $\Omega_i \sim 10^{-11} radS^{-1}$ at the peak value of seed magnetic field generated in time $\tau \sim 10^9 yrs$ after the Big Bang. It may be mentioned here that we need an estimate for ∇T_e to evaluate $|B|$. Here a particular value $T_e \sim 1KeV$ has been used just for an analysis of the scales. Since the gradient scale lengths of density and temperature are of the order of $L_0 \sim 1pc$ (may vary from 1 to $10^3 pc$, for example), therefore the normalization scheme 2 is relevant.

Now we look at the laser-induced plasmas. Let us choose Eq. (30) to look into the scales. When non-linear terms vanish it becomes,

$$\partial_{\tilde{t}} \tilde{\mathbf{B}} = -\frac{\rho_s}{L_0} \left(\tilde{\nabla} \tilde{\psi} \times \tilde{\nabla} \tilde{T}_e \right) \tag{32}$$

Let us assume density gradient along x-axis and temperature gradient along y-axis. If the scale lengths of density and temperature gradients are L_n and L_T respectively and $\frac{T_e}{T_0} \sim 1$, then integrating Eq. (32) for $t = 0 \to \tau$, we can write it as,

$$\mathbf{B} = -\left\{ \frac{c}{e} \left(\frac{T_e}{L_n L_T} \right) \tau \right\} \hat{\mathbf{z}} \tag{33}$$

Note that Eq. (33) is the same as Eq. (5-79) of Ref. [20]. For the laser plasma with $T_e \sim 1KeV, n_0 \sim 10^{20} cm^{-3}, c_s \sim 3 \times 10^7 cmS^{-1}$ and $L_n \sim L_T \sim 0.005cm$, one obtains [20] $|\mathbf{B}| \simeq 0.6 \times 10^6 G$ where $\tau \simeq \frac{L_n}{c_s}$ has been used. We have already explained that Eq. (33) alone must not be used to determine $|\mathbf{B}|$ .But here we use this equation just to understand the scales of applicability. For this plasma we find $\lambda_i \sim 2.25 \times 10^{-4} cm, \epsilon_i \sim 5.6 \times 10^{-2}$ and $\tau = \frac{L_n}{c_s} \sim 1.29 \times 10^{-10} S$ which is smaller than $10^{-9} S$ (the laser pulse duration time in the initial experiments). Since $\frac{m_e}{m_i} < \epsilon_i < 1$, therefore in this case the Hall-Term is a singular perturbation. Our model is applicable to HMHD scale and we may use normalization scheme 2. Note that $1 < \beta \sim 62$ for this plasma.

Important point to note is that in the case $\frac{1}{\beta} \to 0$ the Hall-term is vanishingly small and the model is applicable to MHD scales. But for $L_0 \sim \lambda_i$, our model is applicable to HMHD scale. It may also be mentioned here that we have used the peak value of seed field to estimate ρ_s in case of ISM. But it can be even larger than $1pc$ corresponding to smaller values of B_s. during time $t < \tau$.

Our main aim is to predict the generation of 'seed' magnetic field. Therefore, it seems better to use the equations in physical units without normalization to keep the option open for application to different systems.

4. Seed magnetic field and flow

Let us assume

$$\psi = \psi_0 e^{\mu_1 x} \cos \mu_2 y \tag{34}$$

$$T_e = T_{00e} + T_{0e}^{'}(y - z) f(t) \tag{35}$$

and

$$T_i = T_{00i} + T_{0i}^{'}(y - z) f(t) \tag{36}$$

Here $\mu_1, \mu_2, T_{00j}, T_{0j}^{'}$ are constants whereas T_{0j} are in units of eV and $T_{0j}^{'}$ denote the temperature gradients. Note that Eq. (17) is satisfied with $\lambda = (\mu_2^2 - \mu_1^2)$. Then

$$\nabla T_e \times \nabla \psi = f(t) T_{0e}^{'} \psi_0 e^{\mu_1 x} (-\mu_2 \sin \mu_2 y, -\mu_1 \cos \mu_2 y, -\mu_1 \cos \mu_2 y) \tag{37}$$

and

$$\nabla T_i \times \nabla \psi = f(t) T_{0i}^{'} \psi_0 e^{\mu_1 x} (-\mu_2 \sin \mu_2 y, -\mu_1 \cos \mu_2 y, -\mu_1 \cos \mu_2 y) \tag{38}$$

Now we shall see that all the fields ϕ, ψ, χ and h have the forms similar to ψ. If

$$f(t) = e^{\gamma t} \tag{39}$$

is assumed (where γ is a constant), then equation(22) gives,

$$\chi = \frac{c}{e} \frac{T_{0e}^{'}}{\gamma} \psi \tag{40}$$

and

$$h = -\frac{c}{e} \frac{\mu_1 T_{0e}^{'}}{\gamma} \psi \tag{41}$$

where $\gamma \neq 0$. Similarly Eq.(23) yields,

$$\chi = -\frac{T_{0i}^{'}}{m_i \gamma (a + g^{-1})} \psi \tag{42}$$

and

$$h = \frac{T_{0i}^{'} \mu_1}{m_i \gamma (a + g^{-1})} \psi \tag{43}$$

The relations (40-43) require,

$$T'_{0e} = -\frac{T'_{0i}}{(a+g^{-1})} \tag{44}$$

The fields u and ϕ are related with χ and h through Eqs.(20) and (21). Therefore all the components of $\mathbf{B}$ and $\mathbf{v}_i$ become functions of ϕ which has a linear relationship with ψ. Therefore the conditions (15) and (18) are satisfied along with equation(13). The components of the fields $\mathbf{B}$ and $\mathbf{v}_j$ have the structural forms either of type $F_1 = e^{\mu_1 x}\cos\mu_2 y$ or $F_2 = e^{\mu_1 x}\sin\mu_2 y$. The forms of these functions in xy-plane are shown in Fig. 1 and Fig. 2, respectively, for $\mu_1 x = -1 \to 0$ and $\mu_2 y = 0 \to 2\pi$.

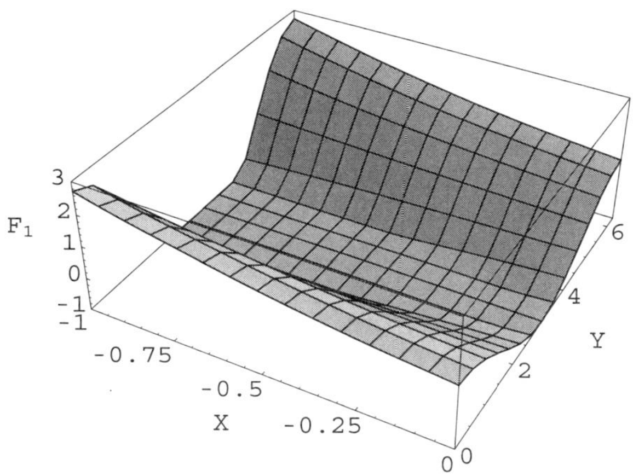

Fig. 1. The function $F_1\,(x,y)$ is plotted for $\mu_1 x = 0 \to -1$ and $0 \leqslant \mu_2 y \leqslant 2\pi$.

5. Applications

We apply the theory to both cosmological and laser-induced plasmas. We may consider any one of the equations (22) and (23) to estimate $|B|$. Let us choose Eq.(22) which for $f(t) = 1$ becomes,

$$\partial_t \mathbf{B} = \frac{cT'_{0e}}{e}\psi_0 e^{\mu_1(x)}\left(-\mu_2\sin\mu_2 y, -\mu_1\cos\mu_2 y, -\mu_1\cos\mu_2 y\right) \tag{45}$$

Integrating this equation from $t=0$ to $t=\tau$, we obtain,

$$\mathbf{B} = \frac{cT'_{0e}}{e}\psi_0 e^{\mu_1(x)}\left(-\mu_2\sin\mu_2 y, -\mu_1\cos\mu_2 y, -\mu_1\cos\mu_2 y\right)\tau \tag{46}$$

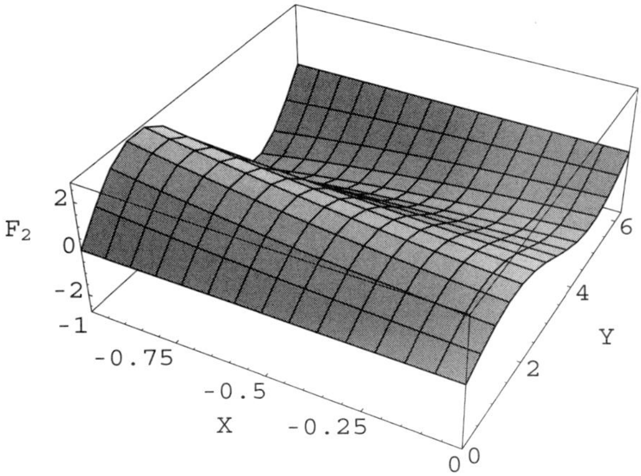

Fig. 2. The function $F_2(x, y)$ is plotted for $\mu_1 x = 0 \rightarrow -1$ and $0 \leqslant \mu_2 y \leqslant 2\pi$.

which gives

$$|B| \simeq \frac{1}{\sqrt{2}} \frac{cT'_{0e}}{e} \psi_0 e^{\mu_1 x} \left(\mu_2^2 + 2\mu_1^2\right)^{\frac{1}{2}} \tau \tag{47}$$

Let us try to find out the origin of galactic magnetic fields. We assume that the seed magnetic field is very weak and it is amplified later on by $\alpha\omega$ dynamo (or by some other mechanism). Here our aim is to generate the magnetic field and flow by the given baroclinic vector. Following Ref. [18] ,we also assume that the galactic cloud is non-uniform and clumpy. The forms of temperature gradients chosen in Eqs.(35) and (36) demand $y \neq z$ and hence $y_1 \neq z_1$ and $y_2 \neq z_2$. Let us choose a parallelogram within the cloud having sides $d = |y_2 - y_1| = |z_2 - z_1|$ while temperatures linearly increase along positive y-axis and negative z-axis. We may choose y_1 and y_2 such that $\mu_2 y_1 = 2\pi$ and $\mu_2(y_2 - y_1) = \mu_2 d = 2\pi$ which gives

$\mu_2 = \frac{2\pi}{d}$. If we set $x_1 = 0$, then

$\psi(x_1, y_1) = \psi_0$ and $\psi(x_2, y) = \psi_0 e^{\mu_1 x_2} cos \mu_2 y$. Since $|x_2 - x_1| = |x_2|$, therefore exponential density increase along positive x-axis requires $\mu_1 = \frac{1}{x_2}$. The density along y axis has been chosen to be sinusoidal. Note that at the points where $cos\mu_2 y = 0$, the density $n(x, y)$ does not vanish. Let N_0 be an arbitrary density and $\bar{n} = \frac{n(x,y)}{N_0}$, such that $\psi = ln\bar{n}$. Hence at $\bar{n} = 1$ we have $\psi = 0$ but $n(x, y) = N_0 \neq 0$. Let us choose N_0 such that $\bar{n} = \frac{n(x_1,y_2)}{N_0} = 20, \psi_{|x=0} = ln 20 \simeq 3$ and $\psi|_{x=x_2} = (\psi_0)e \simeq (2.7)\psi_0$.

Let xy-plane be the galactic disk plane and d be it's width along z-axis. The eight points of this parallelogram are defined as, $P_1(x_1, y_1, z_1), P_2(x_1, y_2, z_1), P_3(x_1, y_2, z_2), P_4(x_1, y_1, z_2), P_5(x_2, y_1, z_2)$, $P_6(x_2, y_1, z_1), P_7(x_2, y_2, z_1), P_8(x_2, y_2, z_2)$.

The side $P_3P_4P_5P_8$ is inside the cloud at lower temperature and the side $P_1P_2P_6P_7$ is the top (or bottom) of galaxy disk at higher temperature.

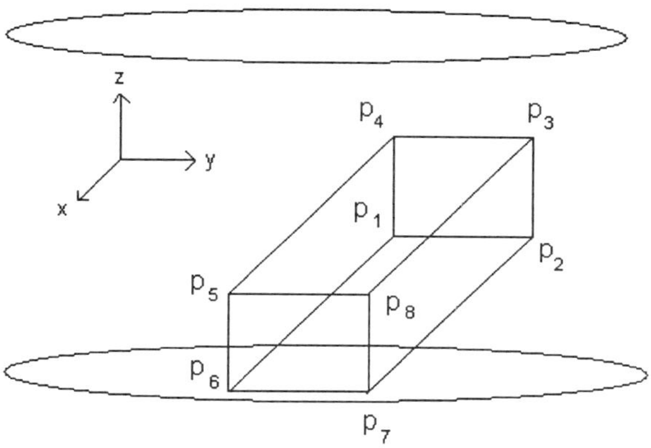

Fig. 3. A parallelogram with sides $P_1P_2P_3P_4$ and $P_5P_6P_7P_8$ within the galaxy cloud is shown. Such clumps of inhomogeneous electron and ion densities can generate three dimensional seed magnetic fields.

The density of electron and ion increases exponentially along x-axis due to ionization processes. It may be mentioned here that the inside region is atomic cloud and galaxy edge region is ionized and consists of plasma [18]. Then we have $0 < \mu_1$. We assume the temperature difference between inside and outside regions of the cloud to be the same as has been used in Ref. [18] which is 10^6 degree Kelvin. Hence we have

$$|T'_{0e}| = \frac{\kappa_\beta(10^6 k^0)}{d} \qquad 48$$

Since $0 < \lambda$ has been assumed in the model and $\lambda = \mu_2^2 - \mu_1^2$, therefore we require $\mu_1 < \mu_2$. Let us choose $x_0 = |x_2 - x_1| = 10d$. Since 1 year $\simeq 3.15 \times 10^7 S$ and 1 parsec $(pc) \simeq 3 \times 10^{18}$ cm therefore for $d = 10^2 pc$, Eq.(47)

gives,

$$|B_s| \sim [(1.05 \times 10^{-22})\tau]G \qquad 49$$

and for $d = 10^3 pc$,

$$|B_s| \sim [(1.05 \times 10^{-24})\tau]G \qquad 50$$

Here the subscript(s) denotes seed field. In Eqs(48) and (49), τ is in years. The galactic evolution time is believed to be one billion (10^9) years [30], in general. If we assume the time of 'seed' magnetic field generation on galactic scale of the order of 10^9 years, then Eq(49) gives,

$$B_s \sim [(1.05 \times 10^{-13})]G \qquad 51$$

and Eq(50) gives,

$$B_s \sim [(1.05 \times 10^{-15})]G \qquad 52$$

It may be mentioned here that $|B_5| \sim 10^{-17}$ G has been estimated assuming $d = 10^4 pc$ and $x_0 = 10^2 pc$ in Ref. [18] and this B-field is unidirectional. It is generated by electron dynamics while the ions are stationary. The electron density and temperature gradients are also assumed to be unidirectional. In our theoretical model, the B_s is three dimensional. However, the present model also needs improvements so that the specific structural forms of gradients are not necessary for the generation of B_s. The increase in the estimated magnitude $|B|$ is due to steeper temperature gradients assumed in the calculation which is necessary to fulfill the condition $0 < \lambda$. This theoretical model can be applied to laser-induced plasma experiments as well. We may define [20], $\kappa_T = \left|\frac{1}{T_e}\frac{dT_e}{dy}\right| = \left|\frac{1}{T_e}\frac{dT_e}{dz}\right|$ and hence $T'_{0e} = T_{0e}\kappa_T$. Let $\mu_1 \simeq \mu_2 = \mu$ and $(\mu_2^2 + 2\mu_1^2)^{\frac{1}{2}} \simeq \sqrt{3}\mu$ where $\mu = \kappa_n = \left|\frac{1}{n_0)}\frac{dn}{dx}\right|$ and $\mu_1 x_0 = -1$. Then we can express equation (47) as,

$$|B| \sim \left\{\sqrt{\frac{3}{2}}\psi_0\left(\frac{cT_e}{e}\right)\kappa_n\kappa_T e^{-1}\right\}\tau \qquad (53)$$

Let $L_n = \frac{1}{\kappa_n}$ and $L_T = \frac{1}{\kappa_T}$ be the density and temperature scale lengths along x-axis and along y and z axes, respectively. Then the above expression can be written as,

$$|B| \sim b_0\left(\frac{cT_{0e}}{e}\frac{1}{L_nL_T}\right)\tau \qquad (54)$$

where $b_0 = \sqrt{\frac{3}{2}}\psi_0 e^{-1}$. Note that equation (54) is the same as equation (5.79) of Ref. [20] and we take the same example of laser-induced plasma as chosen in this reference i.e. $T_{e0} \sim 10^3 eV, c_s \sim 3 \times 10^7 cmS^{-1}, \tau = \frac{L_n}{c_s}$. We assume $\psi_0 = 3$ (to have density positive at the edge region as well) and find $b_0 \simeq 1.36$. Therefore

$$|B_0| \sim (8.7) \times 10^5 G \tag{55}$$

Note that if $b_0 = 1$, we have exactly the same value of $|B| \sim 0.64 \times 10^6 G$ as estimated in Ref. [20]. Note that in the application to laser-plasmas, we assume the density to decrease exponentially along x-axis as has been done in Ref. [26]. But the important point to note is that, the term $\mathbf{v}_e \times \mathbf{B}$ has not been dropped and ion dynamics has also been taken into account because $\tau = \frac{L_n}{c_s} \gg \omega_{pi}^{-1}$.

6. Discussion

A theoretical model for the generation of seed magnetic fields on galactic scales has been presented which can be applicable to laser-induced plasmas as well. The present theoretical model is actually a modified form of the previous work [26]. In Ref. [26], one has to assume poloidal components of the magnetic field and toroidal components of the flow to be static. But in the present theoretical model[27], all plasma fields can grow from their zero values at $t = 0$ due to the source terms $(\nabla\psi \times \nabla T_j)$. An exact analytical solution of a set of highly nonlinear two-fluid partial differential equations of a hot inhomogeneous plasma has been obtained. The plasma is assumed to be produced with electron and ion pressure gradients in a state of non-equilibrium ($T_i \neq T_e$). The baroclinic vectors $(\nabla\psi \times \nabla T_e)$ and $(\nabla\psi \times \nabla T_i)$ then become the source for plasma evolution creating it's flow $\mathbf{v}_i$ and magnetic field $\mathbf{B}$. These vectors are expressed in terms of scalar fields (ϕ, u, χ, h) in Eqs. (11) and (12). A relationship between $\mathbf{v}_i$ and $\mathbf{B}$ is assumed to be given by Eq. (19). The plasma fields are assumed to satisfy the conditions of Eqs. (15) and (18). Then the forms of $\nabla\psi$ and ∇T_j are chosen in Eqs. (34), (35) and (36) in such a way that $(\nabla\psi \times \nabla T_j)$ vectors become proportional to ψ. Then all other fields can also be expressed in terms of ψ in Eqs. (40-43) and hence all the assumed conditions in Eqs. (13), (15) and (18) are satisfied.

This is a particular solution of the plasma equations. But it shows how the plasma can evolve form $t = 0$ with $B = 0$ generating it's magnetic field and flow. It has been shown that the present theoretical model is applicable to both HMHD and MHD scales.

Since $(\nabla\psi \times \nabla T_j)$ are the generating forces therefore the scales of $\mathbf{v}_i$ and $\mathbf{B}$ should be of the same order of magnitude as that of $\nabla\psi$ and ∇T_j

as is clear from equations (22) and (23). The short scale phenomena superimposed upon such fields do not participate in the creation of magnetic field and flow on very large scales. Several effects like dissipation and compressibility have been ignored. It has not been discussed here that how the particular spatial structures of density and temperature gradients are produced. But these simplifications and assumptions have been made to present an exact solution of the complex partial differential equations. The estimates of the magnetic fields at both cosmological and laboratory scales are very close to observations. It is necessary to explain why our estimate of the 'seed' magnetic field is larger than that made in Ref.[18]. The reason for this larger estimate of seed field is that we assume the density gradient scale length μ_1 larger and temperature gradient scale length μ_2 smaller than that used in Ref. [18]. These parameters can be varied because the interstellar cloud can have regions of different magnitudes of the scales of gradients, in our opinion.

It must be pointed out here that our assumption of equal magnitudes of temperature gradients along (+ve) y-axis and (-ve) z-axis forces us to choose $y_0 = |y_2 - y_1| = |z_2 - z_1| = d$. Then the requirement $\mu_2 d = 2\pi$ along with $|\mu_1| < |\mu_2|$ (to have $\lambda = \mu_2^2 - \mu_1^2 > 0$) compels us to assume steeper temperature gradients along both the axes compared to Ref. [18]. However, the future estimates about the gradients can be useful to make the estimate for B-field magnitude more exact.

It is important to point out that the seed magnetic field investigated in Ref. [18] has only non-zero component along z-axis while $B_x = B_y = 0$ is assumed because $(\nabla T_e \times \nabla n_e)$ is along z-axis. In our theoretical model, the baro-clinic vectors $(\nabla T_j \times \nabla \psi)$ have all the three components to be non-zero and hence the galactic seed magnetic field has three components which is in agreement with the observations [11,12].

The different values of the magnitudes of the density and temperature gradients can also be used for the generation of the magnetic and flow fields with $y_0 \neq z_0$ and $\mu_1 < \mu_2$. But in these cases, analytical solution of the set of nonlinear partial differential equations may not be found.

The present theory can have wider applications to many astrophysical situations. However, the numerical simulations will be required, if plasma is considered to be non ideal and if the forms of ∇n and ∇T_j are chosen to be different from the ones used in the present analytical calculations. In the application of the present model to astrophysical and laser plasmas, the density and temperature gradients are assumed to be time-independent. But the time of evolution of the seed galactic field is assumed to be one billion years ($\tau \sim 10^9 yrs$), the time span on galactic scale. During such a long time the gradients of densities and temperatures probably may not remain constant over long spatial scales of the order of $10^2 pc$ or $10^3 pc$. The aim of this investigation is to show that the plasma evolution from a non-equilibrium state can generate microscopic seed magnetic fields due to pressure gradients of ions and electrons. The following three points are important to note about the present model.

1. It considers three dimensional seed magnetic field generation.

2. It points out that the macroscopic field is generated on ion time scale due to nonlinear plasma evolution from a non-equilibrium state. Only electron dynamics can not produce large fields during short times where ions can be assumed to remain stationary

3. The model is applicable to both shorter (HMHD) and larger (MHD) scales.

We think that the computer simulations of the set of model equations presented here can give better estimates about $|B_s|$ assuming different spatial structures of gradients. If we define time-dependent forms of gradients with $f(t) \neq 1$, then it is not easy to find out an exact analytical solution.

Acknowledgement

The author is very grateful to Prof. Z. Yoshida of Tokyo University for several useful discussions on the overall theoretical model at Abdus-Salam International Centre for Theoretical Physics (AS-ICTP), Trieste, Italy during the International Workshop on Frontiers in Plasma Science 21 Aug. -1 Sept. 2006. Discussions with N. Shatashvili and B. Eliasson are also acknowledged.

References

1. L. M., Midrow, *Rev. Mod. Phys.* **74**, 775 (2002).
2. P. P. Kronberg, *Rep. Prog. Phys.* **57**, 325(1994).
3. E. G. Zweibel, and C. Heiles, *Nature (London)* **385**, 131 (1997).
4. J. P. Vallee, *Fundam. Cosmic Phys.* **19**, 1 (1997).
5. R. Beck, *Philos. Trans. Royal Soc. London*, Ser. **A 358**, 777 (2000).
6. W. Dobler, M. Stix and A. Brandenburg, *Astrophys. J.* **638**, 336 (2006).
7. J. T. Frederiksen, C. B. Hededal, T. Haug Bolle and Å. Nordlung, *Astrophys. J.* **608**, L13, (2006).
8. A. A. Schekochihin, and S. C. Cowley, *Magnetohydrodynamics: Historical Evolution and Trends* (Springer Verlag 2006).
9. A. J. Fitt, and P. Alexander, *Mon. Not. R. Astron. Soc.* **261**, 445 (1993).
10. R. Beck in *The Astrophysics of Galactic Cosmic Rays*, edited by R. Diehl et al. (Kluwer, Dordrecht, The Netherlands 2002).

11. M. Dumke, M. Krause, Weilebinski, and U. Klein, *Astron. Astrophys.* **302**, 691 (1995).
12. E. Hummel, R. Beck, and M. Dahlem, *Astron. Astrophys.* **248**, 23 (1991).
13. L. Biermann, *Z. Naturforsch.* **5A**, 65 (1950).
14. T. G. Cowling, *Solar Electrodynamics*, in G. P. Kuiper (ed.), *The Sun* (Chicago: University of Chicago Press 1953).
15. L. Mestel and I. W. Roxburgh, *Astrophys. J.* **136**, 615 (1962).
16. M. R. Kulsrud, R. Gen, J. P. Ostriker, and D. Ryu, *Astrophys. J.* **480**, 481 (1997).
17. M. R. Kulsrud, *Ann. Rev. Astron. Astrophys.* **37**, 37 (1999).
18. A. Lazarian, *Astron. Astrophys.* **264**, 326 (1992).
19. J. A. Stamper, K. Papadopoulos, R. N. Sudan, S. O. Dean, E. A. Mclean and J. W. Dawson, *Phys. Rev. Lett.* **26**, 1012 (1971).
20. K. A. Brueckner and S. Jorna, *Rev. Mod. Phys.* **46**, 325 (1974).
21. A. A. Kingssep, K. V. Chukbar and V. V. Yan'Kov, in *Reviews of Plasma Physics*, edited by B. B. Kadomtsev, (Consultants Bureau, NY, 1990), vol. **16**. p. 243.
22. L. A. Bol'shov, A. M. Dykhne, N. G. Kowalski and A. I. Yudin, in *Handbook of Plasma Physics*, edited by M. N.Rosenbluth and R. Z. Sagdeev, *Physics of Laser Plasma* (Elsevier Science, N. Y. 1991) vol. 3, p. 519-548.
23. H. Saleem, *Phys. Rev. E* **59**, 6196 (1999); H. Saleem, *J. Fusion Energy* **19**, 159 (2002).
24. E. S. Weibel, *Phys. Rev. Lett.* **2**, 83 (1959).
25. P. L. Biermann and P. P. Kronberg,*J. Korean Astron. Soc.* **37**, 527 (2004).
26. H. Saleem and Z. Yoshida, *Phys. Plasmas***11**, 4865 (2004).
27. H. Saleem, *Phys. Plasmas* **14**, 072105 (2007).
28. S. M. Mahajan and Z. Yoshida, *Phys. Rev. Lett.* **81** (22), 4863 (1998).
29. Z. Yoshida, S. M. Mahajan and S. Ohsaki, *Phys. Plasmas* **11** (7), 3660 (2004).
30. J. Bennett, M. Donahue, N. Schneider, and M. Voit, in *The Cosmic Perspective* (Addison Wesley 2004) 3rd Edition Ch.23.

NONLINEAR DYNAMICS OF MIRROR WAVES IN NON-MAXWELLIAN PLASMAS

O. A. POKHOTELOV* and M. BALIKHIN

Department of Automatic Control and Systems Engineering, University of Sheffield, S1 3JD, Sheffield, United Kingdom,
** E-mail: o.a.pokhotelov@sheffield.ac.uk*
www.sheffield.ac.uk

R. Z. SAGDEEV

The University of Maryland, Department of Physics, College Park, MD 20740, USA

O. G. ONISHCHENKO

Institute of Physics of the Earth, 10 B. Gruzinskaya Street, 123995 Moscow, Russia

V. N. FEDUN

Department of Applied Mathematics, University of Sheffield, Hounsfield Road, Sheffield S3 7RH, United Kingdom

A theory of finite-amplitude mirror type waves in non-Maxwellian space plasmas is developed. The collisionless kinetic theory in a guiding center approximation, modified for accounting the effects of the finite ion Larmor radius effects, is used as the starting point. The model equation governing the nonlinear dynamics of mirror waves near instability threshold is derived. In the linear approximation it describes the classical mirror instability with the linear growth rate expressed in terms of an arbitrary ion distribution function. In the nonlinear regime the mirror waves form solitary structures that have the shape of magnetic holes. The formation of such structures and their nonlinear dynamics has been analyzed both analytically and numerically. It is suggested that the main nonlinear mechanism responsible for mirror instability saturation is associated with modification (flattening) of the shape of the background ion distribution function in the region of small parallel particle velocities. The width of this region is of the order of the particle trapping zone in the mirror hole. Near the mirror instability threshold the saturation arises before its width reaches the ion thermal velocity. The nonlinear mode coupling effects in this approximation are smaller and unable to take control over evolution of the space profile of saturated mirror waves or lead to their magnetic collapse. This results in the appearance of quasi-stable solitary mirror structures having the form of deep magnetic depressions. The relevance of the theoretical results to recent satellite observations is stressed.

Keywords: High-beta plasmas; Mirror instability.

1. Introduction

The classical mirror instability (MI) theoretically identified in the end of the 50th[1–3] is a common feature in planetary and cometary magnetosheaths. The early theories were limited by purely linear analysis and thus could not provide a comprehensive comparison with *in situ* observations. Interest to the MI was substantially reinforced by its possible relevance to spacecraft observations of the so-called "magnetic holes" - deep dropouts of the magnetic field in high-β plasmas. The term magnetic hole was first suggested by Turner[4] to describe decreases in the solar wind magnetic field strength. They were found in numerous planetary magnetosheaths[5–10] in the solar wind[4,11–13] and in the vicinity of comets.[14,15] The mirror wave structures were found also in the sheath regions between fast interplanetary coronal mass injections and their preceding shocks.[16] Ultimately they represent a nonlinear phenomenon. It should be noted that besides the magnetic holes some observaions reveal other forms of the mirror structures as quasi-sinusoidal profiles of magnetic humps. For example, they have been found in the Jovian magnetosheath[10] and in numerical simulations.[17]

The cornerstone of the instability mechanism is the pressure anisotropy that provides the free energy necessary for its onset. As the solar wind slams into the magnetosphere it abruptly slows down, and a bow shock and a magnetosheath are formed. In the magnetosheath the pressure anisotropy is generated when the plasma flow crosses the bow shock. The passage of the solar wind through this region leads to an increase in the ion temperature, with the larger proportion of the heating going preferentially into the perpendicular component. This results in the appearance of anisotropic ion populations with stronger pressure anisotropy than in the solar wind. The MI has a number of peculiarities that makes it a possible candidate to fit into a scenario of magnetic hole formation, among these are the nonpropagating nature of the mode and the anti-correlation of plasma density and magnetic field.

The MI corresponds to a resonant type instability that arises due to the interaction of the zero-frequency mode with the ions possessing very small velocities along the external magnetic field. The physical mechanism of the linear MI has been reviewed.[18] The linear mirror mode theory has been extensively developed in a number of previous papers[19] and references therein. Those papers included the effects of a finite electron temperature, a nonvanishing electron temperature anisotropy, multi-ion plasmas, the finite ion Larmor radius (FLR) effect and arbitrary ion or electron distribution functions. The FLR effects in the mirror instability have been recently analyzed by using gyrokinetic theory.[20] A very general compact expression for the linear marginal mirror-mode stability condition and instability growth rate suited for application to a wide class of velocity distribution functions has been derived[19,21,22] and has then been applied to bi-Maxwellian, Dory-Guest-Harris (DGH), Kennel-Ashour-Abdalla (KA) and κ-distributions. However, all these papers have been restricted by the linear approximation.

One of the first attempts of a nonlinear treatment of the MI was made more than four decades ago.[23] These authors using the random phase approximation have reduced the problem of nonlinear saturation of the MI to the study of a quasilinear diffusion equation for the ion distribution function in which all ions were considered to be fully adiabatic. Indeed, the background ion distribution function is shown to be modified which leads to saturation of MI. However, such a scenario that assumes the adiabaticity of particle trajectories does not take into account the effects of resonant trapping, which is truly irreversible collective mechanism. Below we will utilize a fully irreversible approach to nonlinear saturation of the MI. On the other hand, an important conclusion has been made on the special role of ions having small parallel velocities. In what follows, in going beyond the random phase approximation, this feature will remain intact.

It should be noted that some effects related to nonlinear saturation of mirror waves in the magnetosheath have been already discussed.[24,25] Recently the first attempt to develop a nonlinear theory of MI in bi-Maxwellian plasmas has been offered.[26] It should be noted that measured particle distributions in near-Earth space do, in the overwhelming majority of cases, considerably deviate from the bi-Maxwellian shape. They frequently exhibit long suprathermal tails on the distribution function or, in other cases, possess loss cones. In collisionless plasmas, such as planetary magnetosheaths, the decisive role in determining the shape of the particle velocity distribution functions may belong to wave-particle interactions. In the range of resonance these interactions may lead to the formation of diffusion plateaus. For example, such plateaus were recently observed in the velocity distribution of fast solar wind protons.[27] Generally such wave particle interaction at cyclotron resonance is touching particles with energies higher than thermal, unlike the MI which deals mainly with ions below thermal energies. In this respect those two modes could coexist if the background distribution is capable to support both of them.

The main goal of the present paper is to develop a comprehensive nonlinear theory of the MI in a plasma with an arbitrary equilibrium velocity distribution function. It will be shown that a nonlinear process known under the name of wave collapse[28]in the real conditions for the MI is terminated by the effect of particle trapping in the mirror hole that arises in the resonance region.

The paper is organized as follows: Section 2 formulates the derivation of the nonlinear plasma pressure balance condition for the MI in terms of an arbitrary ion distribution function. The generalization of this condition to the case when the FLR effect is included is carried out in Section 3. The linear growth rate of the MI accounting for the FLR effect is derived in Section 4. The mirror mode nonlinear equation is obtained in Section 5. The effects of particle trapping in the mirror holes are discussed in Section 6. Our conclusions and outlook for future research are found in Section 7.

2. Plasma Pressure Balance Condition

We consider a collisionless plasma composed of ions and electrons having the total mass density ρ and macroscopic velocity $\mathbf{v}$, and embedded in a magnetic field $\mathbf{B}$. Our analysis will be limited to the case of most importance when the ion temperature is much larger than the electron temperature. The momentum equation governing the time evolution of this plasma is

$$\rho\frac{d\mathbf{v}}{dt} = \mathbf{J}\times\mathbf{B} - \nabla\cdot\hat{\mathbf{P}}, \tag{1}$$

where $\mathbf{J}$ is the electric current and $d/dt \equiv \partial/\partial t + \mathbf{v}\cdot\nabla$ is the Lagrangian time derivative. The pressure tensor represents the anisotropic pressure tensor $\hat{\mathbf{P}}$ given by

$$\hat{\mathbf{P}} = p_\perp\hat{\mathbf{I}} + (p_\parallel - p_\perp)\hat{\mathbf{z}}\hat{\mathbf{z}}. \tag{2}$$

Here $\hat{\mathbf{I}}$ is the unit dyadic, $p_{\parallel(\perp)}$ is the parallel (perpendicular) plasma pressure and $\hat{\mathbf{z}}$ is the unit vector, $\hat{\mathbf{z}} = \mathbf{B}/B$, along the magnetic field [e.g., Ref. 29].

We make use of a Cartesian coordinate system in which the unperturbed magnetic field $\mathbf{B}_0$ is directed along the z axis. The wave fields exhibit only the y component of the electric field E_y and the z and the x components of the magnetic field, δB_z and δB_x, respectively. The latter are connected through the property of solenoidality of the magnetic field, i.e. $\nabla\cdot\mathbf{B} \equiv \partial B_x/\partial x + \partial B_z/\partial z = 0$. As regards the fluid velocity $\mathbf{v} = (\mathbf{E}\times\mathbf{B})/B^2$, only its x component survives in the leading order, i.e. $v \equiv v_x$. The δB_y component corresponds to the so-called non-coplanar magnetic component, and does not enter our basic equations and can thus be set to zero.

Similarly, Eq. (1) shows that the only nonzero component of the electric current is the y component $J_y \equiv J$ given by

$$J = \frac{\rho}{B}\left(\frac{\partial v}{\partial t} + v\frac{\partial v}{\partial x}\right) + \frac{1}{B}\frac{\partial p_\perp}{\partial x} + \frac{\partial}{B\partial z}\frac{p_\parallel - p_\perp}{B}\frac{\partial}{\partial z}\int \delta B_z dx. \tag{3}$$

The physical meaning of the different terms on the right-hand side of equation (3) is as follows: The first one, containing the Lagrangian time derivative, is due to plasma inertia. This term is important for the fast magnetosonic solitons (MS) in which $\omega^2 \simeq k^2 v_A^2(1+\beta_\perp)$, where ω and k are the characteristic wave frequency and wave number, respectively, and v_A is the Alfvén velocity. We note, that in a high-β plasma the Alfvén velocity is of the order of the ion thermal velocity v_T. The propagation of MS solitons nearly perpendicular to the external magnetic field in non-Maxwellian plasmas has recently been considered in Refs. 30,31. However, for mirror waves, which are the subject of our study, the situation is quite different. These waves are found in the so-called mirror approximation, $\omega \ll k_\parallel \left(n^{-1}\left\langle v_\parallel^2 F\right\rangle\right)^{\frac{1}{2}}$ [cf. Refs. 18,19]. Here $k_\parallel$ is the typical parallel wave number, n is the plasma number density, $v_\parallel$ is the parallel particle velocity and F is the undisturbed particle distribution function. In this limit the inertial term in equation (3) is small as $\omega^2/k^2 v_T^2 \ll 1$, relative to the second term and must thus be neglected. The differences in the mathematical description of the fast MS and mirror waves in the high-β plasmas have been extensively discussed recently.[32] Finally the last term in equation (3) describes the effect of magnetic field line bending. We note that since this term already contains a small parameter $k_\parallel^2/k_\perp^2 \propto \delta B_z/B_0 \propto \varepsilon \ll 1$ one may set here the value of the magnetic field B as well as the difference $p_\parallel - p_\perp$ equal to their unperturbed values.

Thus, in the mirror limit which is the subject of our study the inertial term can be neglected and the electric current J reduces to

$$J \simeq \frac{1}{B}\frac{\partial \delta p_\perp}{\partial x} - \frac{\beta_\perp - \beta_\parallel}{2\mu_0}\frac{\partial^2}{\partial z^2}\int \delta B_z(x')dx', \tag{4}$$

where $\beta_{\perp(\parallel)} = 2\mu_0 p_{\perp(\parallel)}/B_0^2$ is the perpendicular (parallel) plasma beta, and μ_0 is the permeability of free space and delta denotes a perturbation.

Furthermore, the Ampére's law reads

$$\frac{\partial \delta B_x}{\partial z} - \frac{\partial \delta B_z}{\partial x} = \mu_0 J. \tag{5}$$

Substituting (4) into (5) and integrating over x one obtains a nonlinear equation for the perpendicular plasma balance condition

$$\delta p_\perp + \frac{B_0 \delta B_z}{\mu_0} + \frac{\delta B_z^2}{2\mu_0}$$

$$= -\left(1 + \frac{\beta_\perp - \beta_\parallel}{2}\right) \frac{\partial^2}{\partial z^2} \int\int \frac{B_0 \delta B_z}{\mu_0} dx' dx'', \tag{6}$$

where the variation of the plasma perpendicular pressure can be found from the Vlasov equation for the particle distribution function. In the linear approximation equation (6) coincides with the perpendicular plasma balance condition that was usually used for the study of linear MI.[19]

3. FLR Effect

The pressure balance condition (6) can easily be generalized to the case when the FLR effect is taken into consideration. This effect has been extensively studied for mirror waves by many authors.[33–36] It was shown that when the transverse scale size of the mirror wave becomes comparable to the ion Larmor radius, the actual impact of the wave magnetic and electric fields is reduced due to partial cancellation of it integrated over the ion Larmor orbit. This leads to modification of the expression for Fourier transform $(\delta p_\perp(k_\perp) = \int \delta p_\perp(\mathbf{r}_\perp) e^{-i\mathbf{k}_\perp \cdot \mathbf{r}_\perp} d\mathbf{r}_\perp)$ of the variation of the perpendicular plasma pressure $\delta p_\perp \to [I_0(z_i) - I_1(z_i)] e^{-z_i} \delta p_\perp$, where I_0 and I_1 are the modified Bessel functions of the zero and the first order, respectively, and $z_i = -\rho_i^2 \nabla_\perp^2$, with ρ_i being the ion Larmor radius and $\nabla_\perp^2 \equiv \partial^2/\partial x^2$. This effect was first introduced in Ref. 37 for stabilization of flute instability. Later it was widely applied by many authors to other problems of plasma physics. In particular, this effect is similar to the FLR suppression of the perpendicular electric current described in Ref. 38 [cf. Ref. 39] for kinetic Alfvén waves. When the FLR effect is finite but small, i.e. $z_i \ll 1$, we have $\delta p_\perp \to (1 - \frac{3}{2} z_i) \delta p_\perp = (1 + \frac{3}{2} \rho_i^2 \nabla_\perp^2) \delta p_\perp$. Using these considerations Eq. (6) reduces to

$$\delta p_\perp + \frac{B_0 \delta B_z}{\mu_0} - \frac{3}{2}\rho_i^2 \nabla_\perp^2 \frac{B_0 \delta B_z}{\mu_0} + \frac{\delta B_z^2}{2\mu_0}$$

$$= -\left(1 + \frac{\beta_\perp - \beta_\parallel}{2}\right) \nabla_\perp^{-2} \frac{\partial^2}{\partial z^2} \frac{B_0 \delta B_z}{\mu_0}. \tag{7}$$

Here and in what follows we make use of the following hierarchy of small parameters

$$\rho_i^2 \nabla_\perp^2 \simeq \nabla_\perp^{-2} \frac{\partial^2}{\partial z^2} \simeq \frac{\delta B_z}{B_0} \simeq \varepsilon. \tag{8}$$

According to this hierarchy the FLR corrections in the nonlinear term and in the term containing the z derivative are neglected as corrections of higher order on a small parameter ε.

4. MI with FLR Effect

In order to derive the equation governing the dynamics of the MI it is necessary to calculate the variation of the plasma pressure. For that purpose we make use of the drift kinetic equation. Since we are only interested in waves with frequencies much lower than the ion cyclotron frequency we use the distribution function averaged over the ion gyromotion, $f(t,\, \mathbf{r}_g,\, v_\perp,\, v_\parallel)$, where t is the time, $\mathbf{r}_g$ is the guiding center position vector, $v_\perp$ is the gyration speed, and $v_\parallel$ is the velocity along the field lines. The distribution function obeys the Vlasov equation

$$\frac{\partial f}{\partial t} + \dot{\mathbf{r}}_g \cdot \nabla f + \dot{v}_\perp \frac{\partial f}{\partial v_\perp} + \dot{v}_\parallel \frac{\partial f}{\partial v_\parallel} = 0, \tag{9}$$

where $\dot{\mathbf{r}}_g = \mathbf{v}_E + v_\parallel \hat{\mathbf{z}}$, $\mathbf{v}_E$ is the $\mathbf{E} \times \mathbf{B}$ velocity, and $\dot{v}_\perp$ is given by the expression for the conservation of the magnetic moment, $d\mu/dt = 0$, which implies

$$\dot{v}_\perp = \frac{1}{2}\frac{\dot{B}}{B} v_\perp, \tag{10}$$

and $\dot{v}_\parallel$ is given by the parallel component of the equation of the ion motion.[40]

$$m\dot{v}_\parallel = -\mu \hat{\mathbf{z}} \cdot \nabla B, \tag{11}$$

where m is the ion mass.

Let us first analyze MI in the linear approximation. To the first order in the perturbations we have $\dot{v}_\perp \simeq (v_\perp v_\parallel/2)\partial/\partial z(\delta B_z/B_0)$ and $\dot{v}_\parallel \simeq -(v_\perp^2/2)\partial/\partial z(\delta B_z/B_0)$. Substituting these expressions into equation (9), after Fourier transformation and division by the factor $(\omega - k_\parallel v_\parallel)$ in Eq. (9) one finds

$$\delta f^{(1)} = -\mu \frac{\partial F}{\partial \mu}\frac{\delta B_z}{B_0} - \frac{\omega \mu B_0}{\omega - k_\parallel v_\parallel}\frac{\partial F}{\partial W}\frac{\delta B_z}{B_0}, \tag{12}$$

where F is the ion unperturbed distribution function and $\delta f^{(1)}$ is the perturbation of the distribution function in the linear approximation, respectively. In (12) we passed to the new variables, particle magnetic moment and total energy $\mu = mv_\perp^2/2B_0$ and $W = m(v_\parallel^2 + v_\perp^2)/2$. Expression (12) agrees with all previous calculations of the distribution function based on Leuville's theorem.[18,32,41,42] We note, however, that expression for $\delta f^{(1)}$ obtained in Ref. 26 cannot be reduced to that given by equation (12).

The physical content of the different terms in equation (12) is as follows: The first one on the right corresponds to the "mirror effect", i.e. it represents the change in the distribution function that is associated with the expulsion of particles from the regions of increased magnetic field by the quasi-static compressive magnetic field perturbation δB_z. The second term describes the resonant wave-particle interactions contributing to the mirror mode. Note that if ω is small as we expect, at least for marginal stability, then the second term in equation (12) is negligible except for particles with $v_\parallel = 0$. These particles, which we term resonant particles, do not move along the magnetic field and their contribution can be comparable with the hydromagnetic term corresponding to the "mirror effect".

Using Eq. (12) one easily finds the variation of the plasma pressure in the linear approximation

$$\delta p_\perp^{(1)} = \left\langle \mu B_0 \delta f^{(1)} \right\rangle \equiv \sum_{\sigma=\pm 1} \frac{2\pi}{m^{\frac{3}{2}}} \int_0^\infty dW \int_0^{W/B_0} \mu d\mu \frac{B_0^2 \delta f^{(1)}}{[2(W - \mu B_0)]^{\frac{1}{2}}}, \tag{13}$$

where the parentheses $\langle ... \rangle$ denote the averaging process over the full velocity space, the parameter σ has the values ± 1 that corresponds to the direction of particle velocity along ($\sigma = 1$) or opposite ($\sigma = -1$) to the ambient magnetic field.

The substitution of (12) into (13) gives

$$\delta p_{\perp}^{(1)} = -\left\langle \mu^2 B_0 \frac{\partial F}{\partial \mu} \right\rangle \frac{\delta B_z}{B_0} - \left\langle \frac{\omega(\mu B_0)^2}{\omega - k_{\parallel} v_{\parallel}} \frac{\partial F}{\partial W} \right\rangle \frac{\delta B_z}{B_0}. \tag{14}$$

As was already mentioned the mirror modes are found in the mirror approximation, $\omega \ll k_{\parallel}[n^{-1}\left\langle v_{\parallel}^2 F\right\rangle]^{\frac{1}{2}}$. In this case the term containing resonant wave-particle interaction in Eq. (14) can be further simplified, i.e.

$$\left\langle \frac{\omega(\mu B_0)^2}{\omega - k_{\parallel} v_{\parallel}} \frac{\partial F}{\partial W} \right\rangle \simeq -\frac{2\pi^2 i}{m} \frac{\omega}{|k_{\parallel}|} \int_0^{\infty} W^2 \frac{\partial F_{res}}{\partial W} dW$$

$$-\frac{4\pi}{m^{\frac{1}{2}}} \frac{\omega^2}{k_{\parallel}^2} \int_0^{\infty} dW \mathrm{P} \left\{ \int_0^{W/B_0} \frac{\mu^2 B_0^3}{[2(W - \mu B_0)]^{\frac{3}{2}}} \frac{\partial F}{\partial W} d\mu \right\}, \tag{15}$$

where

$$\frac{\partial F_{res}}{\partial W} \equiv \left(\frac{\partial F}{\partial W} \right)_{\mu = W/B_0}. \tag{16}$$

In Eq. (15) P {...} denotes the principal value of the integral. The first term on the right-hand side of this equation describes the contribution of resonant particles with velocity $v_{\parallel} = 0$. These particles do not move along the magnetic field line, and thus for them the condition $\mu = W/B_0$ holds. The second term corresponds to the adiabatic response of the nonresonant particles. In the limit $\omega \ll k_{\parallel}\left(n^{-1}\left\langle v_{\parallel}^2 F\right\rangle\right)^{\frac{1}{2}}$ this term can be neglected. In this case (14) reduces to

$$\delta p_{\perp}^{(1)} = -b p_{\perp} \frac{\langle \mu^2 \partial F / \partial \mu \rangle}{\langle \mu F \rangle} - b \frac{2\pi^2 i \omega D}{m\,|k_{\parallel}|}, \tag{17}$$

where $p_{\perp}$ denotes unperturbed plasma pressure, $b = \delta B_z / B_0$ is the dimensionless wave amplitude and we have introduced the notation

$$D = -\int_0^{\infty} W^2 \frac{\partial F_{res}}{\partial W} dW. \tag{18}$$

Under normal conditions in a plasma the particle distribution function in the range of resonance decreases with increasing energy such that

$\partial F_{res}/\partial W < 0$, corresponding to $D > 0$. The analysis of the particular case when $\partial F_{res}/\partial W > 0$ and hence $D < 0$ requires special considerations. It has been extensively studied in Ref. 43 and corresponds to the case of the so-called halo instability. This, however, is beyond the scope of the present study.

Expression (17) can also be re-written in the form[19]

$$\delta p_{\perp}^{(1)} = -2bAp_{\perp} - b\frac{2\pi^2 i\omega D}{m\left|k_{\parallel}\right|}, \tag{19}$$

where the parameter A appearing in Eq. (17) is defined as

$$A = -\frac{\langle \mu^2 B_0^2 \partial F/\partial W\rangle}{2p_{\perp 0}} - 1. \tag{20}$$

The physical content of this parameter can easily be understood when inserting a specific velocity distribution function into Eq. (20). For a bi-Maxwellian plasma with parallel and perpendicular temperatures $T_{\parallel}$, $T_{\perp}$ relation (20) reduces to the familiar anisotropy factor, $A = T_{\perp}/T_{\parallel} - 1$. Hence A is a generalized anisotropy factor.

Substituting (19) into the Fourier transformed pressure balance condition (7), to the first order in perturbations one obtains

$$\Delta - \frac{3}{2\beta_{\perp}}k_{\perp}^2\rho_i^2 + \frac{i\pi^2 D\omega}{mp_{\perp}\left|k_{\parallel}\right|} = 0, \tag{21}$$

where the parameters Δ and K are given by

$$\Delta = K - \frac{k_{\parallel}^2}{\beta_{\perp}k_{\perp}^2}\left(1 + \frac{\beta_{\perp} - \beta_{\parallel}}{2}\right), \tag{22}$$

$$K = A - \frac{1}{\beta_{\perp}}. \tag{23}$$

From equation (21) one finds that $\mathrm{Re}\omega = 0$ (zero-frequency mode) and

$$\mathrm{Im}\omega = \gamma_L = \frac{mp_{\perp}\left|k_{\parallel}\right|}{\pi^2 D}\left[K - \frac{k_{\parallel}^2}{\beta_{\perp}k_{\perp}^2}\left(1 + \frac{\beta_{\perp} - \beta_{\parallel}}{2}\right) - \frac{3}{2\beta_{\perp}}k_{\perp}^2\rho_i^2\right]. \tag{24}$$

Let us now obtain the condition of maximum growth of the mirror mode as this for certain problems is of utmost importance. For a fixed perpendicular wave number $k_\perp$, maximum growth occurs at a ratio of parallel to perpendicular wave numbers given by

$$\left(\frac{k_\parallel}{k_\perp}\right)^2_{\max} = \frac{\beta_\perp}{3}\frac{K - \frac{3}{2\beta_\perp}k_\perp^2\rho_i^2}{1 + \frac{1}{2}(\beta_\perp - \beta_\parallel)}. \tag{25}$$

Substituting (25) into (24) and maximizing the growth rate over $k_\perp$ we find that it attains its maximum value at

$$(k_\perp\rho_i)^2_{\max} = \beta_\perp\frac{K}{6} \qquad \text{and} \qquad (k_\parallel\rho_i)^2_{\max} = \frac{\beta_\perp^2}{1 + \frac{1}{2}(\beta_\perp - \beta_\parallel)}\frac{K^2}{24}. \tag{26}$$

Using Eq. (26) from (24) we obtain the expression for the maximum growth rate

$$\gamma_{\max} = \frac{\gamma_0}{4\sqrt{6}} \tag{27}$$

where

$$\gamma_0 = \frac{mp_\perp\beta_\perp}{\pi^2\rho_i D}\frac{K^2}{\left(1 + \frac{\beta_\perp - \beta_\parallel}{2}\right)^{\frac{1}{2}}}. \tag{28}$$

For a bi-Maxwellian distribution function we have

$$\gamma_0 = \omega_{ci}\left(\frac{2}{\pi}\right)^{\frac{1}{2}} T_\perp^{\frac{3}{2}} T_\parallel^{-\frac{3}{2}} \frac{K^2}{\left(1 + \frac{\beta_\perp - \beta_\parallel}{2}\right)^{\frac{1}{2}}}. \tag{29}$$

5. MI Nonlinear Equation

Let us now consider how the linear theory is modified in the presence of nonlinear effects. For that we re-write equation (7) in the form

$$\frac{\delta p_\perp^{(1)} + \delta p_\perp^{(2)}}{2p_\perp} + \frac{1}{\beta_\perp}\left(1 - \frac{3}{2}\rho_i^2\nabla_\perp^2\right)b + \frac{b^2}{2\beta_\perp}$$
$$= -\frac{1}{\beta_\perp}\left(1 + \frac{\beta_\perp - \beta_\parallel}{2}\right)\nabla_\perp^{-2}\frac{\partial^2}{\partial z^2}b, \tag{30}$$

where the variation of the perpendicular plasma pressure is decomposed into its linear $\delta p_\perp^{(1)}$ and nonlinear $\delta p_\perp^{(2)}$ parts. Here $\delta p_\perp^{(2)}$ is the variation of the plasma pressure of the second order on the dimensionless wave amplitude $b = \delta B_z/B_0$.

In the second order in the parameter b we have

$$\dot{v}_\perp \simeq \frac{v_\perp v_\parallel}{2}\left(\frac{\partial b}{\partial z} - b\frac{\partial b}{\partial z} + \frac{\delta B_x}{B_0}\frac{\partial b}{\partial x}\right), \tag{31}$$

$$\dot{v}_\parallel = -\frac{v_\perp^2}{2}\left(\frac{\partial b}{\partial z} - b\frac{\partial b}{\partial z} + \frac{\delta B_x}{B_0}\frac{\partial b}{\partial x}\right). \tag{32}$$

Substituting (31) and (32) into the Vlasov equation (9) one obtains

$$\frac{\partial \delta f^{(2)}}{\partial z} = -\frac{\delta B_x}{B_0}\frac{\partial \delta f^{(1)}}{\partial x} - \frac{\partial b}{\partial z}\mu\frac{\partial \delta f^{(1)}}{\partial \mu}$$

$$+\left(b\frac{\partial b}{\partial z} - \frac{\delta B_x}{B_0}\frac{\partial b}{\partial x}\right)\mu\frac{\partial F}{\partial \mu}. \tag{33}$$

Equation (33) is too difficult for analysis since it cannot be integrated over the spatial coordinates z and x. We note that in the similar analysis of Ref. 26 the nonlinear terms containing δB_x were not taken into consideration. This has led to the erroneously conclusion that differential equation for $\delta f^{(2)}$ can be fully integrated in the general case.

Here, we limit our consideration to the case when all perturbed quantities depend solely on one variable $\zeta = x + \alpha z$, where α is the angle of the wave propagation. Since $\delta f^{(1)} \simeq b = \delta B_z/B_0$ and the perturbed magnetic field $\delta\mathbf{B}$ obeys the property of solenoidality, $\partial\delta B_x/\partial x + \partial\delta B_z/\partial z = 0$, in Eq. (33) one can make the following replacements $\delta B_x\partial\delta f^{(1)}/\partial x \to -\alpha\delta B_z\partial\delta f^{(1)}/\partial\zeta$ and $b\partial b/\partial z - (\delta B_x/B_0)\partial b/\partial x \to 2\alpha b\partial b/\partial\zeta$. Then equation (33) can easily be integrated over ς and $\delta f^{(2)}$ reduces to

$$\delta f^{(2)} = b^2\left(\mu\frac{\partial F}{\partial \mu} + \frac{1}{2}\mu^2\frac{\partial^2 F}{\partial \mu^2}\right). \tag{34}$$

Taking the second moment of the distribution function (34) one finds

$$\delta p_\perp^{(2)} = b^2\left\langle \mu B_0\left(\mu\frac{\partial F}{\partial \mu} + \frac{1}{2}\mu^2\frac{\partial^2 F}{\partial \mu^2}\right)\right\rangle, \tag{35}$$

or

$$\frac{\delta p_{\perp}^{(2)}}{p_{\perp}} = 2b^2 A + b^2 \frac{\langle \mu^3 \partial^2 F/\partial \mu^2 \rangle}{2 \langle \mu F \rangle} \equiv 2\Lambda b^2, \tag{36}$$

where A is the generalized anisotropy factor given by equation (20) and Λ is

$$\Lambda = A + \frac{\langle \mu^3 \partial^2 F/\partial \mu^2 \rangle}{4 \langle \mu F \rangle}. \tag{37}$$

For a bi-Maxwellian distribution function $F \propto \exp(\mu B_0 A/T_{\perp} - W/T_{\parallel})$ and $\Lambda = A + (3/2)A^2$, where $A = T_{\perp}/T_{\parallel} - 1$. We note that Λ vanishes in pure Maxwellian plasmas when the ion distribution depends solely on particle energy W.

Substituting (19) and (36) into Eq. (30) and making an inverse Fourier transformation we find

$$\left(\Delta + \frac{3}{2\beta_{\perp}} \rho_i^2 \frac{\partial^2}{\partial \varsigma^2}\right) b - \left(\Lambda + \frac{1}{2\beta_{\perp}}\right) b^2 - \frac{\pi^2 D}{m p_{\perp} \alpha} \hat{k}^{-1} \partial_t b = 0, \tag{38}$$

where Δ is

$$\Delta = K - \frac{\alpha^2}{\beta_{\perp}} \left(1 + \frac{\beta_{\perp} - \beta_{\parallel}}{2}\right). \tag{39}$$

Furthermore, the operator $\hat{k} = \hat{H}\partial/\partial\zeta$ is a positive definite integral operator and $\hat{H}$ is the Hilbert transform [e.g., Ref. 44]

$$\hat{H} f(\zeta) = \frac{1}{\pi} \mathrm{P} \int\limits_{-\infty}^{\infty} \frac{f(\zeta')}{\zeta - \zeta'} d\zeta'. \tag{40}$$

It is convenient to pass to the dimensionless variables ξ and τ according to

$$\xi = \frac{\zeta}{\rho_i} (\beta_{\perp} K)^{\frac{1}{2}} \qquad \text{and} \qquad \tau = \gamma_0 t. \tag{41}$$

Then we have

$$\Delta + \frac{3}{2\beta_{\perp}} \rho_i^2 \frac{\partial^2}{\partial \zeta^2} = K \left(1 - \chi^2 + \frac{3}{2} \frac{\partial^2}{\partial \xi^2}\right), \tag{42}$$

where

$$\chi = \alpha \left(1 + \frac{\beta_\perp - \beta_\parallel}{2}\right)^{\frac{1}{2}} (\beta_\perp K)^{-\frac{1}{2}} . \tag{43}$$

With the help of expression (42) equation (38) reduces to

$$\frac{\partial h}{\partial \tau} = \chi \hat{k}_\xi \left[\left(1 - \chi^2 + \frac{3}{2}\frac{\partial^2}{\partial \xi^2}\right) h - h^2 \right] , \tag{44}$$

where $\hat{k}_\xi = \hat{H}\partial/\partial\xi$ and the normalized amplitude h is

$$h = \frac{\Lambda + (2\beta_\perp)^{-1}}{K} b. \tag{45}$$

Maximizing Eq. (44) over χ we find that the fastest growing mode corresponds to $\chi = 3^{-\frac{1}{2}}$ and thus the equation governing the nonlinear dynamics of this mode is

$$\frac{\partial h}{\partial \tau} = \frac{2\hat{k}_\xi}{3^{\frac{3}{2}}} \left[\left(1 + \frac{9}{4}\frac{\partial^2}{\partial \xi^2}\right) h - \frac{3}{2} h^2 \right] . \tag{46}$$

After rescaling $h = (3/2)h'$, $\tau = (3^{\frac{3}{2}}/2)\tau'$ and $\xi = (3/2)\xi'$ Eq. (46) reduces to

$$\frac{\partial h}{\partial \tau} = \hat{k}_\xi \left[\left(1 + \frac{\partial^2}{\partial \xi^2}\right) h - h^2 \right] , \tag{47}$$

where for the sake of convenience all the primes are omitted.

Eq. (47) describes the nonlinear dynamics of mirror waves in the case when the ion distribution function F does not vary in time. The case when it varies due to the particle trapping will be considered in the next section. Eq. (47) has been solved numerically by using pseudo-spectral scheme.[45] The spatial part of the function h in Eq. (47) has been decomposed into sparse grid by pseudo-spectral method and the time evolution has been given by the fourth-order Runge-Kutta code. The results of numerical simulations are shown in Figures 1 and 2 where the dimensionless amplitude h is plotted as a function of the spatial and temporal variables ξ and τ, respectively. At the initial moment, $\tau = 0$, the wave amplitude has been chosen in the form of a normal Gaussian distribution $h(\tau = 0) = A\exp(-\xi^2/2\sigma^2)$, where $A = -0.2$ is the wave amplitude and $\sigma = 0.05$ stands for the standard deviation.

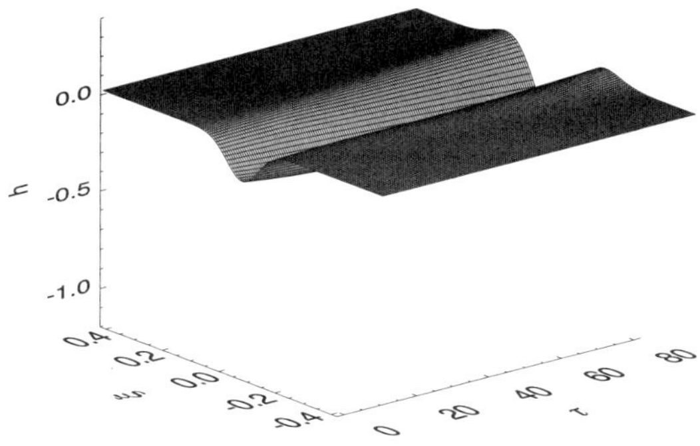

Fig. 1. A plot of h as a function of ξ and τ calculated with the help of equation (47) for the case of the Gaussian initial distribution when the nonlinear term is disregarded. The initial amplitude is $A = -0.2$.

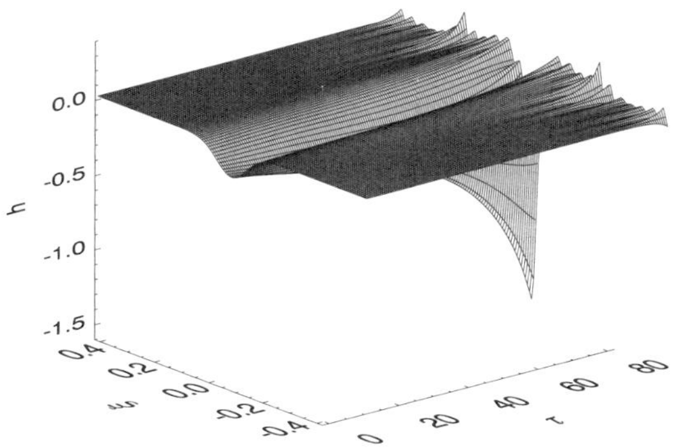

Fig. 2. Same as in Figure 1 but when the nonlinear term is retained. One sees that nonlinear evolution of the MI is accompanied by the formation of a finite-time singularity in the magnetic field.

The latter is connected with the full width at maximum amplitude (FWHA) $\Delta\xi$ through the ordinary relation $\Delta\xi = 2.35\sigma$. In order to stress the role of nonlinear effects this dependence is depicted in two different cases: when the nonlinear term in equation (47) is disregarded (Figure 1) and when both the linear and nonlinear terms are included (see Figure 2). Comparison of these two figures shows that the nonlinearity dramatically changes the evolution of mirror waves. One sees that the exponential increase in the wave amplitude that is inherent to the linear stage of the MI is then replaced by an even faster increase resulting in the formation of a magnetic trough or magnetic hole at the origin of ξ. Figure 2 shows that in a finite time the wave solution blows up. Figure 3 shows the evolution of the mirror mode in case when initial perturbation has the sinusoidal (single-mode) form with the wave amplitude $A = 0.2$. Similar to the previous case the nonlinear evolution possesses the blow-up behavior.

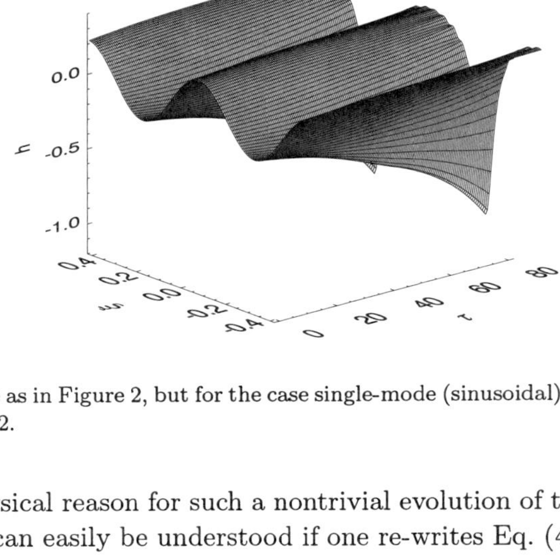

Fig. 3. Same as in Figure 2, but for the case single-mode (sinusoidal) initial perturbation with $A = -0.2$.

The physical reason for such a nontrivial evolution of the magnetic perturbations can easily be understood if one re-writes Eq. (47) in the Hamiltonian form

$$\frac{\partial h}{\partial \tau} = -\hat{k}_\xi \frac{\delta L}{\delta h}, \tag{48}$$

where

$$L = \int \left[-\frac{h^2}{2} + \frac{(\nabla_\perp h)^2}{2} + \frac{h^3}{3} \right] d\xi. \tag{49}$$

The functional L has the meaning of the free energy or the Lyapunov functional.

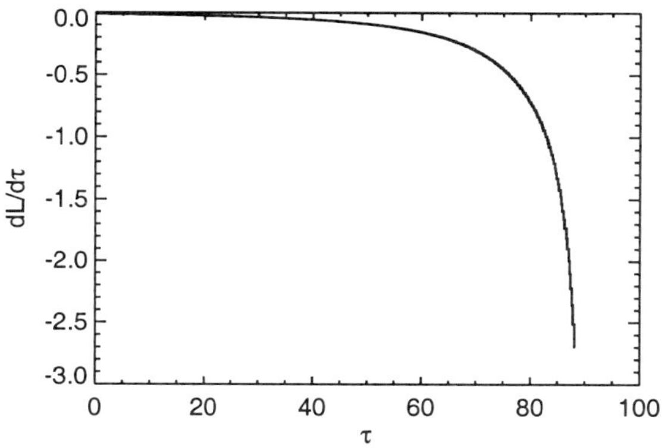

Fig. 4. A plot of the time derivative of the Lyapunov functional $dL/d\tau$ as a function of the dimensionless time τ.

The remarkable feature of this functional is that it cannot grow but solely monotonically decays in time, i.e. $dL/d\tau \leq 0$. This becomes clear from Figure 4 where $dL/d\tau$ is plotted as a function of time. The formation of singularity is connected with the fact that L becomes more and more negative. This effect is produced by the last term in the expression for the Lyapunov functional (49), i.e. by $\int (h^3/3) d\xi < 0$. Physically the blow up behavior arises due to the fast decrease in the free energy shown in Figure 4. We note that in Figures 1-3 the dimensionless amplitude h introduced here for numerical simulations can take large negative values because according to our scaling the dimensionless wave amplitude h introduced here for numerical simulations is in fact $\propto (\delta B_z/B_0)/K$. Since we have limited our consideration by the marginal conditions, i.e. $K \ll 1$ then even for $\delta B_z/B_0 \ll 1$ we have finite values of h.

It should be noted that our numerical modeling shows the appearance of solely magnetic holes. When the magnetic humps as an initial perturbation have been selected they rapidly relax and disappear within relatively short time. This agrees with the particle-in-cell simulations.[17]

The above consideration is valid only for the mirror mode. For the large amplitude fast MS waves the nonlinear scenario is different. These waves can propagate nearly perpendicular to the ambient magnetic field in the form of stable MS solitons.[30,31] Their shapes are sensitive to the details of the ion distribution function. For example, in a bi-Maxwellian plasma the dispersion of the fast MS waves is negative, i.e. the phase velocity decreases with an increase of the wavenumber. This assumes that the solitary solution in this case has the form of a "bright" soliton with the magnetic field increased. On the contrary, in some non-Maxwellian plasmas, such as those with ring-type ion distributions or Dory-Guest-Harris plasmas they can have a positive dispersion and thus solitary solutions may have the form of a magnetic hole ("dark solitons"). We note that analytical and numerical studies of nonlinear structures involving MS waves have also been the subject of previous research.[46,47]

6. Particle Trapping. Collapse Break-up

In the scenario described above it was assumed that unperturbed ion distribution function does not vary in time. This is valid as long as the effects due to particle trapping into the magnetic troughs do not play a role. The effects of particle trapping in the nonlinear mirror modes has been discussed by Pantellini et al.[41,48] These authors argued that the main mechanism that ends the linear phase of the MI has to be particle trapping into the mirror holes. It should be mentioned that in the formation of collapse the nonlinear corrections to the plasma pressure are quadratic in wave amplitude, i.e. $\delta p_\perp^{NL} \propto b^2$. However, the nonlinear corrections to the variation of the particle pressure due to the particle trapping can be as high as $|b|^{3/2}$. Thus, the change in the particle distribution function can develop faster than the wave collapse evolves. As will be shown below this effect can terminate the wave collapse and can result in a collapse break-up. It should be noted that nonlinear FLR effect that has been used in Ref. 26 for similar purposes is of the higher order in small parameter and thus is less important.

In what follows we adopt the following model: - Due to the rapid motion of resonant particles we assume that in the vicinity of small parallel velocities the background ion distribution function would flatten and takes the shape of a quasi-plateau. This to happen does not require the assumption of random phases and is valid even in the single-mode (sinusoidal) regime:

- The width of such flattened region is of the order of the trapping zone in the mirror hole, i.e. $\simeq \Delta v_\parallel \simeq 2v_\parallel^* \simeq 2^{\frac{3}{2}}(\mu\,|\delta B_z|\,/m)^{\frac{1}{2}} = 2v_\perp(|\delta B_z|\,/B_0)^{\frac{1}{2}} = 2v_\perp\,|b|^{\frac{1}{2}}$ (see Figure 5).

- The particle trapping is irreversible, i.e. accompanied by bifurcation and violation of adiabatic invariance since the depth of the mirror hole increases more slowly for the trapped particles that possess the variation of the characteristic velocities $\delta v_\parallel \simeq v_T\,|b|^{\frac{1}{2}}$ whereas for the nonresonant particles $\delta v_\parallel \simeq \delta v_\perp \simeq v_T\,|b|$.

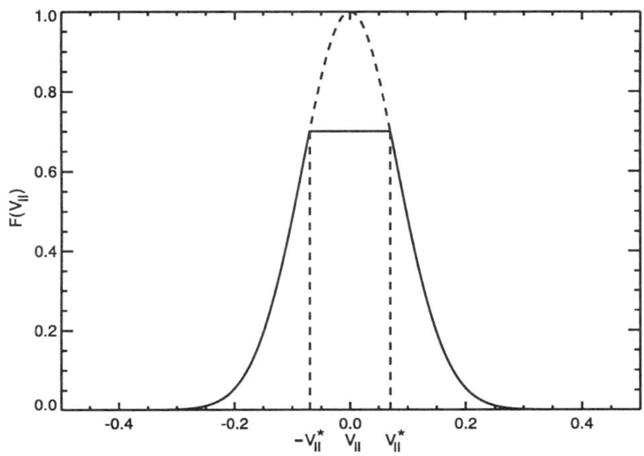

Fig. 5. The ion distribution function with the "plateau" stretched along the parallel velocities.

In order to construct a self-consistent model describing the evolution and final saturation of mirror waves we interpolate the background ion distribution function as a rigorous plateau stretched along $v_\parallel$ from $-v_\perp |b|^{\frac{1}{2}}$ to $v_\perp |b|^{\frac{1}{2}}$ (Figure 5). Outside this region the ion distribution function takes a bi-Maxwellian form. Clearly, such interpolation could be valid up to the numerical coefficient of the order of unity. Its relevant value can be found only by numerical simulation of the time dependent process of particle trapping by the emerging magnetic holes prior to the MI saturation. In somewhat similar problem of saturation of a single mode electrostatic beam instability due to the trapping of the electrons into electrostatic potential wells such computational analysis established that equivalent numerical coefficient appeared to be close to unity.[49]

In our case with the assumed form of the background distribution function, the contribution of the perpendicular plasma pressure variation due to the resonant particles entering the plasma pressure balance condition reduces to

$$\delta p_\perp^{res} = -\pi m \frac{\partial b}{\partial t} \int\limits_0^\infty v_\perp^5 dv_\perp \lim_{\nu \to 0} \int\limits_{v_\parallel^*}^\infty dv_\parallel \frac{\nu}{\nu^2 + k_\parallel^2 v_\parallel^2} \frac{\partial F}{v_\parallel \partial v_\parallel}, \tag{50}$$

where $v_\parallel^* = v_\perp |b|^{\frac{1}{2}}$ and F is given by

$$F = \frac{n}{\pi^{3/2} v_{T_\perp}^2 v_{T_\parallel}} e^{-\frac{v_\perp^2}{v_{T_\perp}^2} - \frac{v_\parallel^2}{v_{T_\parallel}^2}}. \tag{51}$$

Here n is the unperturbed plasma number density, and $v_{T_\perp}$ and $v_{T_\parallel}$ are the thermal perpendicular and parallel velocities, respectively.

When $v_\parallel^* \ll v_{T_\parallel}$, i.e. $|b| \ll 1$, from (50) and (51) one obtains

$$\delta p_\perp^{res} = \frac{2nm}{\pi^{1/2} \left|k_\parallel\right| v_{T_\perp}^2 v_{T_\parallel}^3} \frac{\partial b}{\partial t} \lim_{\nu \to 0} \int_0^\infty v_\perp^5 e^{-\frac{v_\perp^2}{v_{T_\perp}^2}} dv_\perp \left(\frac{\pi}{2} - \arctan \frac{v_\perp \left|k_\parallel\right| |b|^{\frac{1}{2}}}{\nu} \right) \tag{52}$$

If the wave amplitude is small and the plateau is very narrow, $\Delta v_\parallel = 2v_\parallel^* \ll \nu / \left|k_\parallel\right|$, the perturbation of the plasma pressure of the resonant particle reduces to its linear (Landau) value, i.e.

$$\delta p_\perp^{res} = p_\perp \frac{2\pi^{1/2}}{\left|k_\parallel\right| v_{T_\parallel}} \frac{T_\perp}{T_\parallel} \frac{\partial b}{\partial t}. \tag{53}$$

In this case the substitution of $\delta p_\perp^{res}$ into the perpendicular plasma pressure condition recovers the standard expression for the linear MI growth rate when the undisturbed ion distribution function is bi-Maxwellian.[1]

On the contrary, when the plateau is saturated, i.e. $\left|k_\parallel\right| v_\parallel^* \gg \nu$, we expand $\arctan(v_\perp \left|k_\parallel\right| |b|^{\frac{1}{2}} / \nu) \to \pi/2 - \nu / v_\perp \left|k_\parallel\right| |b|^{\frac{1}{2}}$ and thus $\delta p_\perp^{res} \to 0$, i.e. the wave growth is terminated. These features of the resonant particle dynamics should be taken into consideration in the course of the derivation of the corresponding nonlinear equation (47). Taking into account the above considerations we conclude that the left-hand side of this equation should be replaced by a more complex expression that takes into account the effects due to the particle trapping. In the general case it has a quite complex form. For the qualitative conclusions we can use a simplified (model) form for this equation. The simplest extrapolation for the nonlinear equation that takes into account the basic features of the particle trapping effect can be written as

$$\left(1 - \frac{2}{\pi} \arctan \frac{|h|^{1/2}}{\lambda} \right) \frac{\partial h}{\partial \tau} = \hat{k}_\xi \left[\left(1 + \frac{\partial^2}{\partial \xi^2} \right) h - h^2 \right], \tag{54}$$

where λ is a numerical coefficient of the order of unity.

Eq. (54) has been solved numerically. The results are depicted in Figures 6 and 7 for the case of Gaussian and single-mode regimes, respectively.

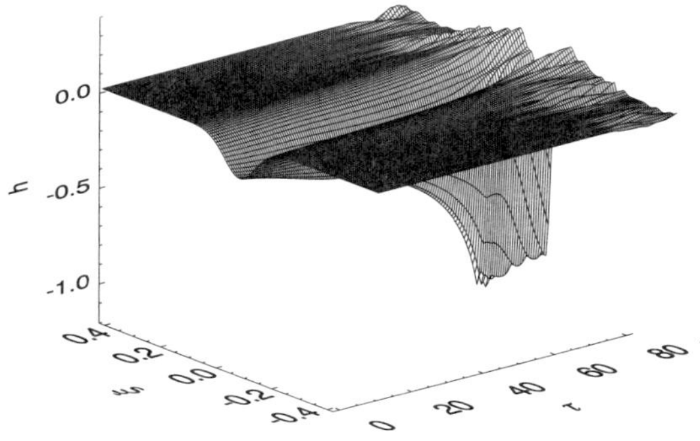

Fig. 6. Same as in Figure 2 but for the case when particle trapping is incorporated.

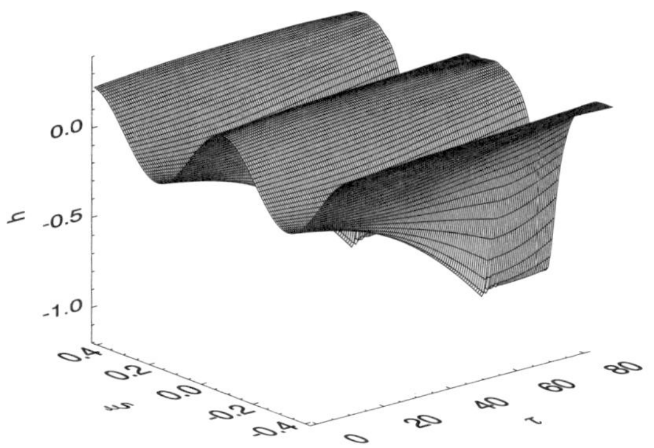

Fig. 7. Same as in Figure 6, but for the one-mode initial regime.

Figure 8 shows the evolution of the maximum depth of the magnetic hole. One sees that the collapse is arrested when $h \simeq -0.8$, then the magnetic hole depth possesses small oscillations and finally is saturated at this value.

We note that numerical simulation of the mirror mode nonlinear dynamics has been recently carried out in the framework of a fluid model that incorporates linear Landau damping and FLR corrections.[50] However, the effects due to nonlinear Landau interactions considered above were not taken into account in this study.

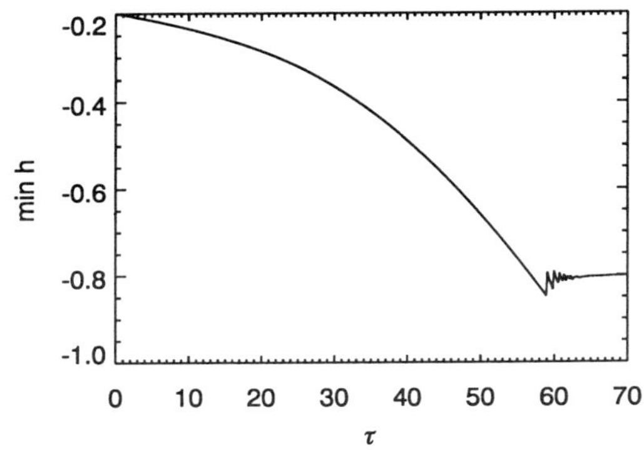

Fig. 8. A plot of maximum depth of the mirror hole as the function of time for single-mode regime.

7. Discussion and Conclusions

We have presented a local analysis of the MI in a high-β non-Maxwellian plasma in the framework of a standard mixed magnetohydrodynamic-kinetic theory accounting for the FLR effect and nonlinearity of the mirror wave perturbations near the instability threshold. It has been shown that nonlinear effects substantially modify the MI dynamics and can lead to the formation of magnetic holes similar to those observed in the satellite data. The characteristic scale of these holes is of the order of a few ion Larmor radii. It has been shown that the resonant ions having small parallel velocities play an important role in the MI nonlinear dynamics. It has been shown that the effect of plateau in the region of small parallel ion velocities can lead to saturation of the magnetic hole depth and results in the appearance of quasi-stationary mirror mode structures. Our paper generalizes previous studies related to the linear MI and provides some additional physical bases for better understanding of large amplitude nonlinear mirror waves in space and astrophysical plasmas.

It is worth mentioning that the model presented in this paper remains oversimplified. So far our consideration is valid solely for the marginal conditions when the parameter $K = A - \beta_{\perp}^{-1}$ is small. An additional analysis is required when the system is far from this state. Furthermore, in the present paper we have considered the case of cold electrons. Actually this is quite common for the magnetosheath where the ion temperature is always larger than the electron temperature. The incorporation of the finite electron temperature in the MI linear theory has been the subject in some previous studies.[41,42] In these papers it has been shown that when the electron temperature becomes of the order of the ion temperature the growth rate of the ion mirror mode is reduced by the presence of the field-aligned electric field. The origin of the electric field is the electron pressure gradient set up as electrons are dragged by the nonresonant ions that have been accelerated as they pass from regions of high magnetic flux into lower flux regions. All these factors have to be taken into account in a more comprehensive nonlinear theory of the MI.

The major and so far unresolved question of the problem at hand regarding the appearance of the MI in space plasmas focuses on the relation between the mirror like and ion-cyclotron instabilities (ICI). Both instabilities compete for the pressure anisotropy. In the anisotropic case it is believed that the latter should grow at faster rate and should therefore appear before the MI can set on. This question was urged in Refs. 51,52, where it was presented the results of numerical simulations of the fully kinetic dispersion relation describing both instabilities, the MI and ICI. The key to the understanding the problem was offered in some previous publications[53,54] in the course of the analysis of the Galileo magnetic field data on the edges of the cold Io wake. They showed that the multispecies content of the Io plasma may create more favorable conditions for the excitation of mirror like instabilities. The more species present in the plasma, the more ion cyclotron modes are possible in a system. However, the growth rate of each individual mode is now smaller than that of a single mode arising in an electron-proton plasma. On the other hand, the mirror like instability exhausts the free energy stored in all components of a multispecies plasma.

The intention of the present approach is to provide deeper insight into the physics of the MI. Hence this paper can be considered as an extension of our previous approach to the study of this instability which was limited by the consideration of a solely linear theory. The results of our study might be useful for a better understanding of the MI properties, as well as for the interpretation of recent satellite observations provided by the Cluster fleet.[55,56]

Acknowledgments

This research was supported by PPARC through grant PP/D002087/1, the Russian Fund for Basic Research grants No 06-05-65176 and 07-05-00774, by ISTC project No 3520 and by the Program of the Russian Academy of Sciences No 16 "Solar activity and physical processes in the Solar-Earth system". The authors are grateful to V. V. Krasnoselskikh, E. A. Kuznetsov and M. S. Ruderman for valuable discussions.

References

1. A. A. Vedenov and R. Z. Sagdeev, *Plasma Physics and the Problem of Controlled Thermonuclear Reactions*, **III** (Pergamon, Tarrytown, New York, 1958), p. 332.
2. L. I. Rudakov and R. Z. Sagdeev, A quasi-hydrodynamic description of a rarefied plasma in a magnetic field, in *Plasma Physics and the Problem of Controlled Thermonuclear Reactions*, **III** (Pergamon, Tarrytown, New York, 1958), p. 321.
3. S. A. Chandrasekhar, A. N. Kaufman, and K. M. Watson (1958), *Proc. R. Soc. London, Ser. A.* **245**, 435 (1958).
4. J. M. Turner, L. F. Burlaga, N. F. Ness, and J. F. Lemaire, *J. Geophys. Res.* **82**, 1921 (1977).
5. R. L. Kaufman, J. T. Horng, and A. Wolfe, *J. Geophys. Res.* **75**, 4666 (1970).
6. B. T. Tsurutani, E. J. Smith, R. R. Anderson, K. W. Ogilvie, J. D. Scudder, D. N. Baker, and S. J. Bame, *J. Geophys. Res.* **87**, 6060 (1982).
7. H. Lühr and N. Klöcker, *Geophys. Res. Lett.* **14**, 186 (1987).
8. R. A. Treumann, L. Brostrom, J. LaBelle and N. Scopke, *J. Geophys. Res.* **95**, 19,099 (1990).
9. M. B. Bavassano Cattaneo, C. Basile, G. Moreno, and J. D. Richarson, *J. Geophys. Res.* **103**, 11,961 (1998).
10. S. P. Joy, M. G. Kivelson, R. J. Walker, K. K. Khurana, C. T. Russell, and W. R. Paterson, *J. Geophys. Res.* **111**, A12212, doi:10.1029/2006JA011985 (2006).
11. Fitzengeiter, R. L., and L. F. Burlaga, *J. Geophys. Res.* **83**, 5579 (1978).
12. D. Winterhalter, D., M. Neugebauer, B. E. Goldstein, E. J. Smith, S. J. Bame and A. Balogh, *J. Geophys. Res.* **99**, 23,371 (1994).
13. M. L. Stevens and J. C. Kasper, *J. Geophys. Res.* **112,** A05109, doi:10.1029/2006JA12116.
14. C. T. Russell, C. T., W. Riedler, K. Schwingenshuh, and Y. Yaroshenko, *Geophys. Res. Lett.* **14**, 644 (1987).
15. K. H. Glassmeier, U. Motschmann, C. Mazelle, F. M. Neubauer, K. Sauer, S. A. Fuselier, and M. H. Acuña, *J. Geophys. Res.* **98**, 20,955 (1993).
16. Y. Liu, J. D. Richardson, J. W. Belcher, J. C. Kasper, and R. M. Skoud, *J. Geophys. Res.* **111**, A09108, doi:10.1029/2006JA011723 (2006).
17. K. Baumg̈artel, K. Sauer, and E. Dubinin, *Geophys. Res. Lett.* **30**, 1761, doi:10.1029/2003GL017373 (2003).
18. D. J. Southwood and M. G. Kivelson, *J. Geophys. Res.* **98**, 9181 (1993).
19. O. A. Pokhotelov, R. Z. Sagdeev, M. A. Balikhin, O. G. Onishchenko, V. P. Pavlenko, and I. Sandberg, *J. Geophys. Res.* **107**, 1312, doi:10.1029/2001JA009125 (2002).
20. H. Qu, Z. Lin and L. Chen, *Phys. Plasmas* **14**, 042108 (2007).
21. M. P. Leubner and N. Schupfer, *J. Geophys. Res.* **105**, 27387 (2000).

22. M. P. Leubner and N. Schupfer, *J. Geophys. Res.* **106**, 12993 (2001).
23. V. D. Shapiro and V. I. Shevchenko, *Soviet Physics-JETP* **18**, 1109 (1964).
24. F. G. E. Pantellini, *J. Geophys. Res.* **103**, 4789 (1998).
25. M. G. Kivelson and D. J. Southwood, *J. Geophys. Res.* **101**, 17,365 (1996).
26. Kuznetsov, E. A., T. Passot, and P. L. Sulem, *Phys. Rev. Lett.* **98**, 235003 (2007).
27. M. Heuer and E. Marsch, *J. Geophys. Res.* **112**, A03102, doi:10.1029/-2006JA011979 (2007).
28. E. A. Kuznetsov, *Chaos* **6**, 381 (1996).
29. N. A. Krall and A. W. Trivelpiece, *Principles of Plasma Physics* (McGraw-Hill, Inc., New York, 1973).
30. O. A. Pokhotelov, M. A. Balikhin, O. G. Onishchenko, and S. N. Walker, *Planet. Space Sci.* **55**, doi:10.1016/j.pss.2007.019 (2007).
31. O. A. Pokhotelov, O. G. Onishchenko, M. A. Balikhin, L. Stenflo and P. K. Shukla, *J. Plasma Phys.* **73**, S0022377807006526 (2007).
32. K. M. Ferrière and N. André, *J. Geophys. Res.* **107**, 1349, doi:10.1029/2002JA009273 (2002).
33. A. Hasegawa, *Phys. Fluids* **12**, 2642 (1969).
34. A. N. Hall, *J. Plasma Phys.* **21**, part 3, 431 (1979).
35. O. A. Pokhotelov, R. Z. Sagdeev, M. A. Balikhin, and R. A. Treumann, *J. Geophys. Res.* **109**, A09213, doi:10.1029/2004JA010568 (2004).
36. P. Hellinger, *Phys. Plasmas* **14**, 082105 (2007).
37. M. N. Rosenbluth, N. A. Krall and N. Rostoker, Nuclear Fusion, Suppl. 1, 143, (1962).
38. B. B. Kadomtsev, *Plasma turbulence*, (Academic, Can Diego, CA, 1965).
39. C. Z. Cheng, *J. Geophys. Res.* **96**, 21,159 (1991).
40. T. G. Northrop, *The Adiabatic Motion of Charged Particles*, (Wiley Interscience, New York, 1963).
41. F. G. E. Pantellini, and S. J. Schwartz, *J. Geophys. Res.* **100**, 3539 (1995).
42. O. A. Pokhotelov, M. A. Balikhin, H. St-C.K. Alleyne, and O. G. Onishchenko, *J. Geophys. Res.* **105**, 2393 (2000).
43. O. A. Pokhotelov, R. Z. Sagdeev, M. A. Balikhin, and R. A. Treumann, *J. Geophys. Res.* **110**, A10206, doi:10.1029/2004JA010933 (2005).
44. R. Bracewell, *The Fourier transform and its applications*, 2nd Edition. (MacGraw-Hill, 1986).
45. J. P. Boyd, *Chebyshev and Fourier Spectral Methods*, 2nd Edition (Dover Publications, New York, 2000).
46. A. N. Kaufman and L. Stenflo, *Phys. Scripta* **11**, 269 (1975).
47. R. P. Sharma and P. K. Shukla, *Phys. Fluids* **26**, (1983).
48. F. G. E. Pantellini, D. Burgess, and S. J. Schwartz, *Adv. Space Res.* **15**(8/9), 341 (1995).
49. B. D. Fried, C. S. Liu, R. Z. Sagdeev, and R. W. Means, *Bull. American Phys. Soc.* **15**, 1421 (1970).
50. Borgogno, D., T. Passot, and P. L. Sulem, *Nonlin. Proc. Geophys.* **14**, 373 (2007).
51. S. P. Gary, *J. Geophys. Res.* **97**, 8523 (1992).

52. S. P. Gary, M. F. Thomsen, L. Yin, and D. Winske, *J. Geophys. Res.* **100**, 21,961 (1995).
53. C. T. Russell, D. E. Huddleston, R. J. Strangeway, X. Blanco-Cano, M. G. Kivelson, K. K. Khurana, L. A. Frank, W. Paterson, D. A. Gurnett, and W. S. Kurth (1999), *J. Geophys. Res.* **104**, 17,471, (1999).
54. D. E. Huddleston, R. J. Strangeway, X. Blanco-Cano, C. T. Russel, M. G. Kivelson, and K. Khurana, K., *J. Geophys. Res.* **104**, 17,479 (1999).
55. Y. Narita, K.-H. Glassmeier, K.-H. Fornaçon, I. Richter, S. Schäfer, U. Motschmann, I. Dandouras, H. Rème, and E. Georgescu, *J. Geophys. Res.* **111**, A01203, doi:10.1029/2005JA011231 (2006).
56. Y. Hobara, S. N. Walker, M. Balikhin, O. A. Pokhotelov, M. Dunlop, H. Nilsson, and H. Remé, *J. Geophys. Res.* **112**, A07202, doi:10.1029/2006JA012142 (2007).

FORMATION OF MIRROR STRUCTURES NEAR INSTABILITY THRESHOLD

E.A. Kuznetsov

P.N. Lebedev Physical Institute RAS, 53 Leninsky Ave., 119991 Moscow, Russia and L.D. Landau Institute for Theoretical Physics RAS, 2 Kosygin str., 119334 Moscow, Russia
E-mail: kuznetso@itp.ac.ru

T. Passot and P.L. Sulem

CNRS, Observatoire de la Côte d'Azur,
PB 4229, 06304 Nice Cedex 4, France
E-mails: passot@oca.eu, sulem@oca.eu

We briefly review recent asymptotic and phenomenological models, aimed to understand the formation of pressure-balanced mirror structures, in the form of magnetic holes and humps, observed in the solar wind and in planetary magnetosheaths, and also obtained by direct numerical simulations of the Vlasov-Maxwell equations.

Keywords: magnetic holes, magnetic humps, mirror modes

1. Introduction

In regions of planetary magnetosheaths close to the magnetopause, and also in the solar wind, satellite observations commonly reveals the presence of quasi-static (in the plasma reference frame) coherent structures, in the form of depressions (magnetic holes) or increase (magnetic humps) of the local magnetic field intensity, elongated in directions making small angles with the ambient field (see e.g. [1] and references therein). These structures that are in pressure balance, correspond to magnetic fluctuations anti-correlated with density and pressure variations. A typical magnitude of the magnetic fluctuations is about 20% of the mean magnetic field value B_0. and can sometimes achieve 50 %. Their characteristic width is of the order of a few ion Larmor radii, and they display an aspect ratio of about 7-10. The origin of these structures is not fully understood, but they are often viewed as associated with the nonlinear development of the mirror instability, a kinetic instability first predicted by Vedenov and Sagdeev.[2]

The linear mirror instability near threshold has been extensively studied both analytically (see[3,4] for recent references), and by means of particle-in-cell (PIC) simulations.[5] This instability develops in a collisionless plasma, when the anisotropy of the ion temperature exceeds a threshold that for cold electrons and a bi-Maxwelian proton distribution is given by

$$\Gamma^* \equiv \beta_\perp \left(\frac{\beta_\perp}{\beta_\parallel} - 1 \right) - 1. \qquad (1)$$

Here $\beta_\perp = 8\pi p_\perp^{(0)}/B_0^2$ (similarly, $\beta_\parallel = 8\pi p_\parallel^{(0)}/B_0^2$), where $p_\perp^{(0)}$ and $p_\parallel^{(0)}$ are the equilibrium perpendicular and parallel plasma pressures respectively, and B_0 the ambient magnetic field. The nonlinear saturation of this instability is still poorly understood, and the origin of the observed structures remains in fact partly unsettled. Magnetic holes are for example also observed in regions where the plasma is linearly stable. Furthermore, in realistic situations, the mirror instability can be competing with the ion cyclotron anisotropic instability, especially at relatively low β and directions making a moderate angle with the ambient field.[6]

Inspection of mirror-like structures recorded by spacecraft missions early suggested that magnetic humps are preferentially present in regions of relatively low magnetic field, while magnetic holes are rather observed in regions of high field.[7] More precisely, the presence of the former or latter structures is strongly correlated with larger or smaller value of the parameter β.[8] A more quantitative picture is obtained when characterizing the nature of the magnetic structures by the skewness of the magnetic fluctuations that appears to be directly related to the distance to threshold.[9,10] Negative skewness (magnetic holes) is observed below or slightly above threshold, while positive skewness (magnetic humps) are measured in more unstable regimes. The phenomenon of bistability (associated with the existence of non trivial solutions in the linearly stable regime), together with the preference of magnetic humps at larger values of β and/or for larger distance from threshold is consistent with a nonlinear stability analysis based on an energy minimization argument, performed in the framework of ordinary anisotropic MHD with a specific equation of state derived from the stationary fluid hierarchy by assuming a bi-Maxwellian distribution function.[11] It is noticeable that this closure where the parallel ion temperature is uniform and the perpendicular one a homographic function of the magnetic fluctuation,[12,13] correctly reproduces the mirror instability threshold, in contrast with bi-adiabatic equations of state or their generalizations.[14] Nevertheless, lacking kinetic effects, this model is not suitable to accurately reproduce the time evolution of linear mirror modes.

The aim of the present paper is to review recent analytical results on the dynamics of nonlinear mirror mode, near the instability threshold. In Section 2, a long-wavelength equation governing the parallel magnetic fluctuations is derived perturbatively from the Vlasov-Maxwell equations. In Section 3, this equation is used to demonstrate the subcritical character of the bifurcation, a property at the origin of the formation of large-amplitude mirror structures and of their bistability. Indeed, for such a bifurcation, non trivial stationary states below threshold are linearly unstable, while above threshold, initially small-amplitude solutions undergo a sharp transition to a large-amplitude state, associated with a blowup behavior within an asymptotic formalism. Such (large-amplitude) bifurcated solutions are not amenable to a perturbative calculation. Section 4 thus discusses a phenomenological equation based on an heuristic modeling of nonlinear finite Larmor radius, that predicts a nonlinear dynamics in very satisfactory agreement with direct numerical simulations of the Vlasov-Maxwell equations. Section 4 is a brief conclusion.

2. Reductive perturbative expansion near threshold

Near the mirror instability threshold, the linearly unstable modes are located at large scales thus permitting the development of a long-wavelength reductive perturbative expansion of the Vlasov-Maxwell equation. We here briefly sketch this derivation that is detailed in [15]. A simplified approach, based on the patching of the linear theory with an estimate of the relevant nonlinear comtributions from the drift-kinetic equation is found in [16].

The equation for the mean proton velocity, as classically derived from the Vlasov equation, reads

$$\frac{d\mathbf{u}}{dt} + \frac{1}{\rho}\boldsymbol{\nabla}\cdot\mathbf{p} - \frac{e}{m_p}(\mathbf{E} + \frac{1}{c}\mathbf{u}\times\mathbf{B}) = 0, \tag{2}$$

where, for cold and massless electrons,

$$\mathbf{E} = -\frac{1}{c}\Big(\mathbf{u} - \frac{\mathbf{j}}{ne}\Big)\times\mathbf{B}, \tag{3}$$

with $\mathbf{j} = (c/4\pi)\boldsymbol{\nabla}\times\mathbf{B}$. The ion pressure tensor is rewritten as the sum of gyrotropic and gyroviscous contributions $\mathbf{p} = p_\perp\mathbf{n} + p_\parallel\boldsymbol{\tau} + \boldsymbol{\Pi}$, with $\mathbf{n} = \mathbf{I} - \widehat{\mathbf{b}}\otimes\widehat{\mathbf{b}}$ and $\boldsymbol{\tau} = \widehat{\mathbf{b}}\otimes\widehat{\mathbf{b}}$, where $\widehat{\mathbf{b}} = \mathbf{B}/|B|$ is the unit vector along the local magnetic field. In order to address the asymptotic regime, we rescale the independent variables in the form $X = \sqrt{\varepsilon}x$, $Y = \sqrt{\varepsilon}y$, $Z = \varepsilon z$,

$T = \varepsilon^2 t$, where ε measures the distance to threshold, and expand any field φ in the form

$$\varphi = \sum_{n=0}^{\infty} \varepsilon^{n/2} \varphi_{n/2}, \tag{4}$$

When retaining the two first nontrivial orders and denoting by $\boldsymbol{\nabla}_\perp = (\partial_X, \partial_Y)$ the transverse gradient, we get

$$\boldsymbol{\nabla}_\perp \Big(p_\perp^{(1)} + \frac{B_0 B_z^{(1)}}{4\pi} \Big) + \varepsilon \Big\{ \boldsymbol{\nabla}_\perp \Big(p_\perp^{(2)} + \frac{B_0 B_z^{(2)}}{4\pi} + \frac{(B_z^{(1)})^2}{8\pi} \Big) - \frac{2}{\beta_\perp} \Big(1 + \frac{\beta_\perp - \beta_\parallel}{2} \Big) p_\perp^{(0)} \partial_z \Big(\frac{\mathbf{B}_\perp^{(3/2)}}{B_0} \Big) + \boldsymbol{\nabla}_\perp \cdot \boldsymbol{\Pi}_\perp^{(2)} + \partial_Z \boldsymbol{\Pi}_{\perp Z}^{(3/2)} \Big\} = 0(\varepsilon^2), \tag{5}$$

that expresses the condition of pressure balance. Using the subdominant character of the longitudinal current together with the divergenceless of the magnetic field, one has

$$\mathbf{B}_\perp^{(3/2)} = (-\Delta_\perp)^{-1} \boldsymbol{\nabla}_\perp \partial_Z B_z^{(1)}. \tag{6}$$

Here, the subscript $\perp$ refers to vector component perpendicular to the ambient field (taken along z). Computing perturbatively the gyrotropic and non-gyrotropic components of the pressure tensors from the Vlasov-Maxwell equations, and also defining $b_z = B_z^{(1)} + \varepsilon B_z^{(2)}$ and $\overline{p}_\perp = p_\perp{}^{(1)} + \varepsilon p_\perp^{(2)}$, we finally obtain the asymptotic equation governing the nonlinear dynamics of mirror modes near the instability threshold in the form[15,16]

$$\begin{aligned} \partial_T \frac{b_z}{B_0} = \frac{v_{th\,\parallel}}{\sqrt{\pi}} \frac{\beta_\parallel}{\beta_\perp} \Big(-\mathcal{H}\partial_Z \Big) \Big\{ \frac{1}{\varepsilon} \Big(\frac{\beta_\perp}{\beta_\parallel} - 1 - \frac{1}{\beta_\perp} \Big) \frac{b_z}{B_0} \\ + \frac{3}{4\beta_\perp} r_L^2 \Delta_\perp \frac{b_z}{B_0} - \frac{1}{\beta_\perp} \Big(1 + \frac{\beta_\perp - \beta_\parallel}{2} \Big) \Delta_\perp^{-1} \partial_{ZZ} \frac{b_z}{B_0} \\ - \frac{3}{2} \Big(\frac{1 + \beta_\perp}{\beta_\perp^2} \Big) \Big(\frac{b_z}{B_0} \Big)^2 \Big\} = 0. \end{aligned} \tag{7}$$

We note that the time derivative and the Hilbert transform $\mathcal{H}$ originates from Landau damping. The parameter r_L is the ion Larmor radius. This equation can be viewed as the linear dispersion relation of large-scale mirror modes retaining leading order finite Larmor radius (FLR) corrections, supplemented by dominant nonlinear contributions. It is noticeable that kinetic effects (such as Landau and FLR effects) contribute only linearly.

We now define $\chi = 1 + (\beta_\perp - \beta_\parallel)/2$ and characterize the regime of linear stability or instability by the parameter $\sigma = \mathrm{sgn}(\beta_\perp/\beta_\parallel - 1 - 1/\beta_\perp)$. The expansion parameter ε is related to the distance to threshold by the condition $|\beta_\perp/\beta_\parallel - 1 - 1/\beta_\perp| = \varepsilon\chi/\beta_\perp$, or in other words $\varepsilon = \Gamma^*/\chi$ with Γ^* defined in (1) as the bi-Maxwellian threshold parameter. We then perform a simple rescaling by introducing the new longitudinal and transverse coordinates $\xi = (2/\sqrt{3})\chi^{1/2} r_L^{-1} Z$, $\mathbf{R}'_\perp = (2/\sqrt{3})\chi^{1/2} r_L^{-1}\mathbf{R}_\perp$, and the new time variable $\tau = (2/\sqrt{3})(\sqrt{\pi}\beta_\perp)^{-1}(\chi\beta_\parallel/\beta_\perp)^{3/2}\Omega T$. We also write

$$b_z/B_0 = 2\chi\beta_\perp\left(1+\beta_\perp\right)^{-1} U. \tag{8}$$

The equation then reduces to

$$\partial_\tau U = -\mathcal{H}\partial_\xi\Big[\sigma U + \Delta_\perp U - \Delta_\perp^{-1}\partial_{\xi\xi}U - 3U^2\Big]. \tag{9}$$

Equation (9) further simplifies when the spatial variations are limited to a direction making a fixed angle with the ambient magnetic field. After a simple rescaling, one gets

$$\partial_T U = \widehat{K}_\Xi\left[\left(\sigma + \partial_{\Xi\Xi}\right)U - 3U^2\right], \tag{10}$$

where Ξ is the coordinate along the direction of variation and $K_\Xi = -\mathcal{H}\partial_\Xi$ is a positive operator whose Fourier transform reduces to the multiplication by the wavenumber absolute value.

Equation (9) (and its one-dimensional reduction (10) as well) possesses the remarkable property of being of the form

$$\frac{\partial U}{\partial T} = -\widehat{K}_z\frac{\delta F}{\delta U},$$

where

$$\begin{aligned} F &= \int\left[-\frac{\sigma}{2}U^2 + \frac{1}{2}U\Delta_\perp^{-1}\partial_Z^2 U + \frac{1}{2}\left(\nabla_\perp U\right)^2 + U^3\right] d\mathbf{R} \\ &\equiv -\sigma N/2 + I_1/2 + I_2/2 + I_3 \end{aligned} \tag{11}$$

has the meaning of a free energy or a Lyapunov functional. The terms $N/2$, $I_1/2$, $I_2/2$ and I_3 correspond to the different contributions in the definition of F. The latter quantity can only decrease in time, since

$$\frac{dF}{dt} = \int\frac{\delta F}{\delta U}\frac{\partial U}{\partial t}d\mathbf{R} = -\int\frac{\delta F}{\delta U}\widehat{K}_z\frac{\delta F}{\delta U}d\mathbf{R} \le 0. \tag{12}$$

This derivative can only vanish at the stationary localized solutions, defined by the equation

$$\frac{\delta F}{\delta U} = \left(\sigma - \Delta_\perp^{-1}\frac{\partial^2}{\partial Z^2} + \Delta_\perp\right)U - 3U^2 = 0. \tag{13}$$

In order to show that non-zero solutions of this equation do not exist above threshold ($\sigma = +1$), we establish relations between the integrals N,

I_1, I_2 and I_3, using the fact that solutions of Eq. (13) are stationary points of the functional F (i.e. $\delta F = 0$). Multiplying Eq. (13) by U and integrating over $\mathbf{R}$ gives the first relation

$$\sigma N - I_1 - I_2 - 3I_3 = 0.$$

Two other relations can be found if one makes the scaling transformations, $Z \to aZ$, $\mathbf{R}_\perp \to b\mathbf{R}_\perp$, under which the free energy (11) becomes a function of two scaling parameters a and b

$$F(a,b) = -\frac{\sigma N}{2} ab^2 + \frac{I_1}{2} b^4 a^{-1} + \frac{I_2}{2} a + I_3 ab^2.$$

Due to the condition $\delta F = 0$, the first derivatives of F at $a = b = 1$ have to vanish:

$$\frac{\partial F}{\partial a} = -\frac{\sigma N}{2} - \frac{I_1}{2} + \frac{I_2}{2} + I_3 = 0,$$

$$\frac{\partial F}{\partial b} = -\sigma N + 2I_1 + 2I_3 = 0.$$

Hence, after simple algebra, one gets the three relations

$$I_1 + \frac{\sigma}{2} N = 0, \quad I_3 = -2I_1, \quad I_2 = 3I_1.$$

For $\sigma = +1$, the first relation can be satisfied only by the trivial solution $U = 0$, because both integrals I_1 and N are positive definite. In other words, above threshold, nontrivial stationary solutions obeying the prescribed scalings do not exist.

In contrast, below threshold, stationary localized solutions can exist. For these solutions, the free energy is positive and reduces to $F_s = N/2$. Furthermore, $I_3 = \int U^3 d^3R < 0$. which means that the structures have the form of magnetic holes. As stationary points of the functional F, these solutions represent saddle points, since the corresponding determinant of second derivatives of F with respect to scaling parameters taken at these solutions, is negative ($\partial_{aa} F \partial_{bb} F - (\partial_{ab} F)^2 = -2N^2 < 0$). As a consequence, there exist directions in the eigenfunction space, for which the free-energy perturbation is strictly negative, corresponding to linear instability of the associated stationary structure. This is one of the properties for subcritical bifurcations.[17]

For the one-dimensional model (10), the proof of instability of stationary solution $U_0 = -\frac{1}{2}\mathrm{sech}^2(\Xi/2)$ (which coincides with the Korteveg-de Vries soliton) is more complicated than in three dimensions. The corresponding free energy turns out to have a minimum relatively to the scaling parameter.

Therefore, one needs to consider the linearized problem for perturbations W ($U = U_0 + W$), which can be formulated as

$$\frac{\partial W}{\partial T} = -\widehat{K}_\Xi \frac{\delta \tilde{F}}{\delta W},$$

where $\tilde{F} = \frac{1}{2}\langle W|L|W\rangle$ is the quadratic part of the free energy and

$$L = 1 - \frac{\partial^2}{\partial \Xi^2} + 6U_0$$

is the 1D Schrödinger operator.

It is easily seen that the operator L has one neutral (shift) mode $\psi_1 = \partial_\Xi U_0$ ($L\partial_\Xi U_0 = 0$) associated with invariance by space translation, which has one node. Thus, according the oscillation theorem, L has one negative energy level with $E = -5/4 < 0$, corresponding to the ground state $\psi_0 = \mathrm{sech}^3(\Xi/2)$ (without nodes), which proves the instability of the stationary solution U_0 with the growth rate equal $\frac{5}{4}\langle\psi_0|\widehat{K}|\psi_0\rangle/\langle\psi_0|\psi_0\rangle > 0$.

As a consequence, starting from general initial conditions, the derivative dF/dt (12) is almost always *negative*, except for unstable stationary points (zero measure) below threshold. In the nonlinear regime, negativeness of this derivative implies $\int U^3 d^3R < 0$, which corresponds to the formation of magnetic holes. Moreover, this nonlinear term (in F) is responsible for collapse, i.e. formation of singularity in a finite time.

3. Saturation of the mirror instability

As discussed in the above sections, the only nonlinearities retained by the near-threshold asymptotics, tend to reinforce the linear instability, leading to a finite time singularity, signature of a sub-critical bifurcation, not amenable to a perturbative calculation. To cope with this situation, a phenomenological model was constructed by supplementing the asymptotic equation with nonlinear FLR effects associated with the local variation of the ion Larmor radius in the coherent structures.[15,16] In regions of weaker magnetic field, the Larmor radius is larger, leading to a more efficient stabilizing effect than within the linear theory. As a consequence, the mirror instability is more easily quenched in magnetic minima than in magnetic maxima, making magnetic humps more likely to form during the saturation phase of the mirror instability. In dimensionless units, this phenomenological model is governed by the equation

$$\partial_\tau U = \widehat{K}_\xi \left[\sigma U - 3U^2 + \frac{1}{1+\alpha U}\partial_{\xi\xi}U - \frac{4\nu}{9(1+\alpha U)^2}\partial_{\xi\xi\xi\xi}U\right], \qquad (14)$$

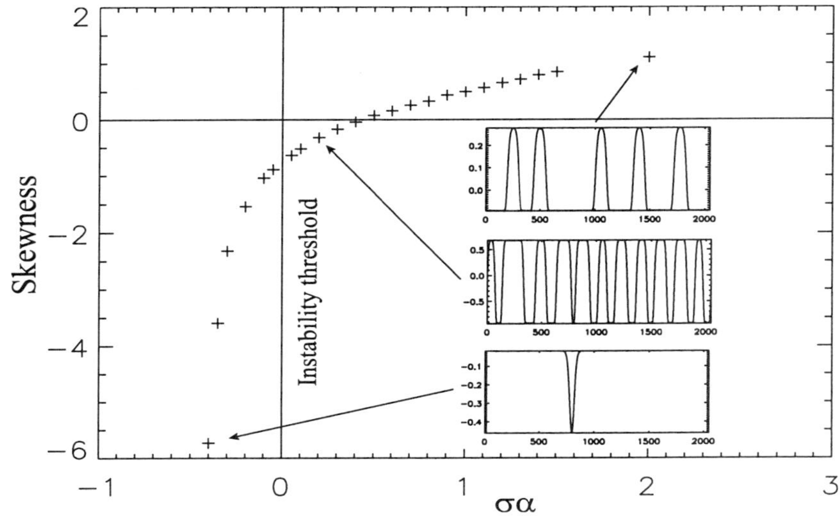

Fig. 1. Variation of the magnetic field skewness with $\sigma\alpha$ predicted by the model equation (14), where α combines the distance to threshold with the value of β, and $\sigma = \pm 1$ characterizes the positive or negative distance to threshold. The insets display typical quasi-stationary solution profiles.

The parameter

$$\alpha = \frac{2\beta_\perp}{1+\beta_\perp}\left[\beta_\perp\left(\frac{\beta_\perp}{\beta_\|} - 1\right) - 1\right] = \frac{2\beta_\perp}{1+\beta_\perp}\Gamma^* \tag{15}$$

scales like the distance to threshold at moderate $\beta_\perp$, while it varies proportionally to $\beta_\perp$) when the latter is small. The coefficient ν fixes the domain size.

In contrast with the previous phenomenological descriptions based on the cooling of a population of trapped particles[18,19] that mostly predict deep magnetic holes (except for very large β) and do not refer to the bistability phenomenon, the present model successfully reproduces spacecraft observations and numerical simulations from Vlasov-Maxwell equations.[15] It indeed predicts the formation of magnetic humps above threshold and also the existence of subcritical magnetic holes (when the system is initialized by large-amplitude perturbations). An interesting quantity also used to analyze satellite data is the skewness of the magnetic perturbations. This quantity is plotted versus $\sigma\alpha$ in Fig. 1. Above threshold ($\sigma = 1$), Eq. (14) is initialized with a small random noise, while in the subcritical regime ($\sigma = -1$) a much larger random initial perturbation is needed. The resulting graph is qualitatively very similar to that designed from Cluster data.[9] The inserted graphs refer to the corresponding typical profiles of quasi-stationary solutions. It is also interesting to note that direct numerical simulations in an extended domain, relatively far from threshold reveals a possible additional origin for magnetic holes. They indeed show that magnetic humps early formed as the saturation of the mirror instability gradually evolve into magnetic holes, an effect that can be related to a decrease of the β of the plasma, as time elapses. This phenomenon that is not obtained in a small computational domain, is also is beyond the capability of the above model.

4. Conclusion

We have presented an asymptotic description of the nonlinear dynamics of mirror modes near the instability threshold. Below threshold, we have demonstrated the existence of unstable stationary solutions. Differently, above threshold, no stationary solution consistent with the prescribed small-amplitude, long-wavelength scaling can exist. For small-amplitude initial conditions, the time evolution predicted by the asymptotic equation (9) leads to a finite-time singularity. These properties are based on the fact that this equation belongs to a class of generalized gradient systems for which it is possible to introduce a free energy or a Lyapunov functional that decreases in time.

The singularity formation as well as the existence of unstable stationary structures below the mirror instability threshold obtained with the asymptotic model, can be viewed as features of a subcritical bifurcation towards a large-amplitude state that cannot be described in the framework of a weakly nonlinear perturbative analysis. In order to model the results of recent numerical simulations[15,20] of the Vlasov-Maxwell equations that display the formation of magnetic humps above threshold together with a phenomenon of bistability, associated with the existence of stable large-amplitude magnetic holes both below and above threshold, we have built a phenomenological model that supplement to the asymptotic equation a heuristic description of the nonlinear FLR corrections. This model reproduces the typical structures observed in the numerical simulations and is also consistent with the the statistics of mirror in the terrestrial magnetosheath structures such as the skewness of the magnetic fluctuations, obtained from CLUSTER satellite data.[9]

An important open question concerns the relation between the present theory of structure formation and the quasi-linear effects[21] that could, in some instances, compete near threshold. An early-time quasi-linear regime could for example modify the onset of coherent structures and, on the other hand, the development of such structures, can also affect the quasi-linear dynamics.

5. Acknowledgments

The work of EK was supported by RFBR (grant no. 06-01-00665) and by the French Ministère de l'Enseignement Supérieur et de la Recherche during his visit at the Observatoire de la Côte d'Azur. TP and PLS acknowledge support from "Programme National Soleil Terre" of CNRS.

References

1. T.S. Horbury, E.A. Lucek, A. Bulogh, I. Dandouras, I., and H. Rème, *J. Geophys. Res.* **109**, A09202 (2004).
2. A.A. Vedenov and R.Z. Sagdeev, Plas. Phys. in *Problem of Controlled Thermonuclear Reactions*, **III**, ed. M.A. Leontovich, 332 (Pergamon Press, NY, 1958).
3. Pokhotelov, O. A., M. A. Balikhin, R. Z. Sagdeev, and R. A. Treumann, *J. Geophys. Res.*, **110**, A10206, (2005) doi:10.1029/2004JA010933.
4. P. Hellinger, *Phys. Plasmas* **14**, 082105 (2007).
5. S.P. Gary, *J. Geophys. Res.* **97**, 8519 (1992).
6. M. E. McKean, D. Winske, and S. P. Gary *J. Geophys. Res.*, **97**, 19421.(1992).
7. E. A. Lucek, M.W. Dunlop, A. Balogh, P. Cargill, W. Baumjohann, E. Georgescu, G. Haerendel, and K. H. Fornacon Geophys. Res. Lett., ***26***, 2159 (1999).
8. S. P. Joy, M. G. Kivelson, R. J. Walker, K. K. Khurana, C. T. Russell, and W. R. Paterson, *J. Geophys. Res.* **111**, A12212,(2006) doi: 10.1029/2006JA011985.
9. V. Génot, E. Budnik, C. Jacquey, J. Sauvaud, I. Dandouras, and E. Lucek, *AGU Fall Meeting Abstracts*, C1412+ (2006).
10. J. Soucek, E. Lucek, and I. Dandouras, Properties of magnetosheath mirror modes observed by Cluster and their responses to changes in plasma parameters, *J. Geophys. Res.*, submitted, doi:10.1029/2007JA012649.
11. T. Passot, V. Ruban, and P.L. Sulem, *Phys. Plasmas*, **13**, 102310 (2006.)
12. T. Passot and P.L. Sulem, *J. Geophys. Res.* **111**, A04203 (2006).
13. T. Chust and G. Belmont, *Phys. Plasmas*, **13**, 012506 (2006),
14. L.N. Hau and B.U.O. Sonnerup, *Geophys. Res. Lett.*, **20**, 1763 (1993).
15. Califano, C., Hellinger, P., Kuznetsov, E., Passot, T., Sulem, P.L., & Travnicek, P. 2007, Nonlinear mirror modes dynamics: simulations and modeling, *J Geophys. Res.* submitted, doi:10.1029/2007JA012898.
16. E.A. Kuznetsov, T. Passot, T. and P.L. Sulem, *Phys. Rev. Lett.*, **98**, 235003 (2007).

17. E.A. Kuznetsov, T. Passot, T. and P.L. Sulem, *Pis'ma ZhETF*, **86**, 725 (2007); *JETP Lett.* in press.
18. M.G. Kivelson and D.S. Southwood, *J. Geophys. Res.*, **101**(A8), 17365 (1996).
19. Pantellini, P.G.E., *J. Geophys. Res.*, **103**(A3), 4789 (1998),
20. K. Baumgärtel, K. Sauer, and E. Dubinin, *Geophys. Res. Lett.* **30** (14), 1761 (2003).
21. V. D. Shapiro, and V. I. Shevchenko, *Sov. Phys. JETP*, **18**, 1109 (1964).

NONLINEAR DISPERSIVE ALFVÉN WAVES IN MAGNETOPLASMAS

P. K. SHUKLA* and B. ELIASSON

Institut für Theoretische Physik IV, Fakultät für Physik und Astronomie, Ruhr-Universität Bochum, D-44780 Bochum, Germany
**E-mail: ps@tp4.rub.de*
www.tp4.rub.de

L. STENFLO

Department of Physics, Umeå University, SE-90187 Umeå, Sweden

R. BINGHAM

Centre for Fundamental Physics, Rutherford Appleton Laboratory, Chilton, Didcot, Oxfordshire OX11 OQX, United Kingdom

Large amplitude Alfvén waves are frequently found in magnetized space and laboratory plasmas. Our objective here is to discuss the linear and nonlinear properties of dispersive Alfvén waves (DAWs) in a uniform magnetoplasma. We first consider the effects of finite frequency (ω/ω_{ci}) and ion gyroradius on inertial and kinetic Alfvén waves, where ω_{ci} is the ion gyrofrequency. Next, we focus on nonlinear effects caused by the dispersive Alfvén waves. Such effects include the plasma density enhancement and depression by the Alfvén wave ponderomotive force, nonlinear interactions among the DAWs, the generation of zonal flows by the DAWs, as well as the electron and ion heating due to wave-particle interactions. The relevance of our investigation to the appearance of nonlinear dispersive Alfvén waves in the Earth's auroral acceleration region, in the solar corona, and in the Large Plasma Device (LAPD) at UCLA is discussed.

1. Introduction

The Alfvén wave[1] is classic in plasma physics. In an Alfvén wave, the restoring force comes from the pressure of the magnetic fields, and the ion mass provides the inertia. The propagation of low-frequency (in comparison with ω_{ci}) nondispersive Alfvén waves is basically governed by the ideal magnetohydrodynamic (MHD) equations. The dispersion of Alfvén waves arises due to the finite ω/ω_{ci} and the parallel electron inertial force in a cold plasma, as well as from the finite ion Larmor radius and the parallel electron pressure gradient in a warm plasma.[2–7] Due to non-ideal effects Alfvén waves also couple to other plasma modes, viz. magnetosonic waves, whistlers, etc.

In laboratory and space plasmas, finite amplitude dispersive Alfvén waves are excited by many sources such as external antennae, energetic charged particle beams, nonuniform plasma parameters, or electrostatic and electromagnetic waves. The dispersive Alfvén waves have wide ranging applications in space, fusion, and laboratory plasmas.[8,9] Specifically, large amplitude Alfvén waves transport a considerable amount of energy from the distant parts of the magnetosphere to the near-Earth space, and play a significant role in the solar corona and in the solar wind.

Large amplitude Alfvén waves can cause a number of nonlinear effects.[4,5,10–21] The latter include harmonic generation for non-circularly polarized waves, parametric processes such as three-wave decay interactions, stimulated Compton scattering, modulational and filamentational interactions, as well as the modification of the background plasma number density by the Alfvén wave ponderomotive force, the formation of Alfvén vortices and current filaments due to the Alfvén wave mode couplings, the generation of zonal flows by the Reynold stress of the kinetic Alfvén waves, as well as the electron and ion heating due to wave-particle interactions.

Understanding the nonlinear properties of dispersive Alfvén waves is a necessary prerequisite in interpreting the numerous observations of large amplitude, low-frequency (in comparison with the electron gyrofrequency) dispersive electromagnetic waves in laboratory and space plasmas.

In the present chapter, we first discuss the linear properties of dispersive Alfvén waves and their relations to other modes in a uniform magnetoplasma. We then include arbitrary ion gyroradius and parallel phase speed effects, as well as ω/ω_{ci} corrections. The finite ion gyroradius effect in the shear Alfvén wave [both inertial Alfvén (IA) and kinetic Alfvén (KA) waves] theory can be considered by means of the kinetic ion response,[22] while for waves with parallel phase speed comparable to the electron thermal speed one has to resort to a gyrokinetic electron response that accounts for the wave-electron interactions. Furthermore, the IA and KA waves have to be treated separately as they have parallel phase speeds $V_p = \omega/k_z$ satisfying opposing limits (viz. $V_p \gg V_{Te}$ for the IA waves and $V_p \ll V_{Te}$ for the KA waves), where ω is the wave frequency, k_z is the magnetic-field aligned wavenumber, and V_{Te} is the electron thermal speed. Second, we go on describing the Alfvén wave ponderomotive force induced density modifications in the background plasma, as well as the localization of magnetic field-aligned circularly polarized Alfvén waves. Third, we consider the mode couplings between dispersive Alfvén waves in plasmas. Fourth, we present a theory[20] for the excitation of zonal flows by the Reynold stresses of the kinetic Alfvén wave. Fifth, we provide mechanism for heating the electrons (ions) by obliquely (magnetic field- aligned) propagating high-frequency electromagnetic ion-cyclotron-Alfvén (left-hand circularly polarized ion-cyclotron) waves, due to the wave-electron (ion) resonance interaction.

2. Linear properties

Let us consider a homogeneous plasma in an external magnetic field $\hat{\mathbf{z}}B_0$, where $\hat{\mathbf{z}}$ is the unit vector along the z direction and B_0 is the strength of the magnetic field. The dynamics of Alfvén waves within the simplest model is governed by the ideal MHD equations (e.g.[23])

$$\frac{dn}{dt} + n\nabla \cdot \mathbf{u} = 0, \tag{1}$$

$$\frac{d\mathbf{u}}{dt} = \frac{(\nabla \times \mathbf{B}) \times \mathbf{B}}{4\pi\rho}, \tag{2}$$

and

$$\frac{\partial \mathbf{B}}{\partial t} = \nabla \times (\mathbf{u} \times \mathbf{B}), \tag{3}$$

where the Ohm's law

$$\mathbf{E} + \frac{\mathbf{u} \times \mathbf{B}}{c} = 0 \tag{4}$$

has been used. Here $d/dt = (\partial/\partial t) + \mathbf{u} \cdot \nabla$, $\rho = nm_i$, where n is the ion number density, $\mathbf{u}$ is the ion fluid velocity, $\mathbf{E}$ and $\mathbf{B}$ are electric and magnetic fields, respectively, m_i is the ion mass, and c is the speed of light in vacuum. The field frozen condition (n/B = constant) follows from Eqs. (1) and (3). The Alfvén wave frequency is $\omega = kV_A \equiv kB_0/\sqrt{4\pi nm_i}$, which shows that the restoring force comes from the magnetic pressure while the ion mass provides inertia to maintain the wave.

The dispersion of the Alfvén wave is here due to the generalized Ohm's law

$$\mathbf{E} + \frac{\mathbf{u} \times \mathbf{B}}{c} - \frac{\nabla \times \mathbf{B}}{4\pi \mathbf{en}} + \frac{\nabla P_e}{en} = 0, \tag{5}$$

which thus replaces Eq. (4), and which is valid for waves with frequencies much smaller than the electron gyrofrequency $\omega_{ce} = eB_0/m_e c$, where e is the magnitude of the electron charge and m_e is the electron mass. Equation (5) takes into account the differential motion between the electrons and ions, as well as the finite electron pressure gradient ∇P_e. Substituting (5) into the ion momentum equation and Faraday's law then yields

$$\frac{d\mathbf{u}}{dt} = \frac{(\nabla \times \mathbf{B}) \times \mathbf{B}}{4\pi\rho} - \frac{\nabla P}{\rho}, \tag{6}$$

and

$$\frac{\partial \mathbf{B}}{\partial t} = \nabla \times \left[\left(\mathbf{u} - \frac{c\nabla \times \mathbf{B}}{4\pi e n} \right) \times \mathbf{B} \right], \tag{7}$$

where $P = P_e + P_i \equiv n(T_e + T_i)$ is the sum of the electron and ion pressures. Equations (1), (6) and (7) are known as the nonlinear Hall-MHD equations. It turns out that the Hall velocity $\mathbf{u}_H = -c\nabla\times\mathbf{B}/4\pi en$ in Eq. (7) introduces a new spatial scale, the ion skin depth c/ω_{pi}, where ω_{pi} is the ion plasma frequency. It also allows us to consider phenomena that are occurring on a timescale smaller than the ion gyroperiod ω_{ci}^{-1}, where $\omega_{ci} = eB_0/m_i c$ is the ion gyrofrequency.

The waves in our Hall-MHD plasma are governed by the dispersion relation[5,14]

$$(\omega^2 - k_z^2 V_A^2) D_m(\omega, \mathbf{k}) = \frac{\omega^2}{\omega_{ci}^2} k_z^2 k^2 (\omega^2 - k^2 V_s^2) V_A^4, \tag{8}$$

where

$$D_m(\omega, \mathbf{k}) = \omega^4 - \omega^2 k^2 (V_A^2 + V_s^2) + k_z^2 k^2 V_A^2 V_s^2, \tag{9}$$

and V_s is the effective sound speed. Equation (8) shows that the Alfvén and magnetoacoustic waves are coupled due to finite ω/ω_{ci} effect. Equation (8) also contains the kinetic Alfvén and long wavelength (in comparison with the electron skin depth $\lambda_e = c/\omega_{pe}$) electron whistlers. The latter usually have $\sqrt{\omega_{ce}\omega_{ci}} \ll \omega \ll kc \ll \omega_{pe}$, where ω_{pe} are the electron plasma frequency.

The dynamics of nonlinear whistlers is governed by the electron-MHD (E-MHD) equation

$$\frac{D\mathcal{B}}{Dt} - \mathcal{B} \cdot \nabla \mathbf{V} = 0, \tag{10}$$

where $\mathcal{B} = \mathbf{B} - \lambda_e^2 \nabla^2 \mathbf{B}$, $D/Dt = (\partial/\partial t) + \mathbf{V} \cdot \nabla$, and $\mathbf{V} = -c\nabla \times \mathbf{B}/4\pi en$ is the electron fluid velocity. The ions are assumed to be immobile. In the linear limit, Eq. (10) admits obliquely propagating whistlers whose frequency is

$$\omega = \frac{k^2 c^2 \omega_{ce} \cos\theta}{\omega_{pe}^2 + k^2 c^2}, \tag{11}$$

where $\cos\theta = k_z/k$. In the long wavelength limit, viz. $kc \ll \omega_{pe}$, we have from (11)

$$\omega = \frac{k_z k c^2 \omega_{ce}}{\omega_{pe}^2}, \tag{12}$$

which reduces to[24]

$$\omega = \frac{k_z^2 c^2 \omega_{ce}}{\omega_{pe}^2} \tag{13}$$

when the perpendicular component of the wavevector is zero.

Next, we discuss the properties of dispersive shear Alfvén waves (DSAWs)[8] whose electric and magnetic fields are $\mathbf{E} = -\nabla\phi - c^{-1}\hat{\mathbf{z}}\partial A_z/\partial t$ and $\mathbf{B}_\perp = \nabla A_z \times \hat{\mathbf{z}}$, respectively, where ϕ is the scalar potential, and A_z is the parallel (to $\hat{\mathbf{z}}$) component of the vector potential. The electron density perturbation n_{e1} associated with low-frequency (in comparison with the electron gyrofrequency) DSAWs is given by

$$n_{e1} = -\frac{k_z J_z}{\omega e}, \tag{14}$$

where the parallel (to $\hat{\mathbf{z}}$) component of the electron current density is[7]

$$J_z = -\frac{\omega n_0 e^2}{k_z^2 T_e} W(\xi) \left(\frac{\omega}{c} A_z - k_z \phi\right), \tag{15}$$

where $W(\xi) = -[1 + \xi Z(\xi)]$, $\xi = V_p/V_{Te}$, $V_{Te} = (T_e/m_e)^{1/2}$ is the electron thermal speed, T_e is the electron temperature, n_0 is the unperturbed value of n, and $Z(\xi)$ is the standard plasma dispersion function of Fried and Conte. In deriving (15) we assumed that $k_\perp^2 \rho_e^2 \ll 1$, where $k_\perp$ is the component of the wavevector perpendicular to $\hat{\mathbf{z}}$ and $\rho_e = V_{Te}/\omega_{ce}$ is the electron gyroradius. From the parallel component of Ampère's law we have

$$J_z = \frac{c}{4\pi} k_\perp^2 A_z, \tag{16}$$

so that from (14)-(16) we obtain[7]

$$n_{e1} = -\frac{n_0 e}{T_e} \frac{k_\perp^2 c^2 k_z^2 \lambda_{De}^2 W(\xi)}{[k_\perp^2 c^2 k_z^2 \lambda_{De}^2 + \omega^2 W(\xi)]} \phi, \tag{17}$$

where $\lambda_{De} = (T_e/4\pi n_0 e^2)^{1/2}$ is the electron Debye radius. The expression (17) incorporates the electron Landau effect.

For finite values of ω/ω_{ci} and $b_i = k_\perp^2 \rho_i^2$, where $\rho_i = V_{Ti}/\omega_{ci}$ is the ion gyroradius, $V_{Ti} = (T_i/m_i)^{1/2}$ is the ion thermal speed and T_i is the ion temperature, the expression for the ion number density perturbation n_{i1} has to be obtained by integrating the ion distribution function calculated from the linearized Vlasov equation. For two-dimensional ion motions across $\hat{\mathbf{z}}$, we have[22]

$$n_{i1} \simeq -n_0 \left[1 - \Gamma_0(b_i) - \frac{2\Gamma_1(b_i)\omega^2}{\omega^2 - \omega_{ci}^2}\right] \frac{e\phi}{T_i}, \tag{18}$$

where $\Gamma_{0,1}(b_i) = I_{0,1}(b_i)\exp(-b_i)$ and $I_0(I_1)$ is the modified Bessel function of zero (first) order.

Invoking the quasi-neutrality $n_{e1} = n_{i1}$, we have from (17) and (18)

$$1 - \Gamma_0(b_i) - \frac{2\Gamma_1(b_i)\omega^2}{\omega^2 - \omega_{ci}^2} - \frac{T_i}{T_e} \frac{k_\perp^2 c^2 k_z^2 \lambda_{De}^2 W(\xi)}{[k_\perp^2 c^2 k_z^2 \lambda_{De}^2 + \omega^2 W(\xi)]} = 0, \tag{19}$$

which is the general dispersion relation for the coupled ion-cyclotron- dispersive Alfvén waves in plasmas.

We now focus on the inertial Alfvén (IA) and kinetic Alfvén (KA) waves for which $\xi \gg 1$ and $\xi \ll 1$, respectively. Thus, the corresponding electron number densities, found from (17), are, respectively,

$$n_{e1}^I = -\frac{k_z^2 c^2 k_\perp^2}{4\pi e(1 + b_e)\omega^2}\phi, \tag{20}$$

and

$$n_{e1}^K = -\frac{k_z^2 c^2 k_\perp^2}{4\pi e(\omega^2 - k_z^2 V_A^2 b_s)}\phi, \tag{21}$$

where $b_e = k_\perp^2 \lambda_e^2$, $b_s = k_\perp^2 \rho_s^2$, $\lambda_e = c/\omega_{pe}$ is the electron skin depth, $\rho_s = C_s/\omega_{ci}$ is the sound gyroradius, and $C_s = (T_e/m_i)^{1/2}$ is the ion sound speed. The superscripts I and K denote the IA and KA waves. The electron Landau damping is neglected in obtaining (20) and (21).

The parallel component of the electric fields associated with the IA and KA waves are, respectively,[7]

$$E_z = i\left[\frac{m_e \omega^2 (1 + b_e)}{b_e e k_z n_0}\right] n_{e1}^I, \tag{22}$$

and

$$E_z = -i\frac{T_e k_z}{e n_0} n_{e1}^K. \tag{23}$$

The wave magnetic field is determined from $\mathbf{B}_\perp = (c/\omega)\mathbf{k} \times \mathbf{E}$.

Noting that $\omega \ll \omega_{ci} \ll \omega_{pi}$ for the shear Alfvén waves in a dense plasma, we obtain from Eq. (19) the dispersion relations

$$\omega^2 = \frac{k_z^2 V_A^2 b_i}{[1 - \Gamma_0(b_i)]\,(1 + b_e)}, \tag{24}$$

for the IA waves, and

$$\omega^2 = k_z^2 V_A^2 \left[b_s + \frac{b_i}{1 - \Gamma_0(b_i)} \right], \tag{25}$$

for the KA waves. For $b_i \ll 1$, we have $\Gamma_0 \approx 1 - b_i + 3b_i^2/4$ so that (24) and (25) give the familiar results for the IA and KA wave frequencies, respectively,

$$\omega = \frac{k_z V_A}{(1 + k_\perp^2 \lambda_e^2)^{1/2}}, \tag{26}$$

and

$$\omega \approx k_z V_A (1 + k_\perp^2 \rho^2)^{1/2}, \tag{27}$$

where $\rho^2 = \rho_s^2 + 3\rho_i^2/4$. For $k_\perp^2 \lambda_e^2 \gg 1$, Eq. (24) gives $\omega = k_z(\omega_{ce}\omega_{ci})^{1/2}/k_\perp$, which is the Okuda-Dawson convective cell frequency. It is obvious that the IA and KA waves appear in plasmas with $\beta \ll m_e/m_i$ and $m_e/m_i \ll \beta \ll 1$, respectively, where $\beta = 8\pi n_0 T/B_0^2$ and $T = T_e + T_i$.

The quasineutrality approximation also allows us to write the parallel electric fields associated with the IA and KA waves as[7]

$$E_z = \frac{[1 - \Gamma_0(b_i)]\omega^2 |\mathbf{E}_\perp|}{k_\perp k_z C_a^2}, \tag{28}$$

and

$$E_z = \frac{c^2 k_z^3 \lambda_{De}^2 |\mathbf{k}_\perp \cdot \mathbf{E}_\perp|}{\omega^2 - k_z^2 V_A^2 b_s}, \tag{29}$$

where $C_a = (T_i/m_e)^{1/2}$ is the electron acoustic speed.

The dispersion relation and the relationship between the parallel and perpendicular electric fields presented above are frequently used to understand the properties of dispersive Alfvén waves in space and laboratory plasmas.

3. Ponderomotive force and plasma density modification

Recent observations by the FREJA and FAST spacecrafts[25] and the LAPD experiments[26] reveal signatures of nonlinear structures consisting of localized DSAWs electric fields and very narrow magnetic-field-aligned density perturbations. Accordingly, in the following, we discuss the ponderomotive force of the DSAWs and the associated plasma density modification. The quasi-stationary plasma slow response in the DSWA fields is given by[23]

$$m_e\left(< \mathbf{v}_{e\perp}\cdot\nabla v_{ez} > +\frac{1}{2} < v_{ez}^2 >\right) + \frac{e}{c} < (\mathbf{v}_{e\perp}\times\mathbf{B}_\perp)_z > \\ = -eE_z^s - T_e\partial_z ln\, n_e^s, \tag{30}$$

$$m_i\left(< \mathbf{v}_{i\perp}\cdot\nabla v_{iz} > +\frac{1}{2} < v_{iz}^2 >\right) - \frac{e}{c} < (\mathbf{v}_{i\perp}\times\mathbf{B}_\perp)_z > \\ = eE_z^s - T_i\partial_z ln\, n_i^s, \tag{31}$$

where the superscript s denotes the quantities associated with the plasma slow motion. The angular bracket denotes averaging over the DSAW period.

We first consider the DSAWs with $k_z V_{Te} \ll \omega \ll \omega_{ce}$, so that the appropriate fluid velocities are

$$\mathbf{v}_{e\perp} \approx \frac{c}{B_0}\mathbf{E}_\perp\times\hat{\mathbf{z}}, \qquad v_{ez} = -i\frac{eE_z}{m_e\omega}, \tag{32}$$

and

$$\mathbf{v}_{i\perp} \approx \frac{e\omega_{ci}}{m_i(\omega_{ci}^2-\omega^2)}\mathbf{E}_\perp\times\hat{\mathbf{z}} - i\frac{c\omega\omega_{ci}\mathbf{E}_\perp}{B_0(\omega_{ci}^2-\omega^2)}. \tag{33}$$

The perpendicular component of the current density is

$$\mathbf{J}_\perp = -\frac{n_0 ec}{B_0}\frac{\omega^2}{(\omega_{ci}^2-\omega^2)}\mathbf{E}_\perp\times\hat{\mathbf{z}} - i\frac{n_0 ec\omega\omega_{ci}\mathbf{E}_\perp}{B_0(\omega_{ci}^2-\omega^2)}, \tag{34}$$

where the asterisk denotes the complex conjugate. Accordingly, we have a balance between the ponderomotive force and the pressure gradient

$$\frac{i}{n_0\omega}\langle \mathbf{J}_\perp^*\cdot\nabla E_z\rangle - \frac{1}{n_0 c}\langle(\mathbf{J}_\perp^*\times\mathbf{B}_\perp)\rangle \\ + \frac{e^2}{4m_e\omega^2}\frac{\partial}{\partial z}|E_z|^2 = -(T_e+T_i)\frac{\partial}{\partial z}\ln n_e^s. \tag{35}$$

By using the wave magnetic field

$$\mathbf{B}_\perp = -i\frac{c}{\omega}(\nabla\times\mathbf{E})_\perp \tag{36}$$

as well as the electric field relationship

$$\frac{\omega^2\omega_{ci}}{(\omega_{ci}^2-\omega^2)\omega_{ce}}\nabla\cdot\mathbf{E}_\perp=\frac{\partial E_z}{\partial z}, \tag{37}$$

and the modified IAW dispersion relation

$$\frac{\omega^2(1+k_\perp^2\lambda_e^2)}{k_z^2\lambda_i^2}=\omega_{ci}^2-\omega^2, \tag{38}$$

we obtain the quasi-neutral density response[23]

$$n_e=n_0\exp\left[\frac{|B_y|^2}{16\pi n_0(T_e+T_i)}\right]. \tag{39}$$

On the other hand, for the modified KAWs we take $v_{ez}=i(e\omega/k_z^2T_e)E_z$ and obtain[23]

$$n_e^s=n_0\exp\left[\frac{e^2|\mathbf{E}_\perp|^2}{4m_i(T_e+T_i)(\omega_{ci}^2-\omega^2)}-\frac{e^2|E_z|^2}{4m_e(T_e+T_i)}\frac{\omega^2}{k_z^4V_{Te}^4}\right]. \tag{40}$$

Since for the modified KAWs we have

$$E_z=\frac{k_zk_\perp C_s^2|\mathbf{E}_\perp|}{\omega_{ci}^2-\omega^2} \tag{41}$$

$$\omega^2=\frac{k_z^2V_A^2}{1+k_z^2\lambda_i^2}(1+k_\perp^2\rho_s^2), \tag{42}$$

the quasi-stationary electron response for the KAWs turns out to be

$$n_e^s=n_0\exp\left[\frac{\alpha e^2|\mathbf{E}_\perp|^2}{4m_i(T_e+T_i)(\omega_{ci}^2-\omega^2)}\right], \tag{43}$$

where $\alpha=1-\omega^2k_\perp^2m_e/(\omega_{ci}^2-\omega^2)k_z^2m_i$.

The expressions (39) and (43) reveal that the ponderomotive force of the DSAWs produces magnetic field-aligned electron density compressions.

4. Modulated circularly polarized dispersive Alfvén waves

Let us now consider the amplitude modulation of circularly polarized dispersive Alfvén waves (CPDAWs) along the external magnetic field direction. The cold plasma dispersion relation for CPDAWs ($\omega\ll\omega_{ce},\quad \mathbf{k}\parallel\hat{\mathbf{z}}$) is

$$\frac{k^2c^2}{\omega^2}\approx\pm\frac{\omega_{pe}^2}{\omega\omega_{ce}}-\frac{\omega_{pi}^2}{\omega(\omega\pm\omega_{ci})}, \tag{44}$$

which for $\omega\sim kV_A\ll\omega_{ci}$ reduces to

$$\omega\approx kV_A\left(1\mp\frac{kV_A}{2\omega_{ci}}\right). \tag{45}$$

Supposing that the nonlinear interaction between CPDAWs and the plasma slow response produces an envelope of waves which varies slowly. Hence, we introduce the eikonal representation[23]

$$\omega \to \omega_0 + i\frac{\partial}{\partial t} - iV_{A0}\frac{\partial}{\partial z}, \qquad k \to k_0 - i\frac{\partial}{\partial z}, \tag{46}$$

in (45) and operate on the wave electric field $\mathbf{E} = E_\perp(\hat{\mathbf{x}} \pm i\hat{\mathbf{y}})$. Assuming that $\partial E_\perp/\partial t \ll \omega_0 E_\perp \equiv \omega_0(E_x \mp iE_y)$ we then obtain the derivative nonlinear Schrödinger equation (DNLSE)

$$\left(\frac{\partial}{\partial t} + V_{A0}\frac{\partial}{\partial z}\right)E_\perp - \frac{V_A}{2n_0}\frac{\partial}{\partial z}(n_1 E_\perp) \pm \frac{iV_{A0}^2}{2\omega_{ci}}\frac{\partial^2 E_\perp}{\partial z^2} = 0. \tag{47}$$

The slow plasma response assumes inertialess electrons

$$0 = -eE^s - \frac{T_e}{n_0}\frac{\partial n_1}{\partial z} \pm \frac{\omega_{pe}^2}{4\pi n_0 \omega\omega_{ce}}\frac{\partial |E_\perp|^2}{\partial z}, \tag{48}$$

inertial ions

$$m_i\frac{\partial v_i^s}{\partial t} = eE^s - \frac{\gamma_i T_i}{n_0}\frac{\partial n_1}{\partial z} - \frac{\omega_{pi}^2}{4\pi n_0 \omega(\omega \pm \omega_{ci})}\frac{\partial |E_\perp|^2}{\partial z}. \tag{49}$$

Eliminating the ambipolar electric field E^s from (48) and (49) and using the ion continuity equation we obtain the driven ion sound waves

$$\left(\frac{\partial^2}{\partial t^2} - C_s^2\frac{\partial^2}{\partial z^2}\right)n_1 = -\frac{n_0 c^2}{B_0^2}\frac{\partial^2 |E_\perp|^2}{\partial z^2}. \tag{50}$$

The quasi-stationary response is then

$$\frac{n_1}{n_0} = \frac{m_i c^2}{(T_e + T_i)}\frac{|E_\perp|^2}{B_0^2}. \tag{51}$$

Subsequently, Eq. (47) takes the form[23]

$$\left(\frac{\partial}{\partial t} + V_{A0}\frac{\partial}{\partial z}\right)E_\perp - \frac{m_i V_{A0} c^2}{2(T_e + T_i)B_0^2}\frac{\partial}{\partial z}(|E_\perp|^2 E_\perp) \pm \frac{iV_{A0}^2}{2\omega_{ci}}\frac{\partial^2 E_\perp}{\partial z^2} = 0 \tag{52}$$

The DNLSE (52) admits a localized DAW electric field envelope accompanied with a background plasma density compression.

5. Nonlinear interactions between DAWs

Satellite (ICB1300) data[27] and recent Cluster observations[28] from the magnetospheric cusp region have unambiguously discovered coherent vortices as a manifestation of Alfvénic turbulence. Accordingly, it is desirable to have the knowledge of the governing nonlinear equations, which admit inverse cascade responsible for the formation of Alfvén vortices.

In the following, we present a set of nonlinear equations[23] for nonlinearly interacting low-frequency (in comparison with ω_{ci}), long wavelength (in comparison with the ion gyroradius) three-dimensional DAWs. For our purposes, the appropriate fluid velocities are then

$$\mathbf{v}_{e\perp} \approx \frac{c}{B_0}\hat{\mathbf{z}} \times \nabla\phi - \frac{c}{eB_0 n_e}\hat{\mathbf{z}} \times \nabla(n_e T_e) + v_{ez}\frac{\nabla_\perp A_z \times \hat{\mathbf{z}}}{B_0}, \tag{53}$$

$$\mathbf{v}_{i\perp} \approx \frac{c}{B_0}\hat{\mathbf{z}} \times \nabla\phi - \frac{c}{B_0\omega_{ci}}\left(\partial_t + \frac{c}{B_0}\hat{\mathbf{z}} \times \nabla\phi \cdot \nabla\right)\nabla_\perp\phi, \tag{54}$$

$$v_{ez} \approx \frac{c}{4\pi e n_0}\nabla_\perp^2 A_z. \tag{55}$$

Substituting (57)-(59) into $\nabla \cdot \mathbf{J} = 0$, where $\mathbf{J} = \mathbf{J}_\perp + \hat{\mathbf{z}}J_z$ is the plasma current density, we obtain

$$\frac{d}{dt}\nabla_\perp^2\phi + \frac{V_A^2}{c}\frac{d}{dz}\nabla_\perp^2 A_z = 0 \tag{56}$$

where $d/dt = (\partial/\partial t) + (c/B_0)\hat{\mathbf{z}} \times \nabla\phi \cdot \nabla$ and $d/dz = (\partial/\partial z) - B_0^{-1}\hat{\mathbf{z}} \times \nabla A_z \cdot \nabla$.

For the IAWs, we neglect the parallel electron pressure gradient in the parallel electron momentum equation and obtain

$$\frac{d}{dt}\left(1 - \lambda_e^2\nabla_\perp^2\right)A_z + c\partial_z\phi = 0. \tag{57}$$

On the other hand, for the KAWs the parallel electron inertia is negligible in comparison with the electron pressure gradient. Thus, we have

$$\frac{\partial A_z}{\partial t} + c\frac{d}{dz}\left(\phi - \frac{T_e n_1}{e n_0}\right) = 0, \tag{58}$$

where the electron continuity equation

$$\frac{dn_1}{dt} + \frac{c}{4\pi e}\frac{d}{dz}\nabla_\perp^2 A_z = 0 \tag{59}$$

determines n_1. Equations (56)–(59) are complex nonlinear partial differential equations containing vector nonlinearities. They admit dual cascades and provide the possibility of self-organization[29] of DAWs in the form of different types of vortical structures[30–32] (e.g. dipolar and tripolar vortices and a vortex street). Equations (56)–(59) can also be expressed in Hamiltonian form, and they are useful for studying collisionless tearing modes and current filaments in plasmas with sheared magnetic fields.[33] Furthermore, in a nonuniform magnetoplasma with the equilibrium density gradient, (56)-(59) have to be modified by including the effects of the density inhomogeneities. In such a situation we will have nonlinearly interacting drift-Alfvén waves.[34] The vortex structures associated with the latter have been identified with those observed in the magnetospheric cusp region.[28]

6. Generation of Zonal Flows by Kinetic Alfvén Waves

The nonlinear couplings[20,35–37] between the dispersive drift and kinetic Alfvén waves and zonal flows/convective cells[38] (also referred to as sheared flows) have been a topic of significant interest in space and laboratory plasmas. It is established that sheared flows can inhibit the transport of plasma particles across the external magnetic field direction. In this section, we present a theory for the generation of convective cells following Ref.[20] We use the two fluid model and account for the combined action of the nonlinear ion polarization drifts and the coupling of the parallel electron flow with sheared magnetic field of KAWs and the convective cell motions/zonal flows.

The electron and ion fluid velocities in the presence of nonlinearly interacting low-frequency ($\ll \omega_{ci}$) DKAWs and zonal flows are, respectively,

$$\mathbf{v}_{e\perp} \approx \frac{c}{B_0}\hat{\mathbf{z}} \times \nabla\phi - \frac{cT_e}{eB_0 n_0}\hat{\mathbf{z}} \times \nabla n_1, \tag{64}$$

and

$$\mathbf{v}_{i\perp} \approx \frac{c}{B_0}\hat{\mathbf{z}} \times \nabla\phi - \frac{c}{B_0\omega_{ci}}\left[\left(\frac{\partial}{\partial t} + \nu_{in} + \mu_i\nabla_\perp^2\right)\nabla_\perp\phi \right.$$
$$\left. + \frac{c}{B_0}(\hat{\mathbf{z}} \times \nabla\psi \cdot \nabla)\nabla_\perp\phi + \frac{c}{B_0}(\hat{\mathbf{z}} \times \nabla\phi \cdot \nabla)\nabla_\perp\psi\right], \tag{65}$$

where ν_{in} is the ion-neutral collision frequency, $\mu_i = 0.3\nu_{ii}\rho_i^2$ is the coefficient of the ion gyroviscosity, $\nu_{in}(\nu_{ii})$ are the ion-neutral (ion-ion) collision frequency, ψ is the potential of zonal flows, and $T_e \gg T_i$ has been assumed.

The appropriate electron and ion velocities involved in zonal flows in the presence of the DKAWs are, respectively,

$$\mathbf{u}_{e\perp} \approx \frac{c}{B_0}\hat{\mathbf{z}} \times \nabla\psi + \frac{< v_{ez}\mathbf{B}_\perp >}{B_0}, \tag{66}$$

and

$$\mathbf{u}_{i\perp} \approx \frac{c}{B_0}\hat{\mathbf{z}} \times \nabla\psi - \frac{c}{B_0\omega_{ci}}\left[\left(\frac{\partial}{\partial t} + \nu_{in} + \mu_i\nabla_\perp^2\right)\nabla_\perp\psi \right.$$
$$\left. + \frac{c}{B_0} < (\hat{\mathbf{z}} \times \nabla\phi \cdot \nabla)\nabla_\perp\phi >\right], \tag{67}$$

where the parallel component of the electron fluid velocity in the DKAW fields is

$$v_{ez} \simeq \frac{c}{4\pi e n_0} \nabla_\perp^2 A_z, \tag{68}$$

which is obtained from the parallel component of Ampere's law, with $\mathbf{B}_\perp = \nabla A_z \times \hat{\mathbf{z}}$. In (68) we have neglected the parallel ion motion, as we are isolating the ion acoustic waves in our intermediate plasma. The last terms in the right-hand side of (66) and (67) are the Reynolds stresses of the KAWs, which reinforce two-dimensional zonal flows.

Substituting (64) and (65) into $\nabla \cdot \mathbf{J} = 0$, where $\mathbf{J} = en_0 (\mathbf{v}_{i\perp} - \mathbf{v}_{e\perp} - v_{ez}\hat{\mathbf{z}})$, we have

$$\frac{\partial}{\partial t}\nabla_\perp^2 \phi + \frac{V_A^2}{c}\frac{\partial}{\partial z}\nabla_\perp^2 A_z + \frac{c}{B_0}(\hat{\mathbf{z}} \times \nabla\psi \cdot \nabla)\nabla_\perp^2 \phi$$

$$+\frac{c}{B_0}(\hat{\mathbf{z}} \times \nabla\phi \cdot \nabla)\nabla_\perp^2 \psi = 0, \tag{69}$$

where we have assumed that $|\partial\phi/\partial t| \gg (\nu_{in} + \mu_i \nabla_\perp^2)\phi$. The quasi-neutrality condition $n_{e1} = n_{i1} = n_1$ holds in a dense plasma with $\omega_{pi} \gg \omega_{ci}$, where $\omega_{pi} = (4\pi n_0 e^2/m_i)^{1/2}$ is the ion plasma frequency.

From the parallel component of the inertialess electron equation of motion, we obtain

$$\frac{\partial}{\partial t} A_z + c\frac{\partial}{\partial z}\left(\phi - \frac{T_e n_1}{e n_0}\right) + \frac{c}{B_0}\hat{\mathbf{z}} \times \nabla\psi \cdot \nabla A_z = 0. \tag{70}$$

On the other hand, the ion continuity equation, together with Eq. (67), yields

$$\left(\frac{\partial}{\partial t} + \frac{c}{B_0}\hat{\mathbf{z}} \times \nabla\psi \cdot \nabla\right)\left(n_1 - \frac{c n_0}{B_0 \omega_{ci}}\nabla_\perp^2 \phi\right) = 0. \tag{71}$$

The equation for two-dimensional zonal flows is obtained by inserting (68) and (69) into the electron and ion continuity equations, respectively, and substituting them into Poisson's equation. We obtain

$$\left(\frac{\partial}{\partial t} + \nu_{in} + 0.3\nu_{ii}\rho_i^2 \nabla_\perp^2\right)\nabla_\perp^2 \psi + \frac{c}{B_0} < (\hat{\mathbf{z}} \times \nabla\phi \cdot \nabla)\nabla_\perp^2 \phi >

$$-\frac{V_A^2}{cB_0} < (\hat{\mathbf{z}} \times \nabla A_z \cdot \nabla))\, \nabla_\perp^2 A_z >= 0, \tag{72}$$

where to lowest order, we must use

$$\frac{\partial A_z}{\partial z} + \frac{c}{V_A^2}\frac{\partial \phi}{\partial t} = 0 \tag{73}$$

into the last term of Eq. (72) to eliminate A_z in terms of ϕ. Equations (69)–(72) form a closed system of equations for studying the excitation of zonal flows by finite amplitude KAWs.

In the following, we derive a general dispersion relation for the modulational instability of a constant amplitude DAW pump against zonal flow perturbations. For this purpose, we decompose the high-frequency potentials into those of the pump and the two sidebands, viz.

$$\phi = \phi_{0+} \exp(-i\omega_0 t + i\mathbf{k}_0 \cdot \mathbf{r}) + \phi_{0-} \exp(i\omega_0 t - i\mathbf{k}_0 \cdot \mathbf{r})$$
$$+ \sum_{+,-} \phi_\pm \exp(-i\omega_\pm t + i\mathbf{k}_\pm \cdot \mathbf{r}), \tag{74}$$

$$A_z = A_{z0+} \exp(-i\omega_0 t + i\mathbf{k}_0 \cdot \mathbf{r}) + A_{z0-} \exp(i\omega_0 t - i\mathbf{k}_0 \cdot \mathbf{r})$$
$$+ \sum_{+,-} A_{z\pm} \exp(-i\omega_\pm t + i\mathbf{k}_\pm \cdot \mathbf{r}), \tag{75}$$

where $\omega_\pm = \Omega \pm \omega_0$ and $\mathbf{k}_\pm = \mathbf{K} \pm \mathbf{k}_0$ are the frequency and wavevector of the upper and lower DAW sidebands. The subscript $0_\pm$ and $\pm$ represent the pump and sidebands, respectively.

Assuming further that $\psi = \varphi \exp(-i\Omega t + i\mathbf{K}\cdot\mathbf{r})$, we insert (74) and (75) into Eqs. (69)-(71) and Fourier transform them and combine the resultant equations to obtain

$$D_\pm \phi_\pm = \pm \frac{ic}{B_0} \hat{\mathbf{z}} \times \mathbf{k}_0 \cdot \mathbf{K} \left(\omega_0 + \omega_\pm \frac{k_{0\perp}^2 - K_\perp^2}{k_{\perp\pm}^2} \right) \varphi \phi_{0\pm}, \tag{76}$$

where $D_\pm = \omega_\pm^2 - k_{z0}^2 V_A^2 (1 + k_{\perp\pm}^2 \rho_s^2) \approx \pm 2\omega_0 (\Omega - \mathbf{K}_\perp \cdot \mathbf{V}_{g\perp} \mp \delta$, with $\omega_0 = k_{z0} V_A (1 + k_{\perp 0}^2 \rho_s^2)^{1/2}$, $\mathbf{V}_{g\perp} = \mathbf{k}_{0\perp} \rho_s^2 k_{z0}^2 V_A^2 / \omega_0$, and $\delta = k_{z0}^2 V_A^2 K_\perp^2 \rho_s^2 / 2\omega_0$.

On the other hand, inserting (74) and (75) into Eq. (72) and Fourier transforming the resultant equation, we have

$$(\Omega + i\Gamma_z)\varphi = i\frac{2c}{B_0}\frac{\hat{\mathbf{z}}\times\mathbf{k}_0\cdot\mathbf{K}}{K_\perp^2}\left(1-\frac{\omega_0^2}{k_{z0}^2V_A^2}\right)$$

$$\times\left(\mathcal{K}_-^2\phi_{0+}\phi_- - \mathcal{K}_+^2\phi_{0-}\phi_+\right), \tag{77}$$

where $\Gamma_z = \nu_{in} + 0.3\nu_{ii}K_\perp^2\rho_i^2$, $\mathcal{K}_\pm^2 = k_{\perp\pm}^2 - k_0^2 \equiv K_\perp^2 \pm 2\mathbf{k}_{0\perp}\cdot\mathbf{K}_\perp$. Equation (77) reveals that the coupling constant on the right-hand side remains finite only if $\omega_0 \neq k_{0z}V_A$. Thus, dispersion to Alfvén waves is required in order for the parametric coupling between convective cells and the KAWs to remain in tact.

Eliminating $\phi_\pm$ from Eq. (77) by using (76) we finally obtain the nonlinear dispersion relation

$$\Omega + i\Gamma_z = \frac{2c^2\omega_0|\phi_0|^2}{B_0^2}\frac{|\hat{\mathbf{z}}\times\mathbf{k}_0\cdot\mathbf{K}|^2}{K_\perp^2}k_{0\perp}^2\rho_s^2\sum_{+,-}\frac{\mathcal{K}_\pm^2\kappa_\pm^2}{k_{\perp\pm}^2D_\pm}, \tag{78}$$

where $\kappa_\pm^2 = K_\perp^2 \pm \mathbf{k}_{0\perp}\cdot\mathbf{K}_\perp$. We see from Eq. (15) that the coupling constant in the right-hand side is proportional to $k_{0\perp}^2\rho_s^2$, which is a feature of the kinetic Alfvén wave dispersion. For long wavelength zonal flows with $|\mathbf{K}_\perp| \ll |\mathbf{k}_{0\perp}|$, Eq. (78) reduces to

$$(\Omega + i\Gamma_z)\left[(\Omega - \mathbf{K}_\perp\cdot\mathbf{V}_{g\perp})^2 - \delta^2\right]$$

$$= -2K_\perp^2c^2\delta\frac{|E_{0\perp}|^2}{B_0^2}k_{0\perp}^2\rho_s^2|(\hat{\mathbf{K}}\cdot\hat{\mathbf{k}}_{0\perp})(\hat{\mathbf{z}}\times\hat{\mathbf{k}}_0\cdot\hat{\mathbf{K}})|^2, \tag{79}$$

where $|E_{0\perp}|^2 = k_{0\perp}^2|\phi_0|^2$ and $\hat{\mathbf{K}}_\perp$ and $\hat{\mathbf{k}}_{0\perp}$ are the unit vectors.

We analyze Eq. (79) in two limiting cases. First, we let $\Omega = \mathbf{K}_\perp\cdot\mathbf{V}_{g\perp} + i\gamma_m$ in Eq. (79) and obtain for $\gamma_m, \Gamma_z \ll |\mathbf{K}_\perp\cdot\mathbf{V}_{g\perp}|$, the growth rate

$$\gamma_m = \left[\frac{2K_\perp^2c^2k_{0\perp}^2\rho_s^2\delta}{|\mathbf{K}_\perp\cdot\mathbf{V}_{g\perp}|}\frac{|E_{0\perp}|^2}{B_0^2}|(\hat{\mathbf{K}}\cdot\hat{\mathbf{k}}_{0\perp})(\hat{\mathbf{z}}\times\hat{\mathbf{k}}_0\cdot\hat{\mathbf{K}})|^2 - \delta^2\right]^{1/2}. \tag{80}$$

The expression (80) shows that the modulational instability sets in if

$$|E_{0\perp}|^2 > \frac{B_0^2\delta|\mathbf{K}_\perp\cdot\mathbf{V}_{g\perp}|}{2K_\perp^2c^2k_{0\perp}^2\rho_s^2|(\hat{\mathbf{K}}\cdot\hat{\mathbf{k}}_{0\perp})(\hat{\mathbf{z}}\times\hat{\mathbf{k}}_0\cdot\hat{\mathbf{K}})|^2}. \tag{81}$$

Second for $\Omega \gg \Gamma_z, \mathbf{K}_\perp\cdot\mathbf{V}_{g\perp}, \delta$, we have from Eq. (79)

$$\Omega^3 \simeq -2K_\perp^2 c^2 \delta \frac{|E_{0\perp}|^2}{B_0^2} k_{0\perp}^2 \rho_s^2, \tag{82}$$

which admits a reactive instability whose growth rate is

$$\gamma_r \approx (K_\perp c)^{2/3} \delta^{1/3} (k_{0\perp}\rho_s)^{2/3} \left(\frac{|E_{0\perp}|}{B_0}\right)^{2/3}. \tag{83}$$

We observe from (83) that the increment is proportional to two-third power of $k_{0\perp}\rho_s$ and the KAW pump electric field strength $|E_{0\perp}|$. For typical laboratory Argon plasmas with $n_0 = 2 \times 10^{12}$ cm^{-3}, $B_0 = 1.5$ kG, $T_e = 10T_i = 10$ eV, we have $\beta m_i/m_e \approx 30$, $V_A = 10^8$ cm/s, $\rho_s = 0.25$ cm. Taking $k_{0\perp}\rho_s = 0.1$, $\omega_0 \sim 10^5$ s^{-5}, $K_\perp/k_{0\perp} \sim 0.1$, and $|E_{0\perp}| \sim 10^{-4}B_0$, we find that $\delta \sim 10$ s^{-1}, $|\mathbf{K}_\perp \cdot \mathbf{V}_g| \sim 100$ s^{-1}, and $\gamma_r \sim 10^3$ s^{-1}. Thus, the reactive instability can produce zonal flows within a millisecond at the expense of the kinetic Alfvén wave energy. It is expected that zonal flows driven by the dispersive kinetic Alfvén waves can play a significant role in regulating the transport of plasma particles in magnetoplasmas, such as those in the Large Plasma Device (LAPD) at University of California Los Angeles,[8,39] as well as in the Earth's magnetosphere and in the solar corona.

7. The electron Joule heating by high-frequency DAWs

It has been known[3,4] for sometime that the dispersive kinetic Alfvén waves can cause heating of the plasma particles in laboratory magnetoplasma. The idea of wave-particle interactions[3] has been utilized to understand the solar coronal electron heating, which is an outstanding central problem in solar physics.[40–46] About seven years ago, Shukla *et al.*[47] proposed that the solar coronal electron heating could be caused by the resonant interaction between the high-frequency electromagnetic ion-cyclotron-Alfven waves (EMICA), which have the magnetic field-aligned electric field. The latter facilitates the electron Joule heating due to the wave-electron interaction, as described below.

Let us present the essential mathematical steps that are required for understanding the origin of the electron Joule heating. We note that the HF-EMICA waves are mixed modes with a magnetic field-aligned electric field component, and consequently there are density perturbations associated with the HF-EMICA waves. The dispersion relation of low parallel phase velocity (in comparison with the electron thermal speed) long wavelength (in comparison with the ion gyroradius) DA waves in a collisionless plasma is[47,48]

$$\omega = \frac{\omega_A}{(1+k_z^2\lambda_i^2)^{1/2}}\left[1+\left(\frac{3}{4}+\frac{T_e}{T_i}\right)k_x^2\rho_i^2\right]^{1/2}, \tag{84}$$

where the wave frequency ω is much smaller than the electron gyrofrequency $\omega_{ce} = eB_0/m_e c$, and k_x is the component of the wavevector across the ambient magnetic field direction. In obtaining (84), we have neglected the parallel ion dynamics, as the parallel phase speed of the EMICA waves is much larger than the ion-sound speed. We note that the parallel (the λ_i-term) dispersive term in (84) arises from the perpendicular ion inertial force, whereas the term involving 3/4 comes from the ion-finite Larmor radius effect, and the term involving T_e/T_i comes from the parallel electron kinetics. In a plasma with $T_e \gg T_i$, the perpendicular dispersion solely arises from the parallel electron pressure gradient force, and the square bracket in (84) is replaced by $1 + k_x^2\rho_s^2$, where $\rho_s = c_s/\omega_{ci}$ is the ion sound gyroradius, $c_s = (T_e/m_i)^{1/2}$ the ion-acoustic speed. When the parallel wavelength is shorter than the ion skin depth, Eq. (84) gives $\omega \approx \omega_{ci}(1+k_x^2\rho_s^2)^{1/2}$, which is the frequency of ion-cyclotron waves in a plasma with $T_e \gg T_i$.

In the following, we shall present the parallel electron current density associated with the HF-EMICA waves, and discuss the associated electron heating rate arising from the dissipation of the perturbed electron current density.

By using the linearized drift kinetic equation[49] for the perturbed electron distribution function, we obtain an expression for the parallel (to **z**) electron current density J_{ez} in a Maxwellian plasma. The result is[50]

$$J_{ez} = i\frac{\omega E_z}{8\pi k_z^2\lambda_{De}^2}Z'(\xi_e), \tag{85}$$

where E_z, is the parallel electric field, $\lambda_{De} = (T_e/4\pi n_0 e^2)^{1/2}$ the electron Debye radius, and Z' the derivative of the standard plasma dispersion function with argument $\xi_e = \omega/\sqrt{2}k_z V_{te}$. Here $V_{te} = (T_e/m_e)^{1/2}$ is the electron thermal speed. For $\xi_e \ll 1$ (85) becomes

$$J_{ez} = -i\frac{\omega E_z}{4\pi k_z^2\lambda_{De}^2}\left[1 + i\sqrt{\pi}\xi_e exp\left(-\xi_e^2\right)\right]. \tag{86}$$

The electron heating rate in the collisionless regime is governed by the dissipation of the magnetic field-aligned electron current, viz.[47,48]

$$\frac{dT_e}{dt} = \frac{1}{2n_0}\mathrm{Re}\left(J_{ez}E_z^*\right) \equiv \sqrt{\pi}\frac{\omega_{ci}^2\omega_{pe}^2|E_z|^2(1+b_s)}{4\pi n_0 k_z^3 v_{te}^3(1+b_i)}, \tag{87}$$

where $b_s = k_x^2\rho_s^2(1+3T_i/4T_e)$, $b_i = k_z^2\lambda_i^2$, $\omega_{pe} = (4\pi n_0 e^2/m_e)^{1/2}$, and the asterisk stands for the complex conjugate. Equation (87) exhibits that the electrons are rapidly heated by intense localized parallel electric fields of the HF-EMICA waves. The localization of the parallel electric field is caused by the self-modulation[13,51] of the HF-EMICA wave packets by non-resonant density perturbations.

We now apply the results of our theoretical model to the electron heating in the solar corona: Accordingly, we use some typical plasma and field parameters that are appropriate for the coronal heating scenario. The electron number density and the electron temperature in the unperturbed state are typically 5×10^9 cm^{-3} and one million K, respectively, whereas the ambient magnetic field is of the order of 10 Gauss. Thus, we have $\omega_{pe} = 4\times10^9$ rad s^{-1}, $\omega_{ci} = 10^5$ rad s^{-1}, $V_{te} = 4\times10^8$ cm s^{-1}, $V_A = 3\times10^7$ cm s^{-1}, $c_s = (T_e/m_i)^{1/2} = 10^7$ cm s^{-1}, and $\beta = 8\pi n_0 T/B_0^2 = 0.2$. The ion Larmor radius is roughly $\rho_s \approx 1$ m and the ion skin depth (λ_i) turns out to be 3 m. Assuming that both the perpendicular and parallel wavelength of the DA waves is one meter, we have $k_x\rho_s \approx 6$ and $k_z\lambda_i = 18$. It then follows from (87) that within $5-10$ s the electron temperature will rise to a value of 6 million K when the parallel electric field of the DA waves is 0.5 V cm^{-1}. For our adopted nominal value of the parallel electric field (0.5 V cm^{-1}), we find that $E_z/B_0 = 1.6\times10^{-5}$. Thus, the electron-HF-EMICA wave interactions are capable of producing the desired electron heating in the solar corona. The heating rate is sufficiently rapid to balance the cooling losses by conduction or radiation.

8. Ion heating by magnetic field-aligned EMICA waves

In the past, many authors[52–54] have discussed stochastic ion heating in the electrostatic and electromagnetic fields that have electric fields perpendicular to the external magnetic field direction. In the following, we consider differential ion heating in the solar corona caused by magnetic field-aligned left-hand circularly polarized EMICA waves that seem to have been observed[55–57] indirectly. The wave magnetic field is of the form

$$\mathbf{B}_w = \sum_k B_k(\hat{\mathbf{x}} - i\hat{\mathbf{y}})\exp(-i\omega_k t + ikz + i\varphi_k) + \text{c.c.} \tag{88}$$

where $\hat{\mathbf{x}}$ and $\hat{\mathbf{y}}$ are the unit vectors along the x and y axes, respectively, $\omega_k = (k^2c^2\omega_{ci}/2\omega_{pi}^2)\left(\sqrt{1+4\omega_{pi}^2/k^2c^2}-1\right)$ is the EMICA wave frequency, $k \equiv k_z$, φ_k is the phase constant, and c.c. stands for the complex conjugate.

In (88) the wave magnetic field is a resultant vector field composed of a large number of discrete components with different wavenumbers.

The ions in the EMICA wave fields obey the equation of motion

$$\frac{d\mathbf{v}}{dt} = \frac{e}{m_i}\left[\mathbf{E}_w + \frac{\mathbf{v}}{c} \times (\hat{\mathbf{z}}B_0 + \mathbf{B}_w)\right], \tag{89}$$

where $\mathbf{v} = d\mathbf{r}/dt$ is the ion velocity. The wave electric field $\mathbf{E}$ is obtained from

$$\hat{\mathbf{z}} \times \mathbf{E}_w = \frac{\omega_k}{kc}\mathbf{B}_w. \tag{90}$$

From (89) and (90) we then obtain[48]

$$\frac{dv_x}{dt} = \frac{e}{m_i c}\sum_k \left(\frac{\omega_k}{k} - v_z\right) B_{ky}\exp(i\phi_k) + \omega_{ci}v_y, \tag{91}$$

$$\frac{dv_y}{dt} = -\frac{e}{m_i c}\sum_k \left(\frac{\omega_k}{k} - v_z\right) B_{kx}\exp(-i\phi_k) - \omega_{ci}v_x, \tag{92}$$

and

$$\frac{dv_z}{dt} = \frac{e}{m_i c}\sum_k (v_x B_{ky} - v_y B_{kx})\exp(-i\phi_k), \tag{93}$$

where we have denoted $\mathbf{v} = (v_x, v_y, v_z)$, $\mathbf{B}_w = (B_{kx}, B_{ky}, 0)$, and $\phi_k = \omega_k t - kz - \varphi_k$.

For left-hand circularly polarized EMICA waves, we introduce $u_\perp = v_x - iv_y$, and $B_{k\perp} = B_{kx} - iB_{ky}$ to obtain from (91)–(93)

$$\frac{du_\perp}{dt} - i\omega_{ci}u_\perp = -i\sum_k \left(v_z - \frac{\omega_k}{k}\right)\Omega_k \exp(-i\phi_k), \tag{94}$$

and

$$\frac{dv_z}{dt} = \operatorname{Im}\left[u_\perp \Omega_k^* \exp(i\phi_k)\right], \tag{95}$$

where $\Omega_k = (e/m_i c)(B_{kx} - iB_{ky})$ is the gyrofrequency associated with the wave magnetic field.

We have numerically solved[48] Eqs. (94) and (95) to demonstrate the ion energization in the magnetic field-aligned EMICA wave magnetic fields.

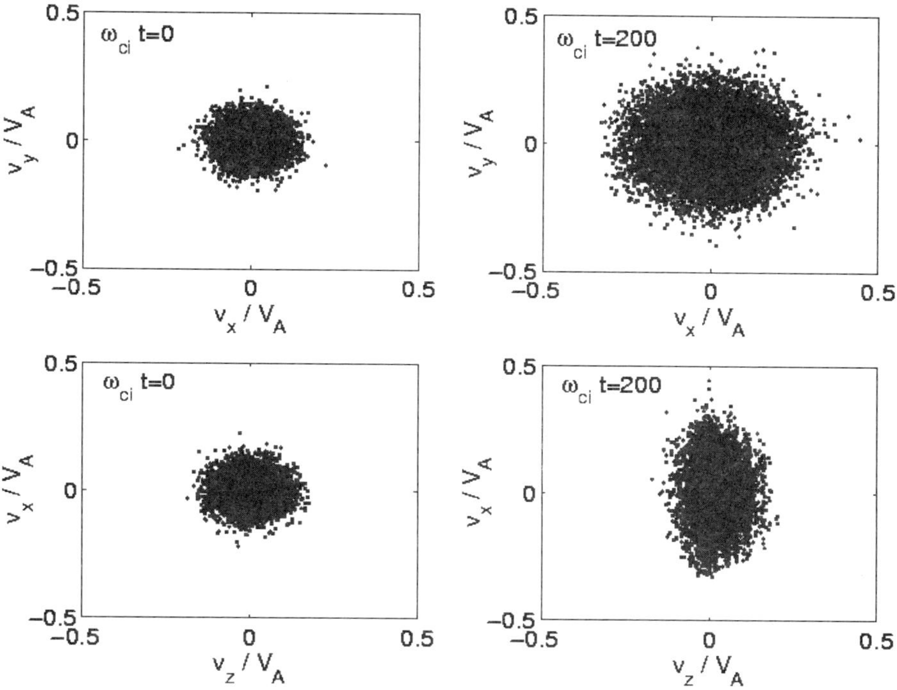

Fig. 1. The ion particle distribution at $t = 0$ (left panels) and $\omega_{ci}t = 200$ (right panels), in the v_x–v_y plane (upper panels) and in the v_z–v_x plane (lower panels). The initially isotropic distribution (left panels) becomes strongly anisotropic due to differential heating of the ions in the perpendicular direction by EMICA waves. After Ref.[48]

In the simulation, we used 10 randomly phased waves, each of amplitude $|\Omega_k| = 0.01\,\omega_{ci}$ and with equally spaced frequencies between $\omega_k = 0.6\omega_{ci}$ and $\omega_k = 1.5\omega_{ci}$. The simulation was performed with 10^4 test particles with initial conditions $u_\perp = u_\perp(0)$, $v_z = v_z(0)$, and $z = z(0)$ that were Maxwellian distributed with an isotropic temperature corresponding to an ion thermal speed of $0.05\,V_A$ in all velocity directions. Figure 1 shows the initial and isotropic ion distribution at time $t = 0$ and the particle distribution at $t = 200\,\omega_{ci}$, where the latter is strongly non-isotropic. The ion heating has been taking place in the perpendicular direction only while the thermal spread in the parallel z direction is the same as at $t = 0$. This differential ion heating is demonstrated in Fig. 2, where we have displayed the parallel and perpendicular ion temperatures as a function of time. We see that the ion temperature, which initially is isotropic, is enhanced strongly

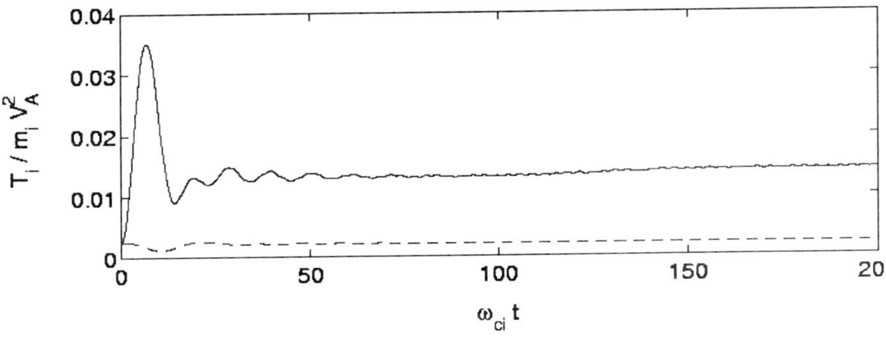

Fig. 2. The perpendicular (solid lines) and parallel (dashed lines) ion temperature as a function of time. The initially isotropic temperature becomes strongly anisotropic due to differential stochastic heating by nonlinear EMICA waves. At $\omega_{ci}t = 200$, the temperature ratio is approximately $T_\perp/T_z = 5$. After Ref.[48]

in the perpendicular direction only. After some time, the ion temperatures stabilize with a temperature ratio $T_\perp/T_z \approx 5$. This is in agreement with the observed ion temperature anisotropy in the solar corona.[58] It should be stressed that the test particle approach used in this section is not self-consistent, since the electromagnetic field is given as an external input and not calculated from the Maxwell equations. Hence, we have not taken into account the damping of the wave via the heating process. Such effects may be important and should be studied in more realistic, self-consistent studies of the wave-ion interactions.

9. Discussion and conclusion

In this paper, we have presented the underlying physics and the dispersion properties of dispersive shear Alfvén waves (DSAWs) in a uniform magnetoplasma. We have obtained a general dispersion relation that includes the finite $\omega/k_z V_{Te}$ and the finite ω/ω_{ci} effects, in addition to the finite ion Larmor radius and electron skin depth corrections. It is shown that these nonideal effects drastically modify the propagation characteristics of the DSAWs in plasmas. Large amplitude DSAWs are shown to produce local plasma density enhancement due to the ponderomotive force. The latter is also responsible for creating nonresonant density perturbation which modulate the magnetic field-aligned dispersive Alfvén waves. The modulated dispersive Alfvén wave packet along the magnetic field direction obeys the DNLSE, which admits localized wave envelopes. The dynamics of nonlinearly interacting three-dimensional dispersive Alfvén waves is governed by a set of equations in which the ion vorticity, the ion number density perturbation and the parallel component of the vector potential are related in a complex manner. Since the governing nonlinear equations contain the vector nonlinearities, they admit dual cascade which ensures the formation of Alfvén vortices. The latter are recently observed in space plasmas. Furthermore, we have presented a nonlinear theory for the generation of sheared plasma flows by the Reynold stresses of the kinetic Alfvén waves. The presence of sheared plasma flows provide the possibility of a better plasma confinement in magnetically confined fusion plasmas. We have also discussed the nonlinear aspects of EMICA waves which are responsible for the heating of plasma particles in the solar corona. Attention is paid on electron Joule heating by the obliquely propagating HF-EMICA waves and ion heating by magnetic field-aligned left-hand circularly polarized EMICA waves. Both mechanisms involve wave-particle interactions and are capable of yielding the desired electron heating and ion temperature anisotropy in the solar corona that are consistent with observations. It should be noted that there is some indirect evidence[55–57] for the existence of the HF-EMICA and left-hand circularly polarized EMICA waves in the solar corona. These waves could be excited due to the ion temperature anisotropy,[58] fast magnetic reconnections,[59] the pre-existing ion and electron beams, sheared plasma flows, etc. In conclusion, we stress that the results presented here are useful for understanding the salient features of dispersive Alfvén waves and some important nonlinear effects (associated with wave-wave and wave-particle interactions) that they produce in the Earth's ionosphere and magnetosphere, in the solar corona, and in LAPD at UCLA.[39]

Acknowledgments This work was partially supported by the Deutsche Forschungsgemeinschaft through the Sonderforschungsbereich 591, Centre for Fundamental Physics (CfFP) at Rutherford Appleton Laboratory, Chilton (UK), as well as by the Swedish Research Council (VR).

References

1. H. Alfvén, *Nature* **150**, 405 (1942).
2. T. J. Stefant, *Phys. Fluids* **13**, 440 (1971).
3. A. Hasegawa and L. Chen, *Phys. Fluids* **19**, 1924 (1976).
4. A. Hasegawa and C. Uberoi, *The Alfvén Wave*, Report DOE/TIC No. 11197, (Technical Center, US Department of Energy, Washington, DC, 1982).
5. G. Brodin and L. Stenflo, *Contrib. Plasma Phys.* **30**, 413 (1990).
6. P. K. Shukla and L. Stenflo, *J Plasma Phys.* **64**, 125 (2000).
7. P. K. Shukla and L. Stenflo, *Geophys. Res. Lett.* **31**, L03810,

doi:10.1029/2003GL018814 (2004).
8. W. Gekelman, *J. Geophys. Res.* **104**, 14417 (1999).
9. C. A. Kletzing *et al.*, *Phys. Rev. Lett.* **90**, 035004 (2003).
10. P. K. Shukla and J. M. Dawson, *Astrophys. J.* **276**, L49 (1984).
11. P. K. Shukla and L. Stenflo, *Phys. Fluids* **28**, 1576 (1985).
12. P. K. Shukla and L. Stenflo, *Phys. Scr.* **T60**, 32 (1995).
13. P. K. Shukla, R. Bingham and R. O. Dendy, *Phys. Lett. A* **239**, 369 (1998).
14. P. K. Shukla and L. Stenflo, in *Nonlinear MHD Waves and Turbulence*, Eds. T. Passot and P. L. Sulem, (Springer, Berlin, 1999) pp 1-30.
15. P. K. Shukla, L. Stenflo and R. Bingham, *Phys. Plasmas* **6**, 1677 (1999).
16. P. K. Shukla and L. Stenflo, *Phys. Plasmas* **6**, 4120 (1999).
17. P. K. Shukla and L. Stenflo, *Phys. Plasmas* **7**, 2738 (2000).
18. T. Drozdenko and G. J. Morales, *Phys. Plasmas* **8**, 3265 (2001).
19. P. K. Shukla and L. Stenflo, *Phys. Rev. Lett.* **85**, 2408 (2000).
20. P. K. Shukla, *Phys. Plasmas* **12**, 012310 (2005).
21. P. H. Diamond and M. A. Malkov, *Astrophys. J.* **654**, 252 (2006).
22. M. Brambilla, *Kinetic Theory of Plasma Waves: Homogeneous Plasmas* (Clarendon, Oxford, 1988).
23. P. K. Shukla, L. Stenflo, R. Bingham and B. Eliasson, *Plasma Phys. Control. Fusion* **46**, B349 (2004).
24. P. K. Shukla, *Nature (London)* **278** 874 (1978).
25. K. Stasiewicz, P. Bellan, C. Chaston *et al.*, *Space. Sci. Rev.* **92**, 423 (2000).
26. W. Gekelman, S. Vincena, N. Palmer *et al.*, *Plasma Phys. Control. Fusion* **42** B15 (2000).
27. V. M. Chmyrev, S. V. Bilichenko, O. A. Pokhotelov, V. A. Marchenko, V. I. Lazarev, A. V. Streltsov and L. Stenflo, *Phys. Scr.* **38**, 841 (1988).
28. D. Sundkvist, V. Krasnoselskikh, P. K. Shukla *et al.*, *Nature (London)* **436**, 825 (2005).
29. A. Hasegawa, *Adv. Phys.* **34**, 1 (1985).
30. V. I. Petviashvili and O. A. Pokhotelov, *Solitary Waves in the Plasma and in the Atmosphere* (Gordon and Breach, London, 1992).
31. P. K. Shukla, G. T. Birk and R. Bingham, *Geophys. Res. Lett.* **22**, 671 (1995).
32. D. Jovanovic, F. Pegoraro and J. J. Rasmussen, *J. Plasma Phys.* **60**, 383 (1998).
33. D. Del Sarto, F. Califano and F. Pegoraro, *Phys. Rev. Lett.* **91**, 235001 (2003).
34. P. K. Shukla, M. Y. Yu and L. Stenflo, *Phys. Rev. A* **34**, 3478 (1986).
35. R. Z. Sagdeev, V. D. Shapiro and V. I. Shevchenko, *Sov. J. Plasma Phys.* **4**, 306 (1978).
36. P. K. Shukla, M. Y. Yu, H. U. Rahman and K. H. Spatschek, *Phys. Rev. A* **23**, 312 (1981).
37. R. Z. Sagdeev, V. D. Shapiro and V. I. Shevchenko, *JETP Lett.* **27**, 340 (1978).
38. H. Okuda and J. M. Dawson, *Phys. Fluids* **16**, 408 (1973).
39. T. A. Carter, R. Brugman, P. Pribyl and W. Lybarger, *Phys. Rev. Lett.* **96**, 155001 (2006).
40. L. Harrison, *Solar Phys.* **175** 467, (1997).

41. B. J. Kellet and V. Tsikoudi, *Mon. Not. R. Astron. Soc.* **288**, 411 (1997).
42. G. B. Laing and P. M. Edwin, *Solar Phys.* **161**, 269 (1995).
43. P. M. Edwin and K. Murawski, *Solar Phys.* **158**, 227 (1995).
44. U. Narain and P. Ulmschneider, *Space Sci. Rev.* **75**, 453 (1996).
45. D. Tsiklauri, *New J. Phys.* **8**, 79 (2006).
46. J. A. Klimchuck, *Solar Phys.* **234**, 41 (2006).
47. P. K. Shukla, R. Bingham, J. McKenzie and W. I. Axford, *Solar Phys.* **186**, 61 (1999).
48. P. K. Shukla, R. Bingham, B. Eliasson, M. E. Dieckmann and L. Stenflo, *Plasma Phys. Control. Fusion* **48** B249 (2006).
49. A. Hasegawa and K. Mima, *J. Geophys. Res.* **83**, 1117 (1978).
50. P. M. Bellan, *Phys. Plasmas* **5**, 3081 (1998).
51. A. B. Mikhailovskii, V. I. Petviashvili and A. M. Friedman, *JETP Lett.* **24**, 43 (1977).
52. J. M. McChesney, R. A. Stern and P. M. Bellan, *Phys. Rev. Lett.* **59**, 1436 (1987).
53. L. Chen, Z. Lin and R. White *Phys. Plasmas* **8**, 4731 (2001).
54. R. White, L. Chen and Z. Lin, *Phys. Plasmas* **9**, 1890 (2002).
55. S. R. Cranmer, G. B. Field and J. L. Kohl, *Astrophys. J.* **518**, 937 (1999).
56. S. R. Cranmer, *J. Geophys. Res.* **106**(A11), 24937 (2001).
57. X. Li, *Astrophys. J.* **571**, L67 (2002).
58. S. P. Gary, R. M. Skoug and C. W. Smith, *Plasma Phys.* **12**, 056501 (2005).
59. J. E. McKenzie, M. Banaszkiewicz and W. I. Axford, *Astron. Astrophys.* **303**, L45 (1995).

PROPERTIES OF DRIFT AND ALFVÉN WAVES IN COLLISIONAL PLASMAS

J. VRANJES[1] and S. POEDTS

Center for Plasma Astrophysics,
Celestijnenlaan 200B, 3001 Leuven, Belgium, and
Leuven Mathematical Modeling and Computational Science Center (LMCC), and
[1]*Faculté des Sciences Appliquées,*
avenue F.D. Roosevelt 50, 1050 Bruxelles, Belgium

B. P. PANDEY

Department of Physics, Macquarie University,
Sydney, NSW 2109, Australia

The stability is discussed of the drift-Alfvén wave which is driven by the equilibrium density gradient, in both unbounded and bounded, collisional plasmas, including the effects of both hot ions and a finite ion Larmor radius. The density gradient in combination with the electron collisions with heavier plasma species is the essential source of the instability of the electrostatic drift mode which is coupled to the dispersive Alfvén mode. In the analysis of modes in an unbounded plasma the exchange of identity between the electrostatic and electromagnetic modes is demonstrated. Due to this, the frequency of the electromagnetic part of the mode becomes very different compared to the case without the density gradient. In the case of a bounded plasma the dispersion properties of modes involve a discrete poloidal mode number, and eigen-functions in terms of Bessel functions with discrete zeros at the boundary. The results are applied to the solar plasma.

A physical background is presented and an analytical description is given for the electrostatic drift mode in partially ionized magnetized plasmas with inelastic collisions. In such plasmas the creation of plasma particles is balanced by a certain number of phenomena in which ions and electrons are lost. Dispersion equation is derived for the drift mode, with the appropriate instability conditions, indicating that inelastic collisions can make the mode unstable.

The physics of Alfvén waves in weakly ionized plasmas like the solar photosphere is discussed. The magnetization and the collision frequencies of the plasma constituents are quantitatively examined. It is shown that the ions and electrons in the photosphere are both un-magnetized, their collision frequency with neutrals is much larger than the gyro-frequency. This implies that eventual Alfvén-type electromagnetic perturbations must involve the neutrals as well. It follows that in the presence of perturbations the whole fluid (plasma

+ neutrals) moves, the Alfvén velocity includes the total (plasma + neutrals) density and is thus considerably smaller compared to the collision-less case, the perturbed velocity of a unit volume, which now includes both plasma and neutrals, becomes much smaller compared to the ideal (collision-less) case, and finally the corresponding wave energy flux for the given parameters becomes much smaller compared to the ideal case.

1. Introduction

The investigation of low frequency (compared to the ion gyro-frequency Ω_i) modes in spatially inhomogeneous and/or bounded plasmas (drift and Alfvén modes being typical representatives) has been the subject of numerous studies in the past. The theory on drift wave has started more than 40 years ago in an attempt of explaining the anomalous diffusion of plasma across the magnetic field lines, which at that time appeared to be due to some instability of unknown origin. According to Chen,[1] the first result on that issue was due to Moiseev and Sagdeev,[2] although already in 1961 a work has been published[3] dealing with waves in inhomogeneous plasmas in the presence of both temperature and density gradients. The term 'drift wave' appears to be used for the first time in the sense we use it nowadays in the work of Kadomtsev and Timofeev.[4] We note that, in the same period, a first report on an experimental verification of low-frequency ion waves propagating perpendicular to the magnetic field lines in the presence of a density gradient has been published by D'Angelo and Motley.[5] In our recent works, we have been dealing with various aspects of the drift wave in both laboratory and space plasmas.[6–13]

The drift wave itself is driven by a density gradient in the direction perpendicular to the magnetic field lines, and the frequency of the mode is close to the diamagnetic frequency. Such a density inhomogeneity is a very common plasma feature and its frequent presence in plasmas implies the possibility for a wide spread appearance of the drift waves. Within the kinetic theory, the mode is unstable (growing) due to the inverse electron Landau damping effect, even for a Maxwellian electron distribution. What is needed is the presence of a density gradient which results in a mode frequency slightly below the diamagnetic frequency. On the other hand, within the fluid theory, the mode is also unstable due to the common effect of the electron collisions and the ion inertia, in the presence of the background density gradient. Hence, the term 'universally unstable mode', used in the early literature on the drift wave. A comparison between the two instabilities[14] reveals that the resistive one is dominant, provided that the electron parallel mean-free path is smaller than the parallel wavelength. This is essentially the reason for the interest in the drift modes with very large parallel wave-lengths and relatively short perpendicular wave-lengths. In fact, such situations occur particularly frequent in magnetized space plasmas with almost unlimited scales in the direction along the magnetic field lines.

The drift mode can be made more unstable in the presence of a plasma current or flow along the magnetic field lines, provided that this background velocity has a gradient in the perpendicular direction. A mode like the drift wave, propagating obliquely with respect to this inhomogeneous equilibrium flow, can become unstable when certain conditions are met. The first demonstration of this instability in a magnetized plasma has been given in the early work of D'Angelo,[15] and more recently in our works.[16,17] An inhomogeneous ion flow with a gradient in the x-direction was considered along an external magnetic field $\vec{B}_0 = B_0\mathbf{z}$. The electron and ion temperatures were equal, i.e. $T_i = T_e$, and the density gradient was assumed to be oriented along the negative x-axis.

In plasmas with not so small plasma β ($> m_e/m_i$), the mode becomes electromagnetic and in such plasmas an interplay between the electrostatic and electromagnetic part of the mode takes place. The electromagnetic component is an Alfvén-type perturbation which is dispersive like the kinetic-Alfvén wave (KAW) or the inertial Alfvén wave (IAW). Which of these modes enter the perturbations depends essentially on the plasma β, yet it also implies a certain range of the wave numbers $k_\perp$ in the direction perpendicular to the magnetic field lines. For the KAW, the determining parameter is $k_\perp \rho_s$, where $\rho_s = c_s/\Omega_i$, with $c_s = (\kappa T_e/m_i)^{1/2}$, and $\Omega_i = eB_0/m_i$. For the solar coronal plasma, ρ_s is typically of the order of 1 m. For the IAW, the relevant parameter is $k_\perp \lambda_e$, where $\lambda_e = c/\omega_{pe}$, and $\omega_{pe} = (e^2 n_0/\varepsilon_0 m_e)^{1/2}$, implying even shorter scales, viz. < 0.5 m. On the other hand, to have modes with reasonably low and detectable frequencies $\sim k_z c_a$, where $c_a = B_0/(\mu_0 n_0 m_i)^{1/2}$, the parallel wave-length is measured in hundreds of kilometers. The huge difference in the perpendicular and parallel scales is, in fact, favorable for the drift wave instability.

Alfvén waves in a weakly ionized but highly collisional medium involve also the motion of the neutrals which are present in the medium. This is due to the friction between charged particles and neutrals. The effect has been first described in the literature[18–20] in 1962 and in many subsequent works ever since.[21–24]

In the present study, we apply the basic theory of the EM drift (i.e. coupled drift and Alfvén) wave to the solar plasma. This implies high and equal electron and ion temperatures, and possibly short scales in the perpendicular direction. Therefore, we derive and include the ion gyro-viscosity stress tensor terms which, for short wave lengths, appear of the same order as the standard polarization drift terms. They both describe the finite ion Larmor radius (FLR) effects, yet the latter appears due to the time dependence of the diamagnetic drift. Some important features, like the exchange of identity of modes and the related frequency change, will be pointed out.

We shall also present a model for the drift modes propagating in a partially ionized magnetized plasma with a relatively high temperature and density, where the negative and/or molecular ions, and the corresponding chemical reactions, are absent. Thus the model includes a three component plasma consisting of singly charged ions, electrons, and neutral parental atoms, and a number of the most important production and loss processes that can take place in such a system. These source/sink terms in the corresponding fluid equations are kept in general form so that the model is applicable to many space and laboratory plasmas.

The Alfvén wave has been a very popular tool in the past in the scenarios and models dealing with the heating of upper solar atmosphere. A necessary ingredient in such models is an efficient and abundant source for the excitation of these waves, which acts permanently and generates waves throughout the solar atmosphere. Very frequently it is assumed that the omnipresent convective motions in the photosphere could serve for this purpose. The amount of thermal energy per unit volume in the solar corona is in fact extraordinary small in comparison with the lower (and much colder) layers of the solar atmosphere. This is due to the rapidly decreasing density with altitude. On the other hand, the complete photosphere is covered by overshooting convective gas motions with typical velocities of about 0.5 km/s, that may go up to 2 km/s. The kinetic energy per cubic meter stored in this macroscopic motion of a mainly neutral gas exceeds for several orders of magnitude the internal energy density in the corona. Clearly, only a tiny fraction of the convective kinetic energy of the neutral gas would be sufficient to heat the higher layers to the given temperatures. Such a scenario is attractive in view of the fact that this macroscopic motion in the lower atmosphere is permanent and widespread throughout the solar surface. However, the photosphere is very weakly ionized and it is also a strongly collisional mixture of the tiny plasma component and the predominantly neutral (uncharged) gas. The energy flux of the Alfvén waves is given by $m_i n_i v_i^2 c_a/2$, where c_a is the Alfvén velocity and v_i is the perturbed velocity of ions involved in the oscillations. Typically, in the estimate of the flux in the photosphere, this perturbed velocity is taken of the same order as the macroscopic convective motion mentioned above.[25] We shall show that in fact this assumption is far from reality.

2. Drift-Alfvén wave in hot plasma

2.1. *Equations*

The momentum equations for ions and electrons which we use here are

$$m_i n_i \left[\frac{\partial \vec{v}_i}{\partial t} + (\vec{v}_i \cdot \nabla)\vec{v}_i\right] = e n_i \left(-\nabla\phi - \frac{\partial A_z}{\partial t}\vec{e}_z + \vec{v}_i \times \vec{B}\right)$$

$$-\kappa T_i \nabla n_i - \nabla \cdot \Pi_i - m_i n_i \nu_i \vec{v}_i, \tag{1}$$

$$m_e n_e \left[\frac{\partial \vec{v}_e}{\partial t} + (\vec{v}_e \cdot \nabla)\vec{v}_e\right] = -e n_i \left(-\nabla\phi - \frac{\partial A_z}{\partial t}\vec{e}_z + \vec{v}_e \times \vec{B}\right)$$

$$-\kappa T_e \nabla n_e - \nabla \cdot \Pi_e - m_e n_e (\nu_e \vec{v}_e - \nu_{ei} \vec{v}_i). \tag{2}$$

Here, $\nu_i \equiv \nu_{in}$ and $\nu_e = \nu_{en} + \nu_{ei}$, i.e., we allow for the presence of neutrals in the plasma and, in view of the huge difference in mass, we neglect the ion momentum change due to their collisions with electrons. The model is consequently applicable to partially ionized as well as to fully ionized plasmas. In order to apply the model to hot plasmas we retain the stress tensor contribution. More precisely, we keep its leading order gyro-viscous part. In practice, provided we deal with relatively short wavelengths, this gives corrections mainly to the ion dynamics. The equilibrium magnetic field is in the z-direction, $B_0\vec{e}_z$, and we have an equilibrium plasma density with a gradient in the perpendicular direction.

In the case of highly magnetized electrons and in the limit when their inertia effects are negligible, the total perpendicular electron velocity in the linear limit reduces to

$$\vec{v}_{e\perp} = \frac{1}{B_0}\vec{e}_z \times \nabla_\perp \phi - \frac{\kappa T_e}{e B_0}\vec{e}_z \times \frac{\nabla_\perp n_e}{n_e}. \tag{3}$$

The parallel electron dynamics, in the limit when ion motion is predominantly polarized in the perpendicular plain, is described by

$$\left(\frac{\partial}{\partial t} + \vec{v}_{e0}\nabla_\perp\right) A_{z1} + \frac{\partial \phi_1}{\partial z} - \frac{\kappa T_e}{n_{e0} e}\frac{\partial n_{e1}}{\partial z} - \frac{m_e \nu_e}{\mu_0 e^2 n_{e0}}\nabla_\perp^2 A_{z1} = 0. \tag{4}$$

Here, v_{e0} denotes the equilibrium electron diamagnetic drift velocity described by the second term in Eq. (3), i.e., $\vec{v}_{e0} = -\kappa T_e \vec{e}_z \times \nabla_\perp n_{e0}/(e B_0 n_{e0})$, and we have also used Ampère law, $\nabla \times \vec{B} = \mu_0 \vec{j}$ yielding $en_{e0}\mu_0 v_{ez1} = \nabla_\perp^2 A_{z1}$.

The electron continuity becomes

$$\frac{\partial n_{e1}}{\partial t} + \frac{1}{B_0}(\vec{e}_z \times \nabla_\perp \phi_1) \cdot \nabla_\perp n_{e0} + \frac{1}{\mu_0 e}\frac{\partial}{\partial z}\nabla_\perp^2 A_{z1} = 0. \tag{5}$$

The ion perpendicular motion, obtained from Eq. (1), is described by the recurrent formula

$$v_{i\perp} = \alpha_i \left[\frac{1}{B_0}\vec{e}_z \times \nabla_\perp \phi + \frac{v_{Ti}^2}{\Omega_i}\vec{e}_z \times \frac{\nabla_\perp n_i}{n_i} - \frac{\nu_i}{\Omega_i}\frac{\nabla_\perp \phi}{B_0} \right.$$
$$- \frac{\nu_i v_{Ti}^2}{\Omega_i^2}\frac{\nabla_\perp n_i}{n_i} + \vec{e}_z \times \frac{\nabla_\perp \cdot \Pi_i}{m_i n_i \Omega_i} + \frac{1}{\Omega_i}\left(\frac{\partial}{\partial t} + \vec{v}_{i\perp} \cdot \nabla_\perp\right)\vec{e}_z \times \vec{v}_{i\perp}$$
$$\left. - \frac{\nu_i}{\Omega_i}\frac{\nabla_\perp \cdot \Pi_i}{e n_i B_0} - \frac{1}{\Omega_i}\frac{\nu_i}{\Omega_i}\left(\frac{\partial}{\partial t} + \vec{v}_{i\perp} \cdot \nabla_\perp\right)\vec{v}_{i\perp}\right]. \tag{6}$$

Here $\alpha_i = 1/(1+\nu_i^2/\Omega_i^2)$. In the case $\nu_i^2 \ll \Omega_i^2$, we may set $\alpha_i \to 1$, the first two terms are the leading order ones and should be used in the polarization drift, while the last two terms are higher order terms and can be omitted.

Eq. (6) is used in the ion continuity equation to calculate the terms $\nabla_\perp(n_i \vec{v}_{i\perp})$. The procedure is straightforward except for the term with the convective derivative in the polarization drift $\vec{v}_p$, i.e., $(\vec{v}_{i\perp} \cdot \nabla_\perp)\vec{e}_z \times \vec{v}_{i\perp}$, and the stress tensor contribution $\vec{v}_\pi$. For a small equilibrium density gradient, the last $\vec{v}_{i\perp}$ in $\vec{v}_p$ from (6) comprises only the leading order perturbed $\vec{E} \times \vec{B}$ and diamagnetic drifts ($\vec{v}_{E1}$ and $\vec{v}_{*i1}$), while the first $\vec{v}_i$ is the equilibrium ion diamagnetic drift $\vec{v}_{i0} = \kappa T_i \vec{e}_z \times \nabla_\perp n_{i0}/(eB_0 n_{i0}) = -\vec{v}_{e0} T_i/T_e$. On the other hand, the stress tensor part after a few steps yields

$$\nabla_\perp \cdot (n\vec{v}_\pi) = -\rho_i^2 \nabla_\perp n_{i0} \cdot \nabla_\perp^2 \vec{v}_{i\perp} - n_{i0}\rho_i^2 \nabla_\perp^2 \nabla_\perp \cdot \vec{v}_{i\perp}$$
$$= -\rho_i^2 \nabla_\perp n_{i0} \cdot \nabla_\perp^2 \vec{v}_{i\perp} + \frac{\rho_i^2 n_{i0}}{\Omega_i}\frac{\partial}{\partial t}\nabla_\perp^4 \left(\frac{\phi_1}{B_0} + \frac{v_{Ti}^2}{\Omega_i}\frac{n_{i1}}{n_{i0}}\right). \tag{7}$$

The first term in this expression, within the second order approximation limit, cancels out with the term $(\vec{v}_{i0} \cdot \nabla_\perp)\vec{e}_z \times \vec{v}_{i\perp}$ from the above discussed convective derivative in the polarization drift which appears in $\nabla_\perp \cdot (n_i \vec{v}_p)$. The second term in Eq. (7) is the FLR contribution, obtained by substituting the velocity (6) into the operator.

Another similar FLR term is obtained from the time derivative part of $\nabla_\perp \cdot (n_i \vec{v}_p)$ reading

$$-n_{i0}\rho_i^2 \frac{\partial}{\partial t}\nabla_\perp^2 \frac{e\phi_1}{\kappa T_i} - \rho_i^2 \frac{\partial}{\partial t}\nabla_\perp^2 n_{i1}. \tag{8}$$

Here, $\rho_i = v_{Ti}/\Omega_i$, and $v_{Ti}^2 = \kappa T_i/m_i$.

The ion continuity equation finally yields

$$\frac{\partial}{\partial t}\left(\frac{n_{i1}}{n_{i0}}\right) + \frac{1}{B_0}\vec{e}_z \times \nabla_\perp \phi_1 \cdot \frac{\nabla_\perp n_{i0}}{n_{i0}} - \nu_i n_{i0}\rho_i^2 \nabla_\perp^2 \left(\frac{e\phi_1}{\kappa T_i} + \frac{n_{i1}}{n_{i0}}\right)$$

$$-n_{i0}\rho_i^2 \frac{\partial}{\partial t}\nabla_\perp^2 \left(\frac{e\phi_1}{\kappa T_i} + \frac{n_{i1}}{n_{i0}}\right) + \frac{\rho_i^2 n_{i0}}{\Omega_i}\frac{\partial}{\partial t}\nabla_\perp^4 \left(\frac{\phi_1}{B_0} + \frac{v_{Ti}^2}{\Omega_i}\frac{n_{i1}}{n_{i0}}\right) = 0. \quad (9)$$

Eqs. (5) and (9) are combined using the quasi-neutrality, yielding a more convenient form of the third equation which closes the system, viz.

$$\left(\frac{\partial}{\partial t} + \nu_i\right)\nabla_\perp^2 \phi_1 + c_a^2 \frac{\partial}{\partial z}\nabla_\perp^2 A_{z1} + \frac{\kappa T_i}{e n_0}\left(\frac{\partial}{\partial t} + \nu_i\right)\nabla_\perp^2 n_1$$

$$-\rho_i^2 \frac{\partial}{\partial t}\nabla_\perp^4 \left(\phi_1 + \frac{\kappa T_i}{e n_0} n_1\right) = 0. \quad (10)$$

The given set of equations (4), (5), and (10) will be used in the description of the drift-Alfvén waves in solar plasma.

2.2. *Dispersion equation*

In Cartesian geometry, for perturbations $\sim \exp(-i\omega t + ik_y y + ik_z z)$, Eqs. (4), (5), and (10) yield

$$\omega^3 - \omega^2 \left[\omega_{*e} + \omega_{*i} - i\left(\delta + \frac{\nu_i}{1 + k_y^2\rho_i^2}\right)\right] + \omega\left\{\omega_{*e}\omega_{*i} - \frac{k_z^2 c_a^2}{1 + k_y^2\rho_i^2}\right.$$

$$-k_y^2 k_z^2 c_a^2(\rho_s^2 + \rho_i^2) - \frac{\nu_i\delta}{1 + k_y^2\rho_i^2} - i\left[\omega_{*i}\delta + \frac{\nu_i(\omega_{*e} + \omega_{*i})}{1 + k_y^2\rho_i^2}\right] + \frac{\omega_{*e}k_z^2 c_a^2}{1 + k_y^2\rho_i^2}$$

$$\left. + \frac{\omega_{*i}\nu_i\delta}{1 + k_y^2\rho_i^2}\right\} + i\frac{\nu_i}{1 + k_y^2\rho_i^2}\left[\omega_{*e}\omega_{*i} - k_z^2 c_a^2 k_y^2\left(\rho_s^2 + \rho_i^2\right)\right] = 0. \quad (11)$$

Here, $\omega_{*e} = k_y v_{e0}$, $v_{e0} = -\kappa T_e \kappa_0/(eB_0)$, $\kappa_0 = n_0'/n_0$, $\omega_{*i} = k_y v_{i0}$, $v_{i0} = \kappa T_i \kappa_0/(eB_0)$, $\delta = m_e \nu_e k_y^2/(\mu_0 n_0 e^2)$.

In the collision-less limit and for $T_e > T_i$, and after the expansion $(1 + k_y^2\rho_i^2)^{-1} \simeq 1 - k_y^2\rho_i^2$, Eq. (11) yields the drift-Alfvén mode derived earlier by using the kinetic theory:[26]

$$\omega^3 - \omega^2(\omega_{*e} + \omega_{*i}) - \omega[k_z^2 c_a^2 - \omega_{*e}\omega_{*i} + k_z^2 c_a^2 k_y^2 \rho_s^2]$$

$$+ k_z^2 c_a^2(\omega_{*e} + k_y^2 \rho_s^2 \omega_{*i}) = 0.$$

In the limit of small $k_y\rho_s$ it reduces to

$$(\omega - \omega_{*e})[\omega^2 - \omega\omega_{*i} - k_z^2 c_a^2] = 0,$$

yielding three obvious solutions, i.e., the electrostatic drift wave and two (accelerated and retarded) Alfvén waves:[27]

$$\omega_1 = \omega_{*e}, \quad \omega_{2,3} = \frac{1}{2}\left[\omega_{*i} \pm \omega_{*i}\left(1 + \frac{4k_z^2 c_a^2}{\omega_{*i}^2}\right)^{1/2}\right]. \tag{12}$$

For small k_z the two latter waves become

$$\omega_2 \simeq \omega_{*i} \equiv -\frac{\omega_{*e} T_i}{T_e}, \quad \omega_3 = \frac{k_z^2 c_a^2}{|\omega_{*i}|}. \tag{13}$$

Hence, the actual frequencies of the modes in a hot plasma with density gradients in the direction perpendicular to the magnetic field lines, may become very different from the frequencies of the standard Alfvén modes propagating in opposite directions, i.e. $\pm k_z c_a$. This fact should be taken into account in fitting observations into the modeling of solar coronal plasmas.

In the limit of negligible ion thermal effects Eq. (11) reduces to

$$\omega^3 - \omega^2\omega_* - \omega k_z^2 c_a^2(1 + k_y^2\rho_s^2) + \omega_* k_z^2 c_a^2 + i\nu_e k_y^2 \lambda_e^2 \omega^2 = 0. \tag{14}$$

In the absence of a density gradient this yields a damped kinetic-Alfvén mode.[28] On the other hand, in the limit $\omega^2 \ll k_z^2 c_a^2$ we have a standard unstable drift mode[9] with frequency $\omega_r = \omega_{*e}/(1 + k_y^2\rho_s^2)$ and increment $\omega_i = \nu_e \omega_{*e}^2 k_y^2 \rho_s^2/(k_z^2 v_{Te}^2)$. This is a unique feature of the drift mode, which is intrinsically unstable in a collisional plasma. The instability appears as a common effect of the electron collisions ν_e, the finite ion mass effect (the term $k_y^2\rho_s^2$), and the equilibrium density gradient (the term ω_{*e}). The two modes are coupled even without collisions, when we have $(\omega - \omega_{*e})(\omega^2 - k_z^2 c_a^2) = \omega k_z^2 c_a^2 k_y^2 \rho_s^2$, and the coupling vanishes in the limit of negligible $k_y^2\rho_s^2$, i.e., for the case of the drift and the non-dispersive Alfvén modes.

Although Eqs. (11) and (14) are relatively simple and cubic only, calculating and discussing exact solutions of these equations appears practically rather inconvenient. This is seen in the case of Eq. (14) which can be easily analytically solved, yielding three complex solutions which, however, are nontrivial to discuss. Yet, to some extent the character of the solutions may be understood even without directly solving the dispersion equations, i.e., by using the generalized Hurwitz method for polynomials with complex coefficients.[29] For a general polynomial of the degree m and with complex coefficients, $x^m + (a_1 + ib_1)x^{m-1} + \ldots + (a_m + ib_m) = 0$, one constructs the sequence of $m+1$ numbers $c_0 = 1$, $c_1 = a_1$, ..., c_r, ..., where r goes to m, and where

$$c_r = (-1)^{r(r-1)/2} \begin{vmatrix} a_1 & 1 & 0 & 0 & 0 & \cdot \\ -b_2 & -b1 & a_1 & 1 & 0 & \cdot \\ a_3 & a_2 & b_2 & b_1 & a_1 & \cdot \\ -b_4 & -b_3 & a_3 & a_2 & -b_2 & \cdot \\ a_5 & a_4 & b_4 & b_3 & a_3 & \cdot \\ \cdot & \cdot & \cdot & \cdot & \cdot & \cdot \\ \cdot & \cdot & \cdot & \cdot & \cdot & \cdot \\ a_{2r-1} & a_{2r-2} & b_{2r-2} & b_{2r-3} & a_{2r-3} & \cdot \end{vmatrix} .$$

According to Giaretta[29] the number of roots with positive real parts equals the number of sign changes in the sequence c_j. A sufficient instability condition is that any of the c_r has a negative sign. As a simple check, we apply the method on Eq. (14). Here, we find $c_0 = 1$, $c_1 = -\omega_*$, $c_2 = -\omega_*^2 k_z^2 k_y^2 c_a^2 \rho_s^2 < 0$, $c_3 = c_a^4 k_z^4 \omega_*^3 (\delta^2 + c_a^2 k_y^4 k_z^2 \rho_s^4) > 0$. Consequently, we have two sign changes in the sequence c_j, i.e., two positive real roots, and c_1 and c_2 are both negative, therefore, there exists at least one unstable mode.

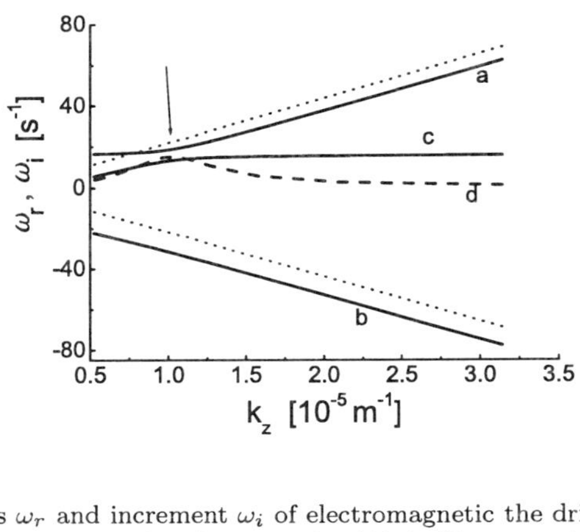

Fig. 1. Frequencies ω_r and increment ω_i of electromagnetic the drift-Alfvén perturbations with the effect of the coupling between the Alfvén (lines a and b) and drift (line c) parts. Dotted lines denote $\pm k_z c_a$. The increment of the electrostatic drift mode (multiplied by 10^3) has a maximum in the region where the retarded kinetic-Alfvén mode and the drift mode change their identities (denoted by arrow).

To discuss the roots and the increments/decrements in detail, we solve Eq. (11) numerically by taking parameters values that are typical for the quiet inner solar corona, viz. $T_e = T_i = 1.5 \cdot 10^6$ K, $n_{e0} = n_{i0} = 10^{14}$ m^{-3}, and taking $B_0 = 10^{-3}$ T, we calculate the spectrum for the two oppositely propagating Alfvén modes and the drift mode. Here, neutrals are absent and the dominant collisions take place between electrons and ions, $\nu_{ei} = (8\pi/m_e)^{1/2}[e^2/(4\pi\varepsilon_0)]^2 n_{e0} L_{ei}/(\kappa T_e)^{3/2}$, $L_{ei} = \log[12\pi\varepsilon_0(\varepsilon_0/n_{e0})^{1/2}(\kappa T_e)^{3/2}/e^3]$. For these parameters we obtain $\nu_{ei} \simeq 2.1$ Hz, $\nu_{ii} \simeq 0.07$ Hz, $c_s = 1.11 \cdot 10^5$ m/s, $c_a = 2.18 \cdot 10^6$ m/s, $v_{Te} = 4.77 \cdot 10^6$ m/s, $\beta = 0.0052 > m_e/m_i = 0.00054$.

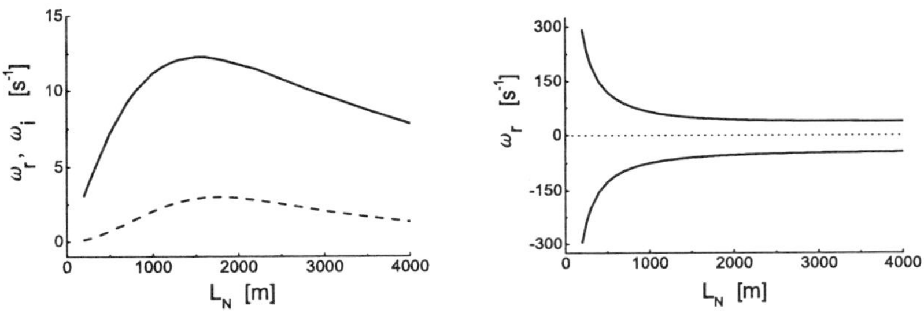

Fig. 2. Left: The drift wave frequency (full line), and its increment multiplied by 100 (dashed line) in terms of the density scale-length, corresponding to the Alfvén modes from the figure right. Right: The frequency of the kinetic-Alfvén modes corresponding to the drift mode (left), for the given coupling parameter $k_y\rho_s = 0.52$ and $k_z c_a = 34.3$ Hz, in terms of the density scale length.

The behavior of the modes in terms of the parallel wave-number k_z is presented in Fig. 1 for $L_n = 10^3$ m and $\lambda_y = 50$ m, while $k_y\rho_s = 0.15$. We thus have a situation similar to that described by Eq. (12), i.e., two (retarded and accelerated) damped kinetic-Alfvén modes, lines *a* and *b*, respectively, and an electrostatic drift mode (line c). The increment of the drift mode (multiplied by 10^3) is presented by line *d*. The drift mode is unstable in the whole range of wave-numbers. Its frequency is nearly constant for large values of k_z. In the area denoted by the arrow, the retarded kinetic-Alfvén mode and the drift mode do not cross each other. Instead, they change identities as typical for an 'avoided crossing'. For observations, the low frequency (small k_z) domain is of particular importance as this parameter can be measured, i.e. spatially resolved. It is clearly seen in Fig. 1 that, in fact, this is the domain in which the Alfvén mode frequency can be very different from what is expected or predicted if the small-scale plasma inhomogeneity, which drives the drift mode, is neglected. Here, for the given density scale length, the two frequency limits at $k_z \to 0$ are ± 16.26 Hz. Clearly, the frequencies can be made arbitrary small by changing the equilibrium density scale length L_N.

The decrement of the Alfvén modes has also been calculated and, in general, the accelerated mode b is less damped. Its damping rate changes between $-3.5 \cdot 10^{-3}$ Hz at $k_z = 3.14$ (in given units), and $-1.1 \cdot 10^{-3}$ Hz at $k_z = 0.5$. The damping rate of the decelerated mode a has a maximum absolute value of about $2.3 \cdot 10^{-2}$ Hz.

The drift mode frequency is normally proportional to $1/L_N$. However, due to the coupling with the Alfvén modes, its behavior is also drastically changed. This is seen in Fig. 2(left) where we fix $k_y \rho_s = 0.52$ and $k_z =$ 400 km. Here, contrary to what may be expected, for small L_N the mode vanishes and the decrement decreases. This is again due to the identity change with the Alfvén mode. As a matter of fact, the Alfvén frequency for the given numbers is constant $k_z c_a = 34.3$ Hz and the KAWs frequencies [see Fig. 2(right)] at large L_N do not change much. For a decreasing L_N the drift mode curve does not intersect with the Alfvén mode. Instead, the Alfvén mode takes over the behavior of the drift mode: it grows while the drift mode decreases.

Note that, for the given parameters, the drift curve changes its direction at frequencies around 12 Hz. This is still far enough from the requirement of a small parallel wave-phase velocity in comparison with the electron thermal velocity used in order to omit the electron inertia. Here, we have $k_z v_{Te} = 75$ Hz. Thus, the inclusion of the electron inertia terms is not expected to considerably change the mode behavior.

We note that a similar sort of identity change of the drift-Alfvén mode, known in the literature,[26] happens also in the case when the ion parallel dynamics is retained, and on the condition that $c_a > c_s$. In this case, the sound part of the drift mode is disconnected and the parallel mode dependence goes along the line $k_z c_a$.

2.3. *Eigen-modes*

In order to present the mode behavior in magnetic structures that are highly elongated along the magnetic field lines and localized in the perpendicular direction, we shall rewrite the starting equations in cylindric coordinates. We use

$$\nabla_{r,\theta} = \vec{e}_r \frac{\partial}{\partial r} + \frac{\vec{e}_\theta}{r} \frac{\partial}{\partial \theta}, \quad \nabla^2_{r,\theta} = \frac{\partial^2}{\partial r^2} + \frac{1}{r}\frac{\partial}{\partial r} + \frac{1}{r^2}\frac{\partial^2}{\partial \theta^2},$$

and consider perturbations of the form $\widehat{f}(r)\exp(-i\omega t + im\theta + ik_z)$, where $\widehat{f}(r)$ denotes the r-dependent amplitude and m the discrete poloidal mode number. In the same frame we have $\vec{v}_{(e,i)0} = \mp\vec{e}_\theta \kappa T_{e,i} n_0'/(eB_0 n_0) = v_{(e,i)0}(r)\vec{e}_\theta$, $n_0' = dn_0/dr$, with the minus sign for electrons. The combined electron dynamics equations (4) and (5) yield:

$$\left(\nabla_\perp^2 + \frac{\dfrac{\omega\omega_2}{c_a^2 k_z^2 \rho_s^2}}{1 - \dfrac{i\nu_e\omega}{k_z^2 v_{Te}^2}}\right)\widehat{A}_{z1} - \frac{1}{k_z c_a^2}\frac{\dfrac{\omega}{\rho_s^2} + \dfrac{m\Omega_i}{r}\dfrac{n_0'}{n_0}}{1 - \dfrac{i\nu_e\omega}{k_z^2 v_{Te}^2}}\widehat{\phi}_1 = 0, \tag{15}$$

where $\omega_2 = \omega - v_{e0}(r)m/r$, and $\nabla_\perp = \partial^2/\partial r^2 + \partial/(r\partial r) - m^2/r^2$. The ion part can be discussed in two limits.

2.4. *Cold ion case*

In this limit, from Eq. (10) we obtain

$$\nabla_\perp^2 \widehat{\phi}_1 - \frac{k_z^2 c_a^2}{\omega}\nabla_\perp^2 \widehat{A}_{z1} = 0, \tag{16}$$

which is used in Eq. (15) yielding

$$\nabla_\perp^2\left[\left(\nabla_\perp^2 - \frac{1}{\omega(1-i\delta_1)}\left(\frac{\omega}{\rho_s^2} + \frac{m\Omega_i}{r}\frac{n_0'}{n_0}\right)\right)\widehat{\phi}_1\right]$$

$$+\nabla_\perp^2\left[\frac{\omega - mv_{e0}/r}{k_z\rho_s^2(1-i\delta_1)}\widehat{A}_{z1}\right] = 0. \tag{17}$$

To further decouple the equations for the potentials we have to assume a density profile. We assume a realistic case, viz. $n_0(r) = N_0\exp(ar^2/2)$, where a can be both positive and negative, and r takes values between 0 and r_0. All the terms under the operators in this case become constant and Eq. (17), after using (16) again, becomes of the form

$$\nabla_\perp^2\left\{\left[\nabla_\perp^2 - \frac{\omega/\rho_s^2 + am\Omega_i}{\omega(1-i\delta_1)} + \frac{\omega(\omega + am\kappa T_e/eB_0)}{k_z^2\rho_s^2 c_a^2(1-i\delta_1)}\right]\widehat{\phi}_1\right\} = 0. \tag{18}$$

or

$$\nabla_\perp^2\psi(r) = 0, \tag{19}$$

Here, consequently

$$\left[\nabla_\perp^2 - \frac{\omega/\rho_s^2 + am\Omega_i}{\omega(1-i\delta_1)} + \frac{\omega(\omega + am\kappa T_e/eB_0)}{k_z^2\rho_s^2 c_a^2(1-i\delta_1)}\right]\widehat{\phi}_1 = \psi(r), \tag{20}$$

and from Eq. (19) $\psi(r)$ is

$$\psi(r) = 0, \quad \text{or} \quad \psi(r) = c_1 \cosh[m \log(r)] + c_2 \sinh[m \log(r)]. \tag{21}$$

Eq. (20) can be readily written in the form

$$\left(\frac{\partial^2}{\partial r^2} + \frac{1}{r}\frac{\partial}{\partial r} - \frac{m^2}{r^2} + \xi^2\right)\left[\widehat{\phi}_1 - \frac{\psi(r)}{\xi^2}\right] = 0, \tag{22}$$

where

$$\xi^2 = \frac{\omega(\omega+\omega_0)}{\rho_s^2(1-i\delta)}\left(\frac{1}{c_a^2 k_z^2} - \frac{1}{\omega^2}\right), \quad \omega_0 = am\frac{\kappa T_e}{eB_0}, \quad \delta_1 = \frac{\nu_e \omega}{k_z^2 v_{Te}^2}.$$

The solutions of Eq. (22) are the Bessel functions of the first and the second kind, $J_n(\xi r)$ and $Y_n(\xi r)$, with a complex argument. For nonsingular eigenfunctions we keep $J_n(\xi r)$. Using the well known theory[30] we have, first, that if $n > -1$, then the zeros of the Bessel function $J_n(z)$ with the complex argument z are all real, and, second, for $n \geq 0$ the functions $J_n(z)$ and $J_{n+s}(z)$ have no common zeros other than the origin, for all $s > 0$. Hence, for vanishing solutions at the boundary, we set that $\xi r_0 = \varepsilon_l$ where ε_l is the real l-th zero of the complex function $J_n(\xi r)$. This allows us to write the dispersion equation for the radially bounded plasma

$$\left(\omega + am\frac{\kappa T_e}{eB_0}\right)(\omega^2 - k_z^2 c_a^2) = \omega k_z^2 c_a^2 \frac{\varepsilon_l^2 \rho_s^2}{r_0^2}(1 - i\delta). \tag{23}$$

This is the equivalent of Eq. (14) in an unbounded plasma. Yet, here both m and ε_l take given discrete values. Eq. (23) describes the global drift-Alfvén wave, with an unstable drift wave part. The poloidal (i.e. in the θ-direction) propagation is the consequence of the drift mode which propagates perpendicular to both the magnetic field lines and the density gradient. Combined with the given z-dependence, this gives the twisting of the global modes.[6,7] The twisting vanishes for $m = 0$ when the two modes decouple. The eigenfunctions $\widehat{A}_{z1}$ can be easily found from Eq. (16), and $\widehat{n}_1$ from Eq. (5), which is rewritten as

$$\widehat{n}_1 = \frac{k_z}{\mu_0 e\omega}\nabla_\perp^2 \widehat{A}_{z1} - \frac{m}{r}\frac{n_0'}{\omega B_0}\widehat{\phi}_1. \tag{24}$$

Mode details are available in our recent work.[10]

2.5. *The hot ion case*

Ion thermal effects enter the equations through (8), the second term in (7), and the collisions. Using Eq. (24) in (10) for the same Gaussian/inverse-Gaussian density profile as before, we obtain an equation containing terms proportional to $\rho_i^4 \nabla_\perp^4 \widehat{\phi}_1$ and $\rho_i^6 \nabla_\perp^6 \widehat{A}_{z1}$, which come from the second part of the stress tensor in Eq. (7). This 6th-order differential equation is to be combined with the 2nd-order Eq. (15) in order to decouple the two potentials. In the case of the global modes studied here and, therefore, for large scales, we have $|\rho_i \nabla_\perp| < 1$ and the high order derivatives yield small terms that can be neglected. Hence, by omitting the stress tensor contribution while still keeping the ion thermal effects through (8) and the collisions, from Eq. (10) we have

$$\nabla_\perp^2 \left\{ \widehat{\phi}_1 - \frac{k_z c_a^2}{(\omega + i\nu_i)(\omega - ma\rho_i^2 \Omega_i)} \left[\omega - (\omega + i\nu_i)\rho_i^2 \nabla_\perp^2 \right] \widehat{A}_{z1} \right\} = 0. \tag{25}$$

This gives

$$\widehat{\phi}_1 = \frac{k_z^2 c_a^2}{(\omega + i\nu_i)(\omega - ma\rho_i^2 \Omega_i)} \left[\omega - \rho_i^2 (\omega + i\nu_i) \nabla_\perp^2 \right] \widehat{A}_{z1} + \psi(r), \tag{26}$$

which is used in Eq. (15) yielding

$$\left(\frac{\partial^2}{\partial r^2} + \frac{1}{r}\frac{\partial}{\partial r} - \frac{m^2}{r^2} + \eta^2 \right) [\widehat{A}_{z1} + c\psi(r)] = 0. \tag{27}$$

Here,

$$\eta^2 = \left[\omega(\omega + i\nu_i) \left(\omega + ma\rho_s^2 \Omega_i \right) \left(\omega - ma\rho_i^2 \Omega_i \right) \right.$$
$$\left. - \omega k_z^2 c_a^2 \left(\omega + ma\rho_s^2 \Omega_i \right) \right] \left\{ k_z^2 c_a^2 (\omega + i\nu_i) \left[\rho_s^2 (1 - i\delta_1) \left(\omega - ma\rho_i^2 \Omega_i \right) \right. \right.$$
$$\left. \left. + \rho_i^2 (\omega + ma\rho_s^2 \Omega_i) \right] \right\}^{-1},$$

and

$$c = \frac{(\omega + ma\rho_s^2 \Omega_i)}{k_z c_a^2 \rho_s^2 \eta^2 (1 - i\delta_1)}.$$

As earlier, the solutions are the Bessel functions of the first kind and the corresponding dispersion equation reads

$$\eta^2 = \frac{\varepsilon_l^2}{r_0^2}. \tag{28}$$

In the cold ion limit, it reproduces Eq. (23). The solutions of Eq. (28) describe the eigen-values of the global eigen-modes in the given cylindrical plasma. Eq. (28) can be easily solved numerically for various harmonics by choosing the appropriate ε_l and m. In the case of solar coronal magnetic structures, the density and the magnetic field have higher values compared to the previous case,[31] $n_{i0} = n_{e0} = 10^{16}$ m^{-3} and $B_0 \simeq 10^{-2}$ T. Taking as an example a magnetic column with the diameter of 200 km and, in the case when the density at its edge is 0.1 of its value at the column axis, we get $a \simeq 7 \cdot 10^{-10}$ m^{-2}, and therefore, the poloidal mode number m takes very high values $\sim 10^5$.

3. Inelastic collisions and the drift wave

In a partially ionized multi-component plasma various inelastic collisions (or chemical reactions) take place, and plasma source-sink terms appear in the equations[32] depending on the temperatures and concentrations of the plasma-gas species. The processes resulting in the production, as well as in the loss of plasma particles are numerous. Typical and most important examples of the former ones are the photo-ionization from the external sources (given by the formal scheme $A + h\nu \rightarrow A^+ + e$; dominant in the D, E, and F regions of the ionosphere and in optically thick systems in general), the ionization by electron impact (of the type $A+e \rightarrow A^+ + e + e$; important in the solar atmosphere, in optically thin and bounded systems, as well as in many dense systems), the ion-atom interchange reactions (the process of creation of molecular ions, given by $A^+ + BC \rightarrow (AB)^+ + C$; dominant in the E and F regions of the ionosphere). The latter ones (losses) include the radiative recombination ($A^+ + e \rightarrow A + h\nu$, insignificant in the ionosphere, dominant in more dense and hot plasmas), the three body recombination ($A^+ + e + e \rightarrow A + e$; important in the D layer of the ionosphere and in the solar photosphere), dissociative recombination (represented by $A_2^+ + e \rightarrow A + A^*$ where A^* is an excited atom; dominant in the E and F layers of the ionosphere, important in the D layer). Consequently in writing the appropriate fluid equations describing the plasma dynamics, the corresponding leading order source-sink terms have to be identified and included in the model. The same holds for the number of plasma-gas species, like the presence and absence of the mentioned molecular ions, or negative ions etc.

The creation of a population of plasma particles with a very narrow distribution function, i.e., nearly mono-energetic beams with a relatively small density compared to the bulk plasma, under laboratory conditions can be achieved in a double layer configuration.[33] In the presence of a large amount of neutrals this results in a directed flux of energetic electrons only, because the ion mean free path for charge exchange is very short. If the electron energy is above the ionization threshold, electrons are capable of ionizing the neutral atoms that are present in the system. The ionization cross section is very much dependent on the energy of the ionizing particles;

a fitting formula for the ionization by electrons is given by[34]

$$\sigma(x) = \frac{\sigma_{00}}{\epsilon_i^2}\frac{1}{x}\left(\frac{x-1}{x+1}\right)^{3/2}\left\{1+\frac{2}{3}\left(1-\frac{1}{2x}\right)\log\left[2.7+(x-1)^1/2\right]\right\}. \tag{29}$$

Here, $x = \epsilon/\epsilon_i$, $\sigma_{00} = 6.56 \cdot 10^{-14}$ eV^2 cm^2, ϵ_i is the binding energy of an orbital electron, and ϵ is the kinetic energy of the bombarding electron. The shape of $\sigma(x)$ in cm^2 (multiplied by 10^{18}) is given in Fig. 3 (line 1). Clearly, in the area slightly above the ionization threshold, small changes of the ionizing electrons may result in a drastic change of the ionization cross section. The cross section changes nearly three orders of magnitude from the threshold (where it is $\approx 10^{-19}$) to the maximum value (where it is $\approx 8 \cdot 10^{-17}$), while in the same domain the electron energy changes for a factor 4 only. Thus, in the presence of a wave, the small energy change of bombarding electrons may result in an increased ionization. This may be formally introduced in calculations by expanding the cross section

$$\sigma_1(x) \approx \sigma_0(x) + (d\sigma_0/dx)x_1 + (d^2\sigma_0/2dx^2)x_1^2 + ..., \tag{30}$$

where $\sigma_0(x)$ is the cross section at the energy of the bombarding electrons. Here we use the fact that the energy ϵ of the bombarding electrons changes only due to electric field effects. An example of that kind has been experimentally verified and analytically explained for an ion sound wave.[33] In the presence of a wave, the cross section changes with the change of the electron energy. The change is most dramatic in a very narrow region of energies, slightly above the threshold. This is presented in Fig. 3 (line 2).

There exist various effects in space plasmas capable of producing fluxes of energetic particles that may further be a source of instability of small accidental perturbations through the explained effect of increased ionization. An example from the magnetosphere[35] shows that electron jets may be generated along magnetic field lines due to the reconnection process in the areas of weak magnetic field. Thus, if we imagine a magnetic arcade or some other magnetic configuration with the magnetic mirror geometry, i.e., with a typical bottle neck configuration at both ends, and assume the input of plasma particles in the area of the weak field, the ions and electrons will flow along the magnetic lines towards the ends in such a way that a potential difference, which accelerates the electrons, is created such that $e\phi_0 = \epsilon_{pi}(1-1/\gamma)\tau/(1+\tau)$. Here, γ is the mirror ratio, $\tau = T_e/T_i$, and $\epsilon_{pi} = m_i v_0^2$ is the ion parallel kinetic energy.

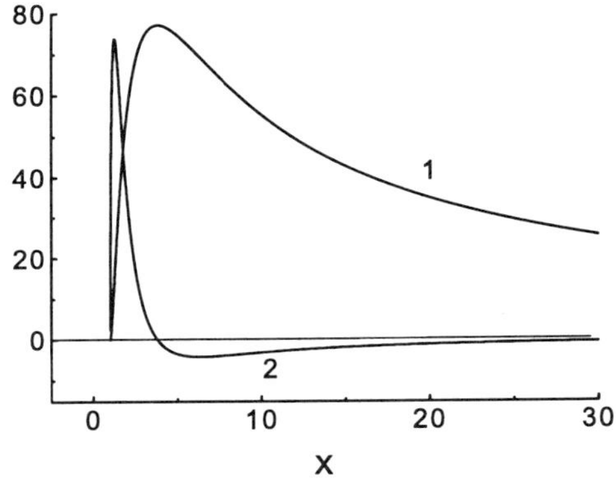

Fig. 3. Line 1 - the ionization cross section of a hydrogen atom vs. the electron energy $x = \epsilon/\epsilon_0$. Line 2 - the derivative of the ionization cross section.

3.1. *Source and sink terms*

In the presence of neutrals (which may imply that inelastic collisions take place in the plasma), the continuity equation becomes

$$\frac{\partial n_\alpha}{\partial t} + \nabla_\perp \cdot (n_\alpha \vec{v}_{\perp\alpha}) + \nabla_z \cdot (n_\alpha \vec{v}_{z\alpha}) = S. \tag{31}$$

Here $\alpha = i, e, n$, the magnetic field is in z-direction, the symbol $\perp$ refers to the perpendicular direction with respect to the magnetic field, and $S = S_C - S_L$ includes source (creation) and sink (loss) terms. As for the source terms, in the plasma comprising ionized and neutral atoms and electrons, the plasma species are predominantly created by the photo-ionization $\sim n_n$, and ionization by electron impact. The latter may include, first, the ionization by the energetic electrons from the tail in the Maxwellian distribution given by $\alpha_{imp} n_e n_n$, where α_{imp} is the ionization rate, and, second, the ionization caused by a flux f of energetic electrons created by some external source described earlier and given by $Q_0 = \sigma_0 f n_n$. Here, σ_0 is the ionization cross section for the given electron energy (which must be above the ionization threshold).

In the same time the plasma particle losses depend on the specific system. It may be due to the radiative and/or three body recombination, or due to transport across the magnetic field lines, and is given by $S_L = a_{rr} n_e n_i + a_{3b} n_i n_e^2 + a_t n$. To quantitatively describe the source and sink terms, as an illustration we take parameters typical for the lower solar atmosphere at several altitudes.[36]

At $h = 1000$ km the corresponding plasma parameters are $n_{i0} \equiv n_0 = 7 \cdot 10^{10}$ cm^{-3}, $n_{n0} = 1.3 \cdot 10^{13}$ cm^{-3}, $T_e = 6000$ K (at $h = 1000$ km). At $h = 500$ km we have $n_0 = 2.6 \cdot 10^{11}$ cm^{-3}, $n_{n0} = 2.3 \cdot 10^{15}$ cm^{-3}, $T_e = 4000$ K, and at $h = 100$ km, we have $n_0 = 10^{13}$ cm^{-3}, $n_{n0} = 6.9 \cdot 10^{16}$ cm^{-3}, $T_e = 5450$ K. Using these numbers we calculate the source/sink terms presented in Table 1, all in the units cm^{-3} s^{-1}. At $h = 1000$ km the impact ionization term appears negligible as well as the three-body recombination term. At the altitude of 500 km the three body recombination becomes important. Finally, at $h = 100$ km the three-body recombination becomes the dominant loss mechanism. Hence, at this altitude all ions in a unit volume are about 26 times recombined (due to the three-body recombination) in one second, while at the altitude $h = 1000$ km about 20 percent of the ions/atoms take part in that process. This must have an impact on the plasma waves because ions and neutrals change their identity so frequently.

Table 1. Numerical values of source/sink terms at several altitudes in solar plasma.

	$\alpha_{imp} n_e n_n$	$a_{rr} n_e n_i$	$a_{3b} n_i n_e^2$
$h = 1000$ km	$2.2 \cdot 10^4$	$2.9 \cdot 10^9 1$	$5.8 \cdot 10^7$
$h = 500$ km	23	$5 \cdot 10^{10}$	$1.8 \cdot 10^{10}$
$h = 100$ km	10^9	$6 \cdot 10^{13}$	$2.6 \cdot 10^{14}$

It can be easily seen that α_{imp} is the only coefficient growing with the temperature (altitude) so that in some regions it becomes the main source term which balances the losses.[37] Here, we have used the following expressions for the rate coefficients:[38]

$$\alpha_{rr}[cm^3/s] = 5.2 \cdot 10^{-14} z \left(\frac{\epsilon_i}{T_e}\right)^{1/2} \left[0.43 + \frac{1}{2}\log\left(\frac{\epsilon_i}{T_e}\right) + 0.469\left(\frac{\epsilon_i}{T_e}\right)^{-1/3}\right],$$

$$\alpha_{3b}[cm^6/s] = 8.75 \cdot 10^{-27} T_e^{-4.5},$$

$$\alpha_{imp}[cm^3/s] = 10^{-5} \exp\left(-\frac{\epsilon_i}{T_e}\right) \frac{(T_e/\epsilon_i)^{1/2}}{\epsilon_i^{3/2}(6 + T_e/\epsilon_i)},$$

where ϵ_i is the ionization energy, z is the ion charge, and the electron temperature is given in eV. The ionization by the external electron flux includes the parameter f which is rather arbitrary. In order to have it of the same order as the radiative recombination term at $h = 1000$ km it turns out that the flux f must be $\sim 10^{17} - 10^{19}$ m^{-2} s, which is still a very small fraction of the local flux due to thermal motions, which can be estimated as $n_0 v_{Te}$, where v_{Te} is the electron thermal velocity; i.e., we have $f/(n_0 v_{Te}) \sim 5 \cdot 10^{-5}$.

Hence, the source-sink term in the equilibrium is

$$S_0 \equiv 0 = \alpha_{ph} n_{n0} + \alpha_{imp} n_{n0} n_0 + Q_0 - a_{rr} n_0^2 - a_{3b} n_0^3 - a_t n_0. \qquad (32)$$

In the perturbed state, after linearization, for a relatively heavy neutral background (or relatively long wavelengths) and a quasi-neutral plasma, and using (32) it becomes

$$S_1 = Q_0 \frac{\sigma'}{\sigma_0} \frac{e\phi_1}{\kappa T_e} + \left(\alpha_{imp} n_{n0} n_0 - 2a_{rr} n_0^2 - 3a_{3b} n_0^3 - a_t n_0\right) \frac{n_1}{n_0}. \qquad (33)$$

Here, $\sigma' = d\sigma/d(e\phi/\kappa T_e)$. Typically, the ionization by ion impact is much smaller compared to the ionization by electron impact. In addition, for ions moving in the background of their own neutral parent atoms the charge exchange may become important. The cross section for the resonant charge exchange (ion and atom of the same kind, the process of the type $A^+ + A \to A + A^+$), in which the ion and the atom exchange their identities, for hydrogen has a maximum at low temperatures, like the usual elastic scattering cross section. However, as a rule, it is larger than the elastic scattering cross section.[39] For example, the charge exchange cross section for hydrogen at a particle energy of 1 eV is $\sim 7 \cdot 10^{-19}$ m^2 (compare this to the usual elastic collision cross section which is roughly speaking $\sim 10^{-20}$ m^2), and it decreases rather slowly so that at 10^4 eV it is $\sim 10^{-19}$ m^2. Consequently it may be taken as constant for the case of the present wave analysis. The curves for the elastic scattering and the charge exchange cross sections, in terms of the particle energy, remain nearly parallel to each other and the total collision cross section may be taken as a multiple (for helium and argon[39] by a factor ~ 3) of the elastic one. The non-resonant charge exchange (between ions and atoms/molecules of different kinds, the process of the type $A^+ + B \to A + B^+$) has a qualitatively different behavior, with a maximum for a certain particle energy, but it is one order of magnitude smaller than the former one.

3.2. *Application to drift waves*

3.2.1. *Boltzmannian electrons.*

Assuming the Boltzmann distribution for electrons, which is satisfied for $\omega/k_z \ll v_{Te}$ the ion continuity can be written[9] as

$$\frac{\partial\Phi}{\partial t} + \frac{c_s^2}{\Omega_i}\vec{e}_z \times \nabla_\perp\Phi\cdot\nabla_\perp \log n_0 - \left(1+\frac{\nu_i}{\Omega_i}\right)\frac{c_s^2+v_{Ti}^2}{\Omega_i}\nabla_\perp^2\Phi = \frac{S_1}{n_0}. \tag{34}$$

Here, $\Phi \equiv e\phi_1/(\kappa T_e)$, and S_1 becomes

$$S_1 = \left(\alpha_{imp}n_{n0}n_0 + Q_0\frac{\sigma'}{\sigma_0} - 2a_{rr}n_0^2 - 3a_{3b}n_0^3 - a_t n_0\right)\frac{e\phi_1}{\kappa T_e}. \tag{35}$$

In the absence of inelastic collisions and for perturbations $\sim \exp(-i\omega t + ik_x x + ik_y y)$, Eq. (34) yields a drift mode damped due to ion-neutral collisions. Inelastic collision terms can make it unstable. This is seen from the dispersion equation which becomes

$$\omega = -\frac{c_s^2 k_y}{\Omega_i}\frac{n_0'}{n_0}\frac{1}{1+k^2\rho_s^2} + \frac{i}{1+k^2\rho_s^2}\left(a_1 - a_2 - \nu_i k^2\rho_s^2\right). \tag{36}$$

Here,

$$a_1 = \frac{1}{n_0}\left(\alpha_{imp}n_{n0}n_0 + Q_0\frac{\sigma'}{\sigma}\right), \quad a_2 = \frac{1}{n_0}(2a_{rr}n_0^2 + 3a_{3b}n_0^3 + a_t n_0), \tag{37}$$

and $v_s^2 = c_s^2 + v_{Ti}^2$, $\rho_s = v_s/\Omega_i$, $k = (k_x^2 + k_y^2)^{1/2}$. The instability condition is obvious, i.e.,

$$\alpha_{imp}n_{n0}n_0 + Q_0\frac{\sigma'}{\sigma_0} > 2a_{rr}n_0^2 + 3a_{3b}n_0^3 + a_t n_0 + \nu_i n_0 k^2\rho_s^2. \tag{38}$$

In application to a laboratory plasma[33] the only important source/sink terms may become $a_t n_0$ and $Q_0\sigma'/\sigma_0$. The instability condition in that case simplifies to

$$f_0\left(\frac{\sigma'}{\sigma_0} - 1\right) > \frac{\nu_i}{\sigma_0}\frac{n_0}{n_{i0}}k^2\rho_s^2. \tag{39}$$

In the solar atmosphere and in the planetary atmospheres the contribution of various terms in the source/sink part in (38) is height dependent, as discussed in the preceding text. On the other hand, the flux f_0 appears as rather arbitrary; in the laboratory it is relatively easily controlled, while in space, as one possibility discussed before, it depends on plasma parameters where the reconnection takes place.[35] The term σ'/σ_0 appears due to the increased ionization caused by the presence of perturbations,[33] i.e., due to the change of the energy of the bombarding electrons. Hence, any value in the range 1 to 15 (or 20) is reasonable. For the electrons in the experiment[33] with the energies close to the ionization threshold and for the wave amplitude $e\phi_1/\kappa T_e \approx 0.1$, the corresponding value used for σ'/σ_0 was 11.

3.2.2. *Non-Boltzmannian electrons.*

The Boltzmann distribution for electrons, which is used in the derivation of Eq. (36) is valid as long as $\omega, \nu_e \ll v_{Te}k_z$. Otherwise, the parallel electron momentum equation should be used yielding

$$v_{ez1} = ik_z D_z \left(\Phi - \frac{n_{e1}}{n_0}\right). \tag{40}$$

The applicability window discussed earlier, in the application to solar and other plasmas for which $T_e \sim T_i$, in that cases becomes larger.

Using the electron continuity (31) and the electron perpendicular equation without the polarization drift, one obtains

$$\frac{n_{e1}}{n_0} = \Phi\left(\frac{\omega_{*e}\omega + c_1c_2}{\omega^2 + c_1^2} - i\frac{c_1\omega_* - c_2\omega}{\omega^2 + c_1^2}\right). \tag{41}$$

Here,

$$c_1 = k_z^2 D_z - \frac{b_2}{n_0}, \quad c_2 = k_z^2 D_z + \frac{b_1}{n_0}, \quad b_1 = Q_0\frac{\sigma'}{\sigma_0},$$

$$b_2 = \alpha_{imp}n_{n0}n_0 - 2a_{rr}n_0^2 - 3a_{3b}n_0^2 - a_t n_0, \quad D_z = v_{Te}^2/\nu_e.$$

The electron collision frequency ν_e includes the electron-ion and electron-neutral elastic collision frequencies, and the inelastic collisions effects. The latter include the same terms as the ion momentum equation, except for the charge exchange term, because ions and electrons are created/lost in the same acts of collisions.

From the ion continuity equation we find

$$\frac{n_{i1}}{n_0} = \frac{\omega_* - \omega k^2\rho_s^2 + i\left(\frac{b_1}{n_0} - \nu_i\rho_s^2k^2\right)}{\omega(1+\rho_i^2k^2) + i\left(\nu_i\rho_i^2k^2 - \frac{b_2}{n_0}\right)}. \tag{42}$$

Eqs. (41), (42) yield the dispersion equation for the collisional drift wave. In order to see more clearly the effects of the collisions, Eq. (41) may be simplified by using the limit $\omega/(k_z^2 D_z) \ll 1$ which is appropriate for the present case. In the case of negligible ion collisions and without inelastic collisions these two equations yield the standard result for the real and imaginary parts of the wave frequency:[26]

$$\omega_r \simeq \frac{\omega_*}{1+\rho_s^2k^2(1+\tau)}, \qquad \omega_i = \omega_*^2\rho_s^2k^2/(k_z^2D_z).$$

The mode is growing due to the common action of the electron collisions and finite ion mass.

The complete dispersion equation is

$$\omega = \frac{\omega_* - \dfrac{\delta_2}{k_z^2D_z}(\omega_* - \omega - \delta_4)}{1+\Gamma} + \frac{i}{1+\Gamma}\left[\frac{(\omega_* - \omega - \delta_4)\omega(1+\rho_i^2k^2)}{k_z^2D_z} + \delta_1 - \delta_2(1+\delta_3)\right]. \tag{43}$$

Here,

$$\delta_1 = \frac{b_1}{n_0} - \nu_i\rho_s^2k^2, \quad \delta_2 = -\frac{b_2}{n_0} + \nu_i\rho_s^2k^2, \quad \delta_3 = \frac{b_1+b_2}{n_0k_z^2D_z},$$

$$\delta_4 = \frac{b_2\omega_* + b_1\omega}{n_0k_z^2D_z}, \quad \Gamma = R^2k^2 + (1+\rho_i^2k^2)\delta_3, \quad R^2 = \rho_i^2 + \rho_s^2.$$

Keeping the leading order terms in the imaginary part we obtain

$$\omega_i \simeq \left(\frac{\omega*^2}{k_z^2D_z} - \nu_i\right)R^2k^2 + \frac{b_1+b_2}{n_0}. \tag{44}$$

For a negligible electron collision term the instability condition becomes identical to Eq. (38). The mode is additionally destabilized by the inelastic collisions and may become unstable even in the case of negligible ion inertia.

4. Alfvén waves in weakly ionized plasma

We focus now on the physics involved in the propagation of the Alfvén wave in a weakly ionized plasma like the solar photosphere. It will be shown that, if we assume the existence of the necessary electromagnetic perturbations in such a weakly ionized medium, the energy flux of the waves is in fact much lower compared to what is usually expected from estimates based on ideal MHD. This is due to the fact that the photospheric gas dynamics is heavily influenced by collisions. More precisely, in the presence of some accidental electromagnetic perturbations, which in the first step involve plasma species (electrons and ions) only, the neutral atoms respond to these electromagnetic perturbations due to the strong friction. This, and the fact that the ionization ratio is rather small (viz. of the order of 10^{-4}), results in very small amplitudes of the perturbed velocity of the *total* plasma-gas fluid.

Table 2. Collision frequencies (in Hz) and magnetization ratio of electrons and protons in the photosphere for two altitudes h (in km) and for the magnetic field $B_0 = 10^{-2}$ T.

h	ν_{in}	ν_{ii}	ν_{en}	ν_{ei}	Ω_i/ν_{it}	Ω_e/ν_{et}
0	$1.6 \cdot 10^9$	$5 \cdot 10^7$	$1.3 \cdot 10^{10}$	$1.5 \cdot 10^9$	$6 \cdot 10^{-4}$	$1.1 \cdot 10^{-1}$
250	$2.6 \cdot 10^8$	$3.8 \cdot 10^6$	$2.2 \cdot 10^9$	$1.2 \cdot 10^8$	$3.6 \cdot 10^{-3}$	$7.3 \cdot 10^{-1}$

Using the data for a quiet Sun model,[36] in Table 2 we summarize the values for the electron and proton elastic scattering collision frequencies at two altitudes (viz. $h = 0$ km, and $h = 250$ km) in the solar photosphere.[40] Here, we have taken $B_0 = 10^{-2}$ T, the corresponding temperatures are respectively $T = 6420$ K and $T = 4780$ K, the electron number densities are $n_0 = 6.4 \cdot 10^{19}$ m^{-3}, $n_0 = 2.7 \cdot 10^{18}$ m^{-3}, the atomic hydrogen number densities are $n_{n0} = 1.17 \cdot 10^{23}$ m^{-3}, $n_{n0} = 2.3 \cdot 10^{22}$ m^{-3} and inelastic collisions are ignored. We assume the proton and electron number densities equal. It is seen that both protons and electrons are un-magnetized. Note that in Table 2 the collision frequencies between the plasma species and neutrals are dominant for both electrons and ions, compared to the frequencies for Coulomb collisions between charged particles.

It is believed[41] that, due to the low temperature, the ions in the lower photosphere are in fact mainly metal ions. Sen and White[41] have assumed that the mean mass of these metal ions is 35 a.u. In that case, due to the rather different masses of (metal) ions and neutral (hydrogen) atoms, in calculating the collision frequency it is appropriate to use a more accurate formula $\nu_{mn} = n_{n0}\sigma_{mn}[m_n/(m_m + m_n)][8\kappa T_m/(\pi\mu)]^{1/2}$, where the index m denotes the metal ion, n denotes the neutrals (hydrogen), and $\mu = m_m m_n/(m_m + m_n)$ is the reduced mass. The calculations may be inaccurate because the collision cross section σ_{mn} is not known. As a guess, we take it as the value for protons multiplied by m_m/m_p. Taking the layer $h = 250$ km, we find $\nu_{mm} = 6.4 \cdot 10^5$ Hz, $\nu_{mn} = 4 \cdot 10^8$ Hz, and $\Omega_m = 2.7 \cdot 10^4$ Hz. Comparing to protons from Table 2, the metal ions appear to be even less magnetized, i.e., $\Omega_m/\nu_m = 6.6 \cdot 10^{-5}$, where $\nu_m = \nu_{mm} + \nu_{mn}$. At $h = 0$ km we have $\nu_{mm} = 1.2 \cdot 10^7$ Hz, $\nu_{mn} = 2 \cdot 10^9$ Hz, and $\Omega_m/\nu_m \simeq 1.3 \cdot 10^{-5}$. The mentioned uncertainty in determining σ_{mn} will clearly not substantially change the fact that the ions are un-magnetized.

The motion of an un-magnetized charged particle is depicted in Fig. 4. Arrows denote the tangential direction at the moment of collision when the particle switches to another gyro-orbit with a possibly different velocity (indicated by different gyro-radius). A collision occurs after a very tiny

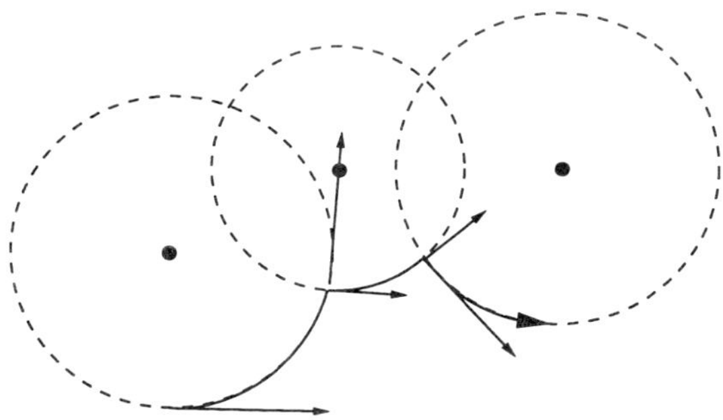

Fig. 4. Schematic presentation of the motion of a charged particle in non-magnetized plasma.

fraction (largely exaggerated here) of the gyro orbit has been traveled. According to numbers from the Table 2, the particle trajectory along a gyro-orbit around one specific magnetic line is only about $1/10^3$ part of the full circle. Hence, in the given case the path of the particle between two collisions is nearly a straight line, like in the case when the magnetic field is absent. In fact, the particle never makes a full rotation. The motion is similar in the fully ionized plasmas, however, there it is related to ion-ion collisions (i.e., viscosity, not to friction).

In the case of the shear Alfvén wave with $\vec{B}_0 = B_0\vec{e}_z$, both ion and electron fluids oscillate in the direction of the perturbed magnetic field vector $\vec{B}_1 = B_1\vec{e}_y$. This is due to the $\vec{E}_1\times\vec{B}_0$ drift, which does not separate neither charges nor masses, and the direction of the electric field is determined by the Faraday law. The wave is in fact sustained by the additional polarization drift $\vec{v}_{pj} = (m_j/q_jB_0^2)\partial\vec{E}_1/\partial t$ and the consequent Lorentz force $j_x\vec{e}_x \times \vec{B}_0$ which is again in the y-direction and has a proper phase shift. Note that the polarization drift appears as a higher order term due to $|\partial/\partial t| \ll \Omega_i$. It introduces the ion inertia effects and if it is neglected, then the Alfvén wave vanishes. The $\vec{E}\times\vec{B}$ term essentially describes the magnetic field frozen-in property of the plasma. The mode is fully described by the wave equation

$$\nabla\times\nabla\times\vec{E}_1 = \frac{\omega^2}{c^2}\vec{E}_1 + \frac{i\omega}{\varepsilon_0 c^2}\vec{j}_1, \tag{45}$$

the modified momentum equations for ions and electrons (1), (2) (due to the friction with neutrals), and the momentum equation for neutrals

$$m_n n_n\left[\frac{\partial\vec{v}_n}{\partial t} + (\vec{v}_n\cdot\nabla)\vec{v}_n\right] = -m_n n_n \nu_{ni}(\vec{v}_n - \vec{v}_i) - m_n n_n \nu_{ne}(\vec{v}_n - \vec{v}_e). \tag{46}$$

In the absence of collisions, the response of a plasma to the magnetic and electric field perturbations is instantaneous, and a volume element of the plasma moves in the previously described manner. In such an ideal case, the energy flux of the Alfvén wave is given by

$$F_{id} = m_i n_0 v_i^2 c_a / 2. \tag{47}$$

Here, v_i is the leading order $\vec{E} \times \vec{B}$ perturbed ion velocity. Its amplitude is given by $v_i = E_1/B_0$. Using the Faraday law we have $E_1 = \omega B_1/k$, hence $v_i = c_a B_1/B_0$. For the estimate only, we assume small perturbations of the magnetic field, viz. around 1 percent (a comment on larger perturbations will be given later on). For the parameters at $h = 250$ km, this yields $c_a = B_0/(\mu_0 n_{i0} m_i) = 1.3 \cdot 10^5$ m/s. Consequently, the perturbed plasma (ion) velocity is $v_i = 10^{-2} c_a = 1.3 \cdot 10^3$ m/s. The wave energy flux in the ideal case, and for $m_i = m_p$, becomes $F_{id} = 5.3 \cdot 10^2$ J/(m^{2}s). Setting $m_i = 35 m_p$ yields $F_{id} \simeq 90$ J/(s m^2).

Table 3. Parameters of waves (wave-lengths λ in km and frequencies in Hz) propagating through the chromosphere for two different altitudes h (in km).

$h = 1065$			
λ	ω	kc_a	ω_i/ω_r
0.1	$311 - 1222i$	44855	3.9
1	$327 - 45i$	4485	0.14
10	$33.1 - 0.45i$	448.5	0.014
100	$3.3 - 0.0045i$	44.85	0.0014
500	$0.66 - 0.0002i$	8.97	0.0003
$h = 1990$			
λ	ω	kc_a	ω_i/ω_r
0.1	$69666 - 732i$	69829	0.01
1	$6891 - 722i$	6983	0.1
10	$371 - 94.5i$	698.3	0.25
100	$36.4 - 0.9i$	69.8	0.025
500	$7.3 - 0.04i$	14	0.005

According to the presently widely accepted physical description of Alfvén waves in partially ionized plasmas, the effects of neutrals are the following. For a relatively small amount of neutrals (or for high frequency short wavelengths) the damping of the mode is proportional to the collision frequency ν_{in}, more collisions increases the friction.[21] In a very weakly ionized plasma the collisions are numerous and the whole fluid moves together. In this case, the stronger collisions the better locking of the gas-plasma fluid, and the damping of the wave (which is now proportional to

$1/\nu_{in}$) vanishes. The Alfvén velocity in such a mixture includes the total fluid density $m_i n_i + m_n n_n$. The dispersion equation of the Alfvén wave in this domain is

$$\frac{\omega}{k} = c_A \left(1 - i\frac{m_n n_n}{m_n n_n + m_i n_i}\frac{\omega}{\nu_{ni}}\right)^{1/2}, \quad c_A = \frac{B_0}{[\mu_0(m_i n_i + m_n n_n)]^{1/2}}.$$

In the same time the perturbed velocity of the gas-plasma mixture may be drastically reduced in a weakly ionized plasma (like the photosphere), and, consequently, the wave energy flux becomes very small. Compared to the ideal case discussed above, in the present case we have the flux[42] (for $m_i = m_p$) given by

$$F = \frac{1}{2}(m_i n_i + m_n n_n) c_A v_c^2 = F_{id}\left(\frac{m_i n_i}{m_i n_i + m_n n_n}\right)^{3/2}. \tag{48}$$

For the given parameters in the photosphere this gives

$$F \simeq 10^{-6} \cdot F_{id} = 5.3 \cdot 10^{-4}\ J/(m^2 s). \tag{49}$$

It is seen that the actual flux is always small for any realistic amplitude of perturbations. For example, even taking exceptionally strong magnetic field perturbations, e.g. $B_1 = B_0$, yields $F \simeq 5$ J/(m^{2}s). Consequently, regardless of the physical mechanism for eventual excitation of the Alfvén waves in the photosphere, the expected amplitude of the perturbed velocity is of the order of 0.1 m/s, and the energy flux of the waves is about one million time smaller than the one obtained from the ideal models.

As an example, assuming the wave propagating towards the chromosphere, the dispersion equation is solved for several wavelengths λ, with all collision frequencies included, at the altitude $h = 1065$ km where[36] $T = 6040$ K, $n_{n0} = 1.71 \cdot 10^{19}$ /m^3, $n_0 = 9.35 \cdot 10^{16}$ /m^3, and at the altitude $h = 1990$ km where $T = 7160$ K, $n_{n0} = 10^{17}$ /m^3, $n_0 = 3.9 \cdot 10^{16}$ /m^3. The results are given in Table 3. It is seen that shorter wavelengths are more damped at lower altitudes. In the same time, longer wavelengths (i.e., those that are presumably better transmitted by the photosphere) are in fact more damped at higher altitudes. This mode behavior is in agreement with the model of Kulsrud and Pierce (1969). However, this trend certainly can not continue because neutrals vanish at still higher altitudes.

5. Conclusions

As seen from Figs. 1, 2, the frequency of the Alfvén modes depends on the density gradient, and on the coupling with the corresponding drift mode which is driven by the density gradient. The drift mode is practically always unstable and its effects are hardly avoidable in any realistic case. Therefore, the change in the frequency of the Alfvén modes should be taken into account in the analysis of observed modes in the solar corona. In fact, this introduces a certain freedom in the fitting of observations into the theoretical modeling. Another consequence of the analysis is that the Alfvén modes are generally damped, and this remains true even if they are coupled to an unstable mode like the drift mode studied here, and even in an environment like the solar corona which is generally assumed as collision-less. Clearly, the coupling of the unstable drift mode to the damped Alfvén mode has some potential for the coronal heating problem. As a matter of fact, the unstable drift modes will always be present in the solar corona and, in turn, excite the Alfvén modes, which are damped and thus contribute to the heating and the acceleration of the coronal plasma. Of course, this is only a qualitative picture and a more detailed modeling and study is required to get a more quantitative idea of the efficiency of this mechanism.

In the case of partially ionized plasmas a physical background is presented for the excitation of electrostatic drift modes. The presence of neutrals introduces a plethora of atomic processes that have been briefly described. It results in the source/sink terms in the mass conservation equations and the appropriate terms in the first momentum equations. The newly created ions and electrons are balanced by some loss mechanisms, so that the plasma is not time evolving in the equilibrium. The source/sink terms presented in the study include several effects, making the model applicable to various situations in space and laboratory plasmas. The model includes also a flux of ionizing electrons which is known to be an instability source for sound type waves in laboratory plasmas[33] and solar plasmas.[37] It has a destabilizing role in a narrow (just a few eV) energy domain between the ionization threshold and the maximum of the cross section curve, cf. Fig. 3. However, the change of the cross section in that region is drastic and its effect on the growth of the sound mode has been experimentally verified.[33] Therefore, it suffices to keep the first term in the expansion (30). For other values of the electron energy the effect may become stabilizing. An interesting case is the electron energy that corresponds to the maximum in Fig. 3. In that case the (nonlinear) second derivative term in (30) becomes dominant and could be used in an eventual numerical study to demonstrate the stabilization of a mode in the presence of the flux of ionizing electrons. These results should be useful in the investigation of plasma waves in any partially ionized plasma. The numbers used for solar plasma show a surprisingly high level of inelastic collision effects, which consequently should not be neglected in the investigation of the majority of plasma modes.

Regarding the properties of Alfvén-type waves in weakly ionized plasmas, we have concluded that their amplitudes are such that the wave energy flux is very small. The main reason for this are ion collisions, which are so frequent that ions almost do not feel the effects of the magnetic field. As seen from Fig. 4, in such an environment the ion motion is very similar to the Brownian motion of atoms and molecules in a gas. The physics presented here should be taken into account in the estimates of the role of the Alfvén waves generated in the solar photosphere in coronal heating scenarios. However, the solar photosphere is only a thin plasma layer and the parameters in the solar atmosphere change with the altitude and so does the physics of the Alfvén waves. Our analysis suggests that if these waves are generated below the chromosphere, they can probably not be generated around the temperature minimum, but perhaps would have to come from lower down, i.e., below the surface where the plasma is again much more ionized and the ion-neutral collisions are not significant.

Acknowledgements: The results presented here are obtained in the framework of the projects G.0304.07 (FWO-Vlaanderen), C 90203 (Prodex), GOA/2004/01 (K.U.Leuven), and the Interuniversity Attraction Poles Programme - Belgian State - Belgian Science Policy.

References

1. F. F. Chen, *Phys. Fluids* **8**, 1323 (1963).
2. S. S. Moiseev and R. Z. Sagdeev, *J. Exp. Theor. Phys. (U.S.S.R.)* **44**, 763 (1963).
3. L. I. Rudakov and R. Z. Sagdeev, *Sov. Phys. Doklady* **6**, 415 (1961).
4. B. B. Kadomtsev and A. V. Timofeev, *Sov. Phys. Doklady* **7**, 826 (1963).
5. N. D'Angelo and R. V. Motley, *Phys. Fluids* **6**, 422 (1963).
6. J. Vranjes and S. Poedts, *Phys. Plasmas* **11**, 891 (2004).
7. J. Vranjes and S. Poedts, *Phys. Plasmas* **11**, 2178 (2004).
8. J. Vranjes and S. Poedts, *Phys. Plasmas* **12**, 064501 (2005).
9. J. Vranjes and S. Poedts, *Phys. Lett. A* **348**, 346 (2006).
10. J. Vranjes and S. Poedts, *Phys. Plasmas* **13**, 032107 (2006).
11. J. Vranjes and S. Poedts, *Astron. Astrophys* **458**, 635 (2006).
12. J. Vranjes, A. Okamoto, S. Yoshimura, S. Poedts, M. Kono, and M. Y. Tanaka, *Phys. Rev. Lett.* **89**, 265002 (2002).
13. A. Okamoto, K. Hara, K. Nagaoka, S. Yoshimura, J. Vranjes, M. Kono, and M. Y. Tanaka, *Phys. Plasmas* **10**, 2211 (2003).
14. R. J. Goldston and P. H. Rutherford, *Introduction to Plasma Physics* (Institute of Physics, Bristol, 1995).

15. N. D'Angelo, *Phys. Fluids* **8**, 1748 (1965).
16. H. Saleem, J. Vranjes and S. Poedts, *Astron. Astrophys* **471**, 289 (2007).
17. H. Saleem, J. Vranjes and S. Poedts, *Phys. Plasmas* **14**, 072104 (2007).
18. B. S. Tanenbaum and D. Mintzer, *Phys. Fluids*, **5**, 10 (1962).
19. L. C. Woods, *J. Fluid Mech.* **13**, 570 (1962).
20. D. F. Jephcott and J. Stocker, *J. Fluid Mech.* **13**, 587 (1962).
21. R. Kulsrud and W. P. Pierce, *Astrophys. J.* **156**, 445 (1969).
22. R. E. Pudritz, *Astrophys. J.* **350**, 195 (1990).
23. G. Haerendel, *Nature* **360**, 241 (1992).
24. C. Watts, and J. Hanna, *J. Phys. Plasmas* **11**, 1358 (2004).
25. J. V. Hollweg, *Sol. Physics* **70**, 25 (1981).
26. J. Weiland, *Collective Modes in Inhomogeneous Plasmas* (Institute of Physics Pub., Bristol, 2000).
27. N. A. Krall and A. W. Trivelpiece, *Principles of Plasma Physics* (McGraw-Hill Kogakusha, Tokyo, 1973).
28. J. Vranjes, D. Petrovic, M. Kono, S. Poedts, and V. Cadez, *Planet. Space Sci.* **54**, 641 (2006).
29. D. L. Giaretta, *Astron. Astrophys.* **75**, 237 (1979).
30. G. N. Watson, *A tretise on the Theory of Bessel Functions* (Cambridge at the University Press, Cambridge, 1962), pp. 482-485.
31. J. P. Goedbloed and S. Poedts, *Principles of Magnetohydrodynamics* (Cambridge Univ. Press, Cambridge, 2004).
32. H. Rishbeth and O. K. Garriott, *Introduction to Ionospheric Physics* (Acad. Press, New York and London, 1969) pp. 87-124.
33. J. C. Johnson, N. D'Angelo, and R. L. Merlino, *J. Phys. D: Appl. Phys.* **23**, 682 (1990).
34. M. Gryzinski, *Phys. Rev. A* **138**, 336 (1965).
35. Y. Serizawa and T. Sato, *Geophys. Res. Lett.* **11**, 595 (1984).
36. J. E. Vernazza, E. H. Avrett, and R. Loeser, *Astrophys. J. Suppl.* **45**, 635 (1981).
37. J. Vranjes, M. Y. Tanaka, M. Kono, and S. Poedts, *Phys. Plasmas* **11**, 4188 (2004).
38. J. D. Huba, *NRL Plasma Formulary* (Naval Research Laboratory, Washington, 2000) p. 54.
39. Y. P. Raizer, *Gas Discharge Physics* (Springer-Verlag, Berlin Heidelberg, 1991), p. 25.
40. J. Vranjes, S. Poedts, and B. P. Pandey, *Phys. Rev. Lett.* **98**, 049501 (2007).
41. H. K. Sen and M. L. White, *Sol. Phys.* **23**, 146 (1972).
42. J. Vranjes, S. Poedts, B. P. Pandey, and B. De Pontieu, Energy flux of Alfvén waves in weakly ionized plasma, in print Astron. Astrophys. (2007).

CURRENT DRIVEN ACOUSTIC PERTURBATIONS IN PARTIALLY IONIZED COLLISIONAL PLASMAS

J. VRANJES[1] and S. POEDTS

Center for Plasma Astrophysics,
Celestijnenlaan 200B, 3001 Leuven, Belgium, and
Leuven Mathematical Modeling and Computational Science Center (LMCC), and
[1]Faculté des Sciences Appliquées,
avenue F.D. Roosevelt 50, 1050 Bruxelles, Belgium

M. Y. TANAKA

Interdisciplinary Graduate School of Engineering Sciences,
Kyushu University Space Environment Research Center,
Kyushu University Kasuga-koen 6-1, Kasuga, Fukuoka 816-8580, Japan

B. P. PANDEY

Department of Physics, Macquarie University,
Sydney, NSW 2109, Australia

A fluid and kinetic analysis is presented of the ion sound mode in a weakly ionized collisional plasma in the regime when the ion collision frequency exceeds the ion gyro-frequency while the electrons remain magnetized. Under these conditions, an ion sound wave can propagate at arbitrary angles with respect to the direction of the magnetic field. In the presence of an electron flow along the magnetic lines the sound mode can grow. Due to the electron collisions the mode is unstable while ion collisions cause an angle dependent instability threshold which is such that the mode is most easily excited at very large angles. Hot ion effects in one part of the work are included by means of an effective viscosity which effectively describes the ion Landau damping effect. In the presence of an additional light ion specie, the mode frequency and increment in a certain parameter range are increased.

Several additional effects are discussed, including the electron-ion collisions, the perturbations of the neutral gas, and the electromagnetic perturbations. The electron-ion collisions are shown to modify the previously obtained angle dependent instability threshold for the driving electron flow. The inclusion of the neutral dynamics implies an additional neutral sound mode which couples to the current driven ion acoustic mode, and these two modes can interchange their identities in certain parameter regimes. The electromagnetic effects, which in the present model imply a bending of the magnetic field lines, result in a further destabilization of an already unstable ion acoustic wave.

In the case when the ion collision frequency is arbitrary the ion species is to be described by a collisional Boltzman kinetic equation. In the same time the electron collision frequency is high enough so that the fluid description is used for the electrons in the presence of an electron drift in the perpendicular direction. This results in the instability of accidentally excited ion sound oscillations, which turn out to be highly unstable for practically all physically acceptable values of the electron drift. In addition, the presence of a population of hotter electrons is shown to reduce the perpendicular electron drift and to increase the instability threshold.

1. Introduction

An ion sound wave in a collision-less hot ion plasma is Landau damped, and waves with wavelengths far exceeding the electron Debye radius may propagate if

$$f(\tau) \equiv (z_i\tau)^{3/2}\exp(-z_i\tau/2) \ll 1.$$

Here, $\tau = T_e/T_i$ and z_i is the ion charge number. For an electron-proton plasma $f(\tau)$ becomes completely negligible for $\tau \geq 15$. However, for higher ion charges, e.g. $z_i = 5$, $f(\tau)$ becomes negligible already at $\tau \geq 3$. Thus, highly charged and hot ions practically do not contribute to the Landau damping as long as $\tau \geq 3$, a condition that may be taken as always satisfied.

In the presence of an electron stream described by a velocity v_{e0}, the ion sound in an ordinary electron-proton plasma may become unstable, with the icrement/decrement given by[1]

$$\omega_{im} = \left(\frac{\pi}{8}\right)^{1/2} kc_s \left[\left(\frac{m_e}{m_i}\right)^{1/2}\left(\frac{v_{e0}}{c_s} - 1\right) - \tau^{3/2}\exp\left(-\frac{\tau}{2}\right)\right]. \tag{1}$$

The mode is unstable provided that

$$v_{e0} > c_s\left[1 + (m_i/m_e)^{1/2} f(\tau)\right]. \tag{2}$$

Here $c_s = (\kappa T_e/m_i)^{1/2}$ is the ion sound speed. For $f(1) = 0.6$ it is seen that the instability develops if $v_{e0} > 27c_s$, while for $f(10) = 0.2$ the threshold is still high, viz. $v_{e0} > 10c_s$.

The physical picture and the stability/instability conditions may become different in a weakly ionized plasma with dominant collisions with neutrals. The ion-neutral collision frequency may exceed the ion gyro-frequency so that the ions are unmagnetized, i.e. they behave as if the magnetic field was absent. On the other hand, the electron-neutral collisions may still occur less frequent than the electron gyro-frequency. In such a situation, acoustic-type oscillations of the ion gas may propagate at an angle with respect to the magnetic field, and the oscillations may have a sound-type character as long as the magnetized electrons are able to follow by moving only along the magnetic field lines.[2–4] Collisions determine the mode behavior in such a partially ionized plasma. The plasma-neutrals collisions, roughly speaking, dominate the collisions between charged particles[5] as long as the ionization ratio n_{i0}/n_{n0} remains below the value $3 \cdot 10^{10}/T_e^2$, where T_e is in units of K.

In the presence of an electron drift/flow, perpendicular[2,3] or parallel[4] to the magnetic lines, the ion sound modes becomes unstable even within the fluid theory. Without collisions and in the electron inertia-less limit, the instability is absent in the fluid model. On the other hand, such an electron-current-driven instability does exist within the *collision-less kinetic* description,[1] where it appears due to the positive slope in the electron distribution function in the domain around the wave phase speed. Consequently, for such an instability it is necessary that the electron macroscopic speed exceeds the sound speed c_s. However, the actual instability threshold is usually much higher, like in the cases $\tau = 10$ and $\tau = 1$, when it is $10\, c_s$ and $27\, c_s$, respectively. In the fluid case[2–4] of a collisional plasma, and for an obliquely propagating sound mode, the necessary instability condition appears higher and given by $c_s k/k_z$, where $k_z = k \cos\varphi$ and φ is the angle between the wave-vector and the direction of the magnetic field. However, the actual instability threshold is angle dependent and it is in fact lower[4] compared to the collision-less case.

In this work we give some details of the parallel and perpendicular electron flow driven instability of the ion acoustic perturbations propagating obliquely to the magnetic field lines in a weakly ionized plasma, with a particular emphasis on collisions, electron inertia and hot ion effects. The latter implies the ion Landau damping which is in the part of the present work described within the fluid theory by including an effective viscosity in the ion momentum equation. Real plasmas frequently involve several ion species, including positive and negative ions, or a specific case of pair ion plasma with equal mass and with charge numbers of the two ion species with an opposite polarity. The behavior of sound waves in such plasmas will be discussed here in view of the angle dependent mode behavior and the instability mentioned above. We investigate also the behavior of an ion acoustic (IA) mode driven in the presence of the ion-neutral, electron-neutral, and electron-ion collisions. Neutrals can also be perturbed due to various external reasons (a common situation in the lower solar atmosphere, in the terrestrial atmosphere, in the molecular clouds etc.), or due to interaction with the perturbed plasma species (the instabilities in the plasma component of a plasma-gas mixture are numerous and are generally more easily excited). This introduces an additional (neutral) gas acoustic (GA) mode. Moreover, these two modes may interact and such effects are studied in the present work in both the electrostatic (ES) and the electromagnetic (EM) regimes.

2. Basic equations and model

In the case when the dominant electron collisions are those with the neutrals, the electron momentum equation in the standard notation is

$$m_e n_e \left[\frac{\partial \vec{v}_e}{\partial t} + (\vec{v}_e \cdot \nabla)\vec{v}_e\right] = en_e \nabla\phi - en_e \vec{v}_e \times \vec{B} - \kappa T_e \nabla n_e - m_e n_e \nu_{en}(\vec{v}_e - \vec{v}_n). \tag{3}$$

For plasma perturbations propagating obliquely with respect to the magnetic field $\vec{B}_0 = B_0 \vec{e}_z$, and with the background of a heavy neutral gas, the perturbed parallel electron velocity is given by

$$v_{ez1} = -\frac{ek_z}{m_e \omega_e}\phi_1 + \frac{v_{Te}^2 k_z}{\omega_e}\frac{n_{e1}}{n_{e0}}. \tag{4}$$

Here, $\omega_e = \omega_0 + i\nu_e$, $\omega_0 = \omega - k_z v_0$, $v_{Te}^2 = \kappa T_e/m_e$, $\vec{v}_0 = v_0 \vec{e}_z$. The ω_0 term appears in ω_e from the left-hand side of the electron momentum equation as the finite electron mass effect. The perpendicular electron motion is described by the following recurrent formula

$$\begin{aligned} v_{e\perp} = {} & \frac{1}{B_z}\vec{e}_z \times \nabla_\perp \phi + \frac{\nu_e}{\Omega_e}\frac{\nabla_\perp \phi}{B_z} - \frac{v_{Te}^2 \nu_e}{\Omega_e^2}\frac{\nabla_\perp n_e}{n_e} - \frac{v_{Te}^2}{\Omega_e}\vec{e}_z \times \frac{\nabla_\perp n_e}{n_e} \\ & - \frac{1}{\Omega_e}\left(\frac{\partial}{\partial t} + \vec{v}_e \cdot \nabla\right)\vec{e}_z \times \vec{v}_{e\perp} - \frac{\nu_e}{\Omega_e^2}\left(\frac{\partial}{\partial t} + \vec{v}_e \cdot \nabla\right)\vec{v}_{e\perp}. \end{aligned} \tag{5}$$

The condition we use here are

$$\omega \ll \nu_e, \quad \nu_e \ll \Omega_e. \tag{6}$$

The first inequality is to be satisfied in order to use the fluid equations for the electrons, implying also negligible electron kinetic effects, i.e., the electron mean free path is much shorter than the parallel wave-length, and the second part implies magnetized electrons. Hence, the small parameters in calculating the higher order terms in (5) are $\epsilon_1 = \omega_0/\nu_e$ and $\epsilon_2 = \nu_e/\Omega_e$. Eqs. (4) and, (5) are used in the electron continuity equation

$$\frac{\partial n_e}{\partial t} + \nabla_\perp \cdot (n_e \vec{v}_{e\perp}) + \nabla_z \cdot (n_e \vec{v}_{ez}) = 0.$$

The leading order vector-product type terms from (5) vanish in $\nabla_\perp(n\vec{v}_\perp)$. The remaining perpendicular and parallel terms together with Eq. (4) yield

$$\frac{n_{e1}}{n_{e0}} = -\frac{ek_z^2}{m_e\omega_0\omega_e\alpha_e}\left(1 - \frac{\omega_e^2 k_\perp^2}{\Omega_e^2 k_z^2}\right)\phi_1. \tag{7}$$

Here,

$$\alpha_e = 1 - \frac{k_z^2 v_{Te}^2}{\omega_0\omega_e} + \rho_e^2 k_\perp^2\left(1 + i\frac{\nu_e}{\omega_0}\right), \quad \rho_e = v_{Te}/\Omega_e. \tag{8}$$

In the absence of collisions and for inertia-less electrons Eq. (7) yields the standard Boltzmann's distribution.

As for the collisional ions, in the proper fluid limit of negligible trapping effect (wave frequency lower than the ion collision frequency) and small Landau damping (the ion mean free path lower than the wave-length), and in the case when they are not magnetized (i.e. when the gyro frequency is much smaller than the ion-neutral collision frequency),

$$\frac{\Omega_i}{\nu_i} \ll 1, \tag{9}$$

the perturbations of the ions may be treated within the fluid theory, and the terms parallel and perpendicular to the magnetic field are irrelevant. Therefore, the ion perturbations are given simply by $\sim \exp(-i\omega t + i\vec{k}\cdot\vec{r})$. Here, $\vec{r}$ denotes an arbitrary direction with respect to $\vec{B}_0 = B_0\vec{e}_z$. The earlier introduced k_z is then $k\cos\varphi$, where φ denotes the angle between $\vec{B}_0$ and $\vec{k}$. The ion momentum equation and the continuity in this case yield

$$\frac{n_{i1}}{n_{i0}} = \frac{ek^2}{m_i}\frac{\phi_1}{\omega^2 + i\nu_i\omega - k^2 v_{Ti}^2}. \tag{10}$$

Using the condition of quasi-neutrality, Eqs. (7),(10) yield the dispersion equation which will be discussed below.

3. Dispersion equation

In the case of an ordinary ion sound mode, the electron inertia is neglected, resulting in a Boltzmann distribution for the electrons, which is always a good approximation because $(\omega/k)^2/v_{Te}^2 \simeq m_e/m_i \ll 1$. In the present case of an obliquely propagating sound wave, and for electrons participating in the perturbations by moving only along the magnetic field lines, the above small factor appears multiplied by $k^2/k_z^2 > 1$, and consequently the electron inertia (the left hand side of the electron momentum equation) may not always be negligible.

The electron mass correction in the perpendicular dynamics may be neglected. After examining the real and the imaginary terms with the mass corrections in Eq. (7) we find out that if

$$\xi \equiv \frac{k_\perp^2}{k_z^2}\frac{\nu_e^2}{\Omega_e^2} < 1, \tag{11}$$

the perpendicular part can be omitted, and only the parallel electron dynamics becomes modified due to the electron mass corrections through the term ω_0 comprised in ω_e. In view of (6) the condition (11) is usually satisfied, otherwise electrons would not be able to follow the nearly perpendicular ion perturbations by moving only in the parallel direction. The conditions (6), (9), and (11) will be checked for every set of parameters used later in the text.

With this simplification, from Eqs. (7),(10) we obtain

$$\omega^2 = k^2c_s^2\left(1+\frac{1}{\tau}\right) - \frac{m_e}{m_i}\frac{k^2}{k_z^2}(\omega - k_z v_0)^2 - i\left[\nu_i\omega + \frac{m_e}{m_i}\frac{k^2}{k_z^2}\nu_e(\omega - k_z v_0)\right]. \tag{12}$$

The term $(\omega - k_z v_0)^2$ is the effect of the left-hand side of the electron momentum equation. In the absence of collisions with neutrals and for perturbations along the magnetic lines, this term yields an electron two stream instability of the sound mode, which sets in provided that

$$v_0^2 > v_{Te}^2\left(1+\frac{1}{\tau}\right). \tag{13}$$

The threshold of this reactive instability is very high.

Another kind of instability, with a much lower threshold, is obtained in the presence of collisions. Setting $\omega = \omega_r + i\omega_i$ and $|\omega_i| \ll |\omega_r|$, from Eq. (12) we find the imaginary part of the frequency

$$\omega_i = -\frac{\nu_i}{2(1-\chi)}\left(1 - \frac{\nu_e}{\nu_i}\chi\right), \quad \chi = \frac{m_e}{m_i}\frac{k^2}{k_z^2}\left(\frac{k_z v_0}{\omega_r} - 1\right). \tag{14}$$

Here, the χ in denominator appears as the electron mass effect. The collision frequency of plasma species with neutrals is given by $\nu_{e,i} = \sigma_{(e,i)n} n_n v_{T(e,i)}$. Note that $\sigma_{in}(m_i, m_n, T_i)$ is generally larger than $\sigma_{en}(m_n, T_e)$, and both cross sections are strongly dependent on the temperature of the colliding plasma particle. Hence, we have

$$\frac{\nu_i}{\nu_e} = \left(\frac{m_e}{m_i\tau}\right)^{1/2}\frac{\sigma_{in}(m_i, m_n, T_i)}{\sigma_{en}(m_n, T_i)} = \beta\left(\frac{m_e}{m_i\tau}\right)^{1/2}, \quad \beta > 1.$$

For e–H collisions[6,7] the cross section at 1 eV is around $2.5 \cdot 10^{-19}$ m^2. For H$^+$–H momentum transfer[8] at 1 eV it is around 350 a.u. (1 a.u. =

$2.8 \cdot 10^{21}$ m^2), i.e., the cross section is 10^{-18} m^2 so that the ratio $\beta = \sigma_{in}/\sigma_{en}$ is around 4, and $\nu_i/\nu_e \approx 0.09/\tau^{1/2}$, or $\nu_e = 10.7\nu_i\tau^{1/2}$. This ratio does not change much, even for lower temperatures that are of particular interest for the partially ionized plasma studied here. Consequently, in most practical situations when $T_e \geq T_i$ or T_i is not very much larger than T_e we have $\nu_e > \nu_i$ and the sign of ω_i changes primarily due to the term in the numerator of Eq. (14). Note that here we are interested in values of τ close to unity, firstly in order to investigate the hot ion case, and secondly to have the conditions (6), (9), and (11) well satisfied.

Setting $\omega_r \approx kv_s$, where $v_s^2 = c_s^2 + v_{Ti}^2$, in Eq. (14) we obtain the instability condition

$$\frac{v_0}{v_s} > \frac{k}{k_z}\left(1 + \frac{m_i}{m_e}\frac{\nu_i}{\nu_e}\frac{k_z^2}{k^2}\right) = \frac{1}{\cos\varphi}\left(1 + \frac{m_i}{m_e}\frac{\nu_i}{\nu_e}\cos^2\varphi\right). \tag{15}$$

Here we have an angle dependent instability threshold for an unstable ion sound mode driven by the constant electron flow directed along the magnetic field lines. It is a purely fluid instability in the presence of electron collisions.

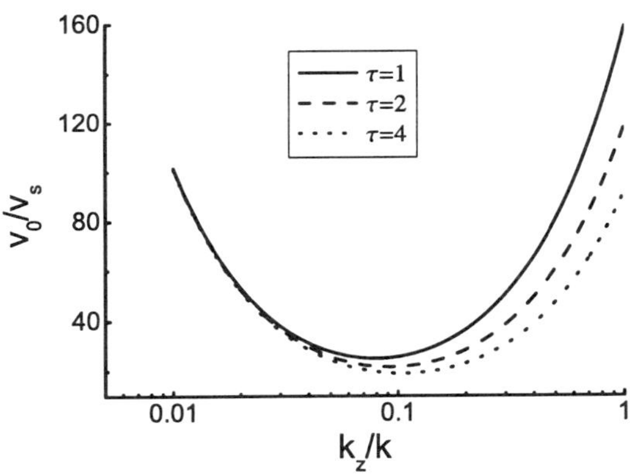

Fig. 1. The angle dependent instability threshold (15) for electron-H^+ plasma in H-gas.

The threshold (15) is presented in Fig. 1 in terms of k_z/k for hydrogen, $m_i/m_e = 1838$, and for the electron energy of 1 eV for which[6] $\sigma_{en} = 2.5 \cdot 10^{-19}$ m^2, and for three values of $\tau = 1$, 2, 4. The corresponding H^+–H collision cross sections for the momentum transfer (in 10^{-19} m^2) are[8] $\sigma_{in} =$

9.24, 9.8, and 10.64, respectively. The collision frequency ratio ν_i/ν_e is 0.086, 0.064, and 0.049, respectively. The mode is most easily excited at given large angles (around $k_z/k = 0.1$) with respect to the driving electron flow which is in the z-direction. At this angle, the minimum instability threshold v_0/v_s for the three lines is 25.1, 21.7, and 19, respectively. Compare this with the threshold for the mode propagating along the magnetic lines which for the given values of τ is 159, 118, and 92, respectively. Note that taking Ω_e/ν_e, in the interval $20 - 80$ we have the condition (9) reasonably satisfied as it takes values from the interval $0.12\tau^{1/2} - 0.48\tau^{1/2}$, while the left-hand side of the condition (11) takes values from the interval $0.25 - 0.015$, respectively. So the parameters used above satisfy the model, particularly for τ close to unity.

The same sort of behavior is obtained for helium ions in a helium gas. In this case at $\tau = 1$ we have[6,9] $\beta = 8.3$, $\nu_i/\nu_e = 0.097$ and the minimum threshold $v_0/v_s = 46$ is around $k_z/k = 0.05$. Here, taking Ω_e/ν_e of the order of 50 or 100 we have un-magnetized ions and condition (11) is satisfied.

In the case of singly ionized cesium ions in a cesium gas at and electron energy of 1 eV and for $\tau = 10$, we have[6] $\sigma_{in} \simeq 7.5 \cdot 10^{-17}$ m^2, $\sigma_{en} = 3 \cdot 10^{-18}$ m^2, hence $\beta = 25$ and $\nu_i/\nu_e = 0.016$. At $T_e = T_i = 1$ eV we have $\beta = 12$ and $\nu_i/\nu_e = 0.024$. Taking Ω_e/ν_e of the order of 50 or larger, it turns out that all conditions discussed above are well satisfied. A plot similar to that in Fig. 1 reveals that the instability threshold curve passes through a minimum at $k_z/k \approx 0.014$ with two high thresholds $v_0/v_s \approx 125$ and 150, respectively. However this is still much below the values for the mode along the electron flow $k_z/k = 1$, viz. 3876 and 5814, respectively.

Table 1. The collision cross sections and collision frequencies for electrons, and several ion species, in helium and argon gasses, in units of 10^{-20} m^2 and at electron and ion temperatures of 0.1 eV, i.e., $\tau = 1$. The values in brackets are for electrons at 1 eV, or $\tau = 10$.

helium gas					argon gas		
σ_{eHe}	σ_{H^+He}	σ_{He^+He}	σ_{Li^+He}	σ_{K^+He}	σ_{eAr}	σ_{Li^+Ar}	σ_{K^+Ar}
5.86 (6.85)	28	50	106	165	0.45 (1.05)	303	580
	$\nu_{H^+}/\nu_e = 0.11(0.03)$ $\nu_{He^+}/\nu_e = 0.097$ $\nu_{Li^+}/\nu_e = 0.16(0.04)$ $\nu_{K^+}/\nu_e = 0.1(0.028)$					$\nu_{Li^+}/\nu_e = 5.9(0.27)$ $\nu_{K^+}/\nu_e = 4.8(0.65)$	

Note however that for some other gasses and ions the sign in Eq. (14) can also change by the term in the denominator. This is seen in Table 1 for several types of ions in gases that are of particular interest for laboratory investigations[6,9,10] i.e., helium and argon. Here, $\nu_i/\nu_e > 1$ for lithium and potassium in the argon gas and for equal electron and ion temperatures. This is due to much larger cross sections for collisions between these large target atoms and large colliding ions, compared to the electron collisions. In this case from (14) the instability threshold is considerably higher and it reads

$$\frac{v_0}{v_s} > \frac{k}{k_z}\left(1 + \frac{m_i}{m_e}\frac{k_z^2}{k^2}\right). \tag{16}$$

We stress the essential difference between the electron flow driven instability (15) and the one obtained from the collision-less kinetic theory (1). In (1) the electron current which drives the instability is associated with the electron Landau-damping term $\omega_r[\pi m_e/(8m_i)]^{1/2}$, and it is a purely collision-less plasma instability. In a strongly collisional plasma it is not expected to play a significant role. On the other hand, in Eq. (15) the driving current term is associated with the electron collisions and by its nature it is a fluid effect.

4. Landau damping in fluid modeling

The ion Landau damping is not expected to play an important role[11] in a strongly collisional plasma as long as the ion mean free path is much shorter that the wave-length. This is verified experimentally[12] even for $\tau \approx 1$, with the strong-weak damping transition observed at $\omega \sim \nu_i$. The inclusion of the ion Landau damping effects implies a proper kinetic domain[11] in which the collisions are neither too strong (too short ion mean free path) nor too weak (implying that ion trapping effects must be included). In this case, Eq. (10) is replaced by its kinetic counterpart[3,4]

$$n_{i1}/n_{i0} = -e\phi_1[1 - J_i(\omega/(kv_{Ti}))]/(\kappa T_i),$$

yielding the dispersion equation. Here, $J_i(\psi)$ denotes the plasma dispersion function.

In 1979 a fluid model was introduced by D'Angelo *et al.*[13] in order to describe the Landau damping effects on the solar wind fast streams with a spatially varying ratio $\tau(= T_e/T_i)$ of the electron and ion temperatures. Within the distances 0.8 - 1 AU from the Sun, where the ratio τ is of the order of unity, the Landau damping on ions is significant and it counteracts the steepening of sound perturbations. Further away, for τ above 4 or 5 it becomes less important and the steepening takes place again. To describe this effect within a collision-less fluid theory, an effective 'viscous' term is introduced in the ion momentum equation. This term is of the form $\mu_L \nabla^2 \vec{v}_i$, where μ_L is chosen in such a way to mimic the known properties of the Landau effect. These include the fact that the ratio δ/λ, between the attenuation length δ and the wavelength, is (i) independent of the wavelength, (ii) independent of the plasma density n, and (iii) dependent in a prescribed way[14] on τ. According to D'Angelo *et al.*[13] these requirements are fulfilled by

$$\mu_L = \frac{m_i n_{i0} v_s \lambda}{2\pi^2 \delta/\lambda}. \tag{17}$$

Here, v_s is the ion sound speed which should include the ion temperature contribution, while δ/λ in terms of τ satisfies a curve which is such that the attenuation is strong at $\tau \approx 1$ and weak for higher values of τ. This fluid model can nicely describe such an essentially kinetic effect, yet it has remained practically unnoticed in the past.

How the model works can easily be demonstrated by applying it to a fluid description of the ion sound, and by then comparing the result with the kinetic theory. Take the ion momentum equation in the form

$$\frac{\partial v_{i1}}{\partial t} = -\frac{e}{m_i}\frac{\partial \phi_1}{\partial r} - \frac{\kappa T_i}{m_i n_0}\frac{\partial n_{i1}}{\partial r} + \frac{\mu_L}{m_i n_0}\frac{\partial^2 v_{i1}}{\partial r^2}.$$

This is combined with the ion continuity and the Boltzmann distribution for electrons yielding

$$\omega^2 + i\mu_0 \omega k^2 - k^2(c_s^2 + v_{Ti}^2) = 0, \quad \mu_0 = \mu_L/(m_i n_{i0}).$$

Setting here $\omega = \omega_r + i\omega_{if}$ we have

$$\omega_{if} = -\mu_0 k^2/2 = -v_s/(\lambda d), \quad \omega_r^2 = k^2 v_s^2 - \mu_0^2 k^4/4, \quad d = \delta/\lambda. \tag{18}$$

From the standard kinetic theory[1] in the limit $T_e \gg T_i$ the Landau damping of the ion acoustic wave for singly charged ions is given by the approximate formula

$$\omega_i = k v_s (\pi/8)^{1/2} \left[(m_e/m_i)^{1/2} + \tau(3+\tau)^{1/2} \exp[-(3+\tau)/2]\right]. \tag{19}$$

In comparison to the exact solution[15] the damping rate (19) in terms of τ has a somewhat different behavior, as at small values of τ it has a local maximum. The exact solution for the decrement is a monotonic, decreasing function of τ in the interval $\tau > 1$ (see Fig. 2). Using the graph of the exact solution for the Landau damping,[15] we find that the exact (normalized) Landau decrement ω_{ik}/ω_r can be well fitted by the following polynomial:

$$-\omega_{ik}/\omega_r = 0.682 - 0.369765\,\tau + 0.0934595\,\tau^2 - 0.012\,\tau^3$$
$$+0.00075245\,\tau^4 - 0.000018\,\tau^5. \tag{20}$$

Using data and a graph from D'Angelo *et al.*[13] we find that the 'fluid' attenuation length d introduced above can be expressed by the following approximate fitting formula to give the same decrement as the kinetic expression (20):

$$d \equiv \delta/\lambda \approx 0.2751 + 0.0421\,\tau + 0.089\,\tau^2 - 0.011785\,\tau^3 + 0.0012186\,\tau^4. \tag{21}$$

As a matter of fact, the graph $-\omega_{if}/(kv_s) = 1/(2\pi d)$ practically coincides with (20), as seen from Fig. 2. In fact, it much better describes the damping on ions than the approximate formula (19) which is obtained after the standard expansion of the plasma dispersion function $J_i(\psi)$. For example, for $\tau = 1$ we have the 'fluid' Landau damping $\omega_{if}/(kv_s) = 0.40$ while the exact kinetic damping is $\omega_{ik}/\omega_r = 0.3944$. For practical purposes a better agreement is usually not needed, although the overlapping of the two lines can easily be improved.

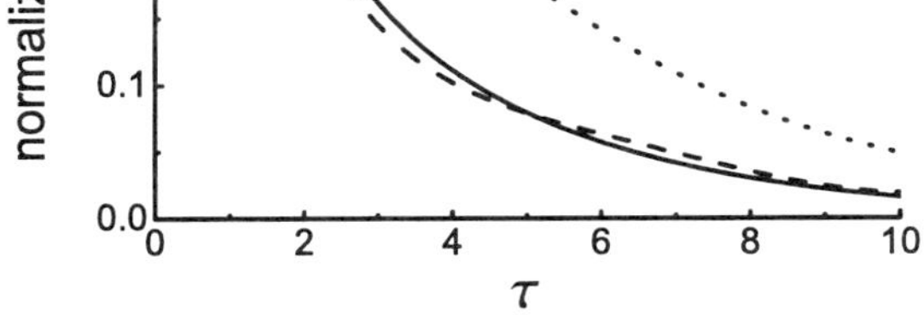

Fig. 2. Comparison of the fluid model decrement ω_{if} and the kinetic decrement ω_{ik} (normalized to ω_r) of the ion acoustic mode in a plasma with hot ions.

Hence, the fluid 'viscosity' term (17) and the corresponding attenuation length (21) can be successfully used as a first approximation in the practical fluid description of the Landau damping, especially in the numerical modeling of ion acoustic waves.

Using the given model we may now modify our dispersion equation (12) by including the effective viscosity term (17) in a plasma where τ is not necessarily much larger than 1. The results should well describe what happens in reality. Using the ion momentum

$$\frac{\partial v_{i1}}{\partial t} = -\frac{e}{m_i}\frac{\partial \phi_1}{\partial r} - \frac{\kappa T_i}{m_i n_0}\frac{\partial n_{i1}}{\partial r} + \frac{\mu_L}{m_i n_0}\frac{\partial^2 v_{i1}}{\partial r^2} - \nu_i v_{i1}$$

it is easily seen that the only change in Eq. (10) is by replacing ν_i with $\delta_i = \nu_i + \mu_0 k^2$. The same change is in the dispersion equation (12).

Eq. (12) is normalized to kc_s and the normalized ion dissipative term becomes of the form

$$\frac{\nu_i}{kc_s} + \frac{1}{\pi d}\left(1 + \frac{1}{\tau}\right)^{1/2}.$$

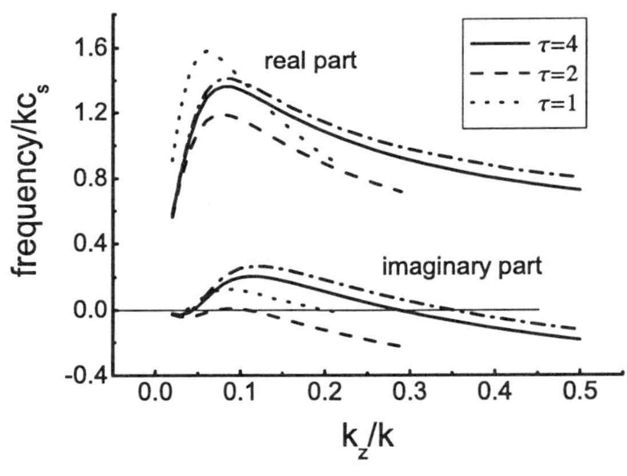

Fig. 3. The angle dependent frequency of the ion acoustic mode for hot ions and Landau damping modeled by (17) and (21), for $v_0 = 30c_s$ and $\nu_e = 30kc_s$. The dotted lines are for $\tau = 1$ and $v_0 = 50c_s$. For comparison with full line, the dash-dot line is for $\tau = 4$ *without* the Landau damping.

The solution is given in Fig. 3 for two values of $\tau = 2$ and 4 and $\nu_e = 30kc_s$ and $v_0 = 30c_s$ (full and dashed lines, respectively). The instability for $\tau = 2$ is not much pronounced, the maximum increment is about $\omega_i/(kc_s) = 0.01$ at the frequency $\omega_r/(kc_s) = 1.18$. On the other hand, the mode is strongly unstable for $\tau = 4$ with the maximum increment $\omega_i/(kc_s) = 0.2$ at the frequency $\omega_r/(kc_s) = 1.3$. The dash-dot line represents the solution for $\tau = 4$ without the Landau damping. In agreement with Eq. (18) the calculated frequency is higher.

Eq. (12) is solved also for $v_0 = 50c_s$, keeping the other parameters the same as before. As a result, for $\tau = 2$ the mode is unstable for k_z/k between 0.04 and 0.26, with the maximum increment $\omega_i = 0.2$ at $k_z/k = 0.1$. For the same electron velocity and for $\tau = 4$ the mode is unstable between 0.04 and 0.58 with the maximum $\omega_i = 0.45$ at $k_z/k = 0.12$.

The case $\tau = 1$ is shown by dotted line in Fig. 3 where $v_0 = 50c_s$ and for the same electron frequency as above.

5. Inhomogeneous electron flow

In the case when the electron flow along the magnetic field has a small gradient in the x direction the perturbations may be taken in the form $\widehat{f}(x)\exp(-i\omega t + ik_y y + ik_z z)$, where $|\partial/\partial x| \ll k_y,\ k_z$. Instead of Eq. (7) for electrons we now have

$$\frac{n_{e1}}{n_{e0}}\left[1 - \frac{\omega_e k_y^2 v_{Te}^2}{\omega_0(\omega_e^2 - \Omega_e^2)} - \frac{k_z^2 v_{Te}^2}{\omega_0\omega_e}\left(1 - \frac{k_y v_0' \Omega_e}{k_z(\omega_e^2 - \Omega_e^2)}\right)\right]$$
$$= -\left[\frac{\omega_e k_y^2}{\omega_0(\omega_e^2 - \Omega_e^2)} + \frac{k_z^2}{\omega_0\omega_e}\left(1 - \frac{k_y v_0'\Omega_e}{k_z(\omega_e^2 - \Omega_e^2)}\right)\right]\frac{e\phi_1}{m_e}. \tag{22}$$

Here $v_{e0}' = dv_0/dx$ and we shall further assume $|\omega_e^2| \ll \Omega_e^2$. The same equation has been derived recently[16] in a study dealing with the ion cyclotron wave excitation by the parallel electron shear flow. Eq. (22) is combined with the corresponding ion equation

$$\frac{n_{i1}}{n_{i0}} = \frac{ek^2}{m_i}\frac{\phi_1}{\omega^2 + i\delta_i\omega - k^2 v_{Ti}^2}, \tag{23}$$

yielding

$$\omega^2 - k^2 v_{Ti}^2 + i\delta_i\omega - k^2 c_s^2 \frac{1 + \dfrac{k_y\Gamma}{k_z}}{1 + \dfrac{k_y\Gamma}{k_z} - \dfrac{k_y^2}{k_z^2}\dfrac{\omega_{e1}^2}{\Omega_e^2}}$$
$$+\frac{m_e}{m_i}\frac{k^2}{k_z^2}\left(\omega_0^2 + i\nu_e\omega_0\right)\frac{1 + \rho_e^2 k_y^2\left(1 + i\dfrac{\nu_e}{\omega_0}\right)}{1 + \dfrac{k_y\Gamma}{k_z} - \dfrac{k_y^2}{k_z^2}\dfrac{\omega_e^2}{\Omega_e^2}} = 0. \tag{24}$$

Here $\Gamma = (dv_0/dx)/\Omega_e$, and $k^2 = k_y^2 + k_z^2$. The perpendicular electron dynamics now can not be omitted. The real part of Eq. (24), i.e., without any dissipation yields the shear flow instability. However this limit in the present model is not physically justified and Eq. (24) should be solved as it is. We note that when $1 > |k_y\Gamma/k_z| > |k_y^2\omega_e^2/(k_z^2\Omega_e^2)|$, setting $\omega = \omega_r + i\omega_i$, from the imaginary part of (24) we have the increment/decrement

$$\omega_i \approx -\frac{\delta_i}{2}\left(1 - \frac{\nu_e}{\delta_i}\frac{\chi}{1+\dfrac{k_y\Gamma}{k_z}}\right)\left(1 - \frac{\chi}{1+\dfrac{k_y\Gamma}{k_z}}\right)^{-1} . \tag{25}$$

If $|\nu_e| > |\delta_i|$ the instability sets in provided that

$$\frac{v_0}{v_s} > \frac{k}{k_z}\left[1 + \frac{\delta_i}{\nu_e}\frac{m_i}{m_e}\left(1 + \Gamma\frac{(1-k_z^2/k^2)^{1/2}}{k_z/k}\right)\frac{k_z^2}{k^2}\right]$$

$$= \frac{1}{\cos\varphi}\left[1 + \frac{\delta_i}{\nu_e}\frac{m_i}{m_e}(1+\Gamma\tan\varphi)\cos^2\varphi\right] . \tag{26}$$

Clearly, except for the mode propagating parallel to the magnetic field, a negative shear flow gradient can considerably reduce the instability threshold.

6. Two ion species

Multi-ion plasmas frequently occur both in the laboratory and space. In the upper atmosphere of the Sun, there is mainly a mixture of hydrogen and helium, while in the solar photosphere it is believed that the main constituents of the plasma are hydrogen and metal ions (with a typical mass[17] of about 35 proton mass). In the terrestrial ionosphere, the most abundant ions[18] in the day-time E-region and lower F-region (i.e. for an altitude h between 120 and 160 km) are molecular ions NO^+ and O_2^+. In a narrow region around $h = 160$ km the third (atomic-ion) component O^+ with the same number density is present. Above this layer, we find mainly a two component electron-O^+ plasma. At night-time, the molecular-atomic ion transition occurs at $h = 220$ km. In the laboratory, classical experiments involve the argon-helium,[19] argon-neon,[20] or xenon-helium plasmas.[21] Each ion specie, in principle, includes an additional branch of acoustic oscillations and various new physical effects emerge, both in the fluid and kinetic domain. In a highly ionized plasma, there is an additional friction between the two ion species.[22] The damping of the heavy ion mode is strongly increased in the presence of even small concentration of light ions.[19,20] The addition of light ions has the same effect as lowering the electron temperature in the kinetic Landau term, i.e., we have $T_e \to T_e/(1+\eta T_e/T_i)$, where $\eta = n_b/n_e$ and n_b is the number density of the added ion specie. Thus the effective value of τ is reduced and as a result the Landau damping is increased. This means that the Landau damping can be controlled, turned on and off. In the presence of a wave driver, like the field aligned electron current discussed in the preceding sections, this implies the possibility for an efficient heating of the heavy ion component. Note that this change of the effective electron temperature is just the opposite to the situation when an additional hot *electron* component is added, which makes the effective ratio τ larger and consequently reduces the ion Landau damping,[3,15] i.e., $T_e/T_i \to \gamma T_e/T_i$, $\gamma > 1$.

For two ion species we have

$$\frac{n_{(a,b),1}}{n_{(a,b)0}} = \frac{q_{a,b}k^2}{m_{a,b}} \frac{\phi_1}{\omega^2 + i\delta_{a,b}\omega - k^2 v_{T(a,b)}^2}, \tag{27}$$

where $\delta_{a,b} = \nu_{a,b} + \mu_{a,b}k^2$, $\mu_{a,b} = \mu_{La,b}/[m_{a,b}n_{0(a,b)}]$, $q_{a,b} = ez_{a,b}$, and $z_{a,b}$ may in principle have also a negative sign for one ion specie. This combined with Eq. (7) with the help of the quasi-neutrality condition $z_a n_a + z_b n_b = n_e$ yields the dispersion equation

$$\frac{m_e n_{a0} z_a^2 k^2}{m_a n_{e0} k_z^2} \frac{1}{\omega^2 + i\delta_a\omega - k^2 v_{Ta}^2} + \frac{m_e n_{b0} z_b^2 k^2}{m_b n_{e0} k_z^2} \frac{1}{\omega^2 + i\delta_b\omega - k^2 v_{Tb}^2}$$
$$+ \frac{1}{\omega_0\omega_e - k_z^2 v_{Te}^2} = 0. \tag{28}$$

In the collision-less limit, for cold ions, and $k_z = k$ and $z_a = z_b = 1$, from (28) we have only one modified sound mode $\omega^2 = k^2 c_{ef}^2$, where $c_{ef}^2 = \kappa T_e/m_{ef}$, $m_{ef} = [m_a + m_e + \eta m_e(m_a/m_b - 1)]/[1 + \eta(m_a/m_b - 1)] \approx m_a m_b/[m_a\eta + m_b(1-\eta)]$, $\eta = n_{b0}/n_{e0}$. For $m_b < m_a$, m_{ef} decreases from m_a at $\eta = 0$ to m_b at $\eta = 1$, so that the phase speed of the mode increases from c_{sa} to c_{sb}. The behavior may change in the case of finite temperatures of the two species[20] when two modes (fast and slow) exist.

In a normalized form, suitable for numerical solution, Eq. (28) becomes

$$\frac{\zeta_1}{\omega^2 + i\delta_a\omega - 1/\tau_a} + \frac{\zeta_2}{\omega^2 + i\delta_b\omega - m_a/(m_b\tau_b)}$$
$$+ \frac{1}{(\omega - k_z v_0/k)[\omega - k_z v_0/k + i\nu_e) - m_a k_z^2/(m_e k^2)]} = 0. \tag{29}$$

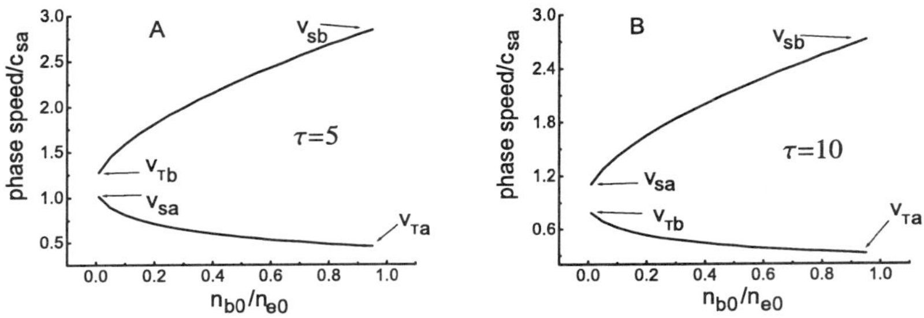

Fig. 4. Phase speed of sound modes in collision-less ideal e-Li$^+$-H$^+$-He plasma without flows and Landau damping, for $k_z = k$. A: the case $v_{sa} < v_{Tb}$, $\tau = 5$. B: the case $v_{sa} > v_{Tb}$, $\tau = 10$. All velocities are normalized to c_{sa}.

Here, ω, $\nu_{a,b}$, $\delta_{a,b}$ are all normalized to kc_{sa}, and v_0 is normalized to c_{sa}, where $c_{sa} = (\kappa T_e/m_a)^{1/2}$, so that

$$\delta_a = \nu_a + \frac{1}{\pi d_a}\left(1+\frac{1}{\tau_a}\right)^{1/2}, \quad \delta_b = \nu_b + \frac{1}{\pi d_b}\left[\frac{m_a}{m_b}\left(1+\frac{1}{\tau_b}\right)\right]^{1/2},$$

$$\zeta_1 = \frac{m_e(1-z_b\eta)z_a k^2}{m_a k_z^2}, \quad \zeta_2 = \frac{m_e \eta z_b^2 k^2}{m_b k_z^2},$$

$$\nu_a = \beta_a\left(\frac{m_e}{m_a\tau_a}\right)^{1/2}\frac{\nu_e}{kc_{sa}}, \quad \nu_b = \beta_b\left(\frac{m_e}{m_b\tau_b}\right)^{1/2}\frac{\nu_e}{kc_{sa}}.$$

For $d_{a,b}$ we use the polynomials (21) with the corresponding $\tau_{a,b} = T_e/T_{a,b}$.

Being interested in ions with low instability threshold, as heavier ions we take Li$^+$, and as light ions we consider H$^+$, both placed in a helium gas. Using data from Table 1, we first check the instability for the two ion species *separately*, i.e., for the H$^+$ sound in an e-H$^+$-He plasma, and the Li$^+$ sound in an e-Li$^+$-He plasma. The instability threshold is calculated with the Landau damping terms (17) and (21) and now it reads

$$\frac{v_0}{c_s} > \frac{k}{k_z}\left[1+\frac{m_i}{m_e}\left(\frac{\nu_i}{\nu_e}+\frac{(1+1/\tau)^{1/2}}{\pi\nu_e d}\right)\frac{k_z^2}{k^2}\right]. \tag{30}$$

Taking $\nu_e = 50$ for $\tau = 10$ from Table 1 we have $\nu_{H^+}/\nu_e = 0.03$ and $\nu_{Li^+}/\nu_e = 0.04$. The minima for H$^+$ and Li$^+$ are $v_0/c_s = 15$, 47 at $k_z/k = 0.13$, 0.04, respectively.

In the plasma with *mixed* ions, i.e., an e-Li$^+$-H$^+$-He plasma, two modes exist with a behavior dependent on the magnitudes of v_{sa} and v_{Tb}. This is

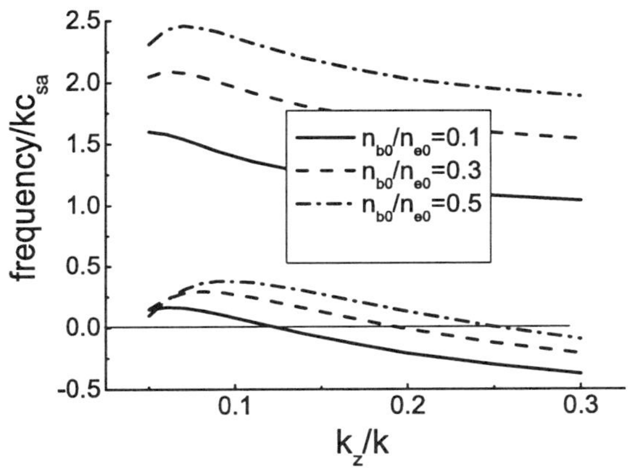

Fig. 5. Real (three upper lines) and imaginary (three lower lines) parts of the frequency (normalized to kc_{sa}) of Li^+ sound wave dependent on the angle of propagation and the number density of H^+ ions. Here $v_0 = 60c_{sa}$, $\nu_e = 50kc_{sa}$, and $\tau = 10$.

presented in Fig. 4 for two values of τ and for a collision-less plasma without Landau damping and flows, and for $k_z = k$. For a large $\tau(= 10)$ (A) and in the case $v_{sb} > v_{sa} > v_{Tb} > v_{Ta}$, the phase speed of the fast mode takes values from v_{sa} (at $\eta = 0$) to v_{sb} (at $\eta = 1$), and a slow mode appears with a phase speed decreasing from v_{Tb} to v_{Ta}. For smaller values of $\tau(= 5)$ (B), when $v_{sb} > v_{Tb} > v_{sa} > v_{Ta}$, the phase speed of the slow mode decreases from v_{sa} to v_{Ta}, while the fast mode changes its phase speed between v_{Tb} (at $\eta = 0$) and v_{sb} (at $\eta = 1$).

In the collisional plasma with the modeled Landau damping we solve Eq. (29) for three different number densities of the H^+ ions. The solutions are presented in Fig. 5 for $v_0 = 60c_{sa}$, $\nu_e = 50kc_{sa}$, and $\tau = 10$. Both real and imaginary parts of the frequency of the Li^+ ion mode are angle dependent, and their magnitude increases with the increased number of H^+ ions. The mode corresponds to the fast mode (upper line) from Fig. 4 (B). In the same time the slow mode has a frequency below 0.01 and it is strongly damped. At the temperature $T_e = 0.1$ eV and $\tau = 1$, the instability threshold is high and requires an electron velocity above 160 at $k_z/k \approx 0.03$.

7. Electron-ion collisions and dynamics of neutrals

In a collisional, weakly ionized three component plasma comprising electrons, ions and neutrals, acoustic perturbations of the neutral gas component include the friction force due to collisions with the ions of the form $-m_n n_{n0}\nu_{ni}(\vec{v}_{n1}-\vec{v}_{i1})$. The momentum conservation due to friction requires $m_\alpha n_{\alpha 0}\nu_{\alpha\beta} = m_\beta n_{\beta 0}\nu_{\beta\alpha}$. The friction due to collisions with the electrons can be omitted for obvious reasons. Using the neutral momentum and continuity equations, one finds that the perturbed velocity of neutrals is coupled to the perturbed ion velocity as

$$v_{n1} = \left(\frac{i\omega\nu_{ni}}{\omega\omega_n - k^2 v_{Tn}^2}\right) v_{i1}. \tag{31}$$

Eq. (31) describes acoustic waves in the neutral gas which is coupled to the ions due to collisions. We have assumed small longitudinal perturbations of the form $\sim \exp(-i\omega t + i\vec{k}\cdot\vec{r})$, propagating in an arbitrary direction $\vec{r}$ which makes an angle ψ with the magnetic field lines $\vec{B}_0 = B_0\vec{e}_z$ that are taken in the z direction. Here, $\omega_n \equiv \omega + i\nu_{ni}$, $v_{Tn}^2 = \kappa T_n/m_n$, and v_{i1} is the perturbed ion velocity in the same $\vec{r}$ direction.

The ion momentum equation is given by

$$m_i n_{i0}\frac{\partial v_{i1}}{\partial t} = -e n_{i0}\frac{\partial \phi_1}{\partial r} - \kappa T_i \frac{\partial n_{i1}}{\partial r} - m_i n_{i0}\nu_{in}(v_{i1} - v_{n1}) + \mu_L \frac{\partial^2 v_{i1}}{\partial r^2}. \tag{32}$$

Eq. (32) and the ion continuity equation yield

$$v_{i1} = \frac{ek}{m_i\omega_2}\left(1 + \frac{k^2 v_{Ti}^2}{\omega\omega_2 - k^2 v_{Ti}^2}\right)\phi_1, \quad \omega_2 = \omega + i(\nu_{in} + \mu_0 k^2) + \frac{\nu_{in}\nu_{ni}\omega}{\omega\omega_n - k^2 v_{Tn}^2}, \tag{33}$$

and

$$\frac{n_{i1}}{n_{i0}} = \frac{ek^2}{m_i(\omega\omega_2 - k^2 v_{Ti}^2)}\phi_1. \tag{34}$$

Electron collisions with both neutrals and ions should be included in the momentum equation which is of the form

$$m_e n_e\left[\frac{\partial \vec{v}_e}{\partial t} + (\vec{v}_e\cdot\nabla)\vec{v}_e\right] = e n_e \nabla\phi - e n_e \vec{v}_e \times \vec{B} - \kappa T_e \nabla n_e$$

$$- m_e n_e \nu_{en}(\vec{v}_e - \vec{v}_n) - m_e n_e \nu_{ei}(\vec{v}_e - \vec{v}_i). \tag{35}$$

The electrons are assumed to be magnetized, i.e., $\Omega_e > \nu_{ei} + \nu_{en}$. In this case, their perpendicular dynamics is negligible, and from Eq. (35) we take the parallel part (to the magnetic field) only. The same holds for the electron continuity which yields

$$\frac{n_{e1}}{n_{e0}} = \frac{\nu_e\omega_0}{\nu_e\omega_0 + ik_z^2 v_{Te}^2}\left(\frac{iek_z^2}{m_e\nu_e\omega_0}\,\phi_1 + \frac{\nu_{en}}{\nu_e}\frac{k_z^2}{k\omega_0}\,v_{n1} + \frac{\nu_{ei}}{\nu_e}\frac{k_z^2}{k\omega_0}\,v_{i1}\right). \qquad (36)$$

Here, $\omega_0 = \omega - k_z v_0$, $\nu_e = \nu_{ei} + \nu_{en}$, and the electron inertia terms are omitted implying a Doppler shifted wave frequency below the electron collision frequency, and a Doppler shifted wave phase velocity below the electron thermal speed. We have assumed a constant equilibrium electron flow $\vec{v}_0 = v_0\vec{e}_z$ along the magnetic lines. Such a flow may be driven by external conditions, e.g., by an electric field E_0 yielding, in the general case[4]

$$v_{i0} = \frac{eE_0}{m_i\nu_i}\left(1 - \frac{m_i}{m_e}\frac{\nu_{ie}}{\nu_e}\right)\left(1 - \frac{\nu_{ie}\nu_{ei}}{\nu_i\nu_e}\right)^{-1}, \qquad \nu_i = \nu_{ie} + \nu_{in}, \qquad (37)$$

$$v_{e0} = \frac{eE_0}{m_e\nu_e}\left[\frac{m_e}{m_i}\frac{\nu_{ei}}{\nu_i}\left(1 - \frac{m_i}{m_e}\frac{\nu_{ie}}{\nu_e}\right)\frac{1}{1 - \frac{\nu_{ie}\nu_{ei}}{\nu_i\nu_e}} - 1\right]. \qquad (38)$$

In a weakly ionized plasma these two velocities become[23] $v_{i0} \approx eE_0/(m_i\nu_{in})$ and $v_{e0} \approx -eE_0/(m_e\nu_{en})$, respectively. The velocity of the ions is usually much smaller and can be omitted, or the equations could be conveniently written in the ion reference frame.

Note that the ambient electric field in purely ionized plasmas may have some interesting effects because of the following. The collisional cross section for Coulomb-type interactions is proportional to $1/v_{T\alpha}^2$, yielding the mean free path λ proportional to $v_{T\alpha}^4$. For λ of the order of the plasma scale, a runaway effect takes place. Without the electric field the effect is not big, otherwise it becomes important because of the acceleration of particles that are already faster (because they feel less collisions). As a result the number of fast particles increases. To have this, the intensity of the applied electric field E_0 must exceed the Dreicer's critical value $E_D = e/(4\pi\varepsilon_0 r_D^2)$, where r_D is the plasma Debye radius. However, we are interested in the application to space plasmas with practically unlimited scales, and u.s. we are dealing with a weakly ionized plasma where the effect in general should have no importance. To have the effect, the collision cross section for plasma-neutral collisions should be proportional to the inverse thermal velocity (with some exponent) which usually is not so, i.e., $\sigma_{\alpha n} \approx const.$

In Eq. (36), we have expressed the ion and neutral velocity components in the z-direction by $v_{j1z} = v_{j1}\cos\psi = v_{j1}k_z/k$. Further, in Eq. (36) we use Eqs. (31) and (33), and then assuming quasi-neutrality, we equate the resulting equation with Eq. (34). This finally yields the following dispersion equation:

$$\left[\omega^2 - k^2 c_s^2 \left(1 + \frac{1}{\tau}\right)\right] (\omega\omega_n - k^2 v_{Tn}^2) = -\nu_{ni}\omega^2 \left(\nu_{in} + \nu_{en}\frac{m_e}{m_i}\right)$$

$$-i(\omega\omega_n - k^2 v_{Tn}^2)\left\{\omega(\nu_{in} + \mu_0 k^2) + \frac{m_e}{m_i}\left[\omega\left(\nu_{ei}\left(\frac{k^2}{k_z^2} - 1\right) + \nu_{en}\frac{k^2}{k_z^2}\right)\right.\right.$$

$$\left.\left. -k_z v_0(\nu_{ei} + \nu_{en})\frac{k^2}{k_z^2}\right]\right\}. \qquad (39)$$

7.1. *Effect of the electron-ion collisions*

We first assume static neutrals and discus the case of perturbations in the plasma species only. In view of the momentum conservation, this may be assumed for very weakly ionized plasmas in which $n_{i0} \ll n_{n0}$ and/or plasmas with a heavy mass of the neutral particles/atoms/molecules in comparison with the ion mass. Setting $\omega = \omega_r + i\omega_i$ into Eq. (39), we find a modified ion sound mode[4,24,25] that is unstable provided that

$$V > \frac{\kappa}{1+\nu}\left(1+\frac{1}{\tau}\right)^{1/2}\left[\nu\left(\frac{1}{\kappa^2} - 1\right) + \frac{1}{\kappa^2} + \frac{\mu}{\widehat{\nu}_{en}}\left(\frac{\widehat{\nu}_{en} b}{(\mu\tau)^{1/2}} + \frac{(1+1/\tau)^{1/2}}{\pi d}\right)\right]. \qquad (40)$$

Here, $\widehat{\nu}_{en}$ is normalized to $\omega_r \simeq kc_s$ and should be chosen in accordance with the assumptions introduced earlier. From $\nu_{en} = \sigma_{en} n_{n0} v_{Te}$ and $\nu_{in} = \sigma_{in} n_{n0} v_{Ti}$, we obtain $\widehat{\nu}_{in} = \widehat{\nu}_{en} b/(\mu\tau)^{1/2}$, where $\mu = m_i/m_e$, $b = \sigma_{in}/\sigma_{en}$ is the ratio of the corresponding collision cross sections, $\kappa = k_z/k$, $\nu = \widehat{\nu}_{ei}/\widehat{\nu}_{en}$, $\widehat{\nu}_{ei}$ is also given in units of kc_s, and $d(\tau)$ is given earlier. For larger values of ν, the second and third terms in Eq. (40) are reduced, the latter implying that the minimum in the threshold velocity profile reduces. This behavior is presented in Fig. 6, for a hydrogen plasma in a neutral hydrogen gas. Here, $\tau = 1$, and[8,26] $\sigma_{en} = 2.5 \cdot 10^{-19}$ m^2, $\sigma_{in} = 9.24 \cdot 10^{-19}$ m^2 at the temperature of 1 eV, and we have chosen $\widehat{\nu}_{en} = 30$. For these parameters $d \simeq 0.4$, and $\widehat{\nu}_{in} = 2.6$. The electron-ion collisions drastically reduce the velocity threshold at small angle of propagation (i.e., for k_z/k close to 1).

7.2. *The dynamics of neutrals*

When the neutral gas is perturbed or when the perturbations in the ionized component induce (due to the friction) perturbations of the neutral

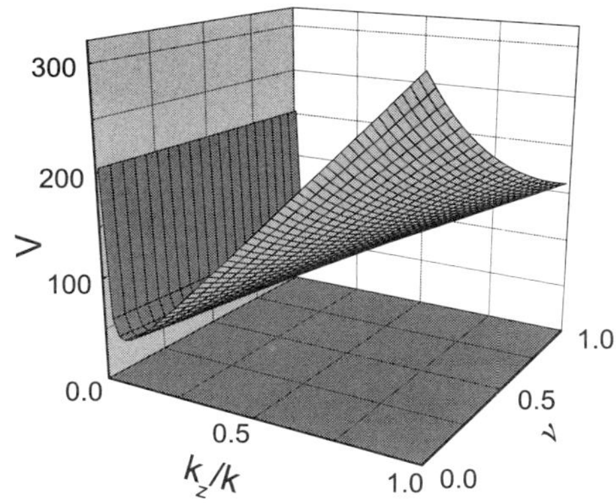

Fig. 6. The normalized threshold velocity $V \equiv v_0/c_s$ for the instability in terms of k_z/k and $\nu \equiv \nu_{ei}/\nu_{en}$. The unstable values are located above the surface.

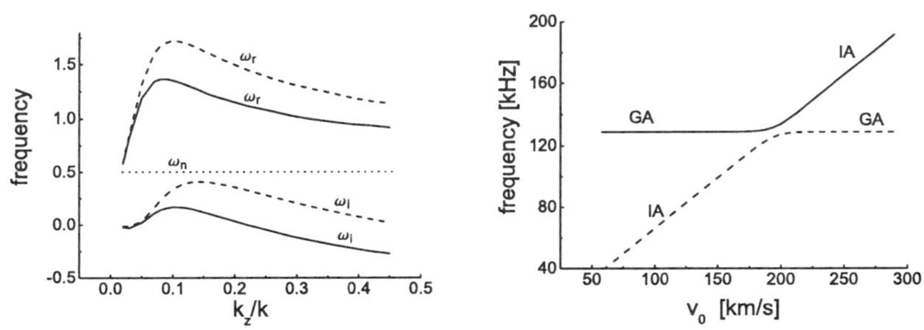

Fig. 7. Left: Normalized real ω_r and imaginary ω_i parts of the angle dependent ion acoustic frequency for $\nu = 0$ (full lines) and $\nu = 0.916$ (dashed lines), in terms of k_z/k. The dotted line ω_n describes the neutral acoustic mode. Right: The frequencies of the electron-flow-driven ion acoustic (IA) mode and the gas acoustic (GA) mode in terms of the electron velocity v_0 for the wave-length $\lambda = 0.3$ m.

background, the full dispersion equation (39) needs to be solved. In dimensionless form it becomes

$$\widehat{\omega}^2 = 1 + \frac{1}{\tau} - \frac{\widehat{\omega}^2 \widehat{\nu}_{ni} \widehat{\nu}_{en} \left[b(\mu\tau)^{-1/2} + \mu^{-1} \right]}{\widehat{\omega}(\widehat{\omega} + i\widehat{\nu}_{ni}) - \frac{\mu_n}{\tau_n}}$$

$$-i\left\{\widehat{\omega}\left[\widehat{\nu}_{in}+\frac{(1+\tau^{-1})^{1/2}}{\pi d}\right]+\frac{1}{\mu}\left[\widehat{\omega}\widehat{\nu}_{en}\left(\nu\left(\frac{1}{\kappa^2}-1\right)+\frac{1}{\kappa^2}\right)-\frac{V\widehat{\nu}_{en}(1+\nu)}{\kappa}\right]\right\}. \tag{41}$$

All frequencies are normalized to kc_s, and we have introduced new parameters $\tau_n = T_e/T_n$, $\mu_n = m_i/m_n$. The number of parameters can be reduced by using as before $\widehat{\nu}_{in} = \widehat{\nu}_{en} b/(\mu\tau)^{1/2}$, and from the momentum conservation in the friction force terms here we have

$$\widehat{\nu}_{ni} = \mu_n X \frac{b}{(\mu\tau)^{1/2}}\widehat{\nu}_{en}, \quad X = n_{i0}/n_{n0}.$$

The parameter ν is dependent of the ionization fraction X because $\nu \equiv \nu_{ei}/\nu_{en} = 2(2\pi/m_e)^{1/2}e^4 L_{ei} X/[3\sigma_{en} v_{Te}(4\pi\varepsilon_0)^2(\kappa T_e)^{3/2}]$. Here, $L_{ei} = \log[12\pi\varepsilon_0(\varepsilon_0/n_{e0})^{1/2}(\kappa T_e)^{3/2}/e^3]$, is the Coulomb logarithm.

As a demonstration Eq. (41) is solved for the parameters $\tau = 4$, $\tau_n = 4$, $\mu = 1838$, $\mu_n = 1$, $\widehat{\nu}_{en} = 30$, and $V = 30$. We have taken $n_{e0} = n_{i0} = 6 \cdot 10^{16}$ m^{-3} and $n_{n0} = 10^{19}$ m^{-3}, which yields $X = 0.006$ and $\nu = 0.916$. The results are presented in Fig. 7(left), with the remarkable angle dependent behavior of the IA mode. The neutral acoustic mode has nearly a constant frequency $\omega_n \simeq 0.5$ and a very small decrement $\simeq -0.005$. The real and imaginary parts of the ion acoustic mode frequency change in the presence of electron-ion collisions ν although the ionization is relatively small. Note that the assumed value of $\widehat{\nu}_{en} = 30$ in principle fixes the wavelength of fluctuations. For example, assuming $T = 5000$ K, one has $c_s = 6.4$ km/s and $\nu_{en}/(kc_s) = 30$ implies a wavelength of 0.7 m.

The dispersion equation Eq. (39) is solved also in terms of the driving velocity v_0 for fixed values $k = 20$ m^{-1}, and $k_z = 0.05k$, and for the following plasma parameters that may be taken as typical for some laboratory plasmas and for the lower solar atmosphere: $T_e = T_i = T_n \simeq 5000$ K, $n_{e0} = n_{i0} = 8 \cdot 10^{16}$ m^{-3}, $n_{n0} = 3 \cdot 10^{20}$ m^{-3}, thus $X = 2.6 \cdot 10^{-4}$. This further yields[8,26] $\nu_{in} = 3.5 \cdot 10^6$ Hz, $\nu_{en} = 3 \cdot 10^7$ Hz, and $\nu_{ei} = 3.8 \cdot 10^6$ Hz. For a magnetic field $B_0 = 0.01$ T this yields un-magnetized ions and magnetized electrons, viz. $\Omega_i/\nu_{it} = 0.27$ while $\Omega_e/\nu_{et} = 53.4$. For these parameters, the plasma β is $1.3 \cdot 10^{-4} < m_e/m_i = 5.4 \cdot 10^{-4}$, implying we are in the proper electrostatic limit. The neutral sound frequency is $kv_{Tn} \simeq 129$ kHz, and the ion sound frequency $kv_s \simeq 182$ kHz. However, in reality for these parameters the IA mode without a source like the electron flow will be completely destroyed by collisions, while the GA mode will exist with the complex frequency $\omega \simeq 129 \cdot 10^3 - i3 \cdot 10^2$ Hz. In the presence of the flow, the two modes exist and this situation is seen in Fig. 7(right). The modes interchange their identities in the vicinity of $v_0 \simeq 190$ km/s, which is around

$v_0/c_s \simeq 30$. It turns out that the imaginary parts of the two frequencies have the opposite peaks in the same domain of close frequencies.[25]

7.3. *Electromagnetic perturbations*

Due to the difference in the parallel motion of electrons and ions, which implies a perturbed parallel current and a perturbed perpendicular magnetic component according to the Ampère law, the magnetic field may be perturbed. In the case of magnetized ions that are tied to the magnetic field lines, such perturbations propagate along the field lines at the Alfvén speed. However, this is not the situation in the present model. For the electromagnetic (EM) perturbations, the dynamics of neutrals is unchanged. The small perturbations of the magnetic field do not change the ion magnetization and thus the Lorentz force in the ion momentum equation can still be omitted. For not so small plasma β, assuming only perpendicular bending of the magnetic field lines, we express the perturbations of the EM field in terms of potentials $\vec{E}_1 = -\nabla\phi_1 - \partial\vec{A}_{z1}/\partial t$ and $\vec{B}_1 = \nabla \times \vec{A}_{z1} = -\vec{e}_z \times \nabla_\perp A_{z1}$. The ion momentum equation (32) now includes the new term $-e\, n_{i0}\, \partial_t A_{z1}\, \vec{e}_z$. Consequently, the ion dynamics in $\vec{k}$-direction comprises the new term $A_k = k_z\, A_{z1}/k$, so that

$$v_{i1} = \frac{ek}{m_i\omega_2}\left(1 + \frac{k^2 v_{Ti}^2}{\omega\omega_2 - k^2 v_{Ti}^2}\right)\left(\phi_1 - \frac{\omega}{k}\frac{k_z}{k}A_{z1}\right). \tag{42}$$

This results in the modified Eq. (34):

$$\frac{n_{i1}}{n_{i0}} = \frac{ek^2}{m_i(\omega\omega_2 - k^2 v_{Ti}^2)}\left(\phi_1 - \frac{\omega}{k}\frac{k_z}{k}A_{z1}\right). \tag{43}$$

The electron perpendicular velocity is calculated from (35) which now comprises an additional term $en_e\vec{e}_z\partial A_{z1}/\partial t$ yielding

$$\vec{v}_{e\perp 1} = \frac{1}{1+\nu_e^2/\Omega_e^2}\left[\frac{1}{B_0}\vec{e}_z \times \nabla_\perp\phi_1 + \frac{\nu_e}{\Omega_e}\frac{\nabla_\perp\phi_1}{B_0} - \frac{v_{Te}^2}{\Omega_e}\frac{\nu_e}{\Omega_e}\frac{\nabla_\perp n_{e1}}{n_0}\right.$$

$$-\frac{v_{Te}^2}{\Omega_e}\vec{e}_z \times \frac{\nabla_\perp n_{e1}}{n_0} - v_0\vec{e}_z \times \nabla_\perp A_{z1} + \frac{\nu_{ei}}{\Omega_e}\vec{e}_z \times \vec{v}_{i\perp 1} + \frac{\nu_{ei}\nu_e}{\Omega_e^2}\vec{v}_{i\perp 1}\Bigg].$$

$$\left. + \frac{\nu_{en}}{\Omega_e}\vec{e}_z \times \vec{v}_{n\perp 1} + \frac{\nu_{en}\nu_e}{\Omega_e^2}\vec{v}_{n\perp 1}\right].$$

This is to be used in the $\nabla_\perp \cdot \vec{v}_{e\perp 1}$ term in the electron continuity equation. Clearly, all non-vanishing terms are multiplied by the small ratio of the collision and gyro-frequencies, so that, like the previous case, the electron perpendicular dynamics can be neglected. The electron parallel dynamics now includes the vector potential, and as a result there appears a new term, viz.

$$-\frac{ie\omega k_z A_{z1}}{m_e \nu_e \omega_0} \tag{44}$$

within the brackets in Eq. (36).

From the Ampère law $\nabla \times \vec{B} = \mu_0 \vec{j}$, we have

$$\frac{\partial^2 A_{z1}}{\partial x^2} = -e\mu_0 n_0 \frac{k_z}{k} v_{i1} + e\mu_0 n_0 v_{ez1} + e\mu_0 n_1 v_0 .$$

Here we shall use Eq. (42) and calculate the electron parallel velocity with the help of modified Eq. (36) [which includes the new term (44)]. The result is given by

$$\frac{k^2 \lambda_i^2 \omega_2}{k_z}\left(\frac{k_z^2}{k^2} - 1\right) A_{z1} = \frac{im_i \omega\omega_2}{m_e(\nu_e\omega_0 + ik_z^2 v_{Te}^2)}\left(\phi_1 - \frac{\omega}{k_z} A_{z1}\right)$$
$$+\left(1 + \frac{k^2 v_{Ti}^2}{\omega\omega_2 - k^2 v_{Ti}^2}\right)\left[-1 + \frac{\omega}{\nu_e\omega_0 + ik_z^2 v_{Te}^2}\left(\nu_{ei} + \frac{i\nu_{en}\nu_{ei}\omega}{\omega\omega_n - k^2 v_{Tn}^2}\right)\right]$$
$$\times\left(\phi_1 - \frac{\omega}{k_z}\frac{k_z}{k} A_{z1}\right). \tag{45}$$

Here, $\lambda_i = c/\omega_{pi}$, where c is the speed of light and ω_{pi} is the ion plasma frequency.

Equating Eq. (43) and the electromagnetically modified Eq. (36), with the help of Eqs. (31) and (42), we obtain the second necessary equation for the two potentials:

$$\frac{k^2}{m_i}\left\{\frac{1}{\omega\omega_2 - k^2 v_{Ti}^2} - \frac{k_z^2}{k^2}\frac{\nu_{ei}}{\omega_2(\nu_e\omega_0 + ik_z^2 v_{Te}^2)}\left(1 + \frac{k^2 v_{Ti}^2}{\omega\omega_2 - k^2 v_{Ti}^2}\right)\right.$$
$$\left.\times\left[1 + \frac{i\omega\nu_{ni}\nu_{en}}{\nu_{ei}(\omega\omega_n - k^2 v_{Tn}^2)}\right]\right\}\left(\phi_1 - \frac{\omega}{k_z}\frac{k_z}{k} A_{z1}\right)$$
$$= \frac{ik_z^2}{m_e(\nu_e\omega_0 + ik_z^2 v_{Te}^2)}\left(\phi_1 - \frac{\omega}{k_z} A_{z1}\right). \tag{46}$$

In the electrostatic limit, Eq. (46) becomes identical to the previously derived Eq. (39). In the electromagnetic collision-less case without the electron flow and Landau damping, Eqs. (45), (46) yield

$$(\omega^2 - k^2 v_{Tn}^2)\left[k^4 k_z^2 v_s^2 \lambda_i^2 - \left(k^2(1 + k_z^2\lambda_i^2) - k_z^2\right)\omega^2\right] = 0.$$

Here, we have the GA mode uncoupled with the IA mode which is modified due to the electromagnetic effects yielding

$$\omega^2 = k^2 c_s^2 \frac{k_z^2 \lambda_i^2}{1 + k_z^2 \lambda_i^2 - \frac{k_z^2}{k^2}}. \tag{47}$$

Hence, in the absence of Alfvén waves (un-magnetized ions), the parallel propagation ($k_z = k$) yields an ordinary ion sound mode. For any other angle of propagation (except for $k_z \to 0$) the ion sound mode is electromagnetically modified and becomes dispersive.

The collisions couple the two modes and the full dispersion equation is given by

$$k^4 \lambda_i^2 v_{Tn}^2 \left(k_z^2 v_s^2 - i\nu_e \omega_0 \frac{m_e}{m_i} \right) + \omega_n \omega^3 \left(1 - \frac{k_z^2}{k^2} \right)$$

$$+ k_z^2 \omega^2 \left[\lambda_i^2 \frac{m_e}{m_i} \left(\nu_{en} \nu_{ni} - i\nu_{ei} \omega_n \right) + v_{Tn}^2 + \omega_n \omega_2 \lambda_i^2 \right]$$

$$-k^2 \omega \left\{ \omega v_{Tn}^2 - i\lambda_i^2 \nu_e \omega_0 \omega_n \frac{m_e}{m_i} + k_z^2 \lambda_i^2 \left[v_{Tn}^2 \left(\omega_2 - i\nu_{ei} \frac{m_e}{m_i} \right) + \omega_n v_s^2 \right] \right\} = 0. \tag{48}$$

$$\omega_0 = \omega - k_z v_0, \quad \omega_n = \omega + i\nu_{ni}, \quad \omega_2 = \omega + i(\nu_{in} + \mu_0 k^2) + \frac{\nu_{in} \nu_{ni} \omega}{\omega \omega_n - k^2 v_{Tn}^2}.$$

In order to compare the ES and EM cases, we use the same set of parameters as in the previous case in Fig. 7. This is only for the sake of comparison because, in principle, the EM effects imply a higher plasma β. It turns out[25] that the most visible difference is in the graph of the IA mode increment, which is now much higher for the velocity v_0 above certain critical value. Physically, here we have the bending of the magnetic lines representing an additional obstacle for the electron motion in the z-direction, and as a result the mode is more unstable. Other effects related to the bending are predictable and will not be discussed in detail.

8. Perpendicular electron drift

Now we assume an electron drift in the perpendicular direction[2,3] and keep the ion-neutral collision frequency arbitrary. Consequently, the ion species should be described by a collisional Boltzman kinetic equation, where we use the Krook's collisional term, which is good enough for ion collisions with neutral having nearly equal mass

$$\frac{\partial f_1}{\partial t}+\vec{v}\frac{\partial f_1}{\partial \vec{r}}+\frac{e}{m_i}(\vec{E}_1+\vec{v}\times\vec{B}_0)\frac{\partial f_1}{\partial \vec{v}}=-\nu_{in}f_1.$$

Performing the usual integration for the ions we obtain[3]

$$\frac{n_{i1}}{n_{i0}}=-\frac{e\phi_1}{m_i v_{Ti}^2}\left[1-J_+\left(\frac{\omega_I}{kv_{Ti}}\right)\right]. \tag{49}$$

Here, $J_+(\alpha)=[\alpha/(2\pi)^{1/2}]\int_c d\zeta\exp(-\zeta^2/2)/(\alpha-\zeta)$ is the plasma dispersion function, $\zeta=v/v_{Ti}$, and $\omega_I=\omega+i\nu_{in}$. Using the quasi-neutrality and the fluid equations for electrons as earlier, we obtain the dispersion equation:

$$\frac{1}{\kappa T_i}\left[1-J_+\left(\frac{\omega_I}{kv_{Ti}}\right)\right]+\frac{ik_z^2}{m_e\nu_{en}(\omega-\vec{k}\cdot\vec{v}_{e0})+ik_z^2\kappa T_e}=0. \tag{50}$$

On condition that $\omega=\omega_r+i\omega_{im}$, $|\omega_{im}|\ll|\omega_r|$ this yields the spectrum:

$$\omega_r^2\approx k^2c_s^2\left(1+\frac{3}{\tau}\right)-6\frac{\nu_{en}\nu_{in}}{\tau}\frac{m_e}{m_i}\frac{k^2}{k_z^2}\left(1-\frac{\vec{k}\cdot\vec{v}_{e0}}{kc_s}\right). \tag{51}$$

In the imaginary part we keep $\omega_r^2\approx k^2c_s^2(1+3/\tau)$ and obtain the increment/decrement approximately:

$$\begin{aligned}\omega_{im}\approx{}&-\nu_{in}\left[1+\frac{3}{\tau(1+3/\tau)^2}\right]\\&-\nu_{en}\frac{m_e}{2m_i}\frac{k^2}{k_z^2}\left[1+\frac{3}{\tau(1+3/\tau)}\right]\left[1-\frac{\vec{k}\cdot\vec{v}_{e0}}{kc_s(1+3/\tau)^{1/2}}\right]\\&-kc_s\left(\frac{\pi}{8}\right)^{1/2}\tau^{3/2}\left(1+\frac{3}{\tau}\right)\exp\left[-\frac{\tau}{2}(1+3/\tau)\right].\end{aligned} \tag{52}$$

It is seen that there can be no instability in the case of a counter propagating wave with respect to the drift velocity, i.e. if $\vec{k}\cdot\vec{v}_{e0}<0$. In the opposite case, the instability sets in if

$$\begin{aligned}\frac{v_{e0}}{c_s}>{}&\frac{(1+3/\tau)^{1/2}}{\cos\psi}\left\{1+\left[\nu_{in}\left(1+\frac{3}{\tau(1+3/\tau)^2}\right)\right.\right.\\&\left.+kc_s\left(\frac{\pi}{8}\right)^{1/2}\tau^{3/2}\left(1+\frac{3}{\tau}\right)\exp\left[-\frac{\tau}{2}\left(1+\frac{3}{\tau}\right)\right]\right]\\&\left.\times\frac{2m_ik_z^2}{m_e\nu_{en}k^2[1+3/(\tau+3)]}\right\}.\end{aligned} \tag{53}$$

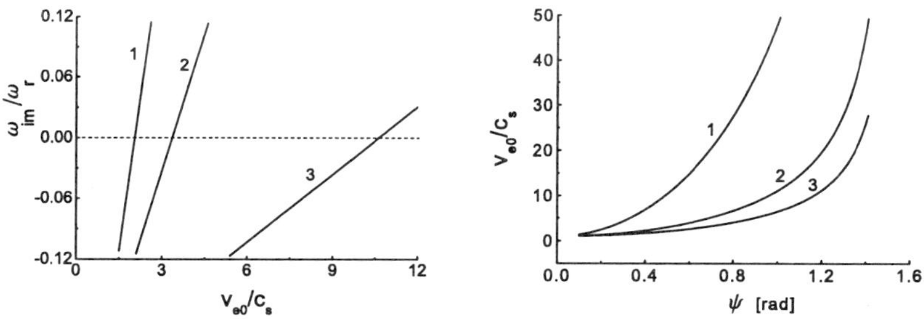

Fig. 8. Left: The imaginary part of the frequency (in units of ω_r) in terms of the electron drift velocity for three values of the angle $\psi = 7.5°$ (line 1), $\psi = 11.25°$ (line 2), and $\psi = 22.5°$ (line 3). Right: Threshold value of the perpendicular electron drift in terms of the angle between the drift velocity vector and the wave-vector. The lines 1-3 are for the ratio kc_s/ν_{en} 1/10, 1/50, 1/100, respectively. Unstable values of the drift are above the lines.

Here, ψ is the angle between $\vec{k}$ and $\vec{v}_{e0}$, and the factor $3/\tau$ can be omitted for larger values of τ. This situation is presented in Fig. 8(left), where the ratio ω_{im}/ω_r is plotted in terms of the normalized electron drift for the normalized value $\nu_{en}/\omega_r = 10$ and consequently for $\nu_{in}/\omega_r = (\nu_{en}/\omega_r)[T_i m_e/(T_e m_i)]^{1/2}$, where $\tau = 10$, and the angle ψ takes values 7.5°, 11.25°, 22.5° for the lines 1-3, respectively. The fast growing mode, driven by the electron current, has a threshold which shifts towards larger values of v_{e0}/c_s, for larger values of the angle ψ. For the lines $1-3$ the mode becomes unstable for v_{e0} exceeding $2c_s$, $3.3c_s$ and $10.7c_s$, respectively. The shift of the instability threshold in the limit $\nu_{in} \ll \omega_r$ and for $\tau = 10$ is presented in Fig. 8(right), for several values of the electron collision frequency and for the angle ψ in the range $0 < \psi < \pi/2$. From (53) one finds that in this limit the instability condition becomes

$$\frac{v_{e0}}{c_s} > \frac{1}{\cos\psi}\left(1 + 377\frac{kc_s}{\nu_{en}}\sin^2\psi\right). \tag{54}$$

Here, the limit $\psi \to 0$, which corresponds to wave propagating along the electron drift, is not justified as it violates the conditions introduced for the electron dynamics. For an electron-proton plasma and for the ion sound, this approximately means that there must be a $k_z > k/43$, or $\psi > 1.33°$. Similarly, the limit $\psi \to \pi/2$ implies propagation perpendicular to the drift and an infinite drift magnitude for the instability. It is seen that, depending on the angle of the propagation, the threshold can be much smaller compared to (2), which according to (53) appears, firstly, due to the oblique propagation because the critical term is multiplied by the square of k_z/k (i.e., by $\sin^2\psi$), and, secondly, due to the collisions (i.e., the term $kc_s/\nu_{en} \ll 1$).

It can be shown[3] that in the presence of an additional hotter electron specie h satisfying $\Omega_h < \nu_{hn}$ the instability threshold is changed reading

$$\frac{v_{e0}}{c_s} > \frac{1}{\cos\psi}\left[\gamma^{1/2} + \frac{m_i c_s k_z^2}{m_e \nu_{en} k}\left(\frac{\pi}{2}\right)^{1/2}(\gamma\tau)^{3/2}\exp\left(-\frac{\gamma\tau}{2}\right)\right]. \quad (55)$$

Here

$$\gamma = \frac{T_h}{T_e}\frac{(n_{h0}/n_{e0}+1)}{n_{h0}/n_{e0}+T_h/T_e}.$$

It is seen that the hotter electron specie makes the electron-ion temperature ratio effectively larger. Since $\gamma > 1$ it is clear that, compared to (53), the instability threshold is higher.

9. Conclusions

In this study some interesting phenomena are presented that are absent in collision-less plasmas. The essential feature exploited here is the fact that electrons may remain magnetized while ions are un-magnetized, and therefore an acoustic wave can propagate at an angle with respect to the magnetic field lines. Electron collisions with neutrals cause a time lag between the density and potential perturbations. In the presence of a driver like the electron flow along the magnetic field lines, this can make the sound mode unstable, provided the magnitude of the flow exceeds some critical threshold. Due to ion collisions, the instability threshold appears to be angle dependent. More precisely, the instability profile has a dip. Hence, in the presence of a driver (flow) along the magnetic lines, the most unstable sound modes are those propagating almost perpendicular to the magnetic field lines. The parameters values considered in the present paper show that the effect should be expected in various realistic plasma configurations both in the laboratory and in space.

In collisional plasmas there exists a clear transition from the (Landau) damping to the non-damping regime, which has been verified experimentally in the past.[12] The essentially kinetic effect called Landau damping, can be modeled within the two-fluid theory by including an effective 'viscosity' term in the ion momentum equation. In the simple linear theory, such a modeling yields good results without missing any essential physics.

The inclusion of the effect of electron-ion collisions implies the applicability of the model to partially ionized plasmas with somewhat higher ionization ratio n_{i0}/n_{n0}. A typical example of such plasma is the upper solar photosphere and the lower chromosphere. Plasmas of the same type are also frequently found in the laboratory environment, like in the ECR-produced plasmas[27–29] where the ionization fraction is maximum at the centrum of the cylindric plasma column and decreases in the radial direction.

In the presence of plasma perturbations, the higher fractional ionization discussed above may involve the dynamics of neutrals due to the friction effects. This is because in such a case, the neutral gas is not necessarily so heavy and may participate in the plasma perturbations. On the other hand, in space plasmas the perturbations of the neutral gas may develop independently due to various external reasons and then involve motion (perturbation) in the plasma component. In either case, the collisions play a crucial role.

The electromagnetic effects may play a role even for the case of unmagnetized ions studied here. We have focused on the perpendicular perturbations only (bending of the magnetic lines) because it is energetically more favorable and is essentially a small-plasma-β effect. The magnetic field is perturbed due to the perturbed parallel current (i.e., the difference between the electron and ion parallel motion) and in the collision-less case yields only a modification of the IA mode. Physically, the bending of the magnetic field lines implies an additional obstacle for electrons and results in their less efficient motion in the z-direction, which increases the instability of the IA mode.

An additional specie of hotter electrons, which is frequently observed in the auroral regions of the terrestrial ionosphere, introduces certain changes in the mode behavior. These hotter electrons are more collisional and, therefore, they perform a motion similar to that of the ions. As a result, the induced electric field becomes smaller due to the fact that a part of the ion charge is shielded by the hotter electron specie moving nearly in the same manner, while in the same time the instability threshold is changed and increased.

We underline an essential difference between the standard electron-drift driven instability condition Eq. (2) and our current results. It is seen that, neglecting the τ contribution, the necessary instability condition in the standard case, $v_{e0} > c_s$, is more easily satisfied compared to the our cases. However, due to the angle dependence, the actual sufficient instability condition is much more easily satisfied in our case.

Acknowledgements: The results presented here are obtained in the framework of the projects G.0304.07 (FWO-Vlaanderen), C 90203 (Prodex), GOA/2004/01 (K.U.Leuven), and the Interuniversity Attraction Poles Programme - Belgian State - Belgian Science Policy.

References

1. R. J. Goldston and P. H. Rutherford, *Introduction to Plasma Physics* (Institute of Physics, Bristol, 1995), p. 447.
2. P. Kaw, *Phys. Lett. A* **44**, 427 (1973).
3. J. Vranjes and S. Poedts, *Eur. Phys. J. D* **40**, 257 (2006).
4. J. Vranjes and S. Poedts, *Phys. Plasmas* **13**, 052103 (2006).
5. J. A. Ratcliffe, *The magneto-ionic theory and its applications to the ionosphere* (Cambridge Univ. Press, Cambridge, 1959), p. 33.
6. M. Mitchner and C. H. Kruger, *Partially Ionized Gasses* (John Willey and Sons, New York, 1973), p. 102.
7. A. Zecca, G. P. Karwasz, and R. S. Brusa, *Riv. Nuovo Cim.* **19**, 128 (1996).
8. P. S. Krstic and D. R. Schultz, *J. Phys. B: At. Mol. Opt. Phys.* **32**, 3485 (1999).
9. Y. P. Raizer, *Gas discharge physics* (Springer-Verlag, Berlin Heidelberg, 1991) p. 25.
10. V. E. Fortov and I. T. Iakubov, *The physics of non-ideal plasma* (World Scientific, Singapore, 2000), p. 380.
11. T. H. Stix, *Waves in plasmas* (AIP, New York, 1992), p. 184.
12. N. D'Angelo, Astrophys. J. **154**, 401 (1968).
13. N. D'Angelo, G. Joyce, and M. E. Pesses, Astrophys. J. **229**, 1138 (1979).
14. H. K. Andersen, N. D'Angelo, V. O. Jensen, P. Michelsen, and P. Nielsen, Phys. Fluids **11**, 1177 (1968).
15. F. F. Chen, *Introduction to plasma physics and controlled fusion* (Plenum Press, New York, 1984), p. 272.
16. B. Eliasson, P. K. Shukla, and J. O. Hall, Phys. Plasmas **13**, 024502 (2006).
17. H. K. Sen and M. L. White, Sol. Phys., **23**, 146 (1972).
18. H. Rishbeth and O. K. Garriott, *Introduction to ionospheric physics* (Acad. Press, New York, 1969), p. 118.
19. B. D. Fried, R. B. White, and T. K. Samec, *Phys. Fluids* **14**, 2388 (1971).
20. Y. Nakamura, M. Nakamura, and T. Itoh, *Phys. Rev. Lett.* **37**, 209 (1976).
21. I. Alexeff, W. D. Jones, and D. Montgomery, *Phys. Rev. Lett.* **19**, 422 (1967).
22. V. Yu. Bychenkov, W. Rozmus, and V. Y. Tikhonchuk, *Phys. Rev. E*, **51**, 1400 (1995).
23. N. D'Angelo, *Phys. Lett. A* **336**, 204 (2005).
24. J. Vranjes, M. Y. Tanaka and S. Poedts, *Phys. Plasmas* **13**, 122103 (2006).
25. J. Vranjes, B. P. Pandey and S. Poedts, *Phys. Plasmas* **14**, 032106 (2007).
26. B. Bederson and L. J. Kieffer, *Rev. Mod. Phys.* **43**, 601 (1971).
27. J. Vranjes, A. Okamoto, S. Yoshimura, S. Poedts, M. Kono, and M. Y. Tanaka, *Phys. Rev. Lett.* **89**, 265002 (2002).

28. A. Okamoto, K. Hara, K. Nagaoka, S. Yoshimura, J. Vranjes, M. Kono, and M. Y. Tanaka, *Phys. Plasmas* **10**, 2211 (2003).
29. M. Kono and M. Y. Tanaka, *Phys. Rev. Lett.* **84**, 4369 (2000).

MULTIFLUID THEORY OF SOLITONS

FRANK VERHEEST

Sterrenkundig Observatorium, Universiteit Gent, Krijgslaan 281, B–9000 Gent, Belgium &

School of Physics, University of KwaZulu-Natal, Private Bag X54001, Durban 4000, South Africa

Email: frank.verheest@ugent.be

After introducing the basic multifluid model equations, this review discusses three different methods to describe nonlinear plasma waves, by giving a rather general overview of the relevant methodology, followed by a specific and recent application. First, reductive perturbation analysis is applicable to waves that are not too strongly nonlinear, if their linear counterparts have an acoustic-like dispersion at low frequencies. It is discussed for electrostatic modes, with a brief application to dusty plasma waves. The typical paradigm for such problems is the well known KdV equation and its siblings. Stationary waves with larger amplitudes can be treated, i.a., via the fluid-dynamic approach pioneered by McKenzie, which focuses on essential insights into the limitations that restrict the range of available solitary electrostatic solutions. As an illustration, novel electrostatic solutions have been found in plasmas with two-temperature electron species that are relevant in understanding certain magnetospheric plasma observations. The older cousin of the large-amplitude technique is the Sagdeev pseudopotential description, to which the newer fluid-dynamic approach is essentially equivalent. Because the Sagdeev analysis has mostly been applied to electrostatic waves, some recent results are given for electromagnetic modes in pair plasmas, to show its versatility.

1. Introduction

Solitons are important nonlinear structures that are frequently observed, not only in laboratory and natural plasmas, as will interest us here, but also in many other physical systems. *In situ* space observations of large-amplitude spikes indicate the need for a nonlinear description, usually in terms of solitary waves or trains of these. Another example among many is the quest by the telecommunications industry for a regular and reliable signal transmission in, e.g., underwater cables, where great store is set by the relation for nonlinear waves between the amplitude and the propagation velocity, so that equal height pulses travel precisely at the same speed, with a minimum of distortion.

In this review, when we speak of solitary waves and/or solitons, we think of hump/dip-like structures whose shape remains unaltered during propagation. For true solitons rather than simple solitary waves, specific interaction properties are needed that confer on them almost particle-like properties, hence the name for these nonlinear entities. On the mathematical side, soliton theory studies completely integrable systems and the many facets of their solutions, while on the experimental side questions arise as to how easily solitons are formed. In the earliest context studied, that of shallow water solitons, the generation is fairly easy, given rather general initial disturbances. There are also rightful questions to be asked about the stability of solitons, in particular in more than one spatial dimension. Here simple answers are not easy to come by, except for solutions of integrable equations.

Since a vast literature exists about nonlinear waves in general and in plasmas in particular, I will quote in this review only those references that are necessary to follow the exposition, without any hope of doing justice to all the interesting papers published in this exciting field. Moreover, not wishing to overload this introduction, I will leave most of the references to the appropriate sections below. After a short section 2 introducing the multifluid model and basic equations, I want to focus, in three sections, on different methods to describe nonlinear waves. Each of these sections will consist of two parts, a rather general overview of the relevant methodology, followed by a specific application involving nonlinear plasma waves.

First, reductive perturbation analysis is widely applicable to waves that are not too strongly nonlinear, if their linear counterparts have an acoustic-like dispersion at low frequencies. This will be dealt with in Section 3 for electrostatic modes in multifluid plasmas, with a brief application to dusty plasma waves. The typical paradigm for such problems is the Korteweg-de Vries (KdV) equation and its siblings, characterized by an infinite number of first integrals and corresponding N-soliton solutions with their inherent stability.

Stationary waves with larger amplitudes can be treated, i.a., via the fluid-dynamic approach pioneered by McKenzie, and I will give in Section 4 the essential insights into the limitations in plasma parameter space, which restrict the range of available solitary electrostatic solutions. As an application I will describe how in a plasma with two-temperature electron species novel electrostatic solutions are possible, that might be relevant in understanding certain magnetospheric plasma observations.

The older cousin of the large-amplitude technique is the Sagdeev pseudopotential description, to which the newer fluid-dynamic approach is essentially equivalent, although the former lacks the deeper insight into the physical restrictions in parameter space. This will be briefly discussed in Section 5, and because the Sagdeev method has traditionally been applied mostly to electrostatic waves, I will in contrast describe some recent results for electromagnetic modes in pair plasmas, to show its versatility.

The conclusions then follow in Section 6.

2. Multifluid plasma model and basic equations

I will use a multifluid model, with a number of species described by standard fluid equations, which only interact through the electromagnetic fields. Although this is a rather simplified model, it is sufficient to bring out the gist of the wave characteristics we are looking for. For waves propagating along the x axis of a reference frame, I can without loss of generality orient the x, z plane so that it contains the static magnetic field $\mathbf{B}_0 = B_{x0}\mathbf{e}_x + B_{z0}\mathbf{e}_z$. Moreover, I will assume that $B_{x0} \geq 0$ and $B_{z0} \geq 0$, and study waves such that the angle ϑ between the direction of wave propagation and $\mathbf{B}_0$ is at most 90°. The standard set of plasma equations includes, per species j, the continuity and momentum equations,

$$\frac{\partial n_j}{\partial t} + \frac{\partial}{\partial x}(n_j v_{jx}) = 0, \tag{1}$$

$$\frac{\partial \mathbf{v}_j}{\partial t} + v_{jx}\frac{\partial \mathbf{v}_j}{\partial x} + \frac{\mathbf{e}_x}{n_j m_j}\frac{\partial p_j}{\partial x} = \frac{q_j}{m_j}(\mathbf{E} + \mathbf{v}_j \times \mathbf{B}), \tag{2}$$

where n_j, p_j and $\mathbf{v}_j$ refer to the number density, scalar pressure and fluid velocity of each plasma species, respectively, with charge q_j and mass m_j, while $\mathbf{E}$ and $\mathbf{B}$ are the electric and (total) magnetic fields, respectively. When I treat electron-ion, electron-positron or pair plasmas, the subscripts $j = e$ will refer to the electrons and $j = i$ to the ions or positrons.

To connect the pressures p_j to the other variables in the system I will introduce polytropic pressure-density relations, with index γ_j, so that $p_j \propto n_j^{\gamma_j}$. This includes the usual adiabatic flow for $\gamma_j = 3$, and by taking the appropriate limits, the special case of isothermal pressures ($\gamma_j = 1$) can also easily be recovered. In many applications, the latter have been described by Boltzmann distributions, obtained by neglecting their inertial effects. As we will see, these restricting regimes are not necessary to derive general results, be it through the reductive perturbation analysis or through the Sagdeev and/or McKenzie method. Specialization to the more common plasma compositions and earlier results from the literature can easily be performed at the last stage.

The system is closed by Maxwell's equations

$$\mathbf{e}_x \times \frac{\partial \mathbf{E}}{\partial x} + \frac{\partial \mathbf{B}}{\partial t} = \mathbf{0}, \tag{3}$$

$$\mathbf{e}_x \times \frac{\partial \mathbf{B}}{\partial x} = \frac{1}{c^2}\frac{\partial \mathbf{E}}{\partial t} + \mu_0 \sum_j n_j q_j \mathbf{v}_j, \tag{4}$$

$$\varepsilon_0 \frac{\partial E_x}{\partial x} = \sum_j n_j q_j, \tag{5}$$

$$\frac{\partial B_x}{\partial x} = 0. \tag{6}$$

The magnetic Gauss's law (6) and the x component of Faraday's law (3) show that $B_x = B_{x0}$ is constant.

3. Reductive perturbation analysis of nonlinear electrostatic modes

3.1. *Reductive perturbation analysis*

3.1.1. *Motivation and linear dispersion*

In this section I will set out the reductive perturbation method to arrive at general evolution equations for electrostatic modes propagating either parallel to an external magnetic field or in unmagnetized multifluid plasmas. These evolution equations express a judicious balance between slow time variations, dispersive effects and nonlinearities that sustain stable solitary waves or double layers. In contrast, for strongly nonlinear phenomena one immediately looks for stationary solutions of the model equations, that can then hopefully be integrated to yield first integrals. The latter allow a reduction of the number of equations that have to be ultimately solved, as will be shown in Sections 4 and 5.

In the context of nonlinear evolution equations and solitary waves, the starting point is now 170 years old, when observations of a single, humped shallow water wave on a Scottish canal forced a rethink of what is understood by wave phenomena. It took some sixty years to derive the appropriate model equation, the KdV equation, the grandfather and archetype of many integrable nonlinear evolution equations. After another seventy years of quasi-dormancy, this part of physics changed dramatically, because it was realized that the KdV equation was equally valid for many other applications, and could be extended in various ways.[1,20]

In particular, the KdV equation and its extensions describe electrostatic longitudinal waves in a wide range of plasma configurations, some with more than one ion species, or with electron species at different temperatures, or a combination of several of these together. Many space plasmas, in the heliosphere or in more distant astrophysical settings, contain more than one ion species, and it is quite remarkable that the KdV and modified KdV (mKdV) equations have shown such a robustness in their applicability.

On top of that, they have generated an intense interest in the properties of integrable equations or systems. The latter have infinitely many constraining first integrals, which ensure the extraordinary stability of solitons that, moreover, have no linear counterpart. These equations also have periodic solutions, called cnoidal waves for the KdV equation, which reduce to the simple harmonic modes known from linear Fourier analysis.[1,20]

For electrostatic plasma modes there are no wave magnetic fields and the electric fields are gradients of the electrostatic potential φ, so that the set of fluid equations (1)–(2) is now complemented by Poisson's equation (5) in the form

$$\varepsilon_0 \frac{\partial^2 \varphi}{\partial x^2} + \sum_j n_j q_j = 0. \tag{7}$$

Before addressing the nonlinear evolution, I briefly discuss linear modes, varying as $\exp[i(kx - \omega t)]$, where k is the wavenumber and ω the angular frequency, for which the dispersion law is

$$\sum_j \frac{\omega_{pj}^2}{\omega^2 - k^2 c_{tj}^2} = 1. \tag{8}$$

Plasma frequencies ω_{pj} are defined through $\omega_{pj}^2 = n_{j0} q_j^2 / \varepsilon_0 m_j$, and thermal velocities c_{tj} through $c_{tj}^2 = \gamma_j p_{j0} / n_{j0} m_j$, corresponding to the general definition of the sound velocity in polytropic (and even in barotropic) fluids.

For small wave numbers the dispersion law (8) can be approximated to lowest order by an acoustic-like dispersion-free propagation along the field, with a dispersive correction of order k^3, i.e.

$$\omega = Vk - \frac{1}{A} k^3 + \dots \tag{9}$$

The phase velocity V in the limit of vanishing wave numbers is determined from

$$D := \sum_j \frac{\omega_{pj}^2}{V^2 - c_{tj}^2} = 0, \tag{10}$$

and the coefficient A is given by

$$A = 2V \sum_j \frac{\omega_{pj}^2}{(V^2 - c_{tj}^2)^2}. \tag{11}$$

As we will see afterwards, $1/A$ is the coefficient of the dispersive term in the KdV equation (21) and in the mKdV equation (23).

If for all species $V > c_{tj}$, i.e. the linear wave velocity V is seen as supersonic by all plasma fluids, then (10) can have no solutions. The same conclusion holds if for all species $V < c_{tj}$, i.e. the linear wave velocity V is now seen as subsonic by all plasma fluids. With respect to V the supersonic species are the cooler ones and the subsonic species the hotter ones. Traditionally, especially in reductive perturbation theory, these concepts have not been used much, so that some care is warranted in the subsequent discussions. In order to obtain a solution to (10), any plasma should have at least one supersonic and one subsonic species. For the simple example of a plasma with one electron and one ion species, the existence of acoustic modes needs an ordering like $c_{ti} < V < c_{te}$ (or very exceptionally the other way around), and then (10) defines the ion-acoustic velocity c_{ia} through

$$V^2 = c_{ia}^2 := \frac{m_i c_{ti}^2 + m_e c_{te}^2}{m_i + m_e}. \tag{12}$$

Under the plausible assumptions that $m_i c_{ti}^2 \ll m_e c_{te}^2 = \kappa_B T_e$ and $m_e \ll m_i$, (12) gives the typical form $c_{ia}^2 \simeq \kappa_B T_e / m_i$ for a wave mode sustained by electron pressure and ion inertia. Here κ_B is Boltzmann's constant and T_j refers to the species' temperatures.

3.1.2. *Coordinate stretching and variable scaling*

The properties of the linear dispersion law (9) for small wave numbers k, translates for the phase argument into

$$kx - \omega t = k(x - Vt) + \frac{1}{A} k^3 t + \cdots \tag{13}$$

Hence we arrive in a natural way at the standard KdV stretching of the independent variables,

$$\xi = \varepsilon^{1/2}(x - Vt), \qquad \tau = \varepsilon^{3/2} t. \tag{14}$$

Basically, V is the linear phase velocity in the limit $k \to 0$, and will later be shown in a natural way to obey (10). I have represented k by $\sqrt{\varepsilon}$, as commonly encountered in the ordinary KdV stretching, but such a choice has evident repercussions on the expansions for the dependent variables.

To be fully systematic and allow for the derivations of both the KdV and mKdV equations in one coherent treatment, I adopt a general expansion for all dependent variables in powers of $\varepsilon^{1/2}$, of the form

$$\begin{aligned} n_j &= n_{j0} + \varepsilon^{1/2} n_{j1} + \varepsilon n_{j2} + \varepsilon^{3/2} n_{j3} + \varepsilon^2 n_{j4} + \dots \\ p_j &= p_{j0} + \varepsilon^{1/2} p_{j1} + \varepsilon p_{j2} + \varepsilon^{3/2} p_{j3} + \varepsilon^2 p_{j4} + \dots \\ v_j &= \varepsilon^{1/2} v_{j1} + \varepsilon v_{j2} + \varepsilon^{3/2} v_{j3} + \varepsilon^2 v_{j4} + \dots \\ \varphi &= \varepsilon^{1/2} \varphi_1 + \varepsilon \varphi_2 + \varepsilon^{3/2} \varphi_3 + \varepsilon^2 \varphi_4 + \dots \end{aligned} \tag{15}$$

Only the densities and pressures have nonzero equilibrium values, n_{j0} and p_{j0}, respectively, and $v_j = v_{jx}$ is shorthand for the only remaining component of the fluid velocities.

Inserting the stretching (14) and the expansions (15) into the basic equations (1)–(2) and (7) gives a series of equations, upon separating out the different orders in $\varepsilon^{1/2}$. When possible, these equations will be integrated with one-sided boundary conditions suitable for the type of solitary wave structures we are interested in,

$$n_j \to n_{j0} \quad \text{and} \quad n_{ji} \to 0, \quad \frac{\partial n_{ji}}{\partial \xi} \to 0 \quad (i = 1, 2, \dots) \tag{16}$$

when $\xi \to \infty$, with similar conditions for the other dependent variables. This procedure is standard and the intermediate steps are well documented in the literature for some of the simpler cases,[13,56] so that there is no need to spell out all the intervening results. The computations here are analogous and straightforward, and I just mention some of the intermediate results,

$$\begin{aligned} n_{j1} &= \frac{n_{j0} q_j}{m_j (V^2 - c_{tj}^2)} \varphi_1, \\ n_{j2} &= \frac{n_{j0} q_j}{m_j (V^2 - c_{tj}^2)} \varphi_2 + \frac{n_{j0} q_j^2 [3V^2 + (\gamma_j - 2) c_{tj}^2]}{2 m_j^2 (V^2 - c_{tj}^2)^3} \varphi_1^2, \\ v_{j1} &= \frac{V q_j}{m_j (V^2 - c_{tj}^2)} \varphi_1, \\ v_{j2} &= \frac{V q_j}{m_j (V^2 - c_{tj}^2)} \varphi_2 + \frac{V q_j^2 (V^2 + \gamma_j c_{tj}^2)}{2 m_j^2 (V^2 - c_{tj}^2)^3} \varphi_1^2, \end{aligned} \tag{17}$$

but will highlight what is obtained from Poisson's equation (7) to various orders in $\varepsilon^{1/2}$.

To order ε^0 there needs to be overall charge neutrality in equilibrium,

$$\sum_j n_{j0} q_j = 0, \tag{18}$$

whereas to order $\varepsilon^{1/2}$ we find $D\varphi_1 = 0$ and the dispersion law (10) determining V is recovered. Next, to order ε, a natural bifurcation point is reached, namely

$$D\varphi_2 + \tfrac{1}{2}B\varphi_1^2 = 0. \tag{19}$$

The term in φ_2 drops out, on account of the linear dispersion law (10), and B is given by

$$B = \sum_j \frac{\omega_{pj}^2 q_j [3V^2 + (\gamma_j - 2)c_{tj}^2]}{m_j(V^2 - c_{tj}^2)^3}. \tag{20}$$

It will be shown that this is the coefficient of the quadratic nonlinearity in a KdV equation.

Although for electron-proton plasmas B can be proved to be strictly positive, for certain more complicated plasma compositions B can become negative,[12,48] and there exist critical densities obeying $B = 0$. Because $\frac{1}{2}B\varphi_1^2 = 0$, the bifurcation means that either the plasma composition is very special, so that $B = 0$, to be discussed further on, or that for generic plasmas $B \neq 0$. Consequently, for $B \neq 0$, $\varphi_1 = 0$ and all variables with subscript 1 then vanish from the expansions (15).

3.1.3. *Generic nonlinear modes: KdV equation*

I first proceed with the generic case where $B \neq 0$, and hence $\varphi_1 = 0$, indicating that there are no terms of order $\varepsilon^{1/2}$ in the expansions (15) of the dependent variables. This is the normal KdV ordering, except that the lowest-order variables now carry a subscript 2 (instead of 1 as usual in the literature) because of our treatment of the KdV and mKdV equations in one coherent derivation.

It turns out that to order $\varepsilon^{3/2}$ Poisson's equation (7) merely duplicates for φ_3 what we learned to order ε for φ_2. This shows that possible intermediate terms with an odd index either vanish from (15) or can be renormalized away.

On the other hand, new information is obtained to order ε^2, and this gives the KdV equation[1,20,52,60]

$$\frac{\partial \varphi_2}{\partial \tau} + \frac{1}{A}\frac{\partial^3 \varphi_2}{\partial \xi^3} + \frac{B}{A}\varphi_2 \frac{\partial \varphi_2}{\partial \xi} = 0. \tag{21}$$

In principle, A and B can change sign. Because A comes from the slow time derivative, a negative sign of A could be absorbed by a time reversal, giving no real new physical insight. However, in plasmas as discussed here, without beam effects, A is always positive. It is the possible sign change of B that leads to physically different situations. Indeed, if critical densities can be exceeded so that B is negative, the solitons will have a negative potential. The transition from positive to negative potentials, of course, occurs at $B = 0$, except that in the immediate vicinity thereof the expansions break down and have to be reconsidered.

To use the more common terminology of "compressive" or "rarefactive" for the solitons is not unambiguous, as we shall see. I note from the expression of n_{j2} in (17), at $\varphi_1 = 0$, that when $\varphi_2 > 0$, the supersonic ($V > c_{tj}$) positive species (think of cool positive ions) are compressed (rarefied when $\varphi_2 < 0$). Conversely, subsonic ($V < c_{tj}$) negative species (think of hot electrons) are also compressed, and in simple plasma models the solitons are thus compressive (rarefactive when $\varphi_2 < 0$). When more species are present, however, such as supersonic species of different signs, the terms "compressive" and "rarefactive" have to be specified for each species.

The standard 1-soliton solution[1,20] of (21) is

$$\varphi_2 = \frac{3MA}{B}\,\mathrm{sech}^2\left[\tfrac{1}{2}\mu(\xi - Mt)\right], \tag{22}$$

where M is the soliton velocity and $\mu = \sqrt{MA}$ a measure of the inverse width. Since we note that φ_2 has the sign of B, and $B > 0$ in a simple electron-proton plasma (prove this!), the upshot is that ion-acoustic solitons are then compressive in both components.

3.1.4. *Nonlinear modes at critical densities: mKdV equation*

Now I return to the bifurcation point encountered before and assume that the plasma is at critical densities, defined here by putting $B = 0$, which implies that we can continue to work with φ_1. The mKdV equation then follows from (7) to order $\varepsilon^{3/2}$ as

$$\frac{\partial \varphi_1}{\partial \tau} + \frac{1}{A}\frac{\partial^3 \varphi_1}{\partial \xi^3} + \frac{C}{A}\varphi_1^2 \frac{\partial \varphi_1}{\partial \xi} = 0. \tag{23}$$

The coefficient of the new, cubic term involves

$$C = \sum_j \frac{\omega_{pj}^2 q_j^2 [15V^4 + (\gamma_j^2 + 13\gamma_j - 18)V^2 c_{tj}^2 + (2\gamma_j^2 - 7\gamma_j + 6)c_{tj}^4]}{2m_j^2 (V^2 - c_{tj}^2)^5}. \tag{24}$$

A change of sign of C (and hence the transition through $C = 0$ under the simultaneous fulfilment of $B = 0$) might be possible, depending on the relative balance between the contributions of the supersonic and the subsonic species, but is not simple to achieve. What is, unfortunately, sometimes overlooked is that the appropriate KdV equation is the valid paradigm when $B \neq 0$, and the question whether C might vanish (for $B \neq 0$) is not of importance.[56]

The 1-soliton solution[1,20] of (23) is

$$\varphi_1 = \pm\sqrt{\frac{6MA}{C}}\,\mathrm{sech}\mu(\xi - Mt), \tag{25}$$

where the parameters have been defined earlier.

In the vicinity of critical densities double layers become possible. For these to occur, one would need that $B\phi_1^2$ become small, of the order of $C\phi_1^3$, so that both quadratic and cubic nonlinearities can be present together in one evolution equation, the mixed KdV equation,

$$\frac{\partial\varphi_1}{\partial\tau} + \frac{1}{A}\frac{\partial^3\varphi_1}{\partial\xi^3} + \frac{B}{A}\varphi_1\frac{\partial\varphi_1}{\partial\xi} + \frac{C}{A}\varphi_1^2\frac{\partial\varphi_1}{\partial\xi} = 0. \tag{26}$$

This has general travelling solitary wave solutions of the form

$$\varphi_1 = \frac{6MA}{B}\cdot\frac{1}{1 \pm \sqrt{1 + \frac{6MAC}{B^2}}\cosh\mu(\xi - Mt)}. \tag{27}$$

Weak double layers are possible if $6MAC + B^2 = 0$, and are of the form

$$\varphi_1 = \frac{3MA}{B}\left[1 \pm \tanh\tfrac{1}{2}\mu(\xi - Mt)\right]. \tag{28}$$

However, the existence of weak double layers involves some tricky discussions about the validity of the expansions assumed in the singular perturbation scheme[18] and will not be pursued here.

3.2. *Dust-acoustic solitons*

In recent years attention has turned to dusty plasmas, where, besides the traditional electrons and ions, one also encounters heavier charged dust grains of different kinds. These mixtures of usual plasmas (electrons, ions) plus dust grains, charged in plasma and radiative environments, occur in the heliosphere, e.g. in noctilucent clouds (in the Earth's polar summer mesosphere), in planetary rings (as spokes and braids) and presumably near comet nuclei and tails. Larger dusty plasmas in molecular interstellar clouds could require also self-gravitation to be taken into account. Other applications range from astrophysics to technology (plasma etching and deposition).

The description of wave processes in dusty plasmas generates interesting difficulties and complications compared to standard plasmas. The typical micron-sized grains of interest are much heavier than protons or other ions, and can collect very many elementary charges, giving frequencies and scales totally outside the usual domain treated by standard plasma textbooks. Contrary to controlled experiments with monodisperse dust, heliospheric and astrophysical dust comes in a range of sizes, and thus masses and charges. As the charges depend on the local plasma potentials, which can be variable, additional complications occur, e.g. with respect to charge fluctuation damping. Further details can be found in review papers and monographs devoted to this challenging subject.[19,33,42,43,51,52]

In view of the space and time scales associated with the charged dust components, which differ vastly from those related to the usual ions and electrons, the simplest modelling has involved treating the charged dust as monodisperse, heavy negative ions, in the presence of hotter electrons and (positive) ions. While in the normal ion-acoustic regime the charged dust can almost be considered as a neutralizing but immobile background, so that the charge imbalance between the electrons and the protons is the main change compared to what happens in normal plasmas, at the lowest end of the frequency spectrum the dust motion has to be taken into account. Here the prime example is the dust-acoustic mode, well studied both in theory[36,49] and in the laboratory.[4]

The simplest model at this very low frequency end of the spectrum is to describe the electrons and ions by Boltzmann distributions, and treat the dust as cold, in view of its great inertia. As should be well known, Boltzmann distributions neglect the species' inertia and treat them as isothermal. This leads to the simplifications $V^2 \ll c_{tj}^2$ and $\gamma_j = 1$ for these hot species, whereas for the cold dust $c_{tj} = 0$. In that case the dispersion law (10) becomes

$$V^2 = \omega_{pd}^2 \lambda_D^2 := c_{da}^2, \tag{29}$$

where the dust-acoustic velocity $c_{da} = \omega_{pd}\lambda_D$ has been introduced. The global Debye length λ_D involving the Boltzmann species is determined by $\lambda_D^{-2} = \lambda_{De}^{-2} + \lambda_{Di}^{-2}$, with the species' Debye lengths given by $\lambda_{Dj} = c_{tj}/\omega_{pj}$. The expression for B thus becomes

$$B = \frac{3q_d}{m_d c_{da}^2 \lambda_D^2} + \frac{e}{\lambda_{Di}^2 \kappa T_i} - \frac{e}{\lambda_{De}^2 \kappa T_e}, \tag{30}$$

and can be rewritten as

$$B = \frac{\varepsilon_0}{n_{d0} q_d} \left[\frac{2}{\lambda_D^4} + \frac{1}{n_{e0} n_{i0}} \left(\frac{n_{e0}}{\lambda_{Di}^2} + \frac{n_{i0}}{\lambda_{De}^2} \right)^2 \right]. \tag{31}$$

This is always nonzero, so that there are no critical densities and the governing paradigm is the KdV equation (21). Moreover, since the one-soliton solution (22) indicates that φ_2 has the sign of B, which itself has the sign of q_d, (17) always gives a compression for the supersonic dust. For negative (positive) dust the subsonic ions are compressed (rarefied) and the subsonic electrons are rarefied (compressed). The solitons hence have a positive or negative potential, depending on whether the dust is positively or negatively charged, respectively. Other labels are ambiguous.

3.3. *Outlook*

To conclude this section, we should mention that the reductive perturbation analysis of nonlinear acoustic-like modes has proven to be a very systematic and versatile technique, with a wide range of applicability and with quasi-algorithmic recipes. Although the mathematical exposition has been given in detail for electrostatic multifluid solitons, other applications run along similar lines. Examples include parallel propagating electromagnetic modes in magnetized plasmas or oblique modes that mix transverse and longitudinal aspects, and this again for many different plasma compositions. Here the nonlinear evolution equations are of the Derivative Nonlinear Schrödinger (DNLS) type, and related model equations like the vector mKdV and the Nonlinear Schrödinger (NLS) equation, although the latter is less general.

Because of its algorithmic character, the reductive perturbation method can easily be implemented via computer algebra software. Finally, generalizations to non-acoustic-like modes (having $\omega \neq 0$ in the limit $k \to 0$) are still controversial and in their infancy, in contrast to the now mature character of the technique for acoustic-like waves.

4. Electrostatic solitons of larger amplitudes

4.1. *Fluid-dynamic treatment of stationary solitary modes*

4.1.1. *McKenzie approach and Bernoulli integrals*

Having detailed in the previous section some aspects of the reductive perturbation techniques to deal with (electrostatic) weakly nonlinear modes, I will in this section discuss solitary waves of larger amplitude, where iterations and expansions are no longer adequate. In what follows, I will use the fluid-dynamic approach,[29,30,32,58] where nonlinear waves are described in a reference frame in which they appear steady ($\partial/\partial t = 0$), in contrast to the Sagdeev pseudopotential theory[37,38] where the nonlinear waves are viewed in an inertial (laboratory) frame. For electrostatic modes the Sagdeev pseudopotential method typically leads, after some manipulation, to an energy-type integral of the form

$$\frac{1}{2}\left(\frac{d\varphi}{dx}\right)^2 + \Psi(\varphi, V) = 0, \tag{32}$$

in terms of the pseudopotential Ψ and the wave electrostatic potential φ, the structure velocity V appearing as a parameter. Such energy-type integrals can be analyzed as in classical mechanics, with φ and x playing the role of a coordinate and of time, respectively, for a pseudoparticle of unit mass. Typical shapes are shown in Figures 4–6 of Section 5, though in terms of different variables.

In order to allow the pseudoparticle to move, an interval in φ is needed where the pseudopotential is negative, leading to typical conditions of the form

$$\begin{aligned} &\Psi(0,V) = \frac{\partial \Psi}{\partial \varphi}(0,V) = 0 \quad \text{and} \quad \frac{\partial^2 \Psi}{\partial \varphi^2}(0,V) < 0 \quad \text{at} \quad \phi = 0, \\ &\Psi(\varphi_m, V) = 0 \quad \text{and} \quad \frac{\partial \Psi}{\partial \varphi}(\varphi_m, V) \gtreqless 0 \quad \text{at} \quad \varphi_m \gtrless 0. \end{aligned} \tag{33}$$

Hence hill- or dip-like structures can exist, when $\partial\Psi/\partial\varphi(\varphi_m, V) \gtrless 0$, respectively, with an extremum at $\varphi_m \gtrless 0$, or even double layers when $\partial\Psi/\partial\varphi(\varphi_m, V) = 0$ and the pseudoparticle transits from a value 0 to $\varphi_m \gtrless 0$. The drawback of the usual pseudopotential method is that the intermediate expressions quickly become complicated when one tries to express all dependent variables in terms of φ, so that the final analysis of the solutions has to be done numerically. Often, physical insight is thus lost on why the parameter ranges that give rise to the existence of solitons are what they are.

The fluid-dynamic point of view, pioneered by McKenzie, on the other hand, uses several first integrals to constrain the existence of solitary waves and the parameter regimes for which they occur. For the clarity of this exposition I recall some of the intermediate steps from earlier work.[8,58] Since in the wave frame there are no time derivatives, the basic equations can easily be integrated, starting with the continuity equations (1) which express conservation of (mass) flux

$$n_j v_j = n_{j0} V. \tag{34}$$

In the case of one-dimensional propagation of electrostatic modes only the x component of $\mathbf{E}$ remains, $E = -d\varphi/dx$, and of Maxwell's equations (3)–(6) only Poisson's equation (5) needs to be considered.

For polytropic pressures, $p_j \propto n_j^{\gamma_j} \propto v_j^{-\gamma_j}$, the last proportionality coming from the conservation of (mass) flux (34). The equations of motion (2) thus become

$$u_j\left(1 - \frac{1}{u_j^{\gamma_j+1}M_j^2}\right)\frac{du_j}{dx} = -\frac{q_j}{m_jV^2}\frac{d\varphi}{dx}, \tag{35}$$

where (34) has been used and dimensionless velocities $u_j = v_j/V$ and Mach numbers $M_j = V/c_{tj}$ have been introduced. In the gas-dynamic description that I am following, the sonic points, defined as

$$u_{js} = M_j^{-2/(\gamma_j+1)}, \tag{36}$$

correspond to where the species' local flow speeds match their local acoustic speeds $(\gamma_j p_j/n_j m_j)^{1/2}$. Hence (35) indicates that these sonic points play a crucial role in limiting the amplitude of the wave, by choking the flow $(du_j/dx \to +\infty)$ before a possible equilibrium point can be reached from the initial values $u_j = 1$ (or $v_j = V$), the beginning of the wave.

After integration of the equations of motion (35) a Bernoulli type integral is obtained per species,

$$E_j := \frac{1}{2}\left(u_j^2 - 1\right) + \frac{1}{(\gamma_j - 1)M_j^2}\left(\frac{1}{u_j^{\gamma_j-1}} - 1\right) = -\frac{q_j}{m_jV^2}\varphi. \tag{37}$$

Here the first term refers to changes in kinetic energy, whereas the second term gives changes in enthalpy. This is clear from the alternative way in which this expression can be written,

$$\frac{1}{2}m_jv_j^2 + \frac{\gamma_j}{\gamma_j - 1}\frac{p_j}{n_j} + q_j\varphi = \frac{1}{2}m_jV^2 + \frac{\gamma_j}{\gamma_j - 1}\frac{p_{j0}}{n_{j0}}. \tag{38}$$

Since a consequence of (37) is that all m_jE_j/q_j are equal, the different species' velocities or the electrostatic potential can be eliminated as functions of one of them. When I now compute

$$\frac{dE_j}{du_j} = u_j\left(1 - \frac{1}{u_j^{\gamma_j+1}M_j^2}\right), \tag{39}$$

I first of all note that at the sonic point given by (36) E_j reaches a minimum, negative value. Initially E_j is zero, and its derivative is then given by

$$\left.\frac{dE_j}{du_j}\right|_{u_j=1} = 1 - \frac{1}{M_j^2}. \tag{40}$$

Thus E_j has a slope at the initial point which is positive if the incoming flow is supersonic ($M_j > 1$) and negative for subsonic flows ($M_j < 1$). If positive particles are discussed first, we learn from the behaviour of the Bernoulli relations that positive particles are decelerated ($u_j < 1$) and driven towards their sonic point in a potential hill ($\varphi > 0$) if the flow is supersonic ($M_j > 1$), whereas subsonic, hotter positive species (with $M_j < 1$) are accelerated ($u_j > 1$) while being driven towards their sonic point. Because of mass conservation (34), this corresponds for supersonic, cooler positive species to a compression, and for subsonic, hotter positive species to a rarefaction.

These conclusions are reversed for negative particles, so that supersonic, cooler negative species are rarefied, whereas subsonic, hotter negative species are compressed, while being driven away from their sonic points. In a potential dip ($\varphi < 0$), all these conclusions are reversed again. This is illustrated schematically in Figure 1. Consequently, if a given plasma species (with index ℓ) is to be driven towards its sonic point, the potential has to obey $q_\ell \varphi > 0$. Thus for a positive species a potential hill ($\varphi > 0$) is needed,

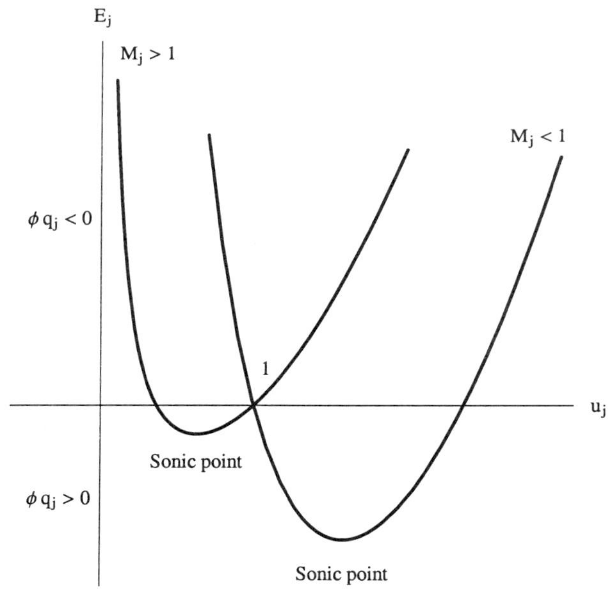

Fig. 1. Schematic representation of Bernoulli integrals for supersonic (cool) and subsonic (hot) species, having a minimum at the respective sonic points.

whereas a negatively charged species would require a potential dip or valley ($\varphi < 0$). This will be of importance when discussing specific solitary waves.

Owing to charge neutrality in the undisturbed state, we also see from (37) that

$$\sum_j n_{j0} m_j E_j = -\frac{\varphi}{V^2} \sum_j n_{j0} q_j = 0, \tag{41}$$

a result that will be useful further on.

4.1.2. *Global structure equation*

Finally, the Poisson equation can be integrated after multiplying it by $d\varphi/dx$, yielding a global structure equation,

$$R := \sum_j n_{j0} m_j P_j = \frac{\varepsilon_0}{2V^2} \left(\frac{d\varphi}{dx} \right)^2 , \tag{42}$$

where the (normalized) particle pressure functions P_j for each species are given by

$$P_j := u_j - 1 + \frac{1}{\gamma_j M_j^2} \left(\frac{1}{u_j^{\gamma_j}} - 1 \right) , \tag{43}$$

and R is the structure function, effectively the negative, up to a multiplicative constant, of the Sagdeev pseudopotential [cf. (32)]. This structure function R is related through $E = -d\varphi/dx$ to the wave electric field. The first term refers to changes in dynamic pressure, and the second term to changes in thermal pressure. Again, (42) obviously is

$$\sum_j V \left(n_j m_j v_j - n_{j0} m_j V \right) + \sum_j \left(p_j - p_{j0} \right) = \frac{\varepsilon_0}{2} \left(\frac{d\varphi}{dx} \right)^2 . \tag{44}$$

Subtracting (41) from (43) gives an alternative expression for the structure equation, namely

$$\sum_j n_{j0} m_j \left(P_j - E_j \right) = \frac{\varepsilon_0}{2V^2} \left(\frac{d\varphi}{dx} \right)^2 , \tag{45}$$

which will allow an easier interpretation of the conditions needed to guarantee the existence of solitary wave solutions. As pointed out already, all $m_j E_j / q_j$ are equal and the different species' velocities can be eliminated as functions of one of them, so that (45) can in principle be reduced to a first-order differential equation in one of the velocities, by using (35) to express also $d\varphi/dx$ in that particular velocity.

Before going to the existence conditions for solitary waves, I briefly digress on how heavier species might be eliminated from the description. Heavy species are in the traditional Sagdeev approach supposed to contribute only charge but no pressure to the wave in the inertial frame and hence have $v_j = V$ or $u_j = 1$ in the wave frame. Contrary to the classic picture, I assume that such heavy species are not quite immobile in the inertial frame, but move very slowly with respect to it,[31] so that their densities also do not change much. Given that for those species the normalized velocity u_j stays close to 1, the pressure and Bernoulli functions may be approximated by the linear term in their Taylor expansions, yielding

$$P_j \simeq (u_j - 1)\left(1 - \frac{1}{M_j^2}\right) \simeq E_j. \tag{46}$$

Consequently, when I split the sums over species into those over heavy and over light elements, we find that (45) is rewritten in terms of the parameters for the lighter species only,

$$\sum_{\text{light}} n_{j0} m_j (P_j - E_j) = \frac{\varepsilon_0}{2V^2}\left(\frac{d\varphi}{dx}\right)^2. \tag{47}$$

4.1.3. *Existence conditions for solitary waves*

Now I come to the conditions for the existence of a solitary wave structure. First, near the initial point I put $u_j = 1 + \Delta_j$ and suppose that $|\Delta_j| \ll 1$, so that with the help of the linearized form of (45) and of the Bernoulli relations (37),

$$\frac{m_j \Delta_j}{q_j}\left(1 - \frac{1}{M_j^2}\right) = \frac{m_\ell \Delta_\ell}{q_\ell}\left(1 - \frac{1}{M_\ell^2}\right), \tag{48}$$

I can linearize (35) to

$$\frac{d\Delta_\ell}{dx} = \pm\,\Delta_\ell \left[\sum_j \frac{\omega_{pj}^2}{c_{tj}^2 - V^2}\right]^{1/2}. \tag{49}$$

The various plasma frequencies ω_{pj} are defined through $\omega_{pj}^2 = n_{j0} q_j^2 / \varepsilon_0 m_j$, and the sum over j again runs over all species.

In these results the index ℓ can refer to any of the species and (49) possesses real, exponential solutions of the form $\Delta_\ell \propto \exp(\pm\kappa x)$, provided the exponent κ is real, or in other words,

$$\sum_j \frac{\omega_{pj}^2}{c_{tj}^2 - V^2} > 0. \tag{50}$$

The soliton condition (50) will determine the different regimes for the speed V of the nonlinear wave, from which it follows that not all species can be supersonic ($c_{tj}^2 < V^2$). As we will later also see, not all species can be subsonic ($V^2 < c_{tj}^2$) if we want a proper solitary wave to exist.

For the second condition I start by noting that R and E have a double root at the initial point, since $dE/dx = -d^2\varphi/dx^2$ also vanishes initially, as follows from the Poisson equation (5) at equilibrium. Moreover, the location of the zeros of R is assisted by computing with the help of (35) its derivative in the form

$$\frac{dR}{du_\ell} = \frac{m_\ell u_\ell}{q_\ell}\left(1 - \frac{1}{u_\ell^{\gamma_\ell+1} M_\ell^2}\right) \sum_j n_j q_j, \tag{51}$$

where it is again understood that all velocities (or densities) have formally been expressed as functions of a chosen u_ℓ with the help of mass conservation (34) and the Bernoulli relations (37). The zeros of R are intertwined with the zeros of the derivative of R, the latter occurring at the charge neutral points, defined as $\sum n_j q_j = \sum n_{j0} q_j = 0$, and at the sonic point in question. A solitary wave is only possible if there is a zero of R located between a charge neutral point (outside the initial point) and one of the sonic or other limiting points, like infinite compression or total rarefaction, or if a double layer is met.

From the Bernoulli integrals (37) it follows that if all species are subsonic ($M_j < 1$), positive species in a potential hill ($\varphi > 0$) are accelerated and rarefied, whereas negative species are decelerated and compressed, so that a charge neutral point outside the initial point cannot be reached before one of the positive species reaches its sonic point, where its flow is choked. In a potential dip the rarefactions and compressions are interchanged, but the conclusion about the unattainability of the charge neutral point remains. Analogous reasoning also applies to the case where all species are supersonic ($M_j > 1$), giving an alternative explanation to the one following from discussing the soliton condition (50). Thus for the proper existence of a solitary wave at least one supersonic and one subsonic species is necessary. Note that such conditions in themselves are not sufficient!

4.2. *Electron-acoustic modes in two-electron species plasmas*

4.2.1. *Magnetospheric electron-acoustic modes*

To illustrate how these general principles work, I address in detail a specific application. Space observations have revealed in several near-Earth environments the presence of positive-potential, large-amplitude electrostatic structures, associated with high-frequency disturbances, and indicative of electron dynamics. They have been found in the Earth's magnetotail by the GEOTAIL satellite,[28] in the polar magnetosphere by the POLAR mission[9,16] and in the mid-altitude auroral zone by the FAST spacecraft.[15,35]

Because of the need to include the electron dynamics in the description, the most obvious model was to seek explanations in terms of solitary electron-acoustic waves, which may exist in plasmas consisting of positive ions and two distinct electron species, one cool and one hot. At high frequencies (but below the plasma frequency) the ion dynamics plays no essential role.[17,61] Kinetic studies[24] have shown that a characteristic of electron-acoustic waves is strong Landau damping at long wavelengths ($k \to 0$) and that there is also a τ-dependent lower cut-off in the fractional cool electron density, f, that can sustain weakly-damped electron-acoustic waves. Here $\tau = c_{tc}^2/c_{th}^2 < 1$ is the square of the ratio of the cool to hot electron thermal velocities, c_{tc} and c_{th}, respectively, and $f = n_{c0}/n_{i0}$ is the fraction of the cool electrons with equilibrium density n_{c0} in terms of the total electron or ion equilibrium density $n_{i0} = n_{c0} + n_{h0}$, n_{h0} being the hot electron equilibrium density.

The weak point of trying to explain the said space observations as being solitary electron-acoustic structures, is that in this model only potential dip solitary waves could be generated.[14,27] Moreover, arguments have also been presented to show that weak electron-acoustic double layers cannot exist.[25] Because the observations point to positive potential structures, the model of a two-electron plasma seemed inadequate, and several attempts have been made to reconcile the space observations with plausible theoretical explanations. Hence, the positive potential observations have been ascribed to localized electrostatic potential structures whose existence is sustained by a trapped electron population, using a Bernstein-Greene-Kruskal technique,[21,28] or explained by the introduction of an additional electron beam into the system, in what is then a three-electron component plasma.[5,26]

4.2.2. *Revisiting large-amplitude electron-acoustic modes*

In a recent paper[8] it was shown that in a simple two-electron species plasma both potential hill solitons and potential dip and hill double layers are possible, besides the known potential dip solitons. This has been achieved without the need for an additional electron beam, but necessitates taking hot electron inertia into account. I will recall in the following review some of the key results of that study[8] and refer to it for further, more technical details and intermediate derivations.

The electron Bernoulli invariants (37) can now be written in terms of a dimensionless electrostatic potential $\psi = e\varphi/mV^2$ as $E_c = E_h = \psi$. The parameter range for the occurrence of solitons is limited by the sonic points, by the total compression or rarefaction ($n_j \to +\infty$ or $n_j \to 0$) of one of the species, or by double layer conditions. Double layers are characterized by the coincidence of a zero of the structure function R with a zero of its derivative, i.e. with a charge neutral point. It is indeed well-known from Sagdeev potential calculations that double layers may represent limiting values for a region in parameter space in which solitons may occur.[3] Use now the Bernoulli relation $E_c = E_h$ to express u_c in terms of u_h, and the hot electron equation of motion to rewrite the simplified structure function equation (47) as

$$R(u_h) := fP_c + (1-f)P_h - E_h = \frac{1}{2}u_h^2\left(1 - \frac{1}{u_h^{\gamma_h+1}M^2}\right)^2\left(\frac{du_h}{d\xi}\right)^2. \quad (52)$$

Here $\xi = x\omega_{pe}/V$ is a dimensionless coordinate, ω_{pe} is the total electron plasma frequency and $M = M_h$ is the hot electron Mach number. Through the use of (47) rather than (45), the neutralizing ion background has been taken into account. In the undisturbed plasma $R(1) = dR/d\psi(1) = 0$ and the soliton condition is

$$\frac{d^2R}{d\xi^2}(1) = \frac{f}{\tau/M^2 - 1} + \frac{1-f}{1/M^2 - 1} > 0, \quad (53)$$

which limits the allowable M to the range

$$f + (1-f)\tau < M^2 \ (< 1) \qquad \text{or} \qquad c_{ea}^2 < V^2 \ (< c_{th}^2), \quad (54)$$

where $c_{ea}^2 = c_{th}^2[f + (1-f)\tau]$ defines the electron-acoustic velocity c_{ea}. This represents the usual soliton requirement of a 'super-acoustic' structure speed.

As an important aside, many traditional treatments of this and related problems describe the hot electrons by Boltzmann distributions. This approach neglects their inertial effects, and is equivalent to $M^2 \ll 1$. The soliton condition is then modified to

$$\frac{f}{1-f} + \tau < M^2 \ (\ll 1), \tag{55}$$

and consistency requires that $f \ll 1$, $1 - f \simeq 1$ and $\tau \ll 1$, conditions that severely constrain the allowable parameter space, as will be pointed out later.

4.2.3. *Results and discussion*

For analytical tractability I consider $\tau = 0$, a simplification which neglects the cool electron thermal pressure and enthalpy, compared to their dynamic pressure and kinetic energy, respectively, and shifts their sonic point to zero. A further argument for so doing is that kinetic theory studies of the linear electron-acoustic wave[24] have shown that there is strong Landau damping unless $\tau < 0.1$. The step to $\tau \to 0$ is thus highly unlikely to give rise to a significantly different physical behaviour and reduces the definition of the electron-acoustic velocity to $c_{ea}^2 = f c_{th}^2$.

Before tackling the general discussion, I look for weakly nonlinear solutions, by expanding (52) (written for $\tau = 0$) around $u_h = 1$ and using $M^2 = f + \delta^2$. This gives

$$\frac{1}{2}\left(\frac{du_h}{d\xi}\right)^2 = \frac{\delta^2}{2(1-f)}(u_h - 1)^2 + \frac{(\gamma+4)f - 3}{6(1-f)}(u_h - 1)^3, \tag{56}$$

with a solitary wave solution,

$$u_h = 1 + \frac{3\delta^2}{3 - (\gamma+4)f} \operatorname{sech}^2\left(\frac{\delta}{2\sqrt{1-f}}\xi\right), \tag{57}$$

which is at the same time the stationary solution of the corresponding KdV equation, obtainable from the reductive perturbation approach of electron-acoustic modes in a two-electron-temperature plasma. It is seen from Figure 1 that potential hill solitons are possible above $f_c = 3/(\gamma + 4)$, a critical cold electron fraction which is not near $f \ll 1$, except for unrealistically large $\gamma = \gamma_h$. This explains why the potential hill solutions cannot be found when the hot electrons are Boltzmann distributed.

The large amplitude treatment proper starts by noting that the existence domain for solitons in the parameter space of M^2 and f is defined, at a given f, by a lower bound on M^2, expressing the soliton condition, $f < M^2$, and by an upper bound coming from one of the limits on the soliton amplitude, imposed by one of the sonic points or by the occurrence of a double layer. At the cold electron sonic point $u_{cs} = 0$ the cold electron flow is choked and u_c as a function of u_h, obtained from the relation $E_c = E_h$ at $\tau = 0$ or explicitly

$$u_c^2 = u_h^2 + \frac{2}{(\gamma-1)M^2}\left(\frac{1}{u_h^{\gamma-1}} - 1\right), \tag{58}$$

becomes imaginary, and hence $R(u_h)$ and $(du_h/d\xi)^2$ are complex. On the other hand, at the hot electron sonic point $u_{hs} = M^{-2/(\gamma+1)}$ the hot electron flow is choked, the structure function $R(u_h)$ has an extremum and $du_h/d\xi$ has an asymptote.

Although it is not possible in general to find an analytic expression for u_{hsc}, the hot species velocity corresponding to the cold electron sonic point, one can decide which sonic point is reached first. It is easily seen that u_{csh}, the cold species velocity corresponding to the hot electron sonic point, can be evaluated from (58) as

$$u_{csh}^2 = \frac{\gamma+1}{\gamma-1}M^{-4/(\gamma+1)} - \frac{2}{\gamma-1}M^{-2}. \tag{59}$$

As $u_c = 1$ initially, it follows (for all f) that the cold electron sonic point ($u_{cs} = 0$) will be reached before the hot electron sonic point, unless $u_{csh}^2 > 0$, or equivalently,

$$M^2 > \left(\frac{2}{\gamma+1}\right)^{(\gamma+1)/(\gamma-1)} := M_A^2. \tag{60}$$

Hence, for the smallest f (and corresponding M^2) it is the cool electron sonic point limitation which is encountered first.

I have plotted in Figure 2 the existence domains in parameter space for $\gamma = 2$.[8] Although $\gamma = 2$ is not a magical number, the analytic results are simpler, lead to expressions that are easier to interpret and can guide us for other values of γ. For the smallest f, the soliton amplitudes are limited by the cold electron sonic point, and at a given f the maximal values of M^2 for which solitons can exist are on the curve OA. Here the point A is defined as that position in the parameter space of M^2 and f at which the amplitude limitation passes from the cold electron sonic point to the hot electron sonic point, with corresponding abscissa f_A and ordinate M_A^2. When $M^2 \gtrsim f$ one has small amplitude solutions. For a given $f < f_A$, the soliton amplitude increases with growing M^2, until the cold electron sonic point is met from below. For higher values of M^2, no solitary waves can be formed. It can be checked that in this range u_{hsc} increases from 1 to higher values, which corresponds to a rarefaction of the hot electrons and thus to an existence range of potential dip solitons.

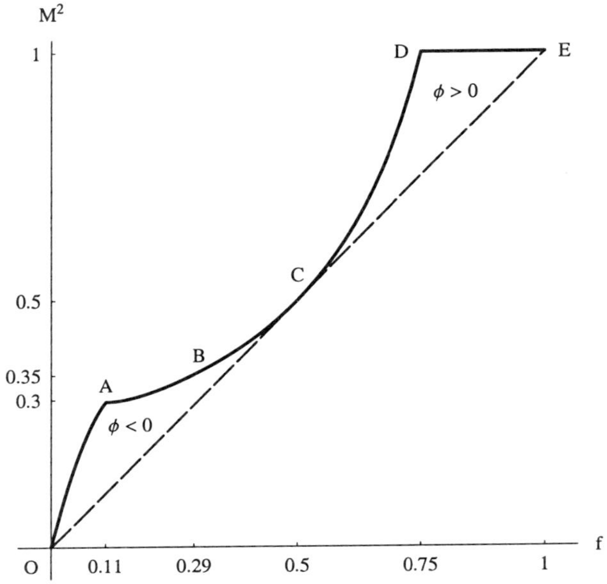

Fig. 2. Graph of bounding curves for $\gamma = 2$. For a given f, allowed M^2 values giving solitary wave solutions lie above the diagonal line $M^2 = f$ and below the curve $OABCDE$. These are potential dips in the $OABC$ region and potential hills in the CDE region. Values of M^2 on BCD itself yield double layers.[8]

For $f_A < f$ the hot electron sonic point becomes the limiting factor for the soliton amplitudes, until the double layer condition comes in at f_B, yielding the bounding curve AB. Now u_{hs} decreases from its value at A to that at B, but we are still in the existence domain of potential dip solitons.

Pursuing our discussion, the curve BCD is obtained from the double layer conditions $R(u_{hdl}) = 0$ and $dR(u_{hdl})/du_h = 0$, and u_{hdl} decreases from its value at B to 1 at C, at which point neither solitons nor double layers occur. The value of f corresponding to C is the critical cool electron density, $f_c = 3/(\gamma+4)$, encountered in the discussion of the weak nonlinear solutions. In the range bounded by BC we have solitons, and on BC itself double layers. An example of R in the range bounded by BC is shown in Figure 3.[8] Hence for $0 < f < f_c$, corresponding to the parameter region for M^2 between the line $M^2 = f$ and the bounding curve $OABC$, we see that $u_h > 1$ and, from mass conservation, that $n_h < n_{h0}$. The solitary waves thus correspond to a rarefaction in the hot (subsonic) electrons, a compression of the cold (supersonic) electrons and a dip for the electrostatic potential.

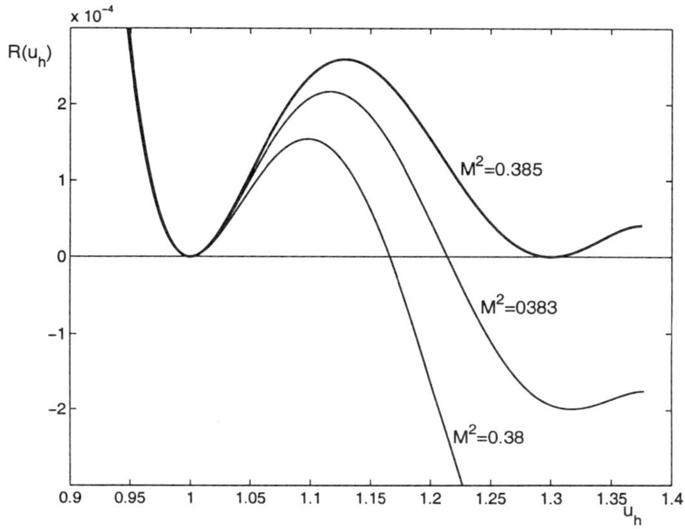

Fig. 3. Potential dip solitons with negative double layer limit for $\gamma = 2$ and $f = 0.35$ in the domain bounded by BC. Corresponding diagrams can be produced for other parts of parameter space with different bounding curves.[8]

However, on CD we can check that u_{hdl} decreases below 1, and now the solitons and double layers are of the potential hill type. From D to E the existence of solitons is solely limited by the condition that $M^2 < 1$. As remarked already, at these high values of f the linear electron-acoustic waves are highly Landau damped and the computed soliton range has no real physical importance. In the regime $f_c < f < 1$, corresponding to the parameter region for M^2 between the line $M^2 = f$ and the bounding curve CDE, $u_h < 1$ and $n_h > n_{h0}$. The solitary waves now correspond to a compression of the hot electrons, a rarefaction in the cold electrons and a hill for the electrostatic potential.

The fact that in this regime both electron species are driven away from their sonic points is rather counter-intuitive, when compared to, e.g., ion-acoustic modes in plasmas with one electron and one ion component, where the ions are driven towards their sonic point and determine the character of the potential.[58] While the fluid-dynamic approach rightly emphasizes the importance of charge neutral and sonic points in delineating the parameter range in which solitons can exist, the method does not give *a priori* indications as to whether a species is driven towards or away from its sonic point. For several simple plasma configurations it turns out that there is always one of the (mobile) species that is driven towards its sonic point, usually, but not necessarily, the heavier one. Because we have treated the ions as almost but not quite immobile, it follows from inspection of Figure 1 that they are actually driven, however slightly, towards their sonic point, located for cold ions at 0.

It is also seen from (37) that the neglect of the hot electron kinetic energy compared to their enthalpy necessitates very small Mach numbers, i.e. the hot electrons have to be highly subsonic. Of course, once $M^2 \ll 1$ we find ourselves in the lower left hand corner of Figure 2, where the solitons are only of the potential dip type and there are no double layers.

Graphs of bounding curves for other values of γ, such as for the traditional Boltzmann case of $\gamma = 1$, all look qualitatively similar. There is only a shift in the critical value of f where the nature of the solitons changes from potential dips to hills, such that f_c increases with $\gamma \to 1$, in the latter case being at $f_c = 3/5$. Historically, most attention has been devoted to the isothermal case, $\gamma = 1$, but as I have commented, that has usually been dealt with by neglecting the inertial effects of the hot electrons. At first sight the logarithmic behaviour, that arises from the use of the isothermal assumption, $\gamma = 1$, might suggest that that case is mathematically and physically significantly different from the general polytropic case $\gamma \neq 1$. Thus it is particularly interesting to note that essentially the same general results have been obtained for $\gamma = 1$ and 2, and indeed for general γ,[8] although then parts of the bounding curves have to be evaluated numerically.

5. Electromagnetic solitons of larger amplitudes

5.1. *General invariants for electromagnetic modes*

Now I turn to nonlinear electromagnetic stationary structures propagating obliquely to a static magnetic field. Before discussing the specific application of propagation in cold pair plasmas in greater detail, I want to derive some of the general multifluid invariants that can be used to reduce the number of equations to be solved. I also want to show that some of the insight gained through the McKenzie approach can be obtained and used in the Sagdeev description,[37,38] although the latter has not really been applied much to the study of electromagnetic waves, contrary to the vast body of papers devoted to large-amplitude electrostatic solitons. Among the very early studies of perpendicularly propagating, large-amplitude hydromagnetic modes in electron-ion plasmas running along similar lines is an interesting and indeed pioneering paper,[2] that, unfortunately, seems to have been largely ignored or forgotten since.

To start with, x and t are transformed to a combined coordinate $\xi = x - Vt$, with V the soliton velocity, and all derivatives are replaced by the usual recipe,

$$\frac{\partial .}{\partial x} = \frac{d.}{d\xi}, \qquad \frac{\partial .}{\partial t} = -V\frac{d.}{d\xi}. \tag{61}$$

This is well know from many studies, in plasma physics mostly of electrostatic modes, leading to their description in terms of a Sagdeev pseudopotential.[37,38]

All derivatives now being with respect to ξ, the continuity equations (1) express conservation of parallel (mass) flux,

$$n_j(V - v_{jx}) = n_0 V, \tag{62}$$

and from Faraday's law (3) there follows that

$$\mathbf{E}_\perp = V(B_z - B_{z0})\mathbf{e}_y - VB_y\mathbf{e}_z, \tag{63}$$

given the boundary conditions at infinity, plus the fact that I want to deal with parallel, oblique and perpendicular wave propagation, by taking $\mathbf{B}_0 = B_{x0}\mathbf{e}_x + B_{z0}\mathbf{e}_z$, as discussed earlier in Section 2. I now multiply the equations of motion (2), written in terms of their ξ dependence, by $n_j m_j$ and sum over all species, use Ampère's law (4), Poisson's equation (5) and flux conservation (62), to obtain expressions which can be integrated with respect to ξ. This yields the momentum invariants

$$\sum_j n_{j0}m_j v_{jx} = \frac{1}{V}\sum_j (p_j - p_{j0}) + \frac{\Gamma}{2\mu_0 V}(B_y^2 + B_z^2 - B_{z0}^2) - \frac{\varepsilon_0}{2V}E_x^2, \tag{64}$$

$$\sum_j n_{j0}m_j v_{jy} = \varepsilon_0 B_{z0}E_x - \frac{\Gamma B_{x0}}{\mu_0 V}B_y, \tag{65}$$

$$\sum_j n_{j0}m_j v_{jz} = -\frac{\Gamma B_{x0}}{\mu_0 V}(B_z - B_{z0}), \tag{66}$$

where $\Gamma = 1 - V^2/c^2$ is the correcting factor that reduces to 1 in the strictly nonrelativistic limit. However, in view of our using the nonrelativistic equations of motion, Γ should stay close to 1.

The sum of the projection of the equations of motion (2) on $n_j m_j \mathbf{v}_j$ can be integrated to yield the energy integral,

$$\sum_j n_{j0} m_j v_j^2 - 2V \sum_j n_{j0} m_j v_{jx} + \frac{2\Gamma B_{z0}}{\mu_0}(B_z - B_{z0}) = 0. \qquad (67)$$

Here $v_j^2 = v_{jx}^2 + v_{jy}^2 + v_{jz}^2$, and I will put $\Gamma = 1$ in what follows.

5.2. *Large-amplitude solitary waves in pair plasmas*

5.2.1. *Motivation and linear waves*

Pair plasmas can differ radically from other plasmas, because the negative and positive charge carriers have the same mass but opposite charges. One of the first examples treated in the literature are electron-positron plasmas, that play an important role in pulsar dynamics and radiation,[23,41,46] but that have also been generated experimentally closer to home.[6,22,47] Very recently, pair plasmas were also created using charged fullerenes, consisting of C_{60}^+ and C_{60}^- in equal numbers.[34] Fullerenes are molecules containing 60 carbon atoms in a very typical geometric arrangement, and a fullerene pair plasma is an exciting novel way of mimicking electron-positron plasma behaviour without having to worry about annihilation, so that longer time scales can be considered.

Since the positive and negative charged particles respond on the same scales, the characteristics of waves in pair plasmas cannot always be translated from what obtains in ordinary plasmas by simply letting $m_i \to m_e$. To give but one simple example, there is no Faraday rotation in pair plasmas,[11] with the immediate consequence that parallel propagating linear electromagnetic waves are not circularly but linearly polarized. Other modes, like the well studied ion-acoustic waves have no counterpart in pair plasmas, where the electrostatic dispersion is of the high frequency Langmuir type.[62]

Before addressing the nonlinear development, I look at linear wave propagation, by linearizing and Fourier transforming the relevant equations (1)–(6) for small disturbances varying as $\exp[i(kx - \omega t)]$, with frequency ω and wave number k. From the equations of motion (2) I deduce that

$$\mathbf{v}_j = \frac{e}{m(\omega^2 - \Omega^2)} \left[\pm i\omega \mathbf{E} \mp i \frac{e^2}{\omega m^2} (\mathbf{E} \cdot \mathbf{B}_0)\mathbf{B}_0 - \frac{e}{m} \mathbf{B}_0 \times \mathbf{E} \right], \qquad (68)$$

where $q_i = +e$, $q_e = -e$, $m_i = m_e = m$, $\Omega = eB_0/m$ is the gyrofrequency in absolute value, for both species, and the upper signs in $\pm$ or $\mp$ refer to the positrons or positive ions and the lower signs to the electrons or negative ions.

The last term between the square brackets is the same for both species and hence will drop out of the current when analyzing Ampère's law (4). Eliminating the wave magnetic field between the linearized and Fourier transformed equations (3) and (4) yields

$$(\omega^2 - c^2k^2)\mathbf{E} + c^2(\mathbf{E}\cdot\mathbf{k})\mathbf{k} + i\frac{en_0\omega}{\varepsilon_0}(\mathbf{v}_i - \mathbf{v}_e) = \mathbf{0}, \tag{69}$$

with $n_0 = n_{e0} = n_{i0}$ the common equilibrium density. Insertion of the velocities (68) then yields the wave equation[57]

$$\begin{aligned} &c^2(\omega^2 - \Omega^2)(\mathbf{E}\cdot\mathbf{k})\mathbf{k} + \frac{\omega_p^2 e^2}{m^2}(\mathbf{E}\cdot\mathbf{B}_0)\mathbf{B}_0 \\ &\quad + \left[(\omega^2 - c^2k^2)(\omega^2 - \Omega^2) - \omega_p^2\omega^2\right]\mathbf{E} = \mathbf{0}. \end{aligned} \tag{70}$$

The total plasma frequency ω_p is defined through $\omega_p^2 = 2n_0e^2/\varepsilon_0 m$.

There is thus at all angles of wave propagation a decoupling of E_y, the component of $\mathbf{E}$ orthogonal to the plane spanned by $\mathbf{k}$ and $\mathbf{B}_0$, from the other components of the electric field. The associated mode has dispersion law

$$\omega^4 - \omega^2(c^2k^2 + \omega_p^2 + \Omega^2) + c^2k^2\Omega^2 = 0, \tag{71}$$

and corresponds at parallel propagation ($B_{z0} = 0$) to the case where the incompressible ($n_i = n_e = n_0$) circularly polarized waves, well known from standard electron-ion plasmas, degenerate into two orthogonal, linearly polarized modes, both with dispersion law (71). At perpendicular propagation ($B_{x0} = 0$) this mode is part of the extraordinary (X) mode, in the Allis terminology.[45] Moreover, the linear $E_y \neq 0$ mode exists at all angles of wave propagation, and is always characterized by charge neutrality ($n_e = n_i = n$ but different from n_0) and linear polarization, with E_y the only nonzero component of the wave electric field. Recall that in ordinary plasmas the X mode cannot be factorized, and that for oblique propagation all modes are mixed. The inclusion of pressures does not substantially modify these conclusions, in the sense that the special X mode remains separate, linearly polarized and charge neutral at all angles of propagation. Further details can be found in a systematic study by Zank and Greaves,[62] although their discussion is not really transparent if one is interested in the polarization and charge density fluctuations of the modes. In addition, it is confusing that several solutions of the dispersion law for parallel propagation refer to the same mode but are catalogued as if they were different.

5.2.2. *Nonlinear modes and charge neutrality*

In this and the following subsections I will study the nonlinear analogue of this special X mode, by imposing charge neutrality for large amplitude stationary structures. In view of the equal charges (in absolute value) and masses in pair plasmas, the momentum invariants (64)–(66) become

$$\begin{aligned} v_{ix} + v_{ex} &= \frac{1}{2\mu_0 n_0 m V}(B_y^2 + B_z^2 - B_{z0}^2) - \frac{\varepsilon_0}{2n_0 m V}\, E_x^2, && (72)\\ v_{iy} + v_{ey} &= \frac{\varepsilon_0 B_{z0}}{n_0 m}\, E_x - \frac{B_{x0}}{\mu_0 n_0 m V}\, B_y, && (73)\\ v_{iz} + v_{ez} &= -\frac{B_{x0}}{\mu_0 n_0 m V}\,(B_z - B_{z0}), && (74) \end{aligned}$$

and the energy integral (67) is

$$v_i^2 + v_e^2 - 2V(v_{ix} + v_{ex}) + \frac{2B_{z0}}{\mu_0 n_0 m}(B_z - B_{z0}) = 0. \qquad (75)$$

For pair plasmas we know from the linear wave behaviour and the weakly nonlinear reductive perturbation results of the special X mode[50,57] that these maintain charge neutrality, not only in equilibrium but also in the wave. Moreover, one can prove for parallel propagation that such a property also holds for stationary nonlinear electromagnetic structures at larger amplitudes, even when pressure effects are taken into account.[7,53]

Hence the assumption of charge neutrality $n_i = n_e = n$ aims at finding the large amplitude generalization of the special X mode discussed in linear and weakly nonlinear theory. In passing, I note that charge neutrality is often assumed for low-frequency nonlinear electromagnetic modes in ordinary electron-ion plasmas, but this is not completely self-consistent[2,59] and needs the invocation of the plasma approximation.[10]

The conservation of mass flux (62) in charge neutral pair plasmas immediately leads to equal parallel velocities, $v_{ix} = v_{ex} = v_x$ and from Poisson's equation (5) also to the vanishing of the parallel electric field, $E_x = 0$, given the conditions at infinity. The two parallel equations of motion should be compatible, in the sense that

$$(v_x - V)\frac{dv_x}{d\xi} = \frac{e}{m}(v_{iy}B_z - v_{iz}B_y) = -\frac{e}{m}(v_{ey}B_z - v_{ez}B_y), \qquad (76)$$

from which I conclude that

$$(v_{iy} + v_{ey})B_z - (v_{iz} + v_{ez})B_y = 0. \qquad (77)$$

When I replace in (77) the sum over the velocities with the help of (73) and (74), at oblique or perpendicular propagation, the upshot is that

$B_{x0}B_{z0}B_y = 0$. For strictly oblique propagation ($\vartheta \neq 0$ or $\vartheta \neq 90°$) this gives $B_y = 0$. In the limit of parallel propagation, we know that the wave magnetic field is transverse and linearly polarized,[54] and there is no harm in taking it along the z axis, so that $B_y = 0$ implies no loss of generality.

Thus for all angles $\vartheta \neq 90°$ I find that the magnetic field has no component outside the plane spanned by the directions of wave propagation and of the external magnetic field, and I will assume, by continuity, that this also holds when $\vartheta \to 90°$. From (63) we learn that only E_y remains, and hence the linear polarization of the wave electric field means that possible solitary wave solutions will occur as true stationary structures without phase oscillations, and hence not as oscillitons,[39,40] as would be the case in the usual electron-ion plasmas.

5.2.3. *Pseudoenergy integral*

From (77) and $B_y = 0$ we see that $v_{iy} = -v_{ey} = v_y$, and what remains of the z component of Ampère's law indicates that $v_{iz} = v_{ez} = v_z$. The y component of Ampère's law (4) reduces to

$$\frac{dB_z}{d\xi} + 2\mu_0 nev_y = 0. \tag{78}$$

Dimensionless variables are introduced as follows: velocities $u_{x,y,z} = v_{x,y,z}/V_A$, a length scale $\zeta = \Omega\xi/V_A$, the Alfvénic Mach number $M = V/V_A$, the only varying component of the wave magnetic field $b = B_z/B_0$, and the direction cosines of the static field $\cos\vartheta = B_{x0}/B_0$ and $\sin\vartheta = B_{z0}/B_0$. Here V_A, the Alfvén velocity in pair plasmas, is given through $V_A^2 = B_0^2/2\mu_0 n_0 m$. Thus we have

$$\begin{aligned} u_x &= \frac{1}{2M}(b^2 - \sin^2\vartheta), \\ u_y &= \frac{1}{M}(u_x - M)\frac{db}{d\zeta}, \\ u_z &= \frac{1}{M}(\sin\vartheta - b)\cos\vartheta, \end{aligned} \tag{79}$$

to be substituted into the energy integral,

$$u_x^2 + u_y^2 + u_z^2 - 2Mu_x + 2(b - \sin\vartheta)\sin\vartheta = 0. \tag{80}$$

This leads to an energy-like integral for a particle with coordinate b and unit mass,

$$\frac{1}{2}\left(\frac{db}{d\zeta}\right)^2 + \psi(b) = 0, \tag{81}$$

moving in the pseudopotential

$$\psi(b) = \frac{M^2(b - \sin\vartheta)^2[(b + \sin\vartheta)^2 - 4(M^2 - \cos^2\vartheta)]}{2(2M^2 + \sin^2\vartheta - b^2)^2} \tag{82}$$

and with ζ playing the role of time. Once b is known, i.e. B_z, the other variables u_x, u_y and u_z and hence v_x, v_y and v_z follow from the existing relations without further integration.

Before going on, I remark that at parallel propagation ($\vartheta = 0$) the pseudopotential reduces to the one computed in an earlier paper.[54] The limit of perpendicular propagation can also be found in the first paper dealing with solitary hydromagnetic waves in electron-ion plasmas,[2] because of their assumption of charge neutrality and when $m_e = m_i$ is taken in their results, although the masses are buried in the nondimensional units used.

5.2.4. *Large amplitude solitons*

A first, necessary condition for the existence of solitons is that $d^2\psi/db^2(\sin\vartheta) < 0$, since we have already that $\psi(\sin\vartheta) = 0$ and $d\psi/db(\sin\vartheta) = 0$. This leads to the requirement that $M^2 > 1$.

Other, single roots of the pseudopotential (82) occur at

$$b_{s\pm} = -\sin\vartheta \pm 2\sqrt{M^2 - \cos^2\vartheta}, \tag{83}$$

and these will give rise to solitons, provided the roots are encountered before hitting the asymptotes,

$$b_{a\pm} = \pm\sqrt{2M^2 + \sin^2\vartheta}. \tag{84}$$

At the latter values the pseudopotential becomes infinite, indicating that $u_x = M$ and giving, from mass conservation (62), an infinite compression of both species. Because of the necessary condition $M^2 > 1$, one can prove that

$$b_{s-} < -3\sin\vartheta \leq \sin\vartheta < b_{s+}, \tag{85}$$

with strict outer inequalities. There are thus two possible ranges to discuss: either $b_{a-} \leq b_{s-} < -3\sin\vartheta$ or $\sin\vartheta < b_{s+} \leq b_{a+}$.

At parallel propagation ($\vartheta = 0$), the negative range is just the mirror image of what obtains for the positive range, and warrants no further attention, since then the orientation of the z axis becomes rather arbitrary. This is illustrated in Fig. 4, where the graph of $\psi(b)$ at moderate Mach numbers ($M^2 = 1.2$) indicates already the existence of symmetric, fairly strong positive and negative solitons.

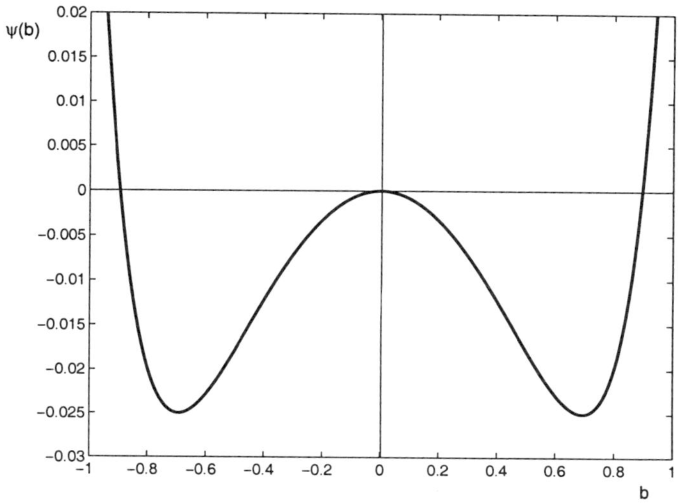

Fig. 4. Pseudopotential $\psi(b)$ for $M^2 = 1.2$ at parallel propagation ($\vartheta = 0°$), yielding symmetric and fairly strong positive and negative solitons.[55]

For oblique or perpendicular propagation, however, the two ranges have a different physical interpretation. For the solutions associated with $b_{s-} < 0$, the transverse magnetic field component starts from the equilibrium value $\sin\vartheta$, goes through zero and attains an extremum $b_{s-} < -3\sin\vartheta < 0$, where $b = B_z/B_0$ is then in the opposite direction to B_{z0}. Moreover, when the magnetic field information is translated to the wave electric field, we see from (63) that $E_y \leq 0$, and I will for reasons of brevity call these the negative solitons. On the other hand, the solutions associated with b_{s+} have a transverse magnetic field component that is always in the same direction as B_{z0}, with $B_z \geq B_{z0}$, and an electric field $E_y \geq 0$. These are then called the positive solitons.

I start with a discussion of the negative solitons, for the existence of which I need, after elimination of the minus signs, that

$$\sin\vartheta + 2\sqrt{M^2 - \cos^2\vartheta} \leq \sqrt{2M^2 + \sin^2\vartheta}. \tag{86}$$

From this I obtain the range of admissible Mach numbers as

$$1 < M^2 \leq 2(1 - \sin\vartheta). \tag{87}$$

The lower bound derives from the necessary condition to have solitons at all, and it is clear that the lower and the upper bound together limit the angles of propagation to below a critical angle of at most $\vartheta_c = 30°$. The

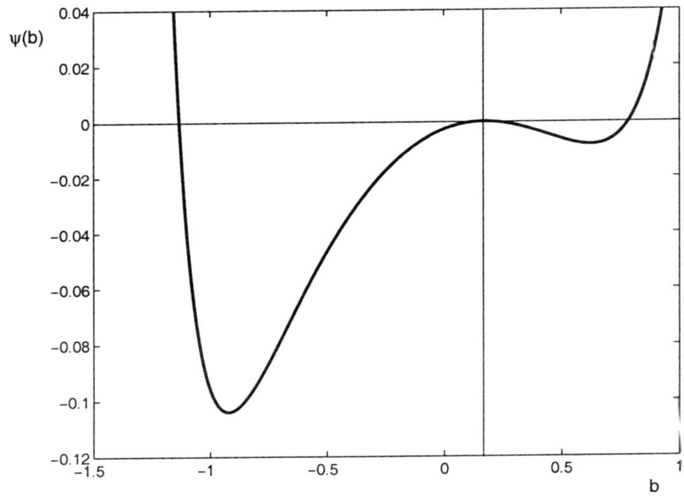

Fig. 5. Pseudopotential $\psi(b)$ for $M^2 = 1.2$ at oblique propagation ($\vartheta = 10°$) below the critical angle $\vartheta_c = 23°$ for this Mach number. The negative soliton has a much stronger amplitude than the positive one.[55]

limiting Mach numbers are $M^2 = 2$ at parallel propagation and diminish to $M^2 = 1$ at $\vartheta_c = 30°$. For $M^2 = 1.2$ at oblique propagation ($\vartheta = 10°$) below the critical angle (for this Mach number $\vartheta_c = 23°$), the graph of $\psi(b)$ in Fig. 5 shows a much stronger negative soliton than the positive one, measured from the equilibrium value $b_0 = \sin\vartheta$. This will have consequences for the weakly nonlinear solutions obtained via reductive perturbation techniques, as indicated in the next subsection.

The discussion of the positive solitons starts from

$$-\sin\vartheta + 2\sqrt{M^2 - \cos^2\vartheta} \leq \sqrt{2M^2 + \sin^2\vartheta}, \tag{88}$$

and runs along analogous lines to arrive at the range of admissible Mach numbers as

$$1 < M^2 \leq 2(1 + \sin\vartheta). \tag{89}$$

The limiting Mach numbers now start from $M^2 = 2$ at parallel propagation and increase to $M^2 = 4$ at perpendicular propagation, without limits on the directions of propagation ϑ. This is shown in Fig. 6, where for $M^2 = 1.2$ at perpendicular propagation the graph of $\psi(b)$ only admits positive solitons. The graphs at angles of propagation above the critical angle ($\vartheta_c = 23°$ for this value of M^2) all look qualitatively similar, but indicate that the

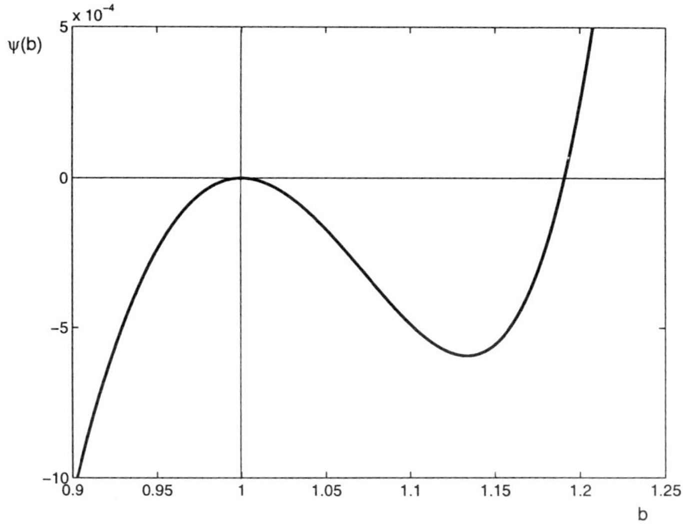

Fig. 6. Pseudopotential $\psi(b)$ for $M^2 = 1.2$ at perpendicular propagation ($\vartheta = 90°$), with only a positive soliton.[55]

soliton amplitude, measured from the initial value $b_0 = \sin\vartheta$, decreases with increasing ϑ. Formally, (81) can be solved for $db/d\zeta$ and integrated for ζ as a function of b.[55] For perpendicular propagation this solution corresponds, for b_{s+}, to the one obtained by Adlam and Allen,[2] since at perpendicular propagation the b_{s-} solution does not exist.

5.2.5. *Weakly nonlinear solitons*

For weakly nonlinear solitons I start from the observation that the undisturbed wave magnetic field is in normalized variables $\sin\vartheta$, and thus I will look for an expansion of (82) to the first significant order, beyond the second, in $w = b - \sin\vartheta$. This gives in general

$$\begin{aligned}\psi(w) = &-\frac{M^2-1}{2M^2}\,w^2 + \frac{(2-M^2)\sin\vartheta}{2M^4}\,w^3 \\ &+\frac{4M^2\cos^2\vartheta + 12\sin^2\vartheta - 3M^4}{8M^6}\,w^4. \end{aligned} \tag{90}$$

For terms higher than second order to play any role, in view of the smallness of w, the coefficient of the second order term has to be sufficiently small.

Consequently, M^2 can only just exceed 1 and this yields

$$\psi(w) \simeq -\frac{M^2-1}{2}\,w^2 + \frac{\sin\vartheta}{2}\,w^3 + \frac{1+8\sin^2\vartheta}{8}\,w^4, \tag{91}$$

where M^2 has been replaced everywhere by 1 except in the numerator of the coefficient of w^2. For $\sin\vartheta$ finite, I can stop the expansion at the cubic term, whereas close to parallel propagation ($\sin\vartheta \ll w$) the cubic term is neglected and the quartic term retained. Hence I put $M^2 = 1+\delta^2$, where δ^2 is of order w or w^2, depending on the case under discussion.

For quasiparallel propagation I determine $w \simeq b$ from

$$\left(\frac{db}{d\zeta}\right)^2 = \delta^2\,b^2 - \frac{1}{4}\,b^4. \tag{92}$$

In the solution,

$$b = 2\delta\,\mathrm{sech}\,(\delta\zeta)\,, \tag{93}$$

δ serves as a measure of the amplitude, while its inverse measures the width of the soliton. This corresponds to the stationary solution of the mKdV equation obtained from reductive perturbation theory at parallel propagation.[50,54] We remark that the mKdV equation is even in b, and thus both positive and negative solutions (of weaker amplitude) can exist.

On the other hand, for oblique propagation I can stop at the cubic term and determine w from

$$\left(\frac{dw}{d\zeta}\right)^2 = \delta^2\,w^2 - \sin\vartheta\,w^3. \tag{94}$$

The solution is now

$$w = b - \sin\vartheta = \frac{\delta^2}{\sin\vartheta}\,\mathrm{sech}^2\left(\frac{\delta\xi}{2}\right) \tag{95}$$

and corresponds to the stationary solution of the KdV equation itself, which results from the reductive perturbation analysis at oblique or perpendicular propagation.[57]

The mKdV solutions are valid very close to parallel propagation, and for all other angles we have to use the KdV equation, which, however, only has positive solutions. It is clear from inspection of the full pseudopotential $\psi(b)$ in Figure 5 that the negative solitons have too large an amplitude, compared to the positive ones, to admit a weakly nonlinear counterpart. Even for $M \to 1$ the negative solitons retain a large amplitude, up to when they disappear, whereas the positive solitons have smaller and smaller amplitudes.

All this indicates that the reductive perturbation techniques would place the transition between parallel and oblique propagation very close to parallel propagation. As we have seen from the full pseudopotential, however, the important transition occurs for finite obliquity, above which the negative, but larger solitons disappear. Quite to the contrary, the change from mKdV to KdV equations close to parallel propagation is thus seen to be an artefact of the expansion of $\psi(b)$, and the discrepancy between the two descriptions should sound a note of caution when relying solely on reductive perturbation analysis. Of course, there are several problems for which that is the only analytic method that works.....

6. Conclusions

In this review the basic multifluid model equations were introduced that have been used to discuss three different methods to describe nonlinear waves. The exposition of each of these methods has consisted of two parts, a rather general overview of the relevant methodology, followed by a specific and recent application.

Reductive perturbation analysis is widely applicable to waves that are not too strongly nonlinear, if their linear counterparts have an acoustic-like dispersion at low frequencies. This has been dealt with for electrostatic modes in multifluid plasmas, with a brief application to dusty plasma waves. The typical paradigm for such problems is the well known KdV equation and its siblings.

Stationary waves with larger amplitudes can be treated, i.a., via the fluid-dynamic approach pioneered by McKenzie, which focuses on essential insights into the limitations that restrict the range of available solitary electrostatic solutions. This has been illustrated for a plasma with two-temperature electron species, where novel electrostatic solutions are possible that might be relevant in understanding certain magnetospheric plasma observations.

The older cousin of the large-amplitude technique is the Sagdeev pseudopotential description, to which the newer McKenzie approach is essentially equivalent, although the former lacks the deeper insight in the physical restrictions in parameter space. Because the Sagdeev method has traditionally been applied mostly to electrostatic waves, some recent results have been given for electromagnetic modes in pair plasmas, to show its versatility.

Acknowledgments

This review has drawn heavily on the many enlightening discussions I have had over the years with T. Cattaert, N.F. Cramer, T.B. Doyle, M.A. Hellberg, G.S. Lakhina, J.F. McKenzie and P.K. Shukla. M.A. Hellberg is also thanked for his critical reading of the manuscript. The ongoing support of the Fonds voor Wetenschappelijk Onderzoek (Vlaanderen) is gratefully acknowledged.

References

1. M.J. Ablowitz and P.A. Clarkson, *Solitons, Nonlinear Evolution Equations and Inverse Scattering* (Cambridge Univ. Press, Cambridge, 1991).
2. J.H. Adlam and J.E. Allen, Philos. Mag. **3**, 448–455 (1958).
3. S. Baboolal, R. Bharuthram and M.A. Hellberg, J. Plasma Phys. **44**, 1–23 (1990).
4. A. Barkan, R.L. Merlino and N. D'Angelo, Phys. Plasmas **2**, 3563–3565 (1995).
5. M. Berthomier, R. Pottelette, M. Malingre, and Yu. Khotyaintsev, Phys. Plasmas **7**, 2987–2994 (2000).
6. H. Boehmer, M. Adams and N. Rynn, Phys. Plasmas **2**, 4369–4371 (1995).
7. T. Cattaert, private communication (2004).
8. T. Cattaert, F. Verheest and M.A. Hellberg, Phys. Plasmas **12**, 042901 (2005).
9. C.A. Cattell, J. Dombeck, J.R. Wygant, M.K. Hudson, F.S. Mozer, M.A. Temerin, W.K. Peterson, C.A. Kletzing, C.T. Russell, and R.F. Pfaff, Geophys. Res. Lett. **26**, 425–428 (1999).
10. F.F. Chen, *Introduction to Plasma Physics* (Plenum Press, New York, 1974), pp. 65–66.
11. F.F. Chen, *Introduction to Plasma Physics* (Plenum Press, New York, 1974), p. 121.
12. G.C. Das and S.G. Tagare, Plasma Phys. **17**, 1025–1032 (1975).
13. K.P. Das and F. Verheest, J. Plasma Phys. **41**, 139–155 (1989).
14. N. Dubouloz, R. Pottelette, M. Malingre and R.A. Treumann, Geophys. Res. Lett. **18**, 155–158 (1991).
15. R.E. Ergun, C.W. Carlson, J.P. McFadden, F.S. Mozer, G.T. Delory, W. Peria, C.C. Chaston, M. Temerin, I. Roth, L. Muschietti, R. Elphic, R. Strangeway, R. Pfaff, C.A. Cattell, D. Klumpar, E. Shelley, W. Peterson, E. Möbius, and L. Kistler, Geophys. Res. Lett. **25**, 2041–2044 (1998).
16. J.R. Franz, P.M. Kintner and J.S. Pickett, Geophys. Res. Lett. **25**, 1277–1280 (1998).
17. S.P. Gary and R.L. Tokar, Phys. Fluids **28**, 2439–2441 (1985).
18. M.A. Hellberg, S. Baboolal, R.L. Mace and R. Bharuthram, IEEE Trans. Plasma Sci. **20**, 695–700 (1992).
19. M. Horányi, Annu. Rev. Astron. Astrophys. **34**, 383–418 (1996).

20. E. Infeld and G. Rowlands, *Nonlinear Waves, Solitons and Chaos* (Cambridge Univ. Press, Cambridge, 2000).
21. V.L. Krasovsky, H. Matsumoto and Y. Omura, J. Geophys. Res. **102**, 22131–22139 (1997).
22. E.P. Liang, S.C. Wilks and M. Tabak, Phys. Rev. Lett. **81**, 4887–4890 (1998).
23. J.G. Lominadze, G.Z. Machabeli and V.V. Usov, Astrophys. Space Sci. **90**, 19–43 (1983).
24. R.L. Mace and M.A. Hellberg, J. Plasma Phys. **43**, 239–255 (1990).
25. R.L. Mace and M.A. Hellberg, J. Plasma Phys. **49**, 283–293 (1993).
26. R.L. Mace and M.A. Hellberg, Phys. Plasmas **8**, 2649–2656 (2001).
27. R.L. Mace, S. Baboolal, R. Bharuthram, and M.A. Hellberg, J. Plasma Phys. **45**, 323–338 (1991).
28. H. Matsumoto, H. Kojima, T. Miyatake, Y. Omura, M. Okada, I. Nagano and M. Tsutsui, Geophys. Res. Lett. **21**, 2915–2918 (1994).
29. J.F. McKenzie, Phys. Plasmas **9**, 800–805 (2002).
30. J.F. McKenzie, J. Plasma Phys. **67**, 353–362 (2002).
31. J.F. McKenzie, J. Plasma Phys. **69**, 199–210 (2003).
32. J.F. McKenzie and T.B. Doyle, New J. Phys. **5**, 26 (2003).
33. D.A. Mendis and M. Rosenberg, Annu. Rev. Astron. Astrophys. **32**, 419–463 (1994).
34. W. Oohara and R. Hatakeyama, Phys. Rev. Lett. **91**, 205005 (2003).
35. R. Pottelette, R.E. Ergun, R.A. Treumann, M. Berthomier, C.W. Carlson, J.P. McFadden, and I. Roth, Geophys. Res. Lett. **26**, 2629–2632 (1999).
36. N.N. Rao, P.K. Shukla and M.Y. Yu, Planet. Space Sci. **38**, 543–546 (1990).
37. R.Z. Sagdeev, in: *Reviews of Plasma Physics* **4** (Ed. M.A. Leontovich, Consultants Bureau, New York, 1966) pp. 23–91.
38. R.Z. Sagdeev and A.A. Galeev, *Nonlinear Plasma Theory* (W.A. Benjamin, New York, 1969).
39. K. Sauer, E. Dubinin and J.F. McKenzie, Geophys. Res. Lett. **28**, 3589–3592 (2001).
40. K. Sauer, E. Dubinin and J.F. McKenzie, Geophys. Res. Lett. **29**, 2226 (2002).
41. P.K. Shukla, Astrophys. Space Sci. **114**, 381 (1985).
42. P.K. Shukla and A.A. Mamun, *Introduction to Dusty Plasma Physics* (IOP Press, London, 2002).
43. P.K. Shukla and A.A. Mamun, New J. Phys. **5**, 17 (2003).
44. K. Stasiewicz, private communication (2005).
45. T.H. Stix, *Waves in Plasmas* (American Institute of Physics, New York, 1992).
46. P.A. Sturrock, Astrophys. J. **164**, 529 (1971).
47. C.M. Surko, M. Leventhal and A. Passner, Phys. Rev. Lett. **62**, 901–904 (1989).
48. F. Verheest, J. Plasma Phys. **39**, 71–79 (1988).
49. F. Verheest, Planet. Space Sci. **40**, 1–6 (1992).
50. F. Verheest, Phys. Lett. A **213**, 177–182 (1996).
51. F. Verheest Space Sci. Rev. **77**, 267–302 (1996).

52. F. Verheest, *Waves in Dusty Space Plasmas* (Kluwer Academic Publishers, Dordrecht, 2000).
53. F. Verheest, Nonlin. Proc. Geophys. **12**, 569–574 (2005).
54. F. Verheest and T. Cattaert, Phys. Plasmas **11**, 3078–3082 (2004).
55. F. Verheest and T. Cattaert, Phys. Plasmas **12**, 032304 (2005).
56. F. Verheest and M.A. Hellberg, Physica Scripta **T82**, 98–105 (1999).
57. F. Verheest and G.S. Lakhina, Astrophys. Space Sci. **240**, 215–224 (1996).
58. F. Verheest, T. Cattaert, G.S. Lakhina and S.V. Singh, J. Plasma Phys. **70**, 237–250 (2004).
59. F. Verheest, T. Cattaert, E. Dubinin, K. Sauer and J.F. McKenzie, Nonlin. Proc. Geophys. **11**, 447–452 (2004).
60. H. Washimi and T. Taniuti, Phys. Rev. Lett. **17**, 996–998 (1966).
61. K. Watanabe and T. Taniuti, J. Phys. Soc. Japan **43**, 1819–1820 (1977).
62. G.P. Zank and R.G. Greaves, Phys. Rev. E **51**, 6079–6080 (1995).

NONLINEAR WAVEPACKETS IN PAIR-ION AND ELECTRON-POSITRON-ION PLASMAS

I. KOURAKIS

Centre for Plasma Physics, Department of Physics and Astronomy, Queen's University Belfast, BT7 1 NN Northern Ireland, UK
E-mail: i.kourakis@qub.ac.uk , www.kourakis.eu

R. ESFANDYARI

Azarbaijan University of Tarbiat Moallem, Faculty of Science, Department of Physics, 51745-406, Tabriz, Iran

P. K. SHUKLA

Institut für Theoretische Physik IV, Fakultät für Physik und Astronomie, Ruhr-Universität Bochum, D-44780 Bochum, Germany

F. VERHEEST

Universiteit Gent, Sterrenkundig Observatorium, Krijgslaan 281, B-9000 Gent, Belgium
and
School of Physics, Howard College Campus, University of KwaZulu-Natal, Durban 4041, South Africa

N. F. CRAMER

School of Physics, University of Sydney, New South Wales 2006, Australia

The occurrence of amplitude-modulated electrostatic and electromagnetic wavepackets in pair plasmas is investigated. A static additional charged background species is considered, accounting for dust defects or for heavy ion presence in the background. Relying on a two-fluid description, a nonlinear Schrödinger type evolution equation is obtained and analyzed, in terms of the slow dynamics of the wave amplitude. Exact envelope excitations are obtained, modelling envelope pulses or holes, and their characteristics are discussed.

1. Introduction

In their widespread textbook picture, charge-neutral electron-ion (e-i) plasmas are modelled as large ensembles of electrons e^- (charge $q_e = -e$, mass

m_e) and positive ions i^+ (charge $q_i = +Z_i e$, mass $m_i \gg m_e$). The small ion-to-electron mass ratio is associated with e-i plasma features, which are most often implicitly taken for granted: for instance, the electron and ion plasma frequency $\omega_{p,s} = (4\pi n_s q_s^2/m_s)^{1/2}$ (for species s; here, $s = e$ or i) and cyclotron frequency $\omega_{c,s} = q_s B/m_s c$ are clearly different, thus allowing for a clear distinction among corresponding time and (e.g., Debye) length) scales and associated wave phenomena.[1] *Pair plasmas* (*p.p.*) are distinct from this picture, in that they consist of two ion populations (say, 1^+ and 2^-) of equal mass and opposite charge (i.e., $q_1 = -q_2 = +Ze$ and $m_1 = m_2 = m$). The pair ion densities at equilibrium, although equal in a symmetric ("pure") *p.p.* configuration, may differ if the charge balance is affected by a 3rd population, e.g. a massive charged defect species $3^{\pm}$ (e.g. dust[2]), assumed present as a stationary background. No plasma or cyclotron frequency separation occurs in *p.p.*; furthermore, a variety of novel physical phenomena (e.g. absence of Faraday rotation) characterizes electrostatic (ES) and electromagnetic (EM) wave propagation in such plasmas.[3] Although this simple description of pair plasmas was originally introduced to model (for $Z = 1$, here) electron-positron (e-p) plasmas[a] (yet overseeing e-p annihilation-recombination processes, here neglected throughout), it may claim to provide a consistent model of fullerene-ion pair plasma configurations which were recently successfully produced in experiments.[4] Significant research effort has recently focused on the properties of linear and nonlinear wave propagation in such plasmas.

Plasma wave observations, both in Space and in the laboratory, provide abundant evidence for the existence of spatially localized propagating wave structures, e.g. in the form of a localized envelope pulse confining (*modulating*) a fast carrier wave.[5] Modulated wavepackets of this form may occur as a result of *modulation instability* (MI), when nonlinearity at the first stages of the amplitude instability is balanced by group dispersion. This physical mechanism is reminiscent of energy localization phenomena in nonlinear optics, hydrodynamics and biophysics.[6] As regards *pair plasmas*, nonlinear modulated wavepackets may occur either as ES[7,8] or electromagnetic EM[9] wavepackets.

The aim of this brief report is to review existing results on linear waves and nonlinear envelope structures propagating in pair plasmas. A two-fluid plasma model is introduced for this purpose, for the pair species; a third massive component is taken to be stationary, referring to three-component

[a]The 3rd species type then represents ions, in electron-positron-ion (e-p-i) plasmas.

pair plasmas of, say, either the type $i^+i^-3^\pm$ (i.e., *p.p.* "doped" with dust defects) or $e^-p^+i^+$. No assumption is made on the density ratio n_+/n_- or the temperature ratio T_+/T_-, since one of our intentions is to point out the role of a possible pair-component asymmetry in the plasma configuration (also, of the fixed background species $3^\pm$) on the properties of nonlinear ES waves. High frequency ES waves in *e-p-i* plasmas are also included in this picture.

2. A two-fluid model for ES wavepackets in pair plasmas

Consider a two-plasma-fluid model, described by the dynamical equations

$$\frac{\partial n_j}{\partial t} + \nabla \cdot (n_j \, \mathbf{u}_j) = 0 \,, \tag{1}$$

$$\frac{\partial \mathbf{u}_j}{\partial t} + \mathbf{u}_j \cdot \nabla \mathbf{u}_j \;=\; -s_j \frac{Ze}{m} \nabla \phi - \frac{1}{m n_j} \nabla p_j \,, \tag{2}$$

for the density n_j and the velocity $\mathbf{u}_j$ of the j-th particle species ($j = 1^+, 2^- \equiv +, -$). The equation of state $p_j \sim n_j^\gamma$ is assumed for the pressure p_j ($\gamma = 1 + 2/f$ is the specific heat ratio, for f degrees of freedom), while $p_{j,0} = n_{j,0} k_B T_j$ is assumed at equilibrium (the Boltzmann constant k_B preceding the temperature T_j will be omitted where obvious). The difference in charge sign is expressed by $s_j = q_j/|q_j| = \pm 1$. The system is closed by Poisson's equation

$$\nabla^2 \Phi \;=\; -4\pi \sum_s q_s \, n_s \;=\; 4\pi \, e \, (Z \, n_- - Z \, n_+ - s_3 Z_3 \, n_3) \tag{3}$$

Note that $s_3 = \pm 1$ may account for either positive or negative background ions[b]. The charge balance, as expressed via the neutrality hypothesis (at equilibrium): $Zn_{+,0} - Zn_{-,0} + s_3 Z_3 \, n_3 = 0$, is obviously affected by the presence of the 3rd background species (of density n_3 =cst.). The case of "pure" pair (two-component, e.g. *e-p*) plasmas is thus recovered for $n_3 = 0$ (i.e. $n_{+,0} = n_{-,0}$), while $n_3 \neq 0$ in general (in $e^-p^+i^+$ or, say, $X^+X^-d^\pm$ type plasmas). A one-dimensional geometry will be adopted here (hence $f = 1$ and $\nabla = \partial/\partial x$), although a (richer) multi-dimensional description may also be found in the bibliography.[8]

In the following, reduced expressions will be implicitly assumed for all quantities, scaling space x and time t units are the Debye length $\lambda_{D,-} = c_s/\omega_{p,-}$ and the inverse plasma frequency $\omega_{p,-}^{-1}$, respectively; also,

[b] i.e., for dust $d^\pm$ (viz. $s_3 = \pm 1$) in "doped *p.p.*", or for ions ($s_3 = +1$) in *e-p-i* plasmas.

the density, velocity and electric potential variables are scaled by $n_{-,0}$, $c_s = k_B T_-/m$ and $k_B T_-/Ze$, respectively.

3. Linear electrostatic wave dispersion properties

The linear (small amplitude) approach consists in assuming perturbations of all elements, say S_l, of the state vector $\mathbf{S} = (\{n_j, u_j;\}; \phi)$ (here $l = 1, ..., 5$; $j = 1, 2$) near equilibrium $\mathbf{S_0} = (\{n_{j,0}, 0;\}; 0)$, in the form $\epsilon S_j \approx S_{1,l}^{(1)} \exp i(kx - \omega t)$; here $S_{1,l}^{(1)}$ is the wave amplitude (assumed constant for a while), while k and ω denote(s) the carrier wavelength and frequency, respectively. Inserting into Eqs. (1)-(3), one obtains the dispersion relation:

$$\frac{1}{\omega^2 - 3k^2} + \frac{\beta}{\omega^2 - 3\sigma\beta^2 k^2} = 1\,, \tag{4}$$

where $\beta = n_{+,0}/n_{-,0}$ is the density ratio[c] and $\sigma = T_+/T_-$ is the temperature ratio, among the pair components. A bi-quadratic polynomial equation is thus obtained and is straightforward to solve, providing two branches, say $\omega = \omega_L(k)$ and $\omega = \omega_U(k)$. The exact expressions are presented and analyzed in Ref.[7] The lower branch ω_L describes an acoustic mode, as $\omega_L(0) = 0$, while the upper one bears a Langmuir-type form, featuring a cutoff frequency $\omega_U(0) = (1+\beta)^{1/2}$ (in units of $\omega_{p,-}$; see above). Interestingly, *no* acoustic mode in principle exists for perfectly symmetric p.p. configurations; to see this, set $\beta = \sigma = 1$ in (4)[d], to obtain $\omega^2 = 2 + 3k^2$ (cf. literature[4]). *Asymmetric p.p.* are henceforth implicitly assumed everywhere. Considering the behavior of both branches near $k = 0$, we obtain

$$\omega_L(k) \approx c_s^2 k^2\,, \qquad \omega_U(k) \approx (\omega_c^2 + c_s^2 k^2)^{1/2}\,. \tag{5}$$

We note that the cutoff frequency $\omega_c = (1+\beta)^{1/2}$ depends on the background (third) species' concentration (via β), while the sound speed $c_s = [3\,\beta(1+\sigma\beta)/(1+\beta)]^{1/2}$ also depends on the pair component temperature asymmetry via σ. This behavior is depicted in Fig. 1. The dispersion reported in fullerene experiments[4] is recovered for $\beta = \sigma = 1$, as expected.

The amplitudes of the linear oscillations are obtained in terms of the electric potential perturbation $\phi_1^{(1)} = \psi$ as

$$n_{+,1}^{(1)} = \frac{\beta k^2}{\omega^2 - 3\sigma\beta^2 k^2}\psi = \frac{\beta k}{\omega}\, u_{+,1}^{(1)}\,, \quad n_{-,1}^{(1)} = -\frac{k^2}{\omega^2 - 3k^2}\psi = \frac{k}{\omega}\, u_{-,1}^{(1)}\,. \tag{6}$$

---

[c]Note that neutrality at equilibrium leads to $\beta = n_{+,0}/n_{-,0} = 1 - s_3 Z_3\, n_3/(Zn_{-,0})$, so a value above (below) unity implies a negatively (positively) charged third species presence in the background, viz. $s_3 = -1$ $(+1)$. Obviously, $\beta = 1$ in "pure" *p.p.*

[d]This seems to suggest an asymmetry among the pair ion species in the experiment of Oohara and Hatakeyama[4] where the observation of an acoustic mode was reported.

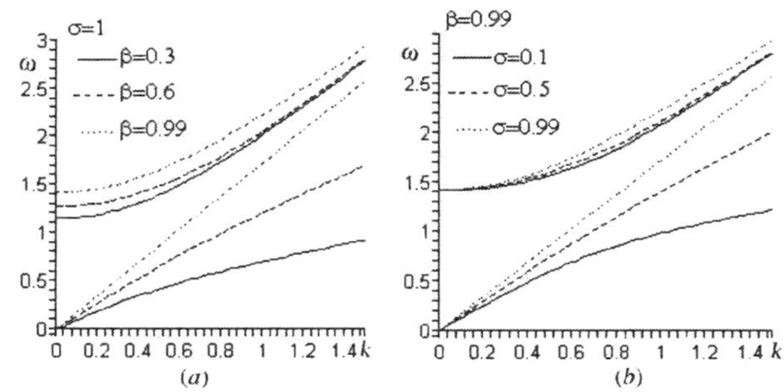

Fig. 1. The two dispersion curves defined by Eq. (4) are depicted: frequency ω vs. wavenumber k (reduced quantities; see in the text). After Ref. 7.

The subscripts +/− will be used, where obvious, to distinguish the positive/negative ions (or, positrons/electrons). These expressions may be used in plasma diagnostics, to trace the presence of charged defects in the background and/or pair-ion asymmetry in real (experimentally produced) *p.p.*

4. Multiple scales theory for modulated ES wavepackets

We shall now consider a small (yet finite) deviation from the equilibrium state $\mathbf{S_0}$, by allowing for a weak time and space dependence (modulation) of the wave's amplitude A_j. All state variables S_l ($l = 1, ..., 5$) are assumed to vary as $S_l(x,t) = S_{0,l} + \sum_{m=1}^{\infty} \epsilon^n \sum_{L=-\infty}^{\infty} S_{L,l}^{(n)}(\xi,\tau)\exp[iL(kx-\omega t)]$ where $\epsilon \ll 1$ is a (real) smallness parameter (the equilibrium state vector $\mathbf{S_0}$ was defined above). Here, the superscript n (subscript L, respectively) denote(s) the expansion order in ϵ (the phase harmonic order m). The reality condition $S_{L,l}^{(n)} = (S_{-L,l}^{(n)})^*$ is implied; the star superscript denotes the complex conjugate. The wave amplitudes $S_{L,l}^{(m)}$ depend on the stretched ("slow") coordinates $\xi = \epsilon\,(x - \lambda\,t)$ and $\tau = \epsilon^2 t$, where $\lambda = v_g = d\omega/dk$ is the wave group (envelope propagation) velocity. We proceed by substituting into Eqs. (1)-(3) and isolating various orders (in ϵ^n), for the $L-$th harmonic contributions; details (omitted here for brevity) can be found elsewhere.[5]

The algebra is tedious yet straightforward. The equations for $n = 1$ reproduce the solution and dispersion characteristics presented in §3. The equations for $n = 2$ produce the (amplitudes of the) 2nd order harmonics, as well as a zeroth order (direct current) term. The solution thus obtained,

$$\phi \approx \epsilon\psi\,\cos\theta_c + \epsilon^2\left[\phi_0^{(2)} + \phi_1^{(2)}\cos\theta_c + \phi_2^{(2)}\cos 2\theta_c\right] + \mathcal{O}(\epsilon^3)\,, \qquad (7)$$

e.g., for the electric potential ϕ, represents a carrier wave (fast phase $\theta_c = kx - \omega t$); a set of similar expressions are obtained for $n_{+/-}$ and $\mathbf{u}_{+/-}$. At this stage, one is after an evolution equation for the potential correction ψ; once this is solved, anticipating a (complex) solution in the form $\psi = \psi_0 \exp i\Theta$, the first-order corrections to all quantities directly follow from (6) above. The real variables ψ_0 and Θ physically represent the potential (wavepacket) real (i.e., measurable) amplitude and a (small) phase correction, leading to a weakly varying total phase $\theta = \theta_c + \epsilon^2\Theta + \mathcal{O}(\epsilon^3)$.

A *nonlinear Schrödinger*-type equation (NLSE) arises, at order $n = 3$, as a compatibility equation (ensuring secular term annihilation). It reads

$$i\,\frac{\partial\psi}{\partial\tau} + P\,\frac{\partial^2\psi}{\partial\xi^2} + Q\,|\psi|^2\,\psi = 0\,. \tag{8}$$

The dispersion coefficient P is related to the dispersion characteristics as $P = d^2\omega/2dk^2$, as obtained from (4) above (for each branch, separately). The lengthy expression for the nonlinearity coefficient $Q = Q(k;\beta,\sigma)$ (here omitted, for brevity) is given by Eq. (19) in Ref. 7. Different expressions are obviously obtained for waves obeying the upper and lower dispersion curves (see in §3 above).

5. Modulational instability & localized envelope excitations

The evolution of the electric potential amplitude ψ depends on the coefficients P and Q,[5] whose analytical behavior is straightforward to investigate in terms of the physical parameters involved. The key element in this analysis, as discussed below, turns out to be the quantity P/Q: its sign (+/-) determines the regions where harmonic oscillations are unstable/stable, as well as the generic type (bright/dark, respectively, to be explained in the following) of envelope excitations, while its magnitude tunes their characteristics (amplitude ψ_0, width L) via a relation in the form $L \sim \psi_0(P/Q)^{1/2}$.

Modulational instability. Summarizing all physically relevant information – yet omitting details[5] – one may consider a harmonic solution of (8) in the form $\psi = \psi_0 \exp(iQ\,|\psi_o|^2\,\tau)$; the standard (linear) stability analysis then leads to the dispersion relation: $\tilde{\omega}^2 = P\tilde{k}^2(P\tilde{k}^2 - 2Q|\tilde{\psi}_0|^2)$, where $\tilde{\omega}$, $\tilde{k}$ and $\tilde{\psi}_0$ denote the frequency, wavenumber and amplitude perturbation(s), respectively. One sees that, for $PQ > 0$, a critical wavenumber threshold $\tilde{k}_{cr} = |\tilde{\psi}_0|(2Q/P)^{1/2}$ exists, below/above which, the envelope is unstable/stable (i.e., for perturbation wavelengths longer/shorter than $2\pi/k_c$); the maximum instability growth rate $|\mathrm{Im}\tilde{\omega}|_{max} = Q|\tilde{\psi}_0|^2$ occurs at $k_m = k_c/\sqrt{2}$. If $PQ < 0$, the amplitude will be stable to perturbations. This is essentially the Benjamin-Feir instability mechanism in hydrodynamics.[6]

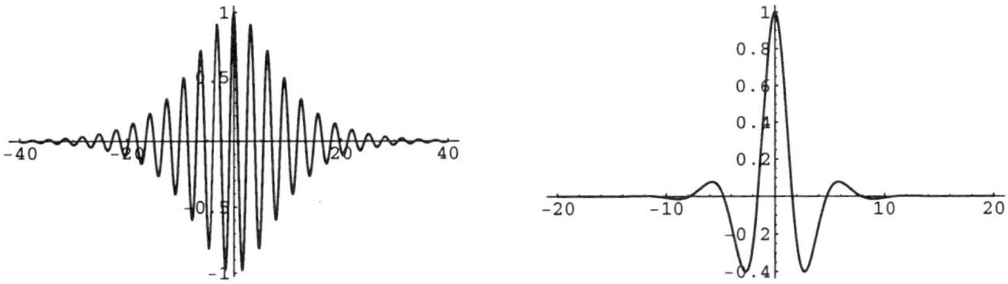

Fig. 2. *Bright* type modulated wavepackets (for $PQ > 0$), for two different (arbitrary) sets of parameter values. After Ref. 5.

Envelope excitations. The NLSE (8) defines an integrable dynamical system, which possesses various types of stationary profile spatially localized solutions, in the form of *envelope solitons*; the analytical form of the latter can be rigorously obtained via the Inverse Scattering Transform (IST) method. The remarkable properties of solitons (e.g. longevity and robustness against perturbations, shape-invariance through collisions), enumerating which goes far beyond our scope here (refer to specialized literature[6]) makes these solutions a loyal working horse for ES wavepacket theories.

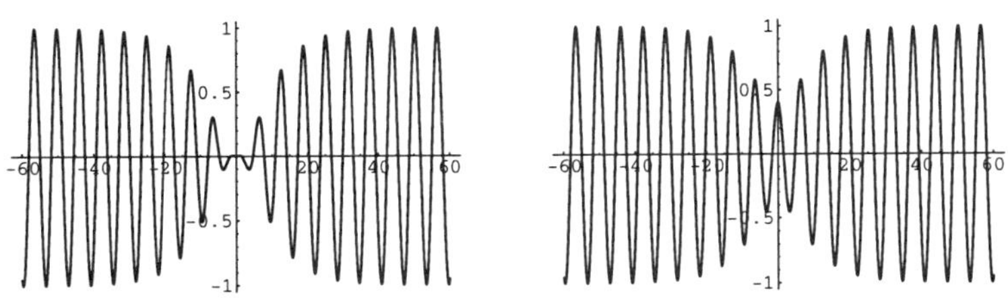

Fig. 3. *Dark*-type modulated wavepackets (for $PQ < 0$) of the *black* (left) and *grey* (right) kind. See that the amplitude never reaches zero in the latter case. After Ref. 5.

The modulated (electrostatic potential) wave amplitude forms resulting from the above analysis may be obtained via an *ansatz* of the form $\psi(\xi, \tau) = \rho(\xi, \tau) \exp[i\Phi(\xi, \tau)]$, where ρ and Φ are analytical (real) functions, to be determined upon substitution into the NLS. A number of exact solutions are thus obtained.[5,10] Omitting algebraic details[5,10] the physically relevant information of importance to us here is summarized in the following.

For positive values of the coefficient product $PQ > 0$ (or, of the ratio

$P/Q)^{e}$, the NLSE (8) possesses *bright*-type soliton solutions representing a propagating potential envelope pulse (vanishing at infinity), which encloses the fast carrier wave oscillation; see Fig. 2. This type of solution is reminiscent of signal pulses in optical fibres.[6]

For negative values of the coefficient product $PQ < 0$, we find *dark*-type soliton solutions, representing a propagating potential envelope void (i.e., a potential hole); these excitations bear a finite (constant) value at infinity; see Fig. 3. Localized envelope solitons of this kind exist in the form of *black* solitons, bearing a vanishing value in the center (see Fig. 3a), or *grey* solitons, bearing a finite potential value everywhere (see Fig. 3b).

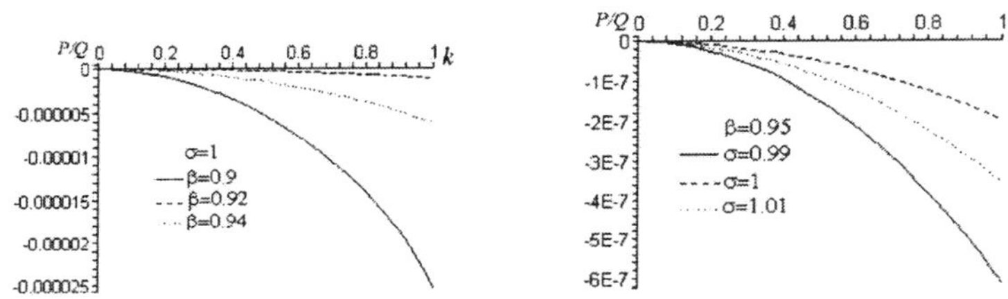

Fig. 4. The NLSE coefficient ratio P/Q corresponding to the *lower* (acoustic) dispersion branch ω_L is depicted against the (reduced) wavenumber k. (a) $\sigma = 1$ and different values of β are considered; (b) $\beta = 0.95$, and σ varies. Note that curves overlap. After Ref. 7.

Pair plasmas – parametric investigation. As regards pair plasmas, the stability profile can be determined along the above guidelines. The analysis shows that the lower (acoustic) mode ω_L is stable (viz. $P/Q < 0$; see Fig. 4) in all cases, for long wavelengths[f], while the upper (optic-type) branch ω_U is generally modulationally unstable (viz. $P/Q < 0$; see Fig. 5). As a consequence, the lower mode favors the propagation of dark-type envelopes (Fig. 3), while bright envelopes (pulses; see Fig. 2) are expected to occur in the upper mode. Increasing β (i.e. for $n_{+,0} > n_{-,0}$, implying a higher concentration of, say, negative background defects, or dust) leads to smaller (less extended) dark excitations (see Fig. 4a), while a positive background leads to the opposite effect. Temperature asymmetry (i.e. β variation) is also seen to affect the characteristics of dark envelopes (see Fig. 4b). On

[e]The coefficient Q is assumed *not to* vanish here. For vanishing Q, nonlinearity does not balance dispersion, so the analysis presented here fails. On the other hand, for vanishing P, one needs to resort to higher-order dispersion effects, here omitted.

[f]Long wavelengths $\lambda = 2\pi/k$ (i.e. small wavenumbers k) are relevant with a fluid model (inevitably invalidated by overseen kinetic effects, e.g. Landau damping, for higher k).

the other hand, bright envelopes, in the upper mode (though more likely to be experimentally observed than the acoustic mode; see comment in §3) bear no significant temperature asymmetry effect (notice the overlapping curves in Fig. 5).

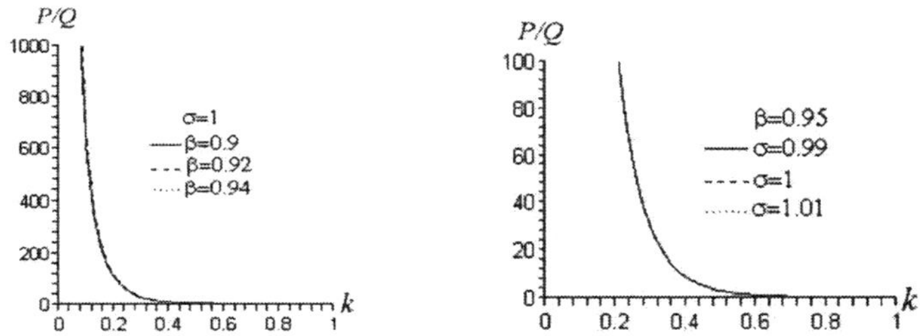

Fig. 5. JPA 4b and 4d. The NLSE coefficient ratio P/Q corresponding to the *upper* (optic) dispersion branch ω_U is depicted against the (reduced) wavenumber k. (a) $\sigma = 1$ and different values of β are considered; (b) $\beta = 0.95$, and σ varies. After Ref. 7.

6. A two-fluid model for EM wavepackets in pair plasmas

We shall now consider electromagnetic excitations propagating in a magnetized pair plasma (in a uniform ambient magnetic field $\mathbf{B_0}$). Retaining the above notation (unless otherwise stated) we adopt, for the pair-component fluids, the set of dynamical equations:

$$\frac{\partial n_j}{\partial t} + \nabla \cdot (n_j\, \mathbf{u}_j) = 0 \tag{9}$$

$$\frac{\partial \mathbf{u}_j}{\partial t} + \mathbf{u}_j \cdot \nabla \mathbf{u}_j = \frac{q_j}{m_j}\left(\mathbf{E} + \frac{1}{c}\,\mathbf{u}_j \times \mathbf{B}\right), \tag{10}$$

The (total) electric and magnetic fields, $\mathbf{E} = -\nabla\,\phi$ and $\mathbf{B}$, obey Maxwell's laws:

$$\frac{1}{c}\,\frac{\partial \mathbf{B}}{\partial t} = -\nabla \times \mathbf{E}\,, \tag{11}$$

$$\frac{1}{c}\,\frac{\partial \mathbf{E}}{\partial t} = \nabla \times \mathbf{B} - \frac{4\pi}{c}\sum_j n_j q_j \mathbf{u}_j\,, \tag{12}$$

and also obey Poisson's equation and Gauss' law(s):

$$\nabla \cdot \mathbf{E} = -\nabla^2 \phi = 4\pi \sum_{j=1,2,3} q_j\, n_j\,, \qquad \nabla \cdot \mathbf{B} = 0\,. \tag{13}$$

At equilibrium, overall charge neutrality is assumed. We take the direction of wave propagation together with $\mathbf{B_0}$ to define the $x-z$ plane, by taking $\mathbf{k} = k\hat{x}$ and $\mathbf{B_0} = B_{0,x}\hat{x} + B_{0,z}\hat{z} = B_0(\cos\theta\hat{x} + \sin\theta\hat{z})$; see Fig. 6. All quantities are assumed to vary along the direction of propagation, i.e. $\nabla \to \partial/\partial x$ (thus $\nabla \times \cdot = \hat{x} \times \partial \cdot /\partial x$). Eqs. (13b) and (11) thus imply a static x-magnetic field component: $B_x = B_{x,0} = B_0 \cos\theta$ =cst.

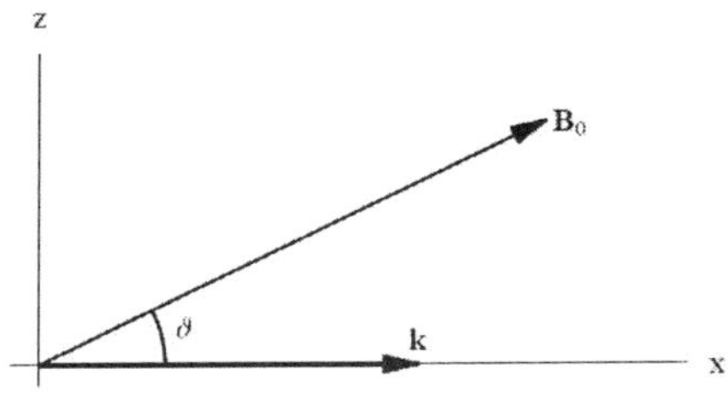

Fig. 6. The reference frame: EM wave propagation takes place along the $x-$axis, while the ambient magnetic field lies in the $xz-$plane.

The generic model described here agrees with previous studies of oblique EM wave propagation,[11,12] as well as for parallel propagation.[13]

7. Linear EM waves in three-component pair plasmas

The (linear wave) dispersion relation obtained from (9)-(13) has the form

$$D(\omega, k; \theta) = d_0(\omega, k) + d_1(\omega, k)\sin^2\theta = 0\,, \tag{14}$$

where d_0 and d_1 are polynomial expressions given by

$$\begin{aligned} d_0(\omega, k) &\equiv D(\omega, k; \theta = 0) \\ &= (\omega^2 - \omega_{p,eff}^2) \\ &\quad \times \left\{ \left[(\omega^2 - c^2k^2)(\omega^2 - \Omega^2) - \omega^2\omega_{p,eff}^2 \right]^2 - \omega^2\Omega^2(\omega_{p,1}^2 - \omega_{p,2}^2)^2 \right\} \\ &= (\omega^2 - \omega_{p,eff}^2) \\ &\quad \times \left\{ (\omega + \Omega)\left[-(\omega^2 - c^2k^2)(\omega - \Omega) + \omega\omega_{p,1}^2 \right] + \omega(\omega - \Omega)\omega_{p,2}^2 \right\} \\ &\quad \times \left\{ (\omega - \Omega)\left[-(\omega^2 - c^2k^2)(\omega + \Omega) + \omega\omega_{p,1}^2 \right] + \omega(\omega + \Omega)\omega_{p,2}^2 \right\}, \end{aligned} \tag{15}$$

and

$$d_1(\omega, k; \theta) = -c^2k^2\Omega^2 \left\{ c^2k^2\omega_{p,eff}^2(\omega^2 - \Omega^2) + \omega^2[4\omega_{p,1}^2\omega_{p,2}^2 - (\omega^2 - \Omega^2)\omega_{p,eff}^2] \right\}. \tag{16}$$

Here we have defined the effective (total) plasma frequency $\omega_{p,eff} = (\omega_{p,1}^2 + \omega_{p,2}^2)^{1/2}$ (cf. definitions in §1) and the (common among 1^+ and 2^-) cyclotron frequency $\Omega = ZeB_0/(mc)$. Note that d_0 is a 10th order polynomial in the frequency ω, while d_1 is a 4th order polynomial in ω; in both quantities, only even terms are present, so that d_0 (d_1) is essentially a quintic (quartic) polynomial in ω^2. Therefore, up to 5 roots for ω^2 may exist, hence 5 distinct propagating modes for the (real part of the) frequency $\omega(k)$ (even in k), depending on the angle θ and relevant parameter values.

Dimensionless form of the dispersion relation. The dispersion relation may be cast into a reduced form, by defining appropriate scales. Following Cramer *et al*,[14] we define the density mismatch parameter

$$\eta = \frac{n_{+,0} - n_{-,0}}{n_{+,0} + n_{-,0}} \tag{17}$$

(see that $\omega_{p,1}^2 - \omega_{p,2}^2 = \eta \omega_{p,eff}^2$), which measures deviation from pair-ion neutrality (the "pure" pair plasma case is recovered for $\eta \to 0$, i.e. $\omega_{p,+} = \omega_{p,-}$; the case $\eta \neq 0$ thus indicates the existence of a third species, or an overall neutrality violation in the plasma composition, at equilibrium). We define the reduced wave frequency, wavenumber and plasma frequency

$$f = \omega/\Omega\,, \qquad \kappa = ck/\Omega\,, \qquad h = \omega_{p,eff}^2/\Omega^2 = (\omega_{p,1}^2 + \omega_{p,2}^2)/\Omega^2\,, \tag{18}$$

respectively; see that $\omega_{p,1/2}^2/\Omega^2 \to (1 \pm \eta)h/2$. Eqs. (15, 16) thus become

$$\begin{aligned}
\hat{d}_0(\omega, k) &\equiv d_0/\Omega^{10} \\
&= (f^2 - h^2)\left\{\left[(f^2 - \kappa^2)(f^2 - 1) - f^2 h^2\right]^2 - f^2 \eta^2 h^2\right\} \\
&= (f^2 - h^2) \\
&\quad \times\left\{(f+1)\left[-(f^2 - \kappa^2)(f-1) + f(1+\eta)h/2\right] + f(f-1)(1-\eta)h/2\right\} \\
&\quad \times\left\{(f-1)\left[-(f^2 - \kappa^2)(f+1) + f(1+\eta)h/2\right] + f(f+1)(1-\eta)h/2\right\},
\end{aligned} \tag{19}$$

and

$$\hat{d}_1(\omega, k; \theta) \equiv d_1/\Omega^4 = -\kappa^2 \left\{\kappa^2 h(f^2 - 1) + f^2[(1-\eta^2)h^2 - (f^2 - 1)h]\right\}, \tag{20}$$

Parallel EM wave propagation. For $\theta = 0$, expression (14) reduces to $d_0 = 0$, implying (temporarily recovering dimensions, for clarity):

$$\omega^4 - \omega^2(\omega_{p,eff}^2 + \Omega^2 + c^2 k^2) \mp \omega\Omega(\omega_{p,1}^2 - \omega_{p,2}^2) + c^2 k^2 \Omega^2 = 0\,, \tag{21}$$

along with trivial plasma oscillations at $\omega = \omega_{p,eff}$. A number of distinct modes are therefore present. Note that the deviation from incompressibility (i.e. for non-zero values of $n_{1,0} - n_{2,0} \sim \omega_{p,1}^2 - \omega_{p,2}^2$), due to the existence of the third (fixed ion) species (e.g. ions in e-p-i plasmas, or "dust" defects in pair-ion, e.g. fullerene, plasmas), leads to the appearance of extra branches (which merge back into one another in the pure *p.p.* limit).

The modes described by Eq. (21) have been briefly analyzed in Ref. 14 (relying on Ref. 13). This equation may be cast in the form[14]

$$(f^2 - 1)(f^2 - \kappa^2) - f^2 h \pm \eta h f = 0 \tag{22}$$

where all (dimensionless) quantities were defined above.

Interestingly, in the pure pair-plasma case (i.e. for $\eta = 0$), (22) can be solved exactly for f^2, leading (apart from $f = \pm h$) to

$$f^2 = \frac{1}{2}(1 + \kappa^2 + h)\{1 \pm \left[1 - 4\kappa^2/(1 + \kappa^2 + h)^2\right]^{1/2}\}\,, \tag{23}$$

i.e.

$$f^2 \approx \frac{1}{2}(1 + \kappa^2 + h)\{1 \pm \left[1 - 2\kappa^2(1 - 2h)\right]\} \tag{24}$$

for small wavenumber κ (and, say, plasma frequency h). One thus obtains (lower branch) acoustic mode:

$$f_-^2 \approx (1 + \kappa^2 + h)\kappa^2(1 - 2h) \approx (1 - 2h)\kappa^2 + \mathcal{O}(\kappa^2)\,, \tag{25}$$

and an (upper branch) optic-type mode:

$$f_+^2 \approx (1 + \kappa^2 + h)\left[1 - \kappa^2(1 - 2h)\right]. \tag{26}$$

Switching back to $\eta \neq 0$, the effect of the density mismatch, which results e.g. from the existence of a third (fixed ion) species, is to split the two *linearly polarized* EM modes (present in *p.p.*[12]) to *four distinct circularly polarized modes.*[14] Focusing on the behavior near $k = 0$, one finds that three out of these modes present a frequency cutoff, i.e. $\omega(k = 0) \neq 0$, below which no wave propagates. In the vicinity of $f \approx 0$, and for small η and h, one finds analytically that the Alfvén type wave which occurs for $\eta = 0$, splits into two modes, one of which presents a cutoff at $f_0 = |\eta| h/(1 + h)$. The behavior presented here is depicted in Fig. 7.

Perpendicular EM wave propagation. For $\theta = \pi/2$, expression (14) reduces to:

$$D(\omega, k; \frac{\pi}{2}) = d_{\perp,1}(\omega, k)\, d_{\perp,2}(\omega, k) = 0\,, \tag{27}$$

where

$$\begin{aligned} d_{\perp,1}(\omega, k) = &-\omega^6 + \omega^4[c^2\, k^2 + 2(\Omega^2 + \omega_{p,eff}^2)] \\ &-\omega^2[(\Omega^2 + \omega_{p,eff}^2)^2 - c^2 k^2(2\Omega^2 + \omega_{p,eff}^2)] \\ &+\Omega^2[c^2 k^2\, (\Omega^2 + \omega_{p,eff}^2) + (\omega_{p,1}^2 - \omega_{p,2}^2)^2] \end{aligned} \tag{28}$$

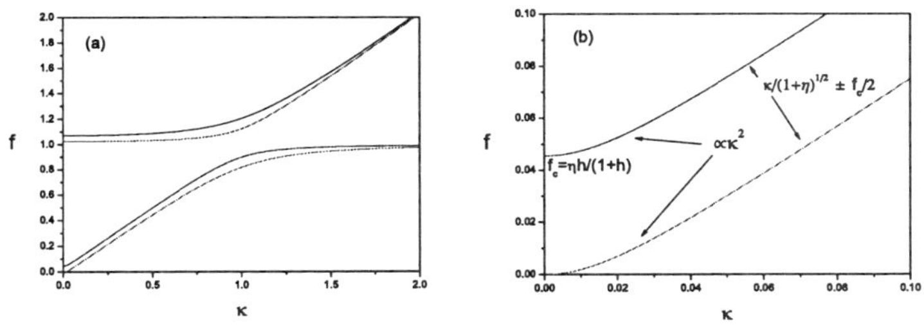

Fig. 7. Linear dispersion relation for EM waves in a three component pair-ion (or *epi*) plasma: the reduced frequency $f = \omega/\Omega$ is depicted against the reduced wavenumber $\kappa = ck/\Omega$; (a) full frequency range; (b) focusing near the origin. Here $\eta = 0.5$, $h = 0.1$ (definitions in the text).

and

$$d_{\perp,2}(\omega, k) = \omega^2 - \omega_{p,eff}^2 - c^2k^2 \tag{29}$$

The latter equation defines the ordinary (or O-) mode;[1] this is a robust perpendicular EM mode, whose dispersion characteristics do not depend on the ambient magnetic field, i.e., it bears the same form for *e-i* plasmas and, in fact, for unmagnetized plasmas. Adopting the O-mode has enabled us to advance in analytical tractability, as regards nonlinear EM wave dynamics,[9] as we shall see below. This study[9] is reminiscent of an earlier study of the O-mode, with respect to modulation effects, yet for unmagnetized plasmas.[15]

The "pure" pair-plasma limit for arbitrary θ. In the absence of the 3rd (fixed ion) species, one recovers, setting $\omega_{p,1} = \omega_{p,2} = \omega_p$) in (14):

$$\begin{aligned} D|_{p.p.}(\omega, \theta) = {} & \left[(\omega^2 - c^2k^2)(\omega^2 - \Omega^2) - 2\omega^2\omega_p^2\right] \\ & \times \left[\omega^2(\omega^2 - c^2k^2 - 2\omega_p^2)(\omega^2 - \Omega^2 - 2\omega_p^2) - 2c^2k^2\Omega^2\omega_p^2\cos^2\theta\right] \end{aligned} \tag{30}$$

Setting the first quantity within brackets to zero, one recovers:

$$\omega^4 - (c^2k^2 + \Omega^2 + 2\omega_p^2)\,\omega^2 + c^2k^2\Omega^2 = 0\,, \tag{31}$$

which coincides[g] with Eq. (9) in Ref. 12 (also see (24)-(26) in Ref. 3a. Eq. (31) represents the dispersion relation of an EM wave which propagates

[g] Upon a trivial difference in notation, though: see that $\omega_{p,eff}^2 = \omega_{p,1}^2 + \omega_{p,2}^2$ here is denoted by ω_p^2 in Refs. 12 and 14.

for any value of the pitch angle θ, and whose only non-vanishing electric field component E_y is perpendicular to the plane spanned by the magnetic field $\mathbf{B_0}$ and the wave vector $\mathbf{k}$ (i.e. $E_x = E_z = 0$); this mode is always characterized by charge neutrality ($n_i = n_e = n \neq n_0$, off equilibrium), for $\theta \neq 0$. For parallel propagation ($\theta = 0$, or $B_z = 0$), this mode corresponds to a splitting of the incompressible ($n_i = n_e = n_0$), circularly polarized EM waves (present in e-i plasmas) into two orthogonal, linearly polarized EM waves, both obeying Eq. (31). For perpendicular propagation ($\theta = \pi/2$, or $B_x = 0$), this mode is part of the extraordinary (X) mode;[1] also see (21)-(22) in Ref. 3a, and the discussion therein. See that both relations (21) merge into (31) for $\omega_{p,1} = \omega_{p,2}$; vice versa, in the presence of a 3rd species (e.g. in *e-p-i* plasmas), this mode splits into the 2 parts in (21).

On the other hand, upon setting the second quantity within brackets, in *rhs*(30) to zero, one recovers exactly[g] Eq. (10) in Ref. 12, representing an EM mode propagating in pair plasmas, for which $E_y = 0$ (i.e. no electric field is generated perpendicular to the plane defined by $\mathbf{B_0}$ and $\mathbf{k}$). Interestingly, for $\theta = \pi/2$, one obtains a pair of (decoupled) dispersion relations, namely $\omega^2 = 2\omega_p^2 + c^2k^2$ (corresponding to an incompressibly, linearly polarized ordinary (O) mode,[1] with $E_z \neq 0$) and $\omega^2 = 2\omega_p^2 + \Omega^2$ (representing a fixed frequency, pure upper-hybrid mode, with $E_x \neq 0$).[12]

The case of parallel EM wave propagation in *p.p.* is obtained either by setting $\theta = 0$ in (30), or by setting $\omega_{p,1} = \omega_{p,2} = \omega_p$ in (15):

$$D_0|_{p.p.}(\omega, \theta = 0) = -i\,\omega^3\,(\omega^2 - 2\omega_p^2)\,[\omega^4 - (c^2k^2 + \Omega^2 + 2\omega_p^2)\,\omega^2 + c^2k^2\Omega^2]^2\,,$$

thus recovering the E_y-mode discussed above, plus trivial (non-propagating, since pressure effects are neglected) plasma oscillations at $\omega = \sqrt{2}\omega_p$.

8. Nonlinear EM harmonic modes

The multiple scales technique presented in §4 above has been employed for EM modes in *p.p.*. The results have been presented in Ref. 9, for the ordinary (O-)mode, while part of the more general results[16] have appeared (yet in a preliminary form) in Ref. 17. In the following, we shall present some essential results of the nonlinear analysis, while an interested reader is referred to the references for details on the tedious calculation.

The algebraic manipulation of the 1st order 1st harmonic ($n = 1$, $l = 1$) amplitude evolution equations (9)-(13) provides a set of equations for the fluid densities and velocities vs. the E/M field components. Furthermore, a 2nd-order correction is obtained, for all quantities, incorporating a 2nd-

and a zeroth-phase-harmonic(s)[h]. For the ordinary mode, the solution (up to $\sim \epsilon^2$) can be summarized as[9]

$$
\begin{aligned}
n_j &= n_{j,0} + \epsilon\, c_j^{(11)} B'_y\, e^{i\phi_c} + \epsilon^2 [c_j^{(22)} B'^2_y e^{i2\phi_c} + n_j^{(20)}] \\
\mathbf{u_j} &= \mathbf{0} + \epsilon\, c_{j,z}^{(11)} B'_y e^{i\phi_c}\, \hat{z} + \epsilon^2 \left\{ c_{j,z}^{(21)} \frac{\partial B'_y}{\partial X_1} e^{i\phi_c} \hat{z} + B'^2_y e^{i2\phi_c} [c_{j,x}^{(22)} \hat{x} + c_{j,y}^{(22)} \hat{y}] + \mathbf{u_j}^{(20)} \right\} \\
\mathbf{E} &= \mathbf{0} + \epsilon\, c_{el,z}^{(11)} B'_y e^{i\phi_c} \hat{z} + \epsilon^2 \left\{ c_{el,z}^{(21)} \frac{\partial B'_y}{\partial X_1} e^{i\phi_c} \hat{z} + B'^2_y e^{i2\phi_c} [c_{el,x}^{(22)} \hat{x} + c_{el,y}^{(22)} \hat{y}] + \mathbf{E}^{(20)} \right\} \\
\mathbf{B} &= B_0 \hat{z} + \epsilon\, B'_y e^{i\phi_c} \hat{y} + \epsilon^2 \left[c_{B,y}^{(21)} \frac{\partial B'_y}{\partial X_1} e^{i\phi_c} \hat{y} + c_{B,z}^{(22)} B'^2_y e^{i2\phi_c} \hat{z} + \mathbf{B}^{(20)} \right] .
\end{aligned}
\tag{32}
$$

Here $j = 1, 2 \equiv +, -)$, $B'_y = B_y^{(11)}/B_0$ and $\phi_c = kx - \omega t$. The *arbitrary* zeroth-order corrections satisfy

$$
u_{1,x}^{(20)} = -u_{2,x}^{(20)} = cE'^{(20)}_y , \qquad u_{1,y}^{(20)} = -u_{2,y}^{(20)} = -cE'^{(20)}_x . \tag{33}
$$

It is worth mentioning that the compatibility conditions, imposed at orders ϵ^2 and ϵ^3 for secular terms to annihilate, respectively take the form

$$
\left(\frac{\partial}{\partial T_1} + v_g \frac{\partial}{\partial X_1} \right) \tilde{B}_\perp = 0 , \tag{34}
$$

[v_g here denotes the EM wave group velocity $\omega'(k) = -(\partial D/\partial k)/(\partial D/\partial \omega)$, as it results from the dispersion relation $D(\omega, k) = 0$ – see (14) – obtained previously [i]] and

$$
i \frac{\partial \tilde{B}_\perp}{\partial \tau} + P \frac{\partial^2 \tilde{B}_\perp}{\partial \xi^2} + Q\, |\tilde{B}_\perp|^2 \tilde{B}_\perp = 0 . \tag{35}
$$

Here, $\tilde{B}_\perp = [B_z^{(11)} + C B_y^{(11)}]/B_0$ is a linear combination of the magnetic field components $\perp$ to the propagation direction, and C is a complex quantity:

$$
C(\omega, k; \theta) = -2\, i\, \omega\, \Omega\, (\omega_{p,1}^2 - \omega_{p,2}^2) \cos\theta\, \frac{C_1}{C_2} = 0 , \tag{36}
$$

where

$$
C_1 = 4c^2k^2(\omega^2 - \Omega^2)(\omega^2 - \omega_{p,eff}^2) + 2\Omega^2 \omega_{p,eff}^2 (\omega^2 - \omega_{p,eff}^2 - c^2k^2) \sin^2\theta
$$

---

[h] The structure of the lengthy algebra and the resulting expressions obtained at orders ϵ^1 and ϵ^2 is (are) reported in Sections V and VI, respectively, of Ref. 17. Details to appear elsewhere.[16]

[i] In specific, differentiating $D(\omega(k), k) = 0$ with respect to k gives:

$$
\frac{\partial D}{\partial k} + \frac{\partial D}{\partial \omega} \frac{d\omega}{dk} = 0 , \qquad \text{hence} \qquad \frac{d\omega}{dk} = - \frac{\frac{\partial D}{\partial k}}{\frac{\partial D}{\partial \omega}} .
$$

and

$$C_2 = 8c^2k^2(\omega^2 - \Omega^2)(\omega^2 - \omega_{p,eff}^2)[\omega^2(\omega^2 - \Omega^2 - \omega_{p,eff}^2) - c^2k^2(\omega^2 - \Omega^2)]$$
$$+ 4\Omega^2\omega_{p,eff}^2 \sin^2\theta$$
$$\times[2c^4k^4(\omega^2 - \Omega^2) - i\,\omega\Omega(\omega_{p,1}^2 - \omega_{p,2}^2)(\omega^2 - \omega_{p,eff}^2 - c^2k^2)\cos\theta]\,. \quad (37)$$

Interestingly, for *parallel* EM wave propagation (i.e. for $\theta = 0$), $C \to \pm i = e^{\pm i\pi/2}$, suggesting a phase difference of $\pm\pi/2$ among B_y and B_z, specifically when the frequency ω obeys the (parallel EM wave) dispersion relation (15). The slowly evolving transverse magnetic field component is then $B_z \pm iB_y = \pm i(B_y \mp iB_z) \equiv \pm iB_\perp^*$. The (anticipated) circular polarization encountered for modulated EM wavepackets propagating parallel to $\mathbf{B_0}$ (see e.g. in Ref.[13]) in multi-component plasmas is thus recovered. Also note, for rigor, that the quantity C vanishes for *perpendicular* EM wave propagation (i.e. for $\theta = \pi/2$), for $\omega_{p,1} \neq \omega_{p,2}$, and also in the *pure* pair-ion plasma case (i.e. for $\omega_{p,1} = \omega_{p,2}$, $\forall\,\theta$).

The dispersion coefficient in Eq. 35 is given by $P = d^2\omega(k)/2dk^2$. The nonlinearity coefficient Q is a complicated function of the angle θ and of the characteristic frequencies $\omega_{p,1/2}$ and Ω. For the O-mode, it has the form $Q = Q_1 + Q_2$, with

$$Q_1 = Q_A/Q_B$$
$$Q_A = 3\omega_{p,eff}^2\{-4\omega^2(\Omega_1^2\omega_{p,1}^2 + Z\Omega_2^2\omega_{p,2}^2) + \Omega_1\omega_{p,1}^2[\Omega_1\Omega_2^2 + (\Omega_1 + \Omega_2)\omega_{p,2}^2] + Z\Omega_2\omega_{p,2}^2[\Omega_2\Omega_1^2 + (\Omega_1 + \Omega_2)\omega_{p,1}^2]\}$$
$$Q_B = \omega(c_4\omega^4 + c_2\omega^2 + c_0)$$
$$c_4 = 48\omega_{p,eff}^2$$
$$c_2 = -4[3\omega_{p,eff}^4 + 3\omega_{p,eff}^2(\Omega_1^2 + \Omega_2^2) + \omega_{p,1}^2\Omega_1^2 + \omega_{p,2}^2\Omega_2^2]$$
$$c_0 = 3(\omega_{p,1}^4\Omega_2^2 + \omega_{p,2}^4\Omega_1^2) + 2\Omega_1\Omega_2\omega_{p,1}^2\omega_{p,2}^2 + 4(\Omega_1^2 + \Omega_2^2)\omega_{p,1}^2\omega_{p,2}^2 + 4\Omega_1^2\Omega_2^2\omega_{p,}^2$$
$$Q_2 = -\frac{1}{2\omega}\left(\frac{n_1^{(20)}}{n_{1,0}}\omega_{p,1}^2 + \frac{n_2^{(20)}}{n_{2,0}}\omega_{p,2}^2\right)$$

($\Omega_1 = \Omega_2$, yet $\omega_{p,1} \neq \omega_{p,2}$, is understood here, for three component *p.p.*). By further assuming $\omega_{p,1} = \omega_{p,2} = \omega_p$ ("pure" *p.p.*, no third component), we obtain the simple expressions:

$$P = \frac{c^2\omega_p^2}{\omega^3}\,, \qquad Q = \frac{3\Omega^2\omega_p^2}{2\omega(\Omega^2 - 3\omega^2)} - \frac{1}{2\omega}\frac{\omega_p^2}{n_0}(n_1^{(20)} + n_2^{(20)})\,. \quad (38)$$

Concluding, the important quantity to deal with, as regards EM wavepackets in pair plasmas, is $\tilde{B}_\perp$ (defined above), which obeys the NLSE (35). The coefficients in the latter may be used for an investigation of the modulation stability profile of EM waves (and associated envelope structures predicted), along the general directions set in §5.

Modulational stability profile of EM waves in pair plasmas. The coefficients of the NLSE (35) are depicted in Fig. 8, for the case of $\theta = 0$, i.e. for EM wave propagation parallel to the external magnetic field. Note the forbidden frequency region (gap) near $f = 1$ (i.e. near $\omega = \Omega$); cf. Fig. 7. We see that, for pure *p.p.* (for $\eta = 0$, i.e. in the absence of a third species), the coefficient product PQ is positive at small frequencies (i.e. for the Alfvén type *p.p.* mode lying below the cyclotron frequency Ω), thus prescribing modulational instability and bright-type envelope excitations. However, PQ becomes negative as one approaches $f = 1$ from below, so high frequency waves will tend to be stable, and propagate as dark-type envelope solitons (envelope holes). A similar alternating (positive, then negative) behavior is obtained by gradually increasing ω (above Ω).

By "switching-on" the existence of the 3rd massive background species, the product PQ ($\alpha\beta$ in Fig. 8), becomes negative for low a frequency, thus apparently stabilizing the two sub-cyclotron modes (see in Fig. 7). This is due to a shift in sign of the dispersion coefficient (right column in Fig. 8, 2nd and 3rd rows) at low ω. We have seen (cf. Fig. 7) that the linearly polarized (pure) *p.p.* EM acoustic mode splits into two modes (one presenting a gap; see Fig. 7b) if a 3rd species is present. Both of these mode, namely a left-hand- and a right-hand-polarized one, exhibit the described behavior.

For perpendicular propagation (O-mode), the result (38) can be employed to investigate the nonlinear amplitude profile. Neglecting $n_j^{(20)}$, we see that the ratio ω/Ω bears a threshold $1/\sqrt{3}$, below (above) which dispersion is anomalous (normal) – borrowing terminology from nonlinear optics, implying bright (dark) type excitations and modulational instability (stability) of the wavepacket amplitude. A detailed investigation of the ordinary mode in pair plasmas is carried out in Ref. 9, so lengthy details (reported therein) were chosen to be omitted here, for brevity.

9. Summary and conclusions

We have considered the propagation of nonlinear amplitude-modulated EM wavepackets in a multi-component plasma. By adopting a multiple scales technique, we have investigated the linear oscillation profile (arising to first order) and have succeeded in showing how secondary harmonic generation, modulational instability and envelope soliton formation may occur in pair plasmas. Both ES and EM waves have been considered.

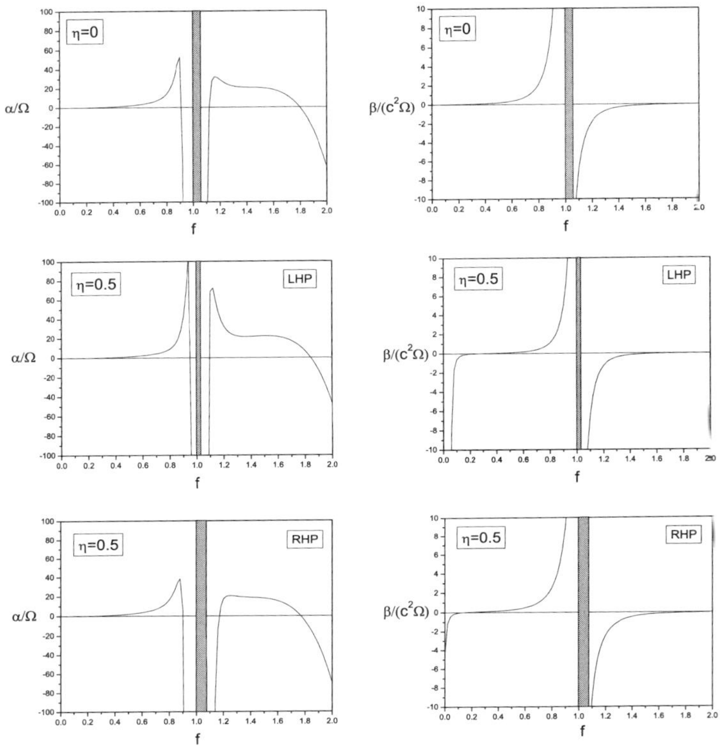

Fig. 8. The nonlinearity (left column) and dispersion (right column) coefficients in the NLS Eq. (35) for parallel EM wave propagation ($\theta = 0$) are plotted against the reduced frequency $f = \omega/\Omega$: (a) Pure pair plasma [$\eta = 0$; recall def. in (17)] - linear polarization (1st row); (b) Three-component pair plasma ($\eta = 0.5$) - left-hand polarization (2nd row); (c) Three-component pair plasma ($\eta = 0.5$) - right-hand polarization (3rd row). Note the frequency gap near $f = 1$, i.e. near $\omega = \Omega$; cf. Fig. 7. From Ref. 14.

We have shown that either a temperature difference among the two pair components or the presence of a third massive species (in *"doped"*, say, pair-ion plasmas, or *e-p-i* plasmas) may affect the stability profile of plasma waves. For instance, parallel EM Alfvén-like waves may be stabilized by the background component, also recovering circular polarization (lost in "pure", symmetric *p.p.*, where the respective mode is linearly polarized).

These results are of relevance with fullerene experiments[4] and *e-p-i* plasma related observations, where they may be tested and (hopefully) confirmed).

Acknowledgments.

This material is based on a lecture given by I.K. in *2007 Summer College on Plasma Physics*, at International Centre for Theoretical Physics (Trieste, Italy, 30 July - 24 August 2007). The Organizers of the event are warmly thanked for the invitation and the hospitality amply offered throughout his stay. The work presented here was carried out during a research visit at Universiteit Gent, Belgium in 2006 (funded by FWO, Belgian Research Fund). I.K. is grateful for that invitation, and vividly reminiscent of a fruitful stay, where multiple interactions with Prof. F. Verheest and a number of visitors have provided a warm scientific environment.

References

1. T. Stix, *Waves in plasmas* (New York: American Institute of Physics, 1992).
2. P. K. Shukla and A. A. Mamun, *Introduction to dusty plasma physics* (Institute of Physics, Bristol, 2002).
3. N. Iwamoto, *Phys. Rev. E* **47**, 604 (1993); G.A. Stewart & E.W. Laing, *J. Plasma Phys.* **47** 295 (1992); G. P. Zank & R. G. Greaves, *Phys. Rev. E* **51**, 6079 (1995).
4. W. Oohara and R. Hatakeyama, *Phys. Rev. Lett.* **91**, 205005 (2003); W. Oohara, D. Date and R. Hatakeyama, *Phys. Rev. Lett.* **95**, 175003 (2005).
5. I. Kourakis and P. K. Shukla, *Nonlin. Proc. Geophys.*, **12**, 407 (2005).
6. M. Remoissenet, *Waves Called Solitons* (Berlin, Springer-Verlag, 1996); T.Dauxois and M. Peyrard, *Physics of Solitons* (Cambridge Univ. Press, 2005).
7. A. Esfandyari-Kalejahi *et al*, *J. Phys. A: Math. Gen.*, **39**, 13817 (2006).
8. A. Esfandyari-Kalejahi *et al*, *Phys. Plasmas*, **13**, 122310 (2006)
9. I. Kourakis, F. Verheest and N. Cramer, *Phys. Plasmas*, **14**, 022306 (2007).
10. R. Fedele and H. Schamel, Eur. Phys. J. **B27**, 313 (2002).
11. H. Hasegawa and Y. Ohsawa, *J. Phys. Soc. Japan* **73**(7), 1764 (2004).
12. F. Verheest and T. Cattaert, Phys. Plasmas **12**, 032304 (2005).
13. S. Irie and Y. Ohsawa, *J. Phys. Soc. Japan* **70**(6), 1585 (2001).
14. N.F. Cramer *et al*, Proc. 28th ICPIG (Prague, 2007), paper 1P04-04.
15. G S Lakhina & B Buti, *Astrophys. Space Sci.* **79**, 25 (1981).
16. I. Kourakis & F. Verheest, unpublished material.
17. *International Workshop on Frontiers of Plasma Science* (2006), Abdus Salam ICTP (activity smr 1765), Trieste, Italy; online Lecture Notes, available at: `http://cdsagenda5.ictp.trieste.it/full_display.php?ida=a05217`.

ELECTRO-ACOUSTIC SOLITARY WAVES IN DUSTY PLASMAS

A. A. MAMUN*

Department of Physics, Jahangirnagar University,
Savar, Dhaka-1342, Bangladesh
**E-mail: mamun_phys@yahoo.co.uk*

P. K. SHUKLA

Institut für Theoretische Physik IV, Fakultät für Physik und Astronomie,
Ruhr-Universität Bochum, D-44780 Bochum, Germany
E-mail: ps@tp4.rub.de

We study two important classes [viz. dust-ion acoustic (DIA) and dust-acoustic (DA)] of electro-acoustic solitary waves in dusty plasmas. We employ the reductive perturbation method for small but finite amplitude solitary waves as well as the pseudo-potential approach for arbitrary amplitude ones. We analyze the effects of positive ions, nonplanar geometry and dust charge fluctuation on DIA solitary waves. On the other hand, we examine the effects of non-isothermal (vortex-like and nonthermal) ion distributions and positive dust on DA solitary waves. It has been reported that the effects, which are included in DIA (DA) solitary waves, do not only significantly modify the basic features of DIA (DA) solitary waves, but also introduce some important new features. The basic features and underlying physics of DIA and DA solitary waves, which are relevant to space and laboratory dusty plasmas, are briefly discussed.

Keywords: Electro-acoustic Waves, Dusty plasmas, Solitary Waves

1. Introduction

The physics of charged dust particles has become an outstanding and challenging research topic not only because dust particles are ubiquitous in most space[1–16] and laboratory plasmas,[17–19] but also because it has introduced a great variety of new phenomena associated with waves and instabilities[13–15,20–30] in a weakly coupled unmagnetized dusty plasma, and it plays a vital role in understanding different interesting phenomena in astrophysical and space environments, such as interplanetary space, interstellar medium, interstellar or molecular clouds, comets, planetary rings, earth's environments, etc.[1–16] The presence or dynamics of charged dust in a plasma does not only modify the existing plasma wave spectra,[20–22] but also introduces a number of new novel eigenmodes. The most important classes of new novel eigenmodes, which are experimentally observed in a weakly coupled unmagnetized duty plasma, are dust-ion acoustic (DIA) waves[23–25] and dust-acoustic (DA) waves.[26–28]

Shukla and Silin[23] have first theoretically shown that due to the conservation of equilibrium charge density $n_{e0} + n_{d0}Z_{d0} = n_{i0}$ (where n_{e0}, n_{i0} and n_{d0} are unperturbed electron, ion and dust number density, respectively, Z_{d0} is number of electrons residing onto the dust grain surface at equilibrium) and the strong inequality $n_{e0} \ll n_{i0}$, a dusty plasma (with negatively charged static dust grains) supports low-frequency DIA waves with a phase speed much smaller (larger) than electron (ion) thermal speed. The dispersion relation (a relation between the wave frequency ω and the wave number k) for the linear DIA waves is[23] $\omega^2 = (n_{i0}/n_{e0})k^2C_i^2/[1 + k^2\lambda_{De}^2(1 + T_i n_{e0}/T_e n_{i0})]$, where $C_i = (k_BT_e/m_i)^{1/2}$ is the ion-acoustic speed, $\lambda_{De} = (k_BT_e/4\pi n_{e0}e^2)^{1/2}$ is the electron Debye-radius, T_e (T_i) is the electron (ion) temperature, m_i is the ion mass, k_B is the Boltzmann constant, and e is the charge of an electron. When we consider a long wavelength limit (viz. $k\lambda_{De} \ll 1$), the dispersion relation for the DIA waves becomes $\omega = (n_{i0}/n_{e0})^{1/2}kC_i$. This form of spectrum is similar to the usual ion-acoustic wave spectrum[31] for a plasma with $n_{i0} = n_{e0}$ and $T_i \ll T_e$. However, in a dusty plasma we usually have $n_{i0} \gg n_{e0}$ and $T_i \simeq T_e$. Therefore, a dusty plasma cannot support the usual ion-acoustic waves, but can do the DIA waves of Shukla and Silin.[23] The phase speed ($v_p = \omega/k$) of the DIA waves is larger than ion-acoustic speed (C_i) because of $n_{i0} \gg n_{e0}$ for negatively charged dust grains. The increase in the phase speed is attributed to the electron density depletion in the background plasma, so that the electron Debye-radius becomes larger. As a result, there appears a stronger space charge electric field which is responsible for the enhanced phase speed of the DIA waves. The DIA waves have been observed in a number of laboratory experiments.[24,25,30]

Rao *et al.*[26] theoretically predicted the existence of extremely low phase velocity (in comparison with the electron and ion thermal speeds) DA waves in an unmagnetized dusty plasma whose constituents are an inertial charged dust fluid and Boltzmann ions and electrons. Thus, in DA waves the inertia is provided by the dust particle mass and the restoring force comes from the pressures of electrons and ions. The dispersion relation for the DA waves with a phase speed $V_p = \omega/k$ much smaller (larger) than the ion (dust) thermal speed is[26] $\omega = kc_d/\sqrt{1+k^2\lambda_D^2}$, where $c_d = \omega_{pd}\lambda_D$ is the dust-acoustic speed, $\omega_{pd} = (4\pi n_{d0}Z_{d0}^2e^2/m_d)^{1/2}$ is the dust plasma frequency, $\lambda_D = \lambda_{De}\lambda_{Di}/\sqrt{\lambda_{De}^2+\lambda_{Di}^2}$ is the effective Debye-radius, and $\lambda_{Di} = (k_BT_i/4\pi n_{i0}e^2)^{1/2}$ is the ion Debye-radius. It is obvious that without the consideration of the dust dynamics, one cannot obtain the DA mode. The theoretical prediction of Rao *et al.*[26] has been conclusively verified by a number of laboratory experiments.[27,28,30]

The linear properties of the DIA and DA waves in dusty plasmas are now well understood and have been reported by a large number of regular and review articles or books during the last few years.[13–15,23–30,32–34] The linear theory is valid only when the wave amplitude is so small that one may neglect the nonlinearities. However, there are numerous processes via which the unstable modes can saturate and attain large amplitudes. When the amplitudes of the waves are sufficiently large, nonlinearities can no longer be ignored. The nonlinearities come from the harmonic generation involving the fluid advection, nonlinear Lorentz force, ponderomotive force, etc. The nonlinearities in plasmas contribute to the localization of waves, leading to different types of interesting nonlinear coherent structures (viz. solitary waves, shock waves, double layers, vortices, etc.) which are important from both theoretical and experimental points of view, and have received a great deal of attention for understanding the basic properties of localized electro-acoustic perturbations in space and laboratory dusty plasmas. Recently, a number of theoretical attempts has been made in order to study the properties of DIA[35–40] and DA[15,26,39–49] solitary waves in unmagnetized dusty plasmas. We, in our present article, systematically study the basic features and underlying physics of small as well as arbitrary amplitude DIA and DA solitary waves in different unmagnetized dusty plasma situations.

The manuscript is organized as follows. We first consider an unmagnetized dusty plasma with static dust, and study the basic features and underlying physics of small as well as arbitrary amplitude DIA solitary waves in Sec. 2. We then consider an unmagnetized dusty plasma with mobile dust, and study the basic features and underlying physics of small as well as arbitrary amplitude DA solitary waves in Sec. 3. We, finally, provide a brief discussion in Sec. 4.

2. Static Dust: DIA Solitary Waves

We confine ourselves, in this section, to the DIA solitary waves in an unmagnetized dusty plasma in which dust particles are stationary and provide only the background charge-neutrality. The basic equations governing the dynamics of one-dimensional DIA waves, whose phase speed v_p is much larger than the ion thermal speed V_{Ti}, but much smaller than the electron thermal speed V_{Ti}, viz. $V_{Ti} \ll v_p \ll V_{Te}$, can be expressed in terms of the normalized variables as

$$\frac{\partial n_i}{\partial t} + \frac{\partial}{\partial x}(n_i u_i) = 0, \tag{1}$$

$$\frac{\partial u_i}{\partial t} + u_i \frac{\partial u_i}{\partial x} = -\frac{\partial \phi}{\partial x}, \tag{2}$$

$$\frac{\partial^2 \phi}{\partial x^2} = \mu \exp(\phi) - n_i + (1 - \mu), \tag{3}$$

where n_i is the ion number density normalized by its equilibrium value n_{i0}, u_i is the ion fluid speed normalized by the ion-acoustic speed C_i, ϕ is the DIA wave potential normalized by $k_B T_e/e$, and $\mu = n_{e0}/n_{i0}$. We have neglected the ion thermal pressure term, and assumed a Boltzmann electron density response, which are valid as long as $V_{Ti} \ll v_p \ll V_{Te}$. The space variable x is normalized by the modified electron Debye-radius $\lambda_{Dem} = (k_B T_e/4\pi n_{i0} e^2)^{1/2}$ and the time variable t is normalized by the ion plasma period $\omega_{pi}^{-1} = (m_i/4\pi n_{i0} e^2)^{1/2}$.

We first study small amplitude DIA solitary waves by the reductive perturbation method.[50] To do so, we express Eqs. (1)–(3) in terms of the stretched coordinates $\zeta = \epsilon^{1/2}(x - v_0 t)$, $\tau = \epsilon^{3/2} t$ (where $v_0 = v_p/C_i$ and ϵ is the expansion parameter, measuring the amplitude of the wave or the weakness of the wave dispersion), expand n_i, u_i and ϕ in power series of ϵ

$$n_i = 1 + \epsilon n_i^{(1)} + \epsilon^2 n_i^{(2)} + \cdots, \tag{4}$$

$$u_i = \epsilon u_i^{(1)} + \epsilon^2 u_i^{(2)} + \cdots, \tag{5}$$

$$\phi = \epsilon \phi^{(1)} + \epsilon^2 \phi^{(2)} + \cdots, \tag{6}$$

and develop equations in various powers of ϵ. The latter can be used to derive a linear dispersion relation ($v_0 = 1/\sqrt{\mu}$) and a Kortweg-de Vries (K-dV) equation

$$\frac{\partial \phi^{(1)}}{\partial \tau} + A_i \phi^{(1)} \frac{\partial \phi^{(1)}}{\partial \zeta} + B \frac{\partial^3 \phi^{(1)}}{\partial \zeta^3} = 0, \tag{7}$$

where $A_i = (3\mu - 1)/\sqrt{2\mu}$ and $B = 1/2\mu^{3/2}$. Now, for a frame moving with speed u_0 (normalized by C_i), the stationary solitary wave solution of Eq. (7) is given by

$$\phi^{(1)} = \left(\frac{3u_0}{A_i}\right) \mathrm{sech}^2 \left[\sqrt{\frac{u_0}{4B}}(\zeta - u_0 \tau)\right]. \tag{8}$$

It is obvious from Eq. (8) that the DIA solitary waves exist with positive (negative) potential for $\mu > (<)1/3$.

We now study arbitrary amplitude DIA solitary waves by the pseudo-potential approach.[51] To do so, we introduce a single independent variable $\xi = x - Mt$, where M is the Mach number (speed of the solitary waves normalized by the ion-acoustic speed C_i), and assume the steady state condition ($\partial/\partial t = 0$). The transformation ($\xi = x - Mt$) along with the steady state condition ($\partial/\partial t = 0$) and the appropriate boundary conditions ($\phi \to 0$, $u_i \to 0$, and $n_i \to 1$ at $\xi \to \pm\infty$) allow us to reduce Eqs. (1)–(3) to an energy integral

$$\frac{1}{2}\left(\frac{d\phi}{d\xi}\right)^2 + V(\phi) = 0, \tag{9}$$

where the pseudo-potential $V(\phi)$ is given by

$$V(\phi) = \mu[1 - \exp(\phi)] + M^2\left(1 - \sqrt{1 - \frac{2\phi}{M^2}}\right) - (1-\mu)\phi. \tag{10}$$

The expansion of $V(\phi)$ around $\phi = 0$ is

$$V(\phi) = C_2^i\phi^2 + C_3^i\phi^3 + \cdots, \tag{11}$$

where

$$C_2^i = \frac{1}{2M^2} - \frac{1}{2}\mu, \tag{12}$$

$$C_3^i = \frac{1}{2M^4} - \frac{1}{6}\mu. \tag{13}$$

To compare the basic features of DIA solitary waves obtained from the reductive perturbation method[50] with those obtained from this pseudo-potential approach,[51] let us first consider the small amplitude DIA solitary waves for which $V(\phi) = C_2^i\phi^2 + C_3^i\phi^3$ holds good. This approximation allows us to write the small amplitude solitary wave solution of Eq. (9) as

$$\phi = \left(-\frac{C_2^i}{C_3^i}\right)\operatorname{sech}^2\left(\sqrt{-\frac{C_2^i}{4}}\,\xi\right). \tag{14}$$

This means that when $C_2^i < 0$, small amplitude solitary waves with positive (negative) potential exist for ($C_3^i < 0$) ($C_3^i > 0$). So, $C_2^i(M = M_c) = 0$, where M_c is the critical value of M above which solitary wave solutions exist, gives the value of $M_c = 1/\sqrt{\mu}$, and $C_3^i(M = M_c, \mu = \mu_c) = 0$, where μ_c is the critical value of μ above (below) which solitary waves with positive (negative) potential exists, gives the value of $\mu_c = 1/3$. Therefore, for

$\mu > 1/3$ ($\mu < 1/3$) the DIA solitary waves exist with positive (negative) potential. This result completely agrees with that obtained from the reductive perturbation method.[50]

We now study the properties of arbitrary amplitude DIA solitary waves by analyzing the general expression for $V(\phi)$ [Eq. (10)]. It is clear that $V(\phi) = dV(\phi)/d\phi = 0$ at $\phi = 0$. Therefore, solitary wave solutions of Eq. (9) exist[52,53] if (i) $\left(d^2V/d\phi^2\right)_{\phi=0} < 0$, i.e. $C_2^i < 0$, so that the fixed point at the origin is unstable, and (ii) $V(\phi) < 0$ when $0 > \phi > \phi_{max}$ for the solitary waves with positive potential and $\phi_{min} < \phi < 0$ for the solitary waves with negative potential, where ϕ_{max} (ϕ_{min}) is the maximum (minimum) value of ϕ for which $V(\phi_{max}) = V(\phi_{min}) = 0$. The condition (i) is satisfied when $M > M_c = 1/\sqrt{\mu}$. To examine whether the condition (ii) is satisfied, we have numerically analyzed Eq. (10), and found that for any set of dusty plasma parameters satisfying $M > 1/\sqrt{\mu}$ and $\mu < 1/3$, potential wells are formed in both positive and negative ϕ-axes, i.e. the DIA solitary waves with both $\phi > 0$ and $\phi < 0$ exist. However, for $M > 1/\sqrt{\mu}$ and $\mu > 1/3$, potential wells are formed in positive ϕ-axis only, i.e. the DIA solitary waves exist with $\phi > 0$ only.

We have discussed the properties of the DIA solitary waves by assuming a single ion-comonent, planar geometry, constant dust charge. However, it is shown that the effects of negative ion-component, nonplanar geometry, and dust grain charge fluctuation introduce new features or significantly modify the properties of DIA solitary waves. We now investigate the effect of negative ions, nonplanar geometry and dust grain charge fluctuation on the properties of the DIA solitary waves.

2.1. *Effect of Negative Ions*

To study the effects of negative ions[54–58] on the properties of the DIA solitary waves, we consider a four-component unmagnetized dusty plasma consisting of negatively charged stationary dust, positive and negative ion fluids and Boltzmann electrons. Therefore, we start with Eqs. (1), (2) and

$$\frac{\partial n_n}{\partial t} + \frac{\partial}{\partial x}(n_n u_n) = 0, \tag{15}$$

$$\frac{\partial u_n}{\partial t} + u_n \frac{\partial u_n}{\partial x} = \alpha_n \frac{\partial \phi}{\partial x}, \tag{16}$$

$$\frac{\partial^2 \phi}{\partial x^2} = \mu \exp(\phi) - n_i + \mu_n n_n + \mu_d, \tag{17}$$

where n_n is the negative ion number density normalized by its equilibrium value n_{n0}, u_n is the negative ion fluid speed normalized by the ion-acoustic speed C_i, $\mu_n = n_{n0}/n_{i0}$ and $\mu_d = 1-\mu-\mu_n$. Since we are interested here to examine the effects of negative ions on DIA solitary waves, the effects of the dust charge fluctuation have been neglected just to avoid the mathematical complexities. As before assuming $\xi = z - Mt$ and $\partial/\partial t = 0$, we can reduce Eqs. (1), (2) and (15)–(17) to an energy integral

$$\frac{1}{2}\left(\frac{d\phi}{d\xi}\right)^2 + U(\phi) = 0, \tag{18}$$

where the pseudo-potential $U(\phi)$ is

$$U(\phi) = \mu[1-\exp(\phi)] + M^2(1-U_i) + \frac{M^2\mu_n}{\alpha_n}(1-U_n) - \mu_d\phi, \tag{19}$$

in which $U_i = (1-2\phi/M^2)^{1/2}$ and $U_n = (1+2\alpha_n\phi/M^2)^{1/2}$. The expansion of $U(\phi)$ around $\phi = 0$ is

$$U(\phi) = C_2^n\phi^2 + C_3^n\phi^3 + \cdots, \tag{20}$$

where

$$C_2^n = \frac{1}{2M^2}(1+\alpha_n\mu_n) - \frac{1}{2}\mu, \tag{21}$$

$$C_3^n = \frac{1}{2M^4}(1-\alpha_n^2\mu_n) - \frac{1}{6}\mu. \tag{22}$$

Now, following the same technique as we used before, the expressions for the critical Mach number (above which the DIA solitary wave exists) and the critical value of μ (below which the DIA solitary wave exists with negative potential), where the effects of negative ions are included, are given by

$$M_c^n = \left(\frac{1+\alpha_n\mu_n}{\mu}\right)^{1/2}, \tag{23}$$

$$\mu_c^n = \frac{(1+\alpha_n\mu_n)^2}{3(1-\alpha_n^2\mu_n)}. \tag{24}$$

$$\tag{25}$$

Therefore, when $M > M_c^n$, small amplitude DIA solitary waves exist with positive (negative) potential for $\mu > \mu_c^n$ ($\mu < \mu_c^n$).

To examine the basic features of arbitrary amplitude DIA solitary waves, we have numerically analyzed Eq. (19), and found that for any set of dusty plasma parameters satisfying $M > M_c^n$ and $\mu < \mu_c^n$, potential wells are formed in both positive and negative ϕ-axes, i.e. positive and negative DIA solitary waves (solitary waves with $\phi > 0$ and $\phi < 0$) coexist. However, for $M > M_c$ and $\mu > \mu_c$, potential wells are formed in positive ϕ-axis only, i.e. DIA solitary waves exist with $\phi > 0$ only.

2.2. *Nonplanar Geometry*

The nonlinear dynamics of the DIA waves, whose phase speed is much smaller (larger) than the electron (ion) thermal speed (viz. $V_{Ti} \ll v_p \ll V_{Te}$), in nonplanar cylindrical and spherical geometries is governed by

$$\frac{\partial n_i}{\partial t} + \frac{1}{r^\nu}\frac{\partial}{\partial r}(r^\nu n_i u_i) = 0, \tag{26}$$

$$\frac{\partial u_i}{\partial t} + u_i \frac{\partial u_i}{\partial r} = -\frac{\partial \phi}{\partial r}, \tag{27}$$

$$\frac{1}{r^\nu}\frac{\partial}{\partial r}\left(r^\nu \frac{\partial \phi}{\partial r}\right) = \mu \exp(\phi) - n_i + (1-\mu), \tag{28}$$

where $\nu = 0$ for a one dimensional planar geometry and $\nu = 1$ (2) for a nonplanar cylindrical (spherical) geometry.

To investigate the basic features of small but finite amplitude nonplanar DIA solitary waves, we express Eqs. (26)–(28) in terms of the stretched coordinates[59] $\zeta = -\epsilon^{1/2}(r + v_0 t)$, $\tau = \epsilon^{3/2} t$, use Eqs. (4)–(6), and develop equations in various powers of ϵ. The latter are used to derive a linear dispersion relation ($v_0 = 1/\sqrt{\mu}$) and a modified K-dV equation

$$\frac{\partial \phi^{(1)}}{\partial \tau} + \frac{\nu}{2\tau}\phi^{(1)} + A_i \phi^{(1)} \frac{\partial \phi^{(1)}}{\partial \zeta} + B_i \frac{\partial^3 \phi^{(1)}}{\partial \zeta^3} = 0, \tag{29}$$

where $A_i = (3\mu - 1)/\sqrt{2\mu}$ and $B_i = 1/2\mu^{3/2}$. We have already mentioned that $\nu = 0$ corresponds to a one dimensional planar geometry, and Eq. (29) is identical to Eq. (7) whose solitary wave solutions have already been discussed. An exact analytic solution of Eq. (29) is not possible. Therefore, we have numerically solved Eq. (29), and have studied the effects of cylindrical and spherical geometries on the time-dependent DIA solitary waves. The initial condition that we have used in our numerical analysis is in the form of $\phi^{(1)} = (3u_0/A)\text{sech}^2(\zeta u_0/\sqrt{4B})$, which is a stationary solitary wave solution of Eq. (29) with $\nu = 0$.

The numerical solutions of Eq. (29) reveal[36] that for a large (negative) value of τ the spherical and cylindrical solitary waves are similar to planar solitary waves. This is because for a large value of τ the term $(\nu/2\tau)\phi^{(1)}$, which is due to the effect of the cylindrical or spherical geometry, is no longer dominant. However, as the value of τ decreases, the term $(\nu/2\tau)\phi^{(1)}$ becomes dominant and both spherical and cylindrical solitary waves differ from planar solitary ones. It is found that as the value of τ decreases, the amplitude of these nonplanar solitary pulses increases. It is also found that the amplitude of cylindrical solitary waves is larger than that of the planar ones, but smaller than that of the spherical ones.

2.3. *Dust Charge Fluctuation*

We study the DIA solitary waves in an unmagnetized dusty plasma where the dust charge is not constant, but varies with space and time. The nonlinear dynamics of one-dimensional DIA waves, whose phase speed is much smaller (larger) than the electron (ion) thermal speed, is governed by Eqs. (1), (2) and

$$\frac{\partial^2 \phi}{\partial x^2} = \mu \exp(\phi) - n_i + (1-\mu) z_d, \tag{30}$$

where z_d is the number of electrons residing onto the dust grain surface normalized by its equilibrium value Z_{d0}. We note that z_d is not constant but varies with space and time. Thus, Eqs. (1), (2) and (30) are completed by the normalized dust grain charging equation[38]

$$\eta \frac{\partial z_d}{\partial t} = \mu\beta \exp(\phi - \alpha_z z_d) - \beta_i n_i u_i \left(1 + \frac{2\alpha_z z_d}{u_i^2}\right), \tag{31}$$

where $\eta = \sqrt{\alpha_z m_e(1-\mu)/2m_i}$, $\alpha_z = Z_{d0}e^2/k_B T_e r_d$, $\beta = (r_d/d)^{3/2}$, $d = n_{d0}^{-1/3}$, and $\beta_i = \beta\sqrt{\pi m_e/8m_i}$. We note that at equilibrium, $\mu\beta\exp(-\alpha_z) = \beta_i u_{i0}(1 + 2\alpha_z/u_{i0}^2)$, where u_{i0} is the ion streaming speed normalized by C_i.

To study small but finite amplitude DIA solitary waves in a dusty plasma with charge fluctuating dust, we express Eqs. (1), (2), (30) and (31) in terms of the stretched coordinates[50] $\xi = \epsilon^{1/2}(x - v_0 t)$, $\tau = \epsilon^{3/2} t$, $\eta = \epsilon\eta_0$, use Eqs. (4)–(6) and $z_d = 1 + \epsilon z_d^{(1)} + \epsilon^2 z_d^{(2)} + \cdots$, and develop equations in various powers of ϵ. The latter are used to derive a linear dispersion relation

$$a w_0^2 - b w_0 - c = 0, \tag{32}$$

where

$$a = \mu + \frac{(1-\mu)\beta_e}{\alpha_z u_\beta}, \tag{33}$$

$$b = \frac{(1-\mu)u_1\beta_i}{\alpha_z u_\beta}, \tag{34}$$

$$c = 1 + \frac{(1-\mu)u_2 u_{i0}\beta_i}{\alpha_z u_\beta}, \tag{35}$$

in which $w_0 = v_0 - u_{i0}$, $u_1 = 1 - 2\alpha_z/u_{i0}$, $u_2 = 1 + 2\alpha_z/u_{i0}^2$, $u_\beta = \beta_e + 2\beta_i/u_{i0}$ and $\beta_e = \mu\beta\exp(-\alpha_z)$, and a K-dV equation

$$\frac{\partial \phi^{(1)}}{\partial \tau} + A_c \phi^{(1)} \frac{\partial \phi^{(1)}}{\partial \xi} + B_c \frac{\partial^3 \phi^{(1)}}{\partial \xi^3} = 0, \tag{36}$$

where

$$A_c = \frac{3c + bw_0 - \beta_2 w_0^4}{w_0(2c + bw_0)}, \tag{37}$$

$$B_c = \frac{w_0^3}{2c + bw_0}, \tag{38}$$

$$\beta_2 = \mu + \frac{\beta_1(1-\mu)}{\alpha_z u_\beta}, \tag{39}$$

$$\beta_1 = \beta_e[1 + (\alpha_z\beta_0 - 2)\alpha_z\beta_0] - \frac{2\beta_i}{w_0^3}\left[1 + \frac{w_\alpha}{u_{i0}^3}\right], \tag{40}$$

$$\beta_0 = \frac{1}{\alpha_z u_\beta}\left[\beta_e - \frac{\beta_i w_2}{w_0} + \frac{2\alpha_z\beta_i}{w_0 u_{i0}}\left(1 - \frac{1}{w_0}\right)\right], \tag{41}$$

in which $w_1 = 1 - u_{i0}/w_0$, $w_2 = 1 + u_{i0}/w_0$ and $w_\alpha = 2\alpha_z w_0 w_1(1 - \beta_0 w_0 u_{i0})$.

Now, for a frame moving with speed u_0 (normalized by C_i), the stationary solitary wave solution of the K-dV equation (36) is

$$\phi^{(1)} = \phi_m^{(1)}\text{sech}^2\left[(\xi - u_0\tau)/\Delta\right], \tag{42}$$

where $\phi_m^{(1)} = 3u_0/A_c$ and $\Delta = \sqrt{4B_c/u_0}$ represent the amplitude and the width of the solitary waves, respectively. It is obvious from Eq. (42) that there exists compressive (rarefactive) solitary waves if $A_c > 0$ ($A_c < 0$). We have numerically analyzed $\phi_m^{(1)}$ and Δ for the parameters corresponding to space dusty plasma parameters:[15] $n_{d0} = 10^{-7}$ cm^{-3}, $r_d = 1$ μm, $k_BT_e = 50$ eV and $Z_{d0} = 10^3$, which correspond to $\alpha_z = 0.0288$ and $\beta = 3 \times 10^{-10}$, as well as laboratory dusty plasma parameters:[24,25] $n_{d0} = 10^5$ cm^{-3}, $r_d = 5$ μm, $k_BT_e = 0.2$ eV and $Z_{d0} = 10^3$, which correspond to $\alpha_z = 1.44$ and $\beta = 3 \times 10^{-4}$. We found that $\phi_m^{(1)}$ is always positive. This means that our present dusty plasma model can support only compressive DIA solitary waves (solitary waves with $\phi > 0$).

3. Mobile Dust: DA Solitary Waves

We, in this section, consider the dynamics of negatively charged dust particles in a weakly coupled unmagnetized dusty plasma, and study the basic features and underlying physics of the DA solitary waves. The nonlinear dynamics of one dimensional DA waves, whose phase speed (V_p) is much smaller than the ion thermal speed, but much larger than the dust thermal speed (V_{Td}), viz. $V_{Td} \ll V_p \ll V_{Ti}$, is governed by[26,46]

$$\frac{\partial n_d}{\partial t} + \frac{\partial}{\partial z}(n_d u_d) = 0, \tag{43}$$

$$\frac{\partial u_d}{\partial t} + u_d \frac{\partial u_d}{\partial z} = \frac{\partial \varphi}{\partial z}, \tag{44}$$

$$\frac{\partial^2 \varphi}{\partial z^2} = n_d + \mu_e \exp(\sigma_i \varphi) - \mu_i \exp(-\varphi), \tag{45}$$

where n_d is the dust number density normalized by its equilibrium value n_{d0}, u_d is the dust fluid speed normalized by $C_d = (Z_{d0} k_B T_i / m_d)^{1/2}$ and φ is the electrostatic wave potential normalized by $k_B T_i/e$, $\mu_e = n_{e0}/Z_{d0} n_{d0} = \mu/(1-\mu)$ and $\mu_i = n_{i0}/Z_{d0} n_{d0} = 1/(1-\mu)$. The time ($t$) and space ($z$) variables are in units of the dust plasma period $\omega_{pd}{}^{-1} = (m_d/4\pi n_{d0} Z_{d0}^2 e^2)^{1/2}$ and the Debye-radius $\lambda_{Dim} = (k_B T_i / 4\pi Z_{d0} n_{d0} e^2)^{1/2}$, respectively. It should be mentioned here that the effects of the dust charge fluctuation have been neglected since the dust charging time (few micro-seconds) is much less than the time period (a fraction of a second) of the DA waves under consideration.[15]

We first study small amplitude DA solitary waves by the reductive perturbation method.[50] To do so, we express Eqs. (43)–(45) in terms of the stretched coordinates $\zeta = \epsilon^{1/2}(z - V_0 t)$, $\tau = \epsilon^{3/2} t$ (where $V_0 = V_p/C_d$), expand n_d, u_d and φ in power series of ϵ

$$n_d = 1 + \epsilon n_d^{(1)} + \epsilon^2 n_d^{(2)} + \cdots, \tag{46}$$

$$u_d = \epsilon u_d^{(1)} + \epsilon^2 u_d^{(2)} + \cdots, \tag{47}$$

$$\varphi = \epsilon \varphi^{(1)} + \epsilon^2 \varphi^{(2)} + \cdots, \tag{48}$$

and develop equations in various powers of ϵ. The latter can be used to derive a linear dispersion relation $[V_0 = 1/\sqrt{\mu_i + \sigma_i \mu_e}]$ and a K-dV equation

$$\frac{\partial \varphi^{(1)}}{\partial \tau} + A_d \varphi^{(1)} \frac{\partial \varphi^{(1)}}{\partial \zeta} + B_d \frac{\partial^3 \varphi^{(1)}}{\partial \zeta^3} = 0, \tag{49}$$

where $A_d = -[1 + (3 + \sigma_i \mu_i)\sigma_i \mu_i + \mu(1 + \sigma_i^2)/2]/[3(1-\mu)^2]$ and $B_d = V_0^3/2$. Now, for a frame moving with speed U_0 (normalized by C_d), the stationary solitary wave solution of Eq. (49) is given by

$$\varphi^{(1)} = \left(\frac{3U_0}{A_d}\right) \operatorname{sech}^2 \left[\sqrt{\frac{U_0}{4B_d}}(\zeta - U_0 \tau)\right]. \tag{50}$$

It is obvious that A_d is always negative for any value of μ or σ_i. Therefore, Eq. (50) clearly implies that the DA solitary waves exist with negative potential ($\varphi < 0$) only.

We now study arbitrary amplitude DA solitary waves by the pseudo-potential approach.[51] To do so, we use a single independent variable $\xi = z - Mt$ and a steady state condition $\partial/\partial t = 0$, and reduce Eq. (43)–(45) to an energy integral

$$\frac{1}{2}\left(\frac{d\varphi}{d\xi}\right)^2 + V(\varphi) = 0, \tag{51}$$

where the pseudo-potential potential $V(\varphi)$ is given by[46]

$$V(\varphi) = \mu_i[1 - \exp(-\varphi)] + \frac{\mu_e}{\sigma_i}[1 - \exp(\sigma_i\varphi)] + M^2[1 - V_d(\varphi)], \tag{52}$$

where $V_d(\varphi) = (1 + 2\varphi/M^2)^{1/2}$. The expansion of $V(\varphi)$ around $\varphi = 0$ is

$$V(\varphi) = C_2^d\varphi^2 + C_3^d\varphi^3 + \cdots, \tag{53}$$

where

$$C_2^d = \frac{1}{2M^2} - \frac{1}{2}(\mu_i + \sigma_i\mu_e), \tag{54}$$

$$C_3^d = -\frac{1}{2M^4} + \frac{1}{6}(\mu_i - \sigma_i^2\mu_e). \tag{55}$$

To compare the basic features of the DA solitary waves obtained from the reductive perturbation method[50] with those obtained from this pseudo-potential approach,[51] let us first consider small amplitude DA solitary waves for which $V(\varphi) = C_2^d\varphi^2 + C_3^d\varphi^3$ holds good. This approximation allows us to write the small amplitude solitary wave solution of Eq. (51) as

$$\varphi = \left(-\frac{C_2^d}{C_3^d}\right)\operatorname{sech}^2\left(\sqrt{-\frac{C_2^d}{4}}\,\xi\right). \tag{56}$$

This means that when $C_2^d < 0$, small amplitude DA solitary waves with positive (negative) potential exist for $(C_3^d > 0)$ $(C_3^d < 0)$. Using $C_2^d(M = M_c) = 0$, where M_c is the critical value of M above which solitary wave solutions exist, we can express M_c as $M_c = 1/\sqrt{\mu_i + \sigma_i\mu_e}$, and at $M = M_c$, we can express C_3^d as $C_3^d = -[1 + (3 + \sigma_i\mu_i)\sigma_i\mu_i + \mu(1 + \sigma_i^2)/2]/[3(1 - \mu)^2]$. This clearly reveals that C_3^d is always negative for any value of σ_i or μ, i.e. small amplitude DA solitary waves with $\varphi < 0$ can only exist. This result completely agrees with that obtained from the reductive perturbation method.[50]

We now study the properties of arbitrary amplitude DA solitary waves by analyzing the general expression [Eq. (52)] for $V(\varphi)$. It is clear that $V(\varphi) = dV(\varphi)/d\varphi = 0$ at $\varphi = 0$. Therefore, solitary wave solutions of Eq. (51) exist[52,53] if (i) $\left(d^2V/d\varphi^2\right)_{\varphi=0} < 0$, i.e. $C_2^d < 0$, so that the fixed point at the origin is unstable, and (ii) $V(\varphi) < 0$ when $0 > \varphi > \varphi_{max}$ for the solitary waves with positive potential and $\varphi_{min} < \varphi < 0$ for the solitary waves with negative potential, where φ_{max} (φ_{min}) is the maximum (minimum) value of φ for which $V(\varphi_{max}) = V(\varphi_{min}) = 0$. The condition (i) is satisfied when $M > M_c = 1/\sqrt{\mu_i + \sigma_i \mu_e}$. We have numerically calculated the critical Mach number M_c for different values of μ and σ_i, and observed[46] that the critical Mach number increases with σ_i, but decreases with μ. To examine whether the condition (ii) is satisfied, we have numerically analyzed the general expression [Eq. (52)] for $V(\varphi)$, and found that for any value of μ or σ_i, the potential wells are formed in the negative φ-axis only. This means that for any value of σ_i or μ, the arbitrary amplitude DA solitary waves with $\varphi < 0$ can only exist.

It is of interest to examine whether or not there exists an upper limit of M for which DA solitary waves can exist. This upper limit of M can be found by the condition $V(\varphi_c) \geq 0$, where $\varphi_c = -M^2/2$ is the minimum value of φ for which the dust number density n_d is real. Thus, the upper limit of M is that maximum value of M for which $S_m \geq 0$, where $S_m = \mu_i + \mu_e/\sigma_i + M^2 - \mu_i \exp(M^2/2) - (\mu_e/\sigma_i)\exp(-\sigma_i M^2/2)$. By numerical analysis of this expression, we have estimated the variation of S_m with M for different values of μ, and found that as we increase μ, the upper limit of M decreases. We note that for $\sigma_i = 0.05$ and $\mu = 0.1$, there exists DA solitary waves with $\varphi < 0$ for $0.95 < M < 1.52$. We have also numerically analyzed $V(\varphi)$ and have found the same results[46] that for $\sigma_i = 0.05$ and $\mu = 0.1$ there exists a potential well on the negative φ-axis for $0.95 < M < 1.52$, i.e. there exists DA solitary waves with $\varphi < 0$ for $0.95 < M < 1.52$.

We have discussed the properties of the DA solitary waves in an unmagnetized dusty plasma by assuming a Maxwellian ion distribution and single (negative) dust component. However, it can be shown that the effects of non-Maxwellian ion distributions and positive dust component introduce new features or significantly modify the properties of the DA solitary waves.[39,42,47–49] We now study the effects of non-Maxwellian (trapped and nonthermal) ion distributions and positive dust component on the properties of the DA solitary waves.

3.1. *Trapped Ion Distribution*

It is well known[60–63] that the electron and ion distribution functions can be significantly modified in the presence of large amplitude waves that are excited by the two-stream instability.[64] Accordingly, the electron and ion number densities depart from a Boltzmann distribution when a phase space vortex distribution appears in a plasma. For the DA waves, the ion trapping in the wave potential is of our interest. To study the effects of non-isothermal ions on the DA solitary waves, we consider the trapped or vortex-like[60,61] ion distribution $f_i = f_{if} + f_{it}$, where

$$f_{if} = \frac{1}{\sqrt{2\pi}} \exp\left[-\frac{1}{2}(v_i^2 + 2\varphi)\right] \tag{57}$$

for $|v_i| > \sqrt{-2\varphi}$ and

$$f_{it} = \frac{1}{\sqrt{2\pi}} \exp\left[-\frac{1}{2}\sigma_{it}(v_i^2 + 2\varphi)\right] \tag{58}$$

for $|v_i| \leq \sqrt{-2\varphi}$. We note that the ion distribution function, as prescribed above, is continuous in velocity space and satisfies the regularity requirements for an admissible BGK solution.[51] Here the ion velocity v_i in Eqs. (57) and (58) is normalized by the ion thermal speed V_{Ti}, and $\sigma_{it} = T_i/T_{it}$, the ratio of the free ion temperature T_i to the trapped ion temperature T_{it}, is a parameter determining the number of trapped ions. Integrating the ion distribution functions over velocity space we readily obtain the ion number density n_i as[62]

$$n_i = I(-\varphi) + \frac{1}{\sqrt{\sigma_{it}}} \exp(-\sigma_{it}\varphi)\, \mathrm{erf}(\sqrt{-\sigma_{it}\varphi}) \tag{59}$$

for $\sigma_{it} > 0$ and

$$n_i = I(-\varphi) + \frac{1}{\sqrt{\pi|\sigma_{it}|}} W_D(\sqrt{\sigma_{it}\varphi}) \tag{60}$$

for $\sigma_{it} < 0$, where

$$I(z_0) = [1 - \mathrm{erf}(\sqrt{z_0})]\exp(z_0), \tag{61}$$

$$\mathrm{erf}(z_0) = \frac{2}{\sqrt{\pi}} \int_0^{z_0} \exp(-y^2)dy, \tag{62}$$

$$W_D(z_0) = \exp(-z_0^2) \int_0^{z_0} \exp(y^2)dy. \tag{63}$$

If we expand n_i in the small amplitude limit (viz. $\varphi \ll 1$) and keep terms up to φ^2, it is found that n_i is the same for both $\sigma_{it} < 0$ and $\sigma_{it} > 0$. Therefore, n_i is expressed as

$$n_i \simeq 1 - \varphi - \frac{4(1-\sigma_{it})}{3\sqrt{\pi}}(-\varphi)^{3/2} + \frac{1}{2}\varphi^2. \tag{64}$$

Now, applying the reductive perturbation technique of Schamel,[61] i.e. using the stretched coordinates $\zeta = \epsilon^{1/4}(z - V_0 t)$, $\tau = \epsilon^{3/4} t$, and Eqs. (43)–(45)

[with the replacement of $\mu_i \exp(-\varphi)$ by right hand side of Eq. (64)] and (46)–(48), we have

$$\frac{\partial \varphi^{(1)}}{\partial \tau} + A_t \sqrt{-\varphi^{(1)}} \frac{\partial \varphi^{(1)}}{\partial \zeta} + B_d \frac{\partial^3 \varphi^{(1)}}{\partial \zeta^3} = 0, \tag{65}$$

where

$$A_t = \frac{V_0^3(1-\sigma_{it})}{\sqrt{\pi}(1-\mu)}, \tag{66}$$

and $B_d = V_0^3/2$. Equation (65) is a modified K-dV equation exhibiting a stronger nonlinearity. As before, for a frame moving with a speed U_0, the stationary solution of Eq. (65) can be expressed as

$$\varphi^{(1)} = -\varphi_m^{(1)} \mathrm{sech}^4[(\zeta - U_0\tau)/\Delta_t], \tag{67}$$

where the amplitude $\varphi_m^{(1)}$ and the width Δ_t are given by $\varphi_m^{(1)} = (15U_0/8a_t)^2$ and $\Delta_t = \sqrt{16b_s/U_0}$, respectively. As $U_0 > 0$ and $\mu < 1$, Eq. (67) reveals that there exist DA solitary waves with negative potential only. It is found that the effect of the trapped ion distribution causes the DA solitary waves of smaller width and larger propagation speed.

3.2. *Nonthermal Ion distribution*

To study the effects of a non-thermal ion distribution on the properties of DA solitary waves, we choose a more general class of ion distribution which includes the population of nonthermal ions.[52] Thus, we take[42]

$$f_i(v_i) = \frac{1+\alpha_i v_i^4}{(1+3\alpha_i)\sqrt{2\pi}} \exp(-v_i^2/2), \tag{68}$$

where v_i is the ion speed normalized by the ion thermal speed V_{Ti} and α_i is a parameter determining the population of nonthermal (fast) ions in our dusty plasma model. The effect of an electrostatic disturbance on the equilibrium ion distribution can easily be introduced by replacing v_i^2 with $v_i^2 + 2\varphi$. The resulting distribution function is then integrated over velocity space, yielding[42]

$$n_i = (1 + \alpha_0\varphi + \alpha_0\varphi^2)\exp(-\varphi), \tag{69}$$

where $\alpha_0 = 4\alpha_i/(1+3\alpha_i)$. Now, using $\xi = z - Mt$, $\partial/\partial t = 0$, Eqs. (43)–(45) [with the replacement of $\mu_i \exp(-\varphi)$ by right hand side of Eq. (69)] we can reduce[42] to an energy integral: $(1/2)(d\varphi/d\xi)^2 + U(\varphi) = 0$, where $U(\varphi)$ is given by

$$U(\varphi) = \mu_i[1 + 3\alpha_0 - (1 + 3\alpha_0 + 3\alpha_0\varphi + \alpha_0\varphi^2)\exp(-\varphi)] + \frac{\mu_e}{\sigma_i}[1 - \exp(\sigma_i\varphi)] + M^2\left[1 - \left(1 + \frac{2\varphi}{M^2}\right)^{1/2}\right]. \tag{70}$$

Again, following the analytical steps or numerical analysis of the pseudo-potential $U(\varphi)$ described as before, we can show that when $\alpha_i > 0.155$ and $M > 1.41$, the potential well develops on both the positive and negative φ-axis.[42] This means that the presence of nonthermal ions ($\alpha_i > 0.155$) supports the coexistence of compressive and rarefactive DA solitary waves (DA solitary waves with $\varphi < 0$ and $\varphi > 0$).

3.3. *Effects of Positive Dust*

To study the effects of positive dust[65–68] on the properties of the DA solitary waves, we consider a four-component unmagnetized dusty plasma system consisting of negatively and positively charged dust particles, and Boltzmann electrons and ions. So, we start with Eqs. (43), (44) and

$$\frac{\partial n_p}{\partial t} + \frac{\partial}{\partial z}(n_p u_p) = 0, \tag{71}$$

$$\frac{\partial u_p}{\partial t} + u_p\frac{\partial u_p}{\partial z} = -\alpha\frac{\partial \varphi}{\partial z}, \tag{72}$$

$$\frac{\partial^2\varphi}{\partial z^2} = n_d - \mu_p n_p + \mu_e e^{\sigma_i\varphi} - \mu_i e^{-\varphi}, \tag{73}$$

where n_p is the positive dust number density normalized by its equilibrium value n_{p0}, u_p is the positive dust fluid speed normalized by C_d. $\alpha = Z_p m_1/Z_{d0}m_d$, $\mu_e = n_{e0}/Z_{d0}n_{d0}$, $\mu_i = n_{i0}/Z_{d0}n_{d0}$, $\mu_p = 1 + \mu_e - \mu_i$, Z_p is the number of protons residing on a positive dust particle, m_p is the positive dust particle mass.

Now, assuming $\xi = z - Mt$ and $\partial/\partial t = 0$, we can reduce Eq. (43), (44), and (71) - (73) to an energy integral

$$\frac{1}{2}\left(\frac{d\varphi}{d\xi}\right)^2 + W(\varphi) = 0, \tag{74}$$

where the pseudo-potential $W(\varphi)$ is[49]

$$W(\varphi) = M^2\left(1 + \frac{\mu_p}{\alpha}\right) + \frac{\mu_e}{\sigma_i} + \mu_i - M^2\left(1 + \frac{2\varphi}{M^2}\right)^{\frac{1}{2}} - \frac{\mu_p M^2}{\alpha}\left(1 - \alpha\frac{2\varphi}{M^2}\right)^{\frac{1}{2}} - \frac{\mu_e}{\sigma_i}e^{\sigma_i\varphi} - \mu_i e^{-\varphi}. \tag{75}$$

The expansion of $W(\varphi)$ around $\varphi = 0$ is

$$W(\varphi) = C_2^p\varphi^2 + C_3^p\varphi^3 + \cdots, \tag{76}$$

where

$$C_2^p = \frac{1}{2M^2}(1 + \alpha\mu_p) - \frac{1}{2}(\mu_i + \sigma_i\mu_e), \tag{77}$$

$$C_3^p = -\frac{1}{2M^4}(1 - \alpha^2\mu_p) + \frac{1}{6}(\mu_i - \sigma_i^2\mu_e). \tag{78}$$

We first consider small amplitude solitary waves for which $W(\varphi) = C_2^p\varphi^2 + C_3^p\varphi^3$ holds good. This approximation allows us to write the small amplitude solitary wave solution of Eq. (74) as

$$\varphi = \left(-\frac{C_2^p}{C_3^p}\right)\operatorname{sech}^2\left(\sqrt{-\frac{C_2^p}{4}}\xi\right). \tag{79}$$

This means that when $C_2^p < 0$, small amplitude DA solitary waves with positive (negative) potential exist for $(C_3^p > 0)$ $(C_3^p < 0)$. So, $C_2^p(M = M_c) = 0$, where M_c is the critical value of M above which solitary wave solutions exist, gives the value of M_c, and $C_3^p(M = M_c, \alpha = \alpha_c) = 0$, where α_c is the critical value of α above (below) which solitary waves with positive (negative) potential exists, gives the value of α_c. We have numerically analyzed M_c and α_c, and found that M_c increases with α and μ, but decreases with σ_i and μ_i.

To examine the basic features of arbitrary amplitude DA solitary waves, we have numerically analyzed Eq. (75), and found that $C_3^p(M = M_c) <0$ for $\alpha < \alpha_c$ and $C_3^p(M = M_c) >0$ for $\alpha > \alpha_c$, and that $\alpha_c \simeq 1$ for $\sigma_i = 0.5$, $\mu_e = 0.2$ and $\mu_i = 0.8$. Therefore, for typical dusty plasma parameters (viz. $\sigma_i = 0.5$, $\mu_e = 0.2$ and $\mu_i = 0.8$) we have the existence of small amplitude DA solitary waves with negative potential for $\alpha = 0.5 < \alpha_c$ and $M > M_c \simeq 1.2$, and we have the existence of small amplitude DA solitary waves with positive potential for $\alpha = 1.5 > \alpha_c$ and $M > M_c \simeq 1.41$. We have used the same sets of parameters, and numerically analyzed the general expression [Eq. (75)] for $W(\varphi)$ to examine the possibility for the coexistence of arbitrary amplitude negative and positive DA solitary potential structures. It has been found from this numerical analysis that for typical dusty plasma parameters (viz. $\sigma_i = 0.5$, $\mu_e = 0.2$ and $\mu_i = 0.8$), we have the existence of solitary waves with negative potential for $\alpha = 0.5$ and $M > M_c \simeq 1.2$, and the coexistence of solitary waves with negative and positive potentials for $\alpha = 1.5$ and $M > M_c \simeq 1.41$.

4. Discussion

We have presented a rigorous theoretical investigation on DIA and DA solitary waves in unmagnetized dusty plasma. The results, which are found in this inverstigation, can be summarized as follows.

We have found that DIA solitary waves with a positive (negative) potential are found to exist for $\mu > (<) 1/3$. It is shown that the presence of negative ions significantly changes the critical values of M (value of M above which the DIA solitary waves exist) and μ (value of μ below which the DIA solitary waves with a negative potential exist).

To examine the effect of a non-planar geometry on DIA solitary waves, we found that the term $(\nu/2\tau)\phi^{(1)}$ in Eq. (29) is due to a non-planar (cylindrical or spherical) geometry. The numerical solutions of Eq. (29) reveal that for a large value of τ the spherical and cylindrical solitary waves are similar to planar ones. This is because for a large value of τ, the term $(\nu/2\tau)\phi^{(1)}$ is no longer dominant. However, as the value of τ decreases, the term $(\nu/2\tau)\phi^{(1)}$ becomes dominant and both spherical and cylindrical solitary waves differ from the planar ones. It is found that as the value of τ decreases, the amplitude of these nonplanar solitary pulses increases, and that the amplitude of cylindrical solitary waves is larger than that of the planar ones, but smaller than that of the spherical ones.

We have shown that the effects of dust grain charge fluctuation modify the properties of the DIA solitary waves. It is found that the effects of dust grain charge fluctuation reduce the speed of the DIA solitary waves. The characteristics of these DIA solitary waves in a space dusty plasma condition are found to be different from those in a laboratory dusty plasma condition. It is seen that for space dusty plasma parameters,[15] as we increase μ, both the amplitude and the width of the DIA solitary waves remains constant for $\mu < 0.5$, but increases very rapidly for $\mu > 0.5$. On the other hand, for laboratory dusty plasma parameters,[24,25] as we increase μ, the amplitude increases, but the width decreases.

The dusty plasma containing mobile dust particles and Boltzmann electrons and ions can support the DA solitary waves with a negative potential only, corresponding to a hump in the dust number density. We have shown that due to the effects of the trapped ion distribution, a dusty plasma admits a modified K-dV equation, exhibiting a stronger nonlinearity, smaller width and larger propagation speed. On the other hand, the presence of nonthermal ions ($\alpha_i > 0.155$) supports the coexistence of compressive and rarefactive DA solitary waves.

To study the effects of positive dust on the basic features of DA solitary waves, it is found that the critical Mach number increases with α and μ_e, but decreases with σ_i and μ_i, and that the presence of positive dust does not only significantly modify the basic properties of DA solitary waves, but also causes the coexistence of positive and negative DA solitary waves.

We finally hope that the basic features and the underlying physics of DIA and DA solitary waves that we have presented here should be useful for understanding the localized electro-acoustic disturbances in space and laboratory dusty plasmas.

References

1. J. R. Hill and D. A. Mendis, *Moon and Planets* **24**, 431 (1981).
2. B. A. Smith *et al.*, *Science* **212**, 163 (1981).
3. B. A. Smith *et al.*, *Science* **215**, 504 (1982).
4. C. K. Goertz and G. E. Morfill, *Icarus* **53**, 219 (1983).
5. C. K. Goertz, *Rev. Geophys.* **27**, 271 (1989).
6. M. Horányi and C. K. Goertz, *Astrophys. J.* **361**, 105 (1990).
7. D. A. Mendis, *Astrophys. Space Sci.* **176**, 163 (1991).
8. U. de Angelis, *Phys. Scripta* **45**, 465 (1992).
9. T. G. Northrop, *Phys. Scripta* **45**, 17 (1992).
10. F. Verheest, *Space Sci. Rev.* **77**, 267 (1996).
11. T. Nakano, *Astrophys. J.* **494**, 587 (2001).
12. E. G. Zweibel, *Phys. Plasmas* **6**, 1725 (1999).
13. F. Verheest, *Waves in Dusty Space Plasmas* (Kluwer, Dordrecht, 2000).
14. P. K. Shukla, *Phys. Plasmas* **8**, 1791 (2001).
15. P. K. Shukla and A. A. Mamun, *Introduction to Dusty Plasman Physics* (Institute of Physics Publishing Ltd., Bristol, 2002).
16. D. A. Mendis, *Plasma Sources Sci. Technol.* **11**, A219 (2002).
17. G. S. Selwyn, *Jpn. J. Appl. Phys.* **32**, 3068 (1993).
18. J. Winter, *Plasma Phys. Control. Fusion* **40**, 1201 (1998).
19. C. Hollenstein, *Plasma Phys. Control. Fusion* **42**, R93 (2000).
20. V. P. Bliokh and V. V. Yaroshenko, *Sov. Astron.* **29**, 330 (1985).
21. U. de Angelis *et al.*, *J. Plasma Phys.* **40**, 399 (1988).
22. P. K. Shukla and L. Stenflo, *Astrophys. Space Sci.* **190**, 23 (1992).
23. P. K. Shukla and V. P. Silin, *Physica Scripta* **45**, 508 (1992).
24. A. Barkan, N. D'Angelo and R. Merlino, *Planet. Space Sci.* **44**, 239 (1996).
25. R. L. Merlino *et al.*, *Phys. Plasmas* **5**, 1607 (1998).
26. N. N. Rao, P. K. Shukla and M. Y. Yu, *Planet. Space Sci.* **38**, 543 (1990).
27. A. Barkan, R. L. Merlino and N. D'Angelo, *Phys. Plasmas* **2**, 3563 (1995).
28. C. Thompson *et al.*, *IEEE Trans. Plasma Sci.* **27**, 146 (1999).
29. M. Rosenberg, *J. Vac. Sci. Technol. A* **14**, 631 (1996).
30. R. L. Merlino and J. Goree, *Phys. Today* **57**, 32 (2004).
31. K. E. Lonngren, *Plasma Phys.* **25**, 943 (1983).
32. P. K. Shukla, *Phys. Scripta* **45**, 504 (1992).
33. R. K. Varma, P. K. Shukla and V. Krishan, *Phys. Rev. E* **47**, 3612 (1993).

34. F. Melandsø , T. Aslaksen and O. Havnes, *Planet. Space Sci.* **41**, 321 (1993).
35. R. Bharuthram and P. K. Shukla, *Planet. Space Sci.* **40**, 647 (1992).
36. A. A. Mamun and P. K. Shukla, *Phys. Plasmas* **9**, 1468 (2002).
37. A. A. Mamun and P. K. Shukla, *Phys. Scripta* **T98**, 107 (2002).
38. A. A. Mamun and P. K. Shukla, *IEEE Trans. Plasma Sci.* **30**, 720 (2002).
39. P. K. Shukla and A. A. Mamun, *New J. Phys.* **5**, 17 (2003).
40. A. A. Mamun and P. K. Shukla, *Plasma Phys. Control. Fusion* **47**, A1 (2005).
41. A. A. Mamun, R. A. Cairns and P. K. Shukla, *Phys. Plasmas* **3**, 702 (1996).
42. A. A. Mamun, R. A. Cairns and P. K. Shukla, *Phys. Plasmas* **3**, 2610 (1996).
43. J. X. Ma and J. Liu, *Phys. Plasmas* **4**, 253 (1997).
44. S. V. Singh and N. N. Rao, *Phys. Lett. A* **235**, 164 (1997).
45. A. A. Mamun, *Phys. Scripta* **57**, 258 (1998).
46. A. A. Mamun, *Astrophys. Space Sci.* **268**, 443 (1999).
47. A. A. Mamun and P. K. Shukla, *Phys. Lett. A* **290**, 173 (2001).
48. A. A. Mamun and P. K. Shukla, *Geophys. Res. Lett.* **29**, 1870 (2002).
49. A. A. Mamun, *Phys. Lett A*, doi:10.1016/j.physleta.2007.07.076 (2007).
50. H. Washimi and T. Taniuti, *Phys. Rev. Lett.* **17**, 996 (1966).
51. I. B. Bernstein, J. B. Greene and M. D. Kruskal, *Phys. Rev.* **108** 546 (1957).
52. R. A. Cairns *et al.*, *Geophys. Res. Lett.* **22**, 2709 (1995).
53. A. A. Mamun, *Phys. Rev. E* **55**, 1852 (1997).
54. H. Amemiya *et al.*, *J. Plasma Phys.* **60**, 81 (1998).
55. H. Amemiya *et al.*, *Plasma Source Sci. Technol.* **8**, 179 (1999).
56. R. N. Franklin, *Plasma Source Sci. Technol.* **9**, 191 (2000).
57. R. N. Franklin, *Plasma Source Sci. Technol.* **11**, A31 (2002).
58. V. Vyas, G. A. Hebner and M. J. Kushner, *J. Appl. Phys.* **92**, 6451 (2002).
59. S. Maxon and J. Viecelli, *Phys. Rev. Lett.* **32**, 4 (1974).
60. H. Schamel, *Plasma Phys.* **14**, 905 (1972).
61. H. Schamel, *J. Plasma Phys.* **13**, 129 (1975).
62. H. Schamel, *Phys. Rep.* **140**, 161 (1986).
63. H. Schamel, *Phys. Plasmas* **7**, 4831 (2001).
64. D. Winske *et al.*, *Geophys. Res. Lett.* **22**, 2069 (1995).
65. V. W. Chow *et al.*, *J. Geophys. Res.* **98**, 19065 (1993).
66. O. Havnes *et al.*, *J. Geophys. Res.* **101**, 1039 (1996).
67. M. Horányi, G. E. Morfill and E. Grün, *Nature* **363**, 144 (1993).
68. V. E. Fortov *et al.*, *J. Exp. Theor. Phys.* **87**, 1087 (1998).

PHYSICS OF DUST IN MAGNETIC FUSION DEVICES

ZHEHUI WANG*

Los Alamos National Laboratory, MS E526,
Los Alamos, New Mexico, USA
**E-mail: zwang@lanl.gov*

CHARLES H. SKINNER

Princeton Plasma Physics Laboratory,
Princeton, NJ 08544, USA

GIAN LUCA DELZANNO

Los Alamos National Laboratory, MS K717,
Los Alamos, New Mexico, USA

SERGEI I. KRASHENINNIKOV

University of California, San Diego,
San Diego, California, USA

GIANNI M. LAPENTA

Los Alamos National Laboratory, MS K717,
Los Alamos, New Mexico, USA
also at
Centre for Plasma Astrophysics, Departement Wiskunde,
Katholieke Universiteit Leuven, Celestijnenlaan 200B,
BE-3001 Heverlee, Belgium

ALEXANDER Yu. PIGAROV

University of California, San Diego,
San Diego, California, USA

PADMA K. SHUKLA

Institut für Theoretische Physik IV,
Fakultät für Physik und Astronomie, Ruhr-Universität Bochum,
D-44780 Bochum, Germany

ROMAN D. SMIRNOV

University of California, San Diego, San Diego, California, USA

CATALIN M. TICOS

National Institute for Laser, Plasma and Radiation Physics, Magurele-Bucharest, Romania

W. PHIL WEST

General Atomics, San Diego, California, USA

Significant amount of dust will be produced in the next generation magnetic fusion devices due to plasma-wall interactions. The dust inventory must be controlled as it can pose a safety hazard and degrade performance. Safety concerns are due to tritium retention, dust radioactivity, toxicity, and flammability. Performance concerns include high-Z impurities carried by dust to the fusion core that can reduce plasma temperature and may even induce sudden termination of the plasma. Questions regarding dust in magnetic fusion devices therefore may be divided into dust safety, dust production, dust motion (dynamics), characteristics of dust, dust-plasma interactions, and most important of all, can dust be controlled in ways so that it will not become a severe problem for magnetic fusion energy production? The answer is not apparent at this time, which has motivated this work. Although dust safety and dust chemistry are important, our discussions primarily focus on dust physics. We describe theoretical frameworks, mostly due to dust research under a non-fusion context, that have already been established and can be used to answer many dust-related questions. We also describe dust measurements in fusion devices, numerical methods and results, and laboratory experiments related to the physics of fusion dust. Although qualitative understanding of dust in fusion has been or can be achieved, quantitative understanding of most dust physics in magnetic fusion is still needed. In order to find an effective way to deal with dust, future research activities include better dust diagnosis and monitoring, basic dusty plasma experiments emulating fusion conditions (for example, by using a mockup facility), numerical simulations bench-marked by experimental data, and development of a new generation of wall materials for fusion, which may include wall materials with engineered nanostructures.

1. Introduction

Dust is a generic name for minute solid particles with diameters less than 0.1 to 0.5 mm. The use of a range rather than a specific number for the cutoff indicates that this upper limit for dust size is somewhat arbitrary and scientifically insignificant. A plausible argument for the range is that particles greater than these sizes are unlikely to float in the air by themselves. Then this range reflects the mass-density difference in dust. There is also a somewhat arbitrary lower limit of a few nanometers to distinguish dust grains from more fundamental particles, such as electrons, protons, deuterons (D), tritons (T), and small molecules (like H_2, H_2O). Therefore, this choice of the 'lower' limit includes small clusters as dust. However, it should be pointed out that this broad definition of dust does not imply all dust particles with the same atomic or molecular composition would be alike. For example, it has been recognized that, when the size of a particle shrinks from a few microns and larger ($\gtrsim 10^{-6}$ m) to $\sim$ 100 nanometers and smaller ($\lesssim 10^{-7}$ m), new material properties that are quite different from bulk 'macroscopic' material properties will arise. As another example, an energetic ion, which can be stopped by and trapped inside a dust with a size greater than a few microns, may penetrate through a dust with a size $\lesssim 10^{-7}$ m rather easily. Strictly speaking, this definition of dust excludes all liquid droplets, which may occur in fusion devices when molten metal (such as liquid lithium) is present. But the physics of dust discussed here should be equally applicable to liquid droplets as long as the appropriate chemistry and physical properties are taken into account.

It is well known that construction and other human presence make it very hard to build a magnetic fusion device free of dust even before a plasma discharge. However, the initial dust content inside a fusion device is less of a concern than dust produced later due to plasma-wall interaction. The fact that the amount (composition) of dust in a fusion device increases (changes) significantly after plasma discharges indicates that additional dust must be produced through plasma-wall interactions, including alpha particle (fusion product)-wall interactions. Direct production of dust by neutron-wall interactions may be less important since neutrons usually penetrate much deeper into the wall than charged ions, well beyond the surface layers of the wall (with the possible exception of grazing incidence). Chemical compositions of dust produced in magnetic fusion devices are therefore determined by the wall materials exposed to the plasma, particles of the fusion plasma itself, and particles of fusion byproducts. Coexistence of many dust-production mechanisms, combined with different kinds of wall materials, can lead to rather diverse distributions in dust composition, dust size, and dust shape in magnetic fusion. Magnetic field can also play a role in dust formation and evolution by affecting the plasma and heat flux to the wall. Versatility of dust production processes and richness of dust species make dust problems in magnetic fusion a challenging and complex one to understand and deal with, to say the least.

We may not ignore dust problems in magnetic fusion devices, however, once dust inventory inside a fusion device exceeds certain limits (there are at least two kinds of dust limits, one is related to the safety, the other is related to performance). Beyond the limits, dust can cause either safety or performance or both kinds of problems that can interrupt the normal operation and therefore must be addressed. Dust problems are particularly pronounced in the next step steady-state burning plasma devices such as the International Thermonuclear Experimental Reactor (ITER), which aims for 1500 MW of fusion power (at least 100 times the output of existing devices) and long-duration operation (pulsed operation of about 400 s), and subsequently, far higher heat, ion, and neutron (≥ 0.5 MW/m^2) loads to the wall are expected [1,2]. The magnetic field for charged particle confinement also redistribute heat flux (except that from neutrons) unevenly on the wall, as much as 20 MW/m^2 is possible on diverter tiles [3]. Such a large amount of heat, if cannot be avoided (the acceptable power exhaust peak load for steady-state operation is below 20 MW/m^2 [3]), will lead to substantially larger dust production rate than existing devices and consequently the dust limit may be reached sooner than we like. Meanwhile, due to inadequate understanding and insufficient data on dust production in fusion devices, reliable and accurate dust limits (which are dependent on dust species) are difficult to obtain [4]. Rough estimates suggest a rate of 0.1 g/s for tungsten (W) dust for ITER, and correspondingly, a total production of 100 kg W dust for one year of operation.

Dust poses serious safety hazard for operation and maintenance of fusion devices because of its radioactivity, chemical reactivity, and toxicity [5,6]. Dust is a repository for radioactive particles of tritium and other activated elements. Dust (metallic dust especially) can also become radioactive by energetic particle (neutrons in particular) bombardment. Dust can spread radiological materials to the environment in the case of an accident. Sufficient amount of dust together can catch a fire and even cause an explosion under certain conditions due to the high flammability and chemical reactivity of the dust. In one scenario, dust of beryllium (Be), graphite (C), or W can react with steam from a broken cooling pipe and produce hydrogen and then initiate an explosion [4]. In another scenario, a sudden loss of vacuum and air ingress mobilizes dust and tritium (in its various chemical forms) around while producing hydrogen. Dust (such as the ones that contain Be) can also be highly toxic. The steady production and accumulation of dust due to plasma operations result in accumulation of radioactive materials (tritium, in particular) inside the vacuum vessel, compounding the 'tritium retention' problem, when tritium content of the wall increases with the time of plasma operation because of the tritium and carbon co-deposition. A guideline for mobilisable tungsten dust is 100 kg inside the ITER vacuum vessel [1]. Less than 6 kg each of beryllium, carbon and tungsten is allowed on each plasma facing component of the diverter to limit the hydrogen potentially generated by chemical reactions following in-vessel coolant spills or air ingress [7,8]. For a W dust production rate of 0.1 g/s as mentioned earlier, this limit can be quickly exceeded and either fusion plasma has to be stopped for in-vessel dust cleaning or some in-situ dust removal technique has to be developed that can continuously operate along with the plasma operation to keep the in-vessel dust inventory below the acceptable limits.

Besides concerns over safety, tritium retention in dust is also a performance issue. Tritium is limited in supply in a fusion device and therefore as much tritium as possible needs to be recycled (if not consumed by fusion) for energy production. The working guideline for the maximum in-vessel inventory of tritium that can be mobilized in the ITER vacuum vessel is about 330 g [9]. Since mobilized dust (graphite dust [a], for example) can trap, spread, and deposit tritium over exposed surfaces of the wall and especially at unintended locations, control of tritium inventory in magnetic fusion devices is a much harder problem because of the dust. Small sizes and the large number of dust particles together make dust removal for the purpose of tritium recovery an nuisance if not worse.

Dust can also compromise the performance of fusion plasma by bringing impurities inside the scrape-off-layer (SOL) and increasing the radiative energy loss from the hot fusion plasma core. Once a dust particle reaches the hot plasma, it is quickly disintegrated into individual atoms and clusters by intense plasma heat (see Fig. 4 in Sec. 3.4.2), in conjunction with or followed by ionization of the individual particles until each ion is fully stripped off its electrons. Since dust contains significant number of high-Z atoms with nuclear charge Z being much greater than the Z_{eff} of the plasma, which is usually within the range of 1 to 2, the resulting high-Z ions will enhance the Bremsstrahlung radiation which is proportional to Z^2 and characteristic line radiation when there are still bounded electrons. If a sufficient amount of dust reaches the hot plasma within certain time, dust cooling may be so large that plasma disruption may even be possible.

[a]The ITER IO has recently made a decision to exclude carbon PFCs from the tritium phase, and therefore, it is very likely that tritium will not be retained by carbon dust in ITER.

Therefore, it is clear that in the next step magnetic fusion devices like ITER, design and operation of the machine that specifically address dust issues will have to be incorporated in order to increase the safety and minimize the adverse dust effects on plasma performance.

Electrons, ions, alpha particles produced by DT fusion, and any combination of the charged-particle fluxes can produce dust (most likely through a multi-step process, details in Sec. 2) inside magnetic fusion devices when the particle fluxes are intercepted by the wall. Under some circumstances such as Marfes [10], electromagnetic radiation may also be sufficient to produce dust by thermal stresses due to overheating. Dust productions are unfortunately not avoidable because of both the need for fusion power exhaust and many kinds of plasma instabilities, up to the extreme case of disruptions, that allow particle or heat fluxes reaching the wall [11]. The amount of dust produced is positively correlated with the number of charged particles and their energy (power and duration) reaching the wall.

While detailed discussion of dust-production process will be postponed to Sec. 2, somewhat independent of their causes, we only give an brief discourse on plasma instabilities that are most important to dust production here. Apparently, any more details regarding the instabilities are beyond the scope of this article and the relevant information should be readily available in the literature if so desired.

In ITER, disruptions are expected to cause 10-100 MJ/m^2 of energy density on surfaces for a duration of 1-10 ms [1], or equivalently, the energy density produced by 2.5 to 25 kg of TNT over one square meter for the same duration. Type I Edge-localized modes (ELMs) is the second most severe energy and plasma loss process from fusion devices that can cause melting and ablations of surfaces, a necessary step for dust production. Type I ELMs are estimated to have an energy density ≤ 1 MJ/m^2 and 0.1-1 ms duration [1].

Even plasma-wall interactions due to instabilities are not necessary to produce dust in magnetic fusion devices. Microscopic processes that induce energetic particle and energy loss to the wall will also generate dust. An example of such a microscopic process is energetic neutral particles produced by charge-exchange. Once these energetic neutrals escape from the hot region of the fusion plasma, they strike the wall and sputter the surface due to their kinetic energies, a process usually known as physical sputtering. Besides physical sputtering due to energetic neutrals and ions, chemical sputtering is another possible contributor to dust production. In comparison, chemical processes require far less energy to erode the surface than a physical sputtering process. However, chemical sputtering process is element-sensitive. The most important chemical process in a machine like ITER is the reaction between graphite and hydrogen isotopes that lead to volatile hydrocarbon molecules.

Magnetic fusion devices are far from being the only plasma environment in which understanding of dust production, dust dynamics, dust-plasma interaction, *etc.* is crucial. As a matter of fact, dust production was recognized as a potential threat to plasma processing of semiconductors, when plasma electrons usually have a temperature of a few eV, ions are essentially at room temperature (except for the ones that are accelerated by sheath potential around a negatively biased object, where an energy gain of keV or larger may be possible). Semiconductor-processing plasma density is usually in the range of 10^{15} to 10^{16} m^{-3}, which is four to five orders of magnitude smaller than the density of an edge plasma in fusion. Besides the laboratory plasmas, ubiquitous presence of dust in the interstellar medium (with a plasma density of $\sim 10^5\ m^{-3}$ and 1 eV temperature) was observed in the 1930's [12]. Dust, as the most common solid matter in the universe, plays an increasingly important role in astrophysics [13]. Dust is being used as a diagnostic of its surroundings, complementary to conventional electromagnetic-radiation based astronomy. Dust is now understood to play critical roles in galactic evolution through, for example, its catalyzing effect on molecular hydrogen formation. Many solar phenomena, such as comets and planetary rings, can be better understood within the framework of solar plasma and dust interaction. It is believed that the very origin of the solar system depends on dust physics [14].

Understanding of dust in plasmas has made a tremendous progress [15–23], in particular, in laboratory dusty plasmas when the plasma condition is similar to the semiconductor-processing plasma condition specified above. Such an impressive progress has been possible partially because theoretical predictions have been routinely compared with and meanwhile, used to guide laboratory studies, or vice versa. For example, we saw the theoretical prediction of dust crystal formation (in so-called strongly coupled dusty plasmas) which was later confirmed by a number of laboratory experiments independently [18–20] and, in another case, the theoretical prediction of dust acoustic waves and its later experimental verification [24–26]. On other occasions, however, experimental observations came before theoretical predictions. The examples are the discoveries of dust voids [27] and 3-D dust Coulomb balls [28] in laboratory. In addition to being one of the most vibrant branches of basic plasma physics, strongly coupled laboratory dusty plasmas have also been recognized as uniquely suitable for '*simulating*' kinetic scale physics in conventional fluids because both individual dust motion and collective behavior of an assembly of dust are readily measurable simultaneously.

While it is hard to exaggerate how valuable the contributions from past and on-going dusty plasmas research have been to the understanding of dust physics in magnetic fusion, particularly, from the point of view of a framework for dust physics in plasmas in general [15], it is still appropriate to point out that semiconductor-processing plasmas and alike are too cold and too dilute compared with edge plasmas in magnetic fusion. Recognition of this drastic parameter change from the '*mild*' and '*benign*' semiconductor processing plasmas to the '*extreme*' and '*hostile*' fusion plasmas is necessary to appreciate the distinctively complex dust physics in edge plasmas as well as in hot and core plasmas of magnetic fusion. We first illustrate this distinction by a brief discussion on dust size with respect to the dust sheath thickness and the likelihood of dust erosion under two plasma conditions.

When a dust grain is immersed in a plasma, the difference in electron and ion mobility will cause the grain to charge up and a sheath to form around the grain. The dust charge and the sheath around it will force the net electric current to the grain to become zero at a time scale that is usually much shorter than the time for dust position change, in other words, a dust grain will have a well-defined charge and a sheath for its instantaneous spatial location, even if the grain moves. When additional effects, such as secondary electron emission, thermionic emission, photoemission, field emission, and radioactive decay, *etc.* are neglected, then the dust is expected to be charged negatively. The sheath thickness is characterized by the Debye length (λ_D), which is a function of electron density, ion density, electron temperature, and ion temperature as $\lambda_D^{-2} = \sum_j \lambda_j^{-2}$, and $\lambda_j = \sqrt{\epsilon_0 k_B T_j/(n_j e^2)}$ in MKS units. The subscripts $j = e$, i are for electron and ion (only one predominant ion species is assumed here to simplify the discussion) respectively. The symbols have their usual meanings. $\epsilon_0 = 8.85 \times 10^{-12}$ F/m is permittivity, $k_B = 1.38 \times 10^{-23}$ J/K is the Boltzmann constant, and $e = 1.60 \times 10^{-19}$ C is the elementary charge. We assume that the plasma is quasineutral, $n_i = n_e$, or equivalently that dust charge does not affect the quasineutral condition. In semiconductor processing plasmas, since $T_i \ll T_e$, therefore, $\lambda_D^l \sim \lambda_i < \lambda_e$, where we used the superscript 'l' for semiconductor processing plasma. While in fusion plasmas, $T_i = T_e$, therefore, $\lambda_D^h = \lambda_e/\sqrt{2} = \lambda_i/\sqrt{2}$, where the superscript 'h' symbolizes edge fusion plasmas. Quantitatively, we have

$\lambda_D^l \sim \lambda_i = 34\ \mu$m for $T_i = 1200$ K (0.1 eV) and $n_e = 5 \times 10^{15}$ m^{-3}. $\lambda_D^h = \lambda_e/\sqrt{2} = \lambda_i/\sqrt{2} = 2.4\ \mu$m for $T_i = 1.2 \times 10^5$ K (10 eV) and $n_e = 5 \times 10^{19}$ m^{-3}. The observation that the dust sheath thickness is usually much larger in semiconductor processing plasmas than in edge fusion plasmas, that is, $\lambda_D^l \gg \lambda_D^h$, can also be better appreciated through dust-to-dust Coulombic coupling, which is usually measured by the parameter $\Gamma^{Coul.}$, a function of dust charge (Q_d), dust density (n_d), and temperature (T_d), $\Gamma^{Coul.} = Q_d^2 n_d^{1/3}/(\epsilon_0 k_B T_d)\exp[-1/(n_d^{1/3}\lambda_D)]$ [15]. Theoretically, when $\Gamma^{Coul.} \geq \Gamma_0^{Coul.} \sim 17$, dust-to-dust Coulombic coupling is possible, although so far strongly coupled dust systems have only been observed experimentally when $\Gamma^{Coul.} \gg \Gamma_0^{Coul.}$. The exponential dependence of $\Gamma^{Coul.}$ on λ_D as $\Gamma^{Coul.} \propto \exp[-1/(n_d^{1/3}\lambda_D)]$ due to Debye shielding means it is much more difficult to achieve strongly coupled dust conditions in edge fusion plasmas than in semiconductor processing and alike plasmas, if not entirely impossible.

Dust erosion by fusion plasma is another process that is normally neglected in other laboratory dusty plasmas. We estimate dust erosion rate here to shed some light on the significance of the process in fusion plasmas. In an edge plasma with density ($\sim 10^{20}$ m^{-3}) and temperature ($\gtrsim 10$ eV), the least amount of erosion on a dust grain would come from heating by thermal motion of charged particles with a Maxwellian velocity distribution. Charge-exchange, fusion byproducts, waves and instabilities that produce energetic particles or reshuffle the particle velocity distribution, will only increase the erosion rate.

For an uncharged dust[b], the heating rate (Γ_j^H) due to jth plasma species ($j = e$, i for electrons and ions as above) onto a surface (with an area A) is given by $\Gamma_j^H = \frac{1}{4}An_jk_BT_jv_{\infty j}$, and $v_{\infty j} = \sqrt{8k_BT_j/\pi m_j}$ is the mean velocity of the species (with mass m_j) at a distance far away from the dust and its sheath. When ions and electrons are at the same temperature, it is apparent that the electron heating is more severe than ion heating by a factor of $\sqrt{m_i/m_e}$, the square-root ratio of the ion mass to the electron mass, which is 60 for a deuterium-ion-dominated plasma. Negative charge on the dust will reduce electron heating and enhance ion heating (and vice versa for a positively charged dust). Even in this 'conservative estimation' scenario, the total heating rate can not be smaller than ion heating alone,

[b]This is rarely the case in plasmas, see Sec. 4.1. In fact, in ambipolar diffusion, when the electron and ion current/flux to the dust are equal, ion heating can exceed electron heating due to energy gain from the sheath potential. We use a neutral dust for a baseline estimate only.

$\Gamma^H_{tot.} \equiv \sum_j \Gamma^H_j > \Gamma^H_i$. To the zeroth order, the erosion rate, defined as dr_d/dt for a spherical dust grain with radius r_d, does not depend on the surface area nor dust size, since the heating rate increases with the surface area linearly, $\Gamma^H_{tot.} \propto A$. That is, $dr_d/dt > (\mu_d/\rho_d)(\Gamma^H_i/AE_0)$, with ρ_d being the dust mass density, μ_d being the 'mean' mass of atoms that form the dust, and E_0 being the amount of energy needed to strip it from the dust (a sum of electronic bonding energy and any other inter-atomic energy). In the case when clusters of atoms instead of individual atoms are stripped at a time, less energy is required instead since fewer chemical bonds are broken. Therefore, the estimate gives the lowest dust-erosion rate. For a graphite dust particle in a 5×10^{19} m^{-3} and 10 eV edge plasma, $\rho_d = 2.25 \times 10^3$ kg/m^3, $\mu_d = 12m_p$ with m_p being the proton mass, $E_0 \sim 5$ eV, we found that $dr_d/dt > 10$ μm/s.

Another distinction comes from dust dynamics. Although the framework for dust dynamics in plasmas has mostly been established [15,29–38], it is necessary to reassess dust dynamics in fusion plasmas because a.) many forces can affect dust motion in plasmas simultaneously; b.) relative importance of different forces may change under vastly different plasma conditions; c.) dust shapes are not necessarily perfect spheres; d.)accurate description of some forces may be difficult due to the collective plasma effects. These forces include gravity, Lorentz force (collective effect possible), ion drag force (or plasma-flow drag force in quasineutral plasmas, collective effect possible), neutral particles drags, radiation pressure, 'rocket' force due to non-uniform ablation across dust surface, and forces due to gradients and waves (collective effect possible). For example, gravity is important to the formation of levitated dust crystals in laboratory plasmas. However, gravitational acceleration can be neglected in fusion plasmas, see Sec. 5. Since dust erosion is severe and up to the point of complete dust destruction, analysis of dust dynamics is more involving than in dusty plasmas when the 'constant' dust approximation is valid [35–38].

Below, we will discuss in more details about dust production and possible dust removal mechanisms (Sec. 2), experimental observation, diagnostic, and analysis of dust in magnetic fusion devices (Sec. 3), basic properties of dust in plasmas and numerical approach to study dust-plasma interaction (Sec. 4), dust dynamics, transport and impact on fusion plasmas (tokamaks mostly) (Sec. 5). Safety issues and dust chemistry, although they are very important, will not be discussed much further. We will summarize our discussion and give some prospective for dust problem in magnetic fusion devices towards the end (Sec. 6).

2. Dust production and removal in fusion devices

In magnetic fusion, since the initial forms of matter are either macroscopic solid (wall and other plasma facing components which are much larger than dust) or atoms and small molecules (gases used for plasma production or glow discharge cleaning, and residue gases inside the vacuum chamber) and possibly liquid lithium, there are, in principle, two possible paths to form dust. One is the *bottom-up* path, the other is the *top-down* path. Dust is formed through condensation (or coagulation) process by the bottom-up path, when atoms, molecules or small (nm in size) clusters condensate on each other and grow to larger particulates. While dust can also form by breaking down relatively larger pieces of material like flakes in one or multiple steps by the top-down path. The bottom-up path is the most discussed pathway for dust formation in plasmas (including magnetic fusion) according to the literature [21–23].

Formation of solid (dust included) from its vapor (atoms, small molecules) is a *phase transition* that happens at certain temperatures (usually below 3000 K [39]) and vapor pressures. Although phase diagrams for many pure materials are readily available, a phase diagram for a mixture of different materials may not be as well defined as a pure material because such a diagram also depends on relative concentrations of different materials. Under a certain pressure, the temperature at which a vapor turns into a liquid or a solid is called the condensation temperature. The hydrogen condensation temperature is about 20 K, a temperature that is too low compared with temperatures inside the vacuum chambers of the existing and near-future fusion devices. These 'icy' dust grains of pure hydrogen and its isotopes are unlikely in this context. Therefore, bottom-up formed dust would either come from wall materials which have higher condensation temperatures, or wall material is at least needed to participate in the coagulation process.

Phase transitions under other contexts (the Wilson chamber, rain formation, and many other examples in astrophysics, geochemistry, biology, medicine, metallurgy, engineering, aerodynamics, crystallography) can be useful references to understand dust formation in plasmas. A phase transition can be divided into four stages [40], that is, development of a supersaturated state, generation of nuclei of the new phase (nucleation), the growth of nuclei to form larger particles or domains of the new phase, and relaxation processes such as agglomeration by which the texture of the new phase alters. As far as the dust formation is concerned, the fourth step does not apply. Therefore, we need to consider the other three stages for dust formation in magnetic fusion (and in other plasmas).

Development of a supersaturated state is possible due to sputtering, evaporation and sublimation of the wall in plasmas (particularly in plasma afterglow or training plasmas). Nucleation can also happen for several different scenarios. In an ideal case, when only one kind of atoms/molecules is involved in the process, so-called *homogenous* nucleation happens when the nuclei of the new phase forms in absence of catalytic agents [c]. Homogenous nucleation is rarely the case in plasmas since many different particle species (including electrons and positively charged ions) are present. Therefore, *heterogeneous* nucleation is more common. In either case, formation or existence of 'nucleation centers' (or simply 'nuclei' [40]) is necessary for dust formation. We do not discuss particle growth due to surface chemistry or any other chemical process here because chemical processes depend on the particular chemicals involved [41]. Nucleation centers could be as trivial as the wall itself. Besides the wall, nucleation centers can also come from the evaporated wall materials during a normal operation (fusion products, such as alpha particles, do not easily form larger molecules either in room temperature or above). We briefly go over the sputtering, melting, evaporation and sublimation processes that create small atoms and molecules in fusion devices before discussion of nucleation, condensation, and then dust growth.

2.1. *Sputtering, melting, evaporation and sublimation*

The conditions to produce the atoms, ions and small cluster for dust coagulation can be rather easily met in fusion devices by a.) physical sputtering, when a sufficient fraction of the kinetic energy from the impinging particles (particularly important ones are H, D, T, He^{2+} and n^0) is transferred to surface atoms that allows the surface atoms to overcome the surface potential and escape from the surface [42]; b.) chemical sputtering, when plasma ions or atoms at excited states chemically react with surface atoms and form unstable compounds that escape from the surface easily [42,43]; or c.) sublimation and evaporation of the wall material resulted from sudden arrival of a large amount of plasma energy (due to disruptions or ELM's) that causes rapid temperature rise above the evaporation and/or sublimation point of the surface. Physical and chemical sputtering are essentially caused by individual particles, while contributions from other particles, plasma heating, and surface condition can significantly modify the individual particle sput-

[c]page 70 of the ref. [40]

tering yield, a measure of number of atoms sputtered per incidental particle. Sublimation and evaporation are caused by the collective heating of many particles simultaneously. Besides supplying the necessary wall particles for dust formation, all of these processes also lead to wall erosion and structure damage.

There is a typical kinetic energy threshold about 10 (9 eV for D on Be) to a few hundred eV (200 eV for D on W) for physical sputtering to take place, below which, the sputtering yield vanishes. Besides the incidental particle energy, physical sputtering yield also depends on the incidental particle mass, wall particle mass, and incidental angle of the bombardment. For the most frequently considered wall materials of fusion, Be, C, and W, sputtering yield database for H, D, and He has been established for energies up to 10 keV [43,44] and were extrapolated to even higher energies that are relevant to fusion. Physical sputtering yields for metallic compounds or potentially useful nanostructured materials [45], however, are not as much as known as for pure materials.

Unlike physical sputtering, chemical sputtering essentially does not have a threshold for the incidental particle energy. But chemical sputtering is particle-species sensitive/selective. The most important pairing are H/D/T with C(graphite) which lead to volatile or loosely-bound hydrocarbon production within the vacuum chamber because of the their large chemical sputtering yield. Graphite could be eroded by hydrogen ions with a maximum chemical erosion yield of $Y \sim 0.1\mathrm{C/D}^+$, a yield several times higher than the maximum physical sputtering yield [46]. At surface temperatures below 400 K, all surface carbon atoms are essentially hydrated but no hydrocarbon release occurs because binding energy of around 1 eV for hydrocarbon and radicals. This energy is much less than binding energy carbon atoms in regular graphite lattice of 7.4 eV. Heating of the surface and particle bombardment can both result in release of hydrocarbons. Although many aspects of chemical sputtering remain to be understood, one of the best solution for now seems to replace graphite wall by other materials such as Be, so that chemical sputtering can be substantially mitigated. But chemical sputtering does not go away completely since graphite tiles used in diverter region (in ITER) can still be chemically eroded and hydrocarbons can be redeposited elsewhere.

Evaporation following melting (for Be, W, and other metals) or sublimation (for C) of the wall happens during ELMs and disruptions, when the wall temperature rises abruptly above the evaporation or sublimation point of the wall material since wall heat load can substantially exceed the cooling. Although the liquid phase of carbon does exist under high temperatures and high pressures (the triple point of carbon occurs at 100 times the atmospheric pressure and a temperature of 4700 K) [47], graphite normally bypasses its liquid phase going from solid to vapor phase under magnetic fusion conditions, a process known as sublimation.

The transient temperature rise ΔT for an originally uniform-temperature surface can be calculated from the 1-D heat conducting equation

$$C_p\rho\frac{\partial\Delta T}{\partial t} = k\frac{\partial^2\Delta T}{\partial x^2} + q_i, \qquad t > 0, \quad -\infty \leq x < \infty, \tag{1}$$

for a surface with a mass density ρ, a heat capacity C_p, and a thermal conductivity k. The heat influx q_i satisfies

$$q_i(x) = \begin{cases} q_0\delta(x), & \text{for } \ 0 \leq \mathrm{t} < \mathrm{t}_0. \\ 0, & \text{for } \ \mathrm{t} \geq \mathrm{t}_0. \end{cases} \tag{2}$$

The solution to Eq. (1) is readily available as

$$\Delta T = \frac{1}{C_p\rho}\int_0^t \frac{d\tau}{2\sqrt{\pi\alpha(t-\tau)}}\int_{-\infty}^{\infty} d\xi q_i(\xi)\exp\left[-\frac{(x-\xi)^2}{4\alpha(t-\tau)}\right], \tag{3}$$

with thermal diffusivity $\alpha = \frac{k}{C_p\rho}$. An estimate for transient temperature rise ($t_0 \to 0$) right after the heat flux corresponds to $t \leq t_0$. Therefore, only near the surface $x \to 0$, the temperature rise is significant. Eq. (3) can be Taylor expanded for small x as

$$\Delta T = \frac{q_0}{C_p\rho}\sqrt{\frac{t}{\pi\alpha}} + O(x^2). \tag{4}$$

This estimate is smaller than the one given in ref. [11] by a factor of two. Based on the lowest order estimation, the first term in Eq. (4), using the melting/boiling temperature of tungsten is 3695/5828 K, the thermal conductivity of tungsten k =170 W$\cdot$ K$^{-1}\cdot$ m^{-1} , the enthalpy of fusion/evaporation is 35/800 kJ$\cdot$ mol^{-1}, $C_p \sim 130$ J kg$^{-1}\cdot$ K^{-1}, $\rho = 1.92\times10^4$ kg/m^3, one obtains the threshold q_0 for tungsten melting is about 130 MW/m^2 for a one-second heat pulse. This estimate is about three times larger than the result given in Ref. [3] for ITER. Such heating can be delivered by disruptions and type I ELMs.

2.2. *Nucleation*

Nucleation, or generation of nuclei of a new phase, happens in the early stage of dust formation when 'clusters of atoms' are relatively small, up to a few nm in size, or $\lesssim 10^5$ atoms per cluster [22]. These small clusters may grow or fragment by collisions with other particles (Although vapor chemistry, electric charges, Van der Waals force, boundary conditions, deviation of clusters from perfect spheres can all modify nucleation, we focus here solely on collisions, which still allow us to gain considerable insight about nucleation [48–50,52]). An elementary process may be illustrated by the formation of a larger cluster (AB) from two smaller ones A and B,

$$A + B + (M) \rightleftharpoons AB + (M), \tag{5}$$

where the extra particle M may help to form a metastable AB in the forward direction and cause AB to disintegrate in the backward direction. From the energy point of view, two clusters A and B initially have kinetic (translational) energies E^k_A, E^k_B and internal energies (E^I_A, E^I_B, which include electron binding energy, vibration, and rotational energy for clusters with multiple atoms), while the cluster AB has E^k_{AB} and E^I_{AB} correspondingly. The forward process is favorable energy-wise if $E^I_{AB} < E^I_A + E^I_B$, and the opposite is true if E^I_{AB} is larger. The absolute value of the internal energy change $|E^I_A + E^I_B - E^I_{AB}|$ is smaller or comparable to the energy absorption available ($\sim k_BT$) from a collision. Therefore, the collision may dictate the direction of the reaction, until a critical cluster size is reached. When larger than the critical size, particle growth (condensation) is favored since larger particles corresponds to a lower internal energy state (intuitively, a large enough cluster would develop sufficiently deep enough potential wells that can absorb all the energy from collisions without fragmentation). The critical particle size thus divides the nucleation phase from the particle growth phase.

A cluster may also carry electric charge in a plasma. Each cluster may be regarded as a small electric capacitor. The maximum amount of charge (q_c^{max}) can be estimated to be

$$\frac{(q_c^{max})^2}{2C_c} \leq k_BT, \tag{6}$$

where C_c is the cluster capacitance. Basically, Eq. (6) treats the total energy due to charging as a destabilizing source. Approximate each cluster by a sphere, $C_c = 4\pi\epsilon_0 r_1 i^{1/3}$, with r_1 being the radius of the vapor molecule and i is the number of vapor molecules in the cluster (for a homogenous vapor, see the paragraph below), we observed that $q_c^{max} \propto i^{1/6}$, which is a very weak dependence on the number of atoms/molecules. For a carbon cluster of $i = 10^5$ atoms, with $r_1 \sim 1$ Å and a temperature of $T \sim 1000$ K, one obtains $q_c^{max} \leq 0.7\ e$ (e is the elementary charge). Therefore, the number of charges on a carbon cluster in the plasma is about one (due to the quantization of charge), consistent with the result given elsewhere [22]. Knowing that the number of charges on a smaller clusters can only be one (both positive and negative are allowed) can reduces the number of reactions like Eq. (5) for nucleation models.

Two basic questions need to be answered by any nucleation model. What are the critical cluster sizes? What are the nucleation rates for neutral and ionic clusters? While the first question can be answered within the scope of thermodynamic equilibrium nucleation, the second question has to be answered by a kinetic theory, when the microscopic details [collision rates, collision cross sections, possible collision reactions like Eq. (5)] of collisions are needed.

2.2.1. *Homogenous nucleation*

One of the simplest nucleation models is the so-called *classical* homogeneous nucleation model [22,40,48–51,53]. The classical model assumes one type of vapor molecule (monomer) A and its clusters A_i's, which have $i \times A$ molecules (i-mer). In addition, the only ways by which an i-mer A_i forms or fragments are through monomer condensation and its reverse process $A_{i-1} + A \rightleftharpoons A_i$, and $A_i + A \rightleftharpoons A_{i+1}$. The net rate of the i-mer formation is given by [40]

$$\frac{dn_i}{dt} = J_i - J_{i+1}, \qquad \text{for } i \geq 2, \tag{7}$$

where [53]

$$J_i = \beta \alpha_{i-1} s_{i-1} n_{i-1} - E_i n_i \tag{8}$$

is for the reversible reaction $A_{i-1} + A \rightleftharpoons A_i$ with n_{i-1} and n_i being the density of A_{i-1} and A_i, $\beta = (\frac{k_B T}{2\pi m_1})^{1/2} n_1$ being the monomer flux onto a unit surface area of A_{i-1}, $0 \leq \alpha_{i-1} \leq 1$ being the sticking coefficient. s_{i-1} being the surface area of A_{i-1}, and E_i being the i-mer evaporation rate.

One may use the *detailed balance* assumption at an equilibrium to obtain E_i in Eq. 8, $J^e = 0$ gives [53],

$$E_i = \beta^e \alpha_{i-1} s_{i-1} \frac{n_{i-1}^e}{n_i^e}, \tag{9}$$

where the superscript e standards for the values at a thermal equilibrium. In a steady state, Eq.(7) implies that $J_i = J_{i+1} \equiv J$. Summation of Eq. (8)

with subscript i from 2 and up to a number N, one obtains

$$J = \left[\sum_{i=1}^{N-1} (\beta \alpha_i s_i n_i^e S^i)^{-1}\right]^{-1} \left(1 - \frac{n_N}{S^N n_N^e}\right), \tag{10}$$

where

$$S \equiv \frac{\beta}{\beta^e} = \frac{n_1}{n_1^e} \tag{11}$$

is the ratio of the vapor pressure to the pressure at equilibrium when a liquid phase is present. When $S > 1$ and for sufficiently large N, the last term in Eq. (10) is small compared with unity and may be neglected. Therefore, the nucleation rate J becomes

$$\frac{1}{J} = \sum_{i=1}^{\infty} \exp\left(-\frac{H_i}{k_B T}\right), \tag{12}$$

with

$$\frac{H_i}{k_B T} = ln(\beta \alpha_i s_i n_i^e) + i \ln S. \tag{13}$$

Further calculation of H_i requires the knowledge of β, α_i, s_i, and n_i^e. n_i^e is related to n_1^e through the Gibbs free energy (ΔG_i) as

$$n_i^e = n_1^e \exp\left(-\frac{\Delta G_i^e}{k_B T}\right) \tag{14}$$

with ΔG_i given by [22,40,54]

$$\Delta G_i = 4\pi r_1^2 \gamma (i^{2/3} - 1) - (i-1) k_B T \ln S, \tag{15}$$

where γ is the surface energy per unit area. This formulation is different from classical formulation slightly as discussed in [54]. Therefore, $\Delta G_i^e = 4\pi r_1^2 \gamma (i^{2/3} - 1)$ since by definition, $S^e = 1$ [see Eq. (11)]. A critical nucleation size $i_c(> 1)$ is obtained from $\partial H_i / \partial i = 0$, that is

$$i_c = \left(\frac{2}{3} \frac{\Theta}{\ln S}\right)^3, \tag{16}$$

while α_i is assumed to be i in-dependent in Eq. (13), $\Theta i_c^{2/3} - 1 \sim \Theta i_c^{2/3}$, and the dimensionless surface energy $\Theta = \frac{4\pi r_1^2 \gamma}{k_B T}$. Approximate the infinite series in Eq. (12) by an integration, the nucleation rate J can be approximated by [53]

$$J \simeq \left[\frac{H''(i_c)}{2\pi k_B T}\right]^{1/2} \exp\left[\frac{H(i_c)}{k_B T}\right], \tag{17}$$

which is obtained by Taylor expansion of H(i) around $i = i_c$ and keep only the zeroth order and the second order term (the first order vanishes according to the definition of i_c). Combining Eqs. (13) to (17), one obtains

$$J = (\frac{4\pi}{3}r_1^3)(\frac{2\gamma}{\pi m_1})^{1/2}(n_1^e)^2 S \exp\left[\Theta - \frac{4\Theta^3}{27(\ln S)^2}\right], \tag{18}$$

in which, the approximation $\frac{H''(i_c)}{k_B T} \sim \frac{4}{9}\Theta i_c^{-4/3}$ is used since $\Theta i_c^{2/3} \gg 1$. The sticking coefficient α_i is chosen to be 1. This formulation is different from the classical theories by a factor of e^{Θ}/S as found by Girshick and Chiu [53,54]. The nucleation rate depends on saturation pressure, pressure of supersaturated state, surface energy of the clusters.

2.2.2. *Heterogeneous nucleation*

Heterogeneous nucleation may happen due to charged particles, wall (boundary) effects, chemical reactivity, and spatial non-uniformity of the particle distribution. We only briefly discuss heterogeneous nucleation due to charged particles here. Similar approach may be applied to other types of heterogeneous mechanisms. Basically, each heterogeneous mechanism modifies the homogeneous Gibbs free energy, Esq. (15). The procedure given above for the homogenous nucleation still applies to derive the nucleation rates for heterogeneous nucleation.

Since charged particles are readily available in a plasma, one immediate question regarding dust formation through nucleation in plasmas would be the possible effects of electrostatic force on nucleation due to cluster charging. Cluster charging was estimated by Eq. (6) to be about one. Therefore, one may focus on singly charged i-mers, $A_i^{\pm}$. The Gibbs free energy for a charged cluster is [22,55,56]

$$\Delta G_i^{q\pm} = \Delta G_i - \frac{q^2}{8\pi\epsilon_0 r_1}(1 - \frac{1}{\epsilon})(1 - \frac{1}{i^{1/3}}) \tag{19}$$

with $q = 1$ for $A_i^{\pm}$ and ΔG_i given by Eq. (15). Even in one of the simplest cases, say a model for A_i^+ formation with the presence of neutrals A_i, the number of reactions to produce A_i^+ is more than its neutral counter part A_i. Besides $A_{i-1}^+ + A \rightleftharpoons A_i^+$, $A_i^+ + A \rightleftharpoons A_{i+1}^+$, one also has to include $A_{i-1} + A^+ \rightleftharpoons A_i^+$ [d]. Correspondingly, at least three terms are needed to calculate $\frac{dn_i^+}{dt}$, while Eq. (7) has only two terms on the right-hand side. We

[d] $A_i + A^+ \rightleftharpoons A_{i+1}^+$ is not needed, since it does not involves A_i^+.

leave out further discussion of heterogeneous nucleation rates, which can be found in literature for different models [22,55,56].

It should be pointed out, although the generic framework for nucleation is mostly established and understood, however, specifics of nucleation process as a part of dust formation mechanism (for fusion in particular) remain open questions. It is anticipated that both numerical simulations and laboratory experiments through sophisticated diagnostics will be needed in the future. In particular, although in-situ measurement of dust in fusion experiment have been achieved (Sec. 3.4). Experiments to understand nucleation phase of dust formation in fusion devices remain to be seen and would be more difficult because of the small sizes of nanoparticles.

2.3. *Redeposition, condensation, and coagulation*

Sputtering, wall evaporation and sublimation supply both nucleation 'nuclei' and vapor for dust formation and growth. When at sufficiently low temperatures, formation and growth of dust particles ('macroparticles') continues beyond the nano-clusters in the nucleation phase. Since the wall has the lowest temperature around a fusion device, dust formation usually happens at or near the wall. Dust formation may also happen inside the plasma, for example, when the electrostatic force draws oppositely charged particulates together, or when the plasma is adiabatically cooled through a nozzle or a nozzle-like converging-expanding structure [57–59]; however, dust formation through the adiabatic cooling is unlikely in fusion. Experimentally, formation of dust particles in discharges due to physical sputtering of cathode was first observed by Langmuir and colleagues [60]. Suspended dust particles were directly observed by laser scattering technique in the cathode sheath of RF plasmas for semiconductor processing [61,62]. More extensive study of dust formation in semiconductor processing plasmas can be found in ref. [22].

Most of the sputtered atoms, molecules, radicals, clusters, and dust, if not ionized by the high-temperature plasma or removed by the vacuum pumping system, will redeposit or condensate on the wall again, thus the name *reposition*, which is used interchangeably with *codeposition*. Therefore, dust can be formed on the wall due to *redeposition*. Although sputtering-deposition is a well-known technique to form thin films on substrates and sometimes being used (such as to produce diamond-like thin films) for deposit protective coatings onto the substrates surfaces, the self-rising thin-film deposition in tokamaks, usually observed in the form of loosely-hanging 'flakes' on the wall, causes more concerns, such as tritium retention, than delivering any real benefit to the inner wall of fusion devices.

Dust may also grow away from the wall. Two scenarios are possible. One is the condensation of molecules and atoms onto dust grains that are already present, this is *condensation* pathway for dust growth. The other is when two or more smaller dust grains stick together to form a larger grain, this is *coagulation* pathway for dust growth [16]. Dust growth due to condensation would be more important if the monomer concentration is much higher than dust particles. The rate for dust mass (m_d) change is relatively simple,

$$\frac{1}{m_1}\frac{dm_d}{dt} = \alpha\beta s_d - E_1 s_d, \tag{20}$$

where α is the monomer sticking coefficient, $\beta = (\frac{k_B T}{2\pi m_1})^{1/2} n_1$ is the monomer flux on the dust, s_d is the dust surface area, and E_1 is the monomer evaporation rate per unit area.

If most of the particles in the plasma are small dust grains and clusters, coagulation will become the primary mechanism for dust growth. Perrin and Hollenstein [22] have pointed out the similarity of dust formation in plasmas through coagulation to aerosol formation in the air [63], and colloid formation in liquid solvents. Since a dust grain can hold multiple charges (significantly different from one if the grain is large enough, see Sec. 4.1), coagulation collisions are similar to Eq. (5) but need to take into account charge, mass, and momentum conservations, that is

$$A(m', q', \mathbf{v}') + B(m - m', q - q', \mathbf{u}) + (M) \rightleftharpoons AB(m, q, \mathbf{v}) + (M). \tag{21}$$

The formation rate of the charged dust grains with mass m and charge q can be described by a Boltzmann (kinetic) equation [16]

$$\begin{aligned}\frac{\partial}{\partial t} f(m, q, \mathbf{v}, t) = \frac{1}{2}\int_0^m dm' \int_{-\infty}^{\infty} dq' \iiint_{-\infty}^{\infty} d\mathbf{v}' \left(\frac{m}{m - m'}\right)^3 K(m', q', \mathbf{v}';\\ m - m', q - q', \mathbf{u}) \times f(m - m', q - q', \mathbf{u}, t) f(m', q', \mathbf{v}', t)\\ - f(m, q, \mathbf{v}, t) \int_0^{\infty} dm' \int_{-\infty}^{\infty} dq' \iiint_{-\infty}^{\infty} d\mathbf{v}' K(m', q', \mathbf{v}';\\ m, q, \mathbf{v}) \times f(m', q', \mathbf{v}', t),\end{aligned} \tag{22}$$

where K is called the coagulation kernel function that is the coagulation probability for the collisions, and $\mathbf{u}$ satisfies $m'\mathbf{v}' + (m - m')\mathbf{u} = m\mathbf{v}$. The integration limits for q' does not have to go to infinity due to physical constraints such as field emission of electron and ions, see Sec. 4.1. Eq. (22)

is different from the usual collision terms in the Boltzmann equation for a fully ionized plasma [64]. Eq. (22) only includes two-body coagulations (A and B) in Eq. (21), and collisions of different bodies (AB, M) for fragmentation, which strictly would correspond to a three-body coagulation process (A, B and M). Therefore, Eq. (22) is valid when two-body coagulations dominate over three-body coagulations, and all coagulations and fragmentations are due to collisions among clusters (effects due to radiation, electron and ion, and monomers are neglected). The 'artificial' term $(\frac{m}{m-m'})^3$ above, also found in ref. [16], is needed to obtain the well-known Smoluchowski equation, as discussed below. Although there is no apparent reason why it is should be included.

If one assumes that the kernel function K and the distribution functions only depend on the mass of the coagulation particles, then by integrating both sides of the Eq. (22) over the $\mathbf{v}$ space (or equivalently over $\mathbf{u}$ space) and over q-space, the integral form of the Smoluchowski equation for coagulation can be obtained [65–67]

$$\frac{\partial}{\partial t}n(m,t) = \frac{1}{2}\int_0^m dm' K(m'; m-m')n(m-m',t)n(m',t) \\ -n(m,t)\int_0^m dm' K(m'; m)n(m',t), \tag{23}$$

If one assumes that the kernel function K and the distribution functions depend on the mass and charges of the coagulation particles but not velocities, then by integrating both sides of the Eq. (22) over the $\mathbf{v}$ space (or equivalently over $\mathbf{u}$ space), the following coagulation equation can be derived [23,68,69]

$$\frac{\partial}{\partial t}n(m,q,t) = \frac{1}{2}\int_0^m dm' \int_{-\infty}^{\infty} dq' K(m',q'; q-q', m-m') \times \\ n(m-m',q-q',t)n(m',q,t) \\ -n(m,q,t)\int_0^m dm' \int_{-\infty}^{\infty} dq' K(m',q'; m,q)n(m',q',t). \tag{24}$$

We do not pursue the discussion of the solutions to the coagulation equations further since a vast number of articles are available, although most of them are not directly related to magnetic fusion. Therefore, just like nucleation, dust formation through coagulation needs to be understood better for magnetic fusions. Among three possible mechanisms for dust formation, whether redeposition, condensation, and coagulation are equally important, or some are more important than others remain unknown.

2.4. *Dust removal mechanisms and techniques*

Dust may be removed by either destructive or non-destructive means. Here the destructive means (dust destruction was also called 'dust disruption' in ref. [70], but we do not use this terminology here to avoid confusion with 'disruption' of a plasma) are ways that turn dust into finer particles up to fully ionized plasmas. Non-destructive means are methods that do not destroy dust. Mechanisms to destroy dust include heating of dust to dust evaporation or sublimation temperature, energetic particle bombardment, collision among dust grains themselves [70] (the relative velocity ranging from 3 km/s for icy dielectrics up to 8 km/s for Fe particles [71], chemical decomposition, plasma-assisted chemical decomposition, dust collision with the wall, Coulomb explosion, or laser evaporation. The non-destructive mechanisms include ways to get the dust moving, either mechanical (such as a brush), aerodynamic (such as a vacuum cleaner), sonic, electric, or magnetic energy may be used.

Versatility of the dust removal mechanisms does not imply there is already an *effective* method to remove dust from the vacuum chamber of a magnetic fusion devices at this moment. Particularly useful methods would be ones that can be used in parallel with the plasma operation, since dust is continuously produced along with fusion energy. Because of the tritium retention and dust radioactivity, another requirement of an 'ideal' method is that it can also recycle tritium back for fusion, and separate the radioactive elements from the bulk of the dust that is non-radioactive.

3. Experimental observation, diagnostic, and analysis of dust in fusion devices.

Dust particles have long been observed to coat the inside surfaces of fusion devices after plasma operations (Fig. 1). Dust can be produced by the disassembly of plasma facing tile surfaces or of plasma-grown co-deposited layers under the impact of ELMs or disruptions, or by the chemical agglomeration of sputtered C_n clusters (sect. 2). In next-step devices, the increase in duty cycle and erosion levels will cause a large scale-up in the amount of dust particles produced. This has important safety consequences as the dust particles may be radioactive from tritium or activated metals, toxic and /or chemically reactive with steam or air. Tritium and dust are related but not necessarily identical source terms in safety analyses. One can have a tritium inventory in codeposited layers separate from the dust inventory, and one can have a hazardous dust inventory (for example of activated tungsten) that is independent of tritium inventory. Previous reviews of dust in fusion devices may be found in refs. [1,4,5,9,72].

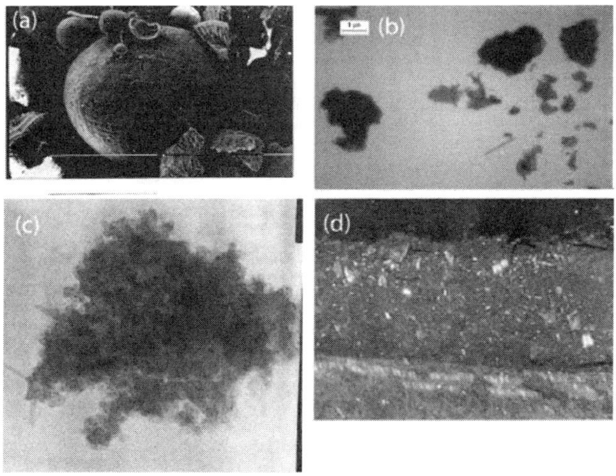

Fig. 1. Dust in contemporary tokamaks (a) Iron spheres from TEXTOR-94 with the large sphere showing a regular surface texture [72]; (b) TEM microphotograph of dust retrieved from TFTR [94]; (c) TEM image of flakes from Tore Supra: globular and elongated structures [87]; (d) Dust in NSTX under lower divertor tile.

To provide a technical basis for assessing the dust inventory limits in next-step machines, dust collection from contemporary tokamaks was begun as part of the Engineering Design Activity of the International Thermonuclear Tokamak Reactor (ITER) and is now an ongoing activity. These tokamaks include Alcator C-Mod [73,74], Asdex-Upgrade [75], DIII-D [73,76], JET [1,77,78], JT60-U [79], LHD [75], NOVA (an ICF facility) [80], NSTX [79], TEXTOR [81–83], TFTR [73,84–86], Tore Supra [87,88], and TRIAM-1M [89]. The dust is typically vacuumed from various areas in the vacuum vessel and trapped in filters with 0.02 μm pore size. In some cases a cyclone vacuum cleaner collected dust and debris down to 2 μm in size [77]or cotton swipes were used.

The disruption heat loads anticipated in next step devices (up to 100 MJ/m^2 on the divertor area in 10 ms) are not attainable in current tokamaks. Disruption simulators are used instead to reproduce the heat loads and study the resulting dust generation [90–92]. The physics of dust production under such extreme conditions has been modeled by Hassanein [93].

3.1. *Amount of dust in contemporary fusion devices*

The amount of dust found in tokamaks depends on the history of the plasma parameters in the discharges and the overall duration of plasma operations since the surfaces were last cleaned. Typically the highest concentrations are found on horizontal areas in the lower part of the vacuum vessel. Estimates of the total dust inventory are listed in Table 1 and range from 0.5 g (NSTX) to 90-120 g (DIII-D). The surface mass density is derived from the mass of dust collected divided by the collection area. Estimates range from 0.03 g/m^2 (NSTX) to 10 g/m^2 (C-mod) (see Table 1).

3.2. *Size distribution and composition*

The radiological and/or toxic hazard of dust depends on how well it is confined in accidental situations and whether it is small enough to remain airborne as an aerosol and be respirable($\lesssim$ 10 μm). Particle size is an important factor in the deposition pattern of particles in the respiratory tract. Tritium bound to metal or carbon particles can have a much longer residence time in the human body (and concomitant radiation dose) than the 10 d biological half life of HTO [e]. A study of the dissolution rate of carbon tritide particles from TFTR in simulated lung fluid found that > 90% of the tritium remained in the particles after 110 d [94]. A further complication with tritiated dust is that it can self charge through the emission of beta electrons and spontaneously levitate in electric fields [95]. Such levitation was observed experimentally in tritiated particles retrieved from TFTR [96] and adds to the challenge of confining dust in accident situations.

The tritium content of dust will depend strongly on whether it originates from crumbling codeposited layers, or under thermal overload of tritiated layers by off-normal events. Dust that is produced during ELMs and disruptions may be heated sufficiently to outgas tritium as T_2 reducing the radiological hazard of the dust. Early studies of dust produced by a plasma gun disruption simulator found a retention level of 6×10^{19} D per gram of carbon or D/C of 1.2×10^{-3} [97]. The D content in dust retrieved from TFTR MIRI diagnostic windows [74] was measured to be D/C $\sim 5.8\times10^{-3}$ and the T/C ratio was 26× lower, as expected from the T/D fuelling ratio. Analysis of dust vacuumed from the TFTR vacuum vessel also showed a

[e] a tritiated form of H_2O.

Machine	Estimate of total dust (g)	Surface mass density (g/m^2)	Count median diameter (μm)	BET surface area (m^2 /g)	Elemental composition
Alcator C-mod [74,88]		≤ 10	0.78-2.89	0.77	Mo, B
ASDEX-U [75]		≤ 1.5	3.33	≤ 3.7	Cu, Fe Cr, Ni
DIII-D [74,88]	90-120	≤ 1	0.46-1.0	2.44	C
JET [1,78]		< 4	4	4.7	C, Fe Cr, Ni, T
JT60-U [79]	7.5	0.037	3.08	≤ 1.18	C, Si
LHD [75]	16.2	0.04	9.6		C, Fe, Cr, Mn
NOVA [80]	44	36	≤ 8.85	0.43	Si, Cu, O
NSTX [79]	0.5	0.033	3.27		C,B
TEXTOR [5]		5.2			
TFTR [74,88]			0.76-3.0	0.82	C
Tore Supra [88]	31	≤ 1.1	3.0	1.32	C, O Fe, Cr, Ni
TRIAM-1M [89]			1-5		Mo, Fe Ni, Cr

Table 1. Parameters of dust collected from various tokamaks. Note that this table samples a diverse data set and the references in column 1 should be consulted for important experimental details.

low D/C ratio of 8.1×10^{-3} and T/C 4.4×10^{-4}. Baking of 0.24 g of flakes from TFTR codeposits at 773 K for 1 h released 0.72 Ci or 75 μg of tritium, a T/C ratio of 3×10^{-4} [86] broadly consistent with the dust results. It should also be noted that vessel access to obtain these samples was only possible after substantial detritiation activities [98,99]. A low H/C ratio of

0.04 was found in dust from JT-60 [100] and interpreted in terms of high wall temperatures. In contrast ion beam analysis of JET flakes showed a D/C ratio of 0.75 [77], which is about two orders of magnitude higher than the TFTR deuterium fraction.

The particle size distribution of tokamak dust has been measured by analysis of optical and SEM & TEM microscope images of the dust and results are listed in Table 1. The particle sizes most often follow a log normal distribution. The count median diameter ranges from 0.46 μm (DIII-D) to 9.6 μm (LHD). Note that particles smaller than the 0.02 μm pores of the filters used to vacuum up the dust are not collected or included in the above estimates. In particular, nano-scale particles have been observed in Tore Supra [87] and TEXTOR [81]. Particles generated by the SIRENS disruption simulator have smaller count median diameters - 0.1 μm in ref. [90] than found in tokamaks and this can be understood by modeling of the condensation of a vapor cloud [91]. The QSPA disruption simulator facility produced a significant number of nano-scale particles together with particles of 0.1 - 3 μm [92].

The dust particle composition can be measured by energy dispersive X-ray (EDX) analysis and where applicable, the tritium content measured by thermal desorption spectroscopy (Table 1). The elemental composition generally reflects that of the plasma facing components. The dust chemical reactivity depends on the effective surface area. This is measured by the BET technique [101] which employs krypton gas absorption to determine the total surface area of the dust sample. The total surface area is then divided by the sample mass to obtain the specific surface area. High specific surface areas are typically observed $\sim$ few m^2/g (Table 1).

The different formation mechanisms of carbonaceous dust will leave their signatures in the macroscopic and microscopic structure and this may be accessed through Raman scattering. A strong increase in structural disordering was found in graphite samples irradiated with high energy D^+ and He^+ ions [102] and by sputtering [103]. Raman analysis of dust from Tore Supra [104] and NSTX [105] also showed structural modifications induced by the plasma.

3.3. *Dust production rate*

The control of dust inventory in next-step devices will depend on the expected location of dust and its production rate. The total estimated mass of carbon dust collected in Tore Supra for 986 discharges was 31 g and in the same range as the 27 g estimate of the mass of eroded plasma facing material [87]. The high duty cycle of ITER will result in thick codeposits, up to 0.4 mm over a 10 day operational period [106]. These will have internal stress, be less thermally and mechanically stable than the thin layers in existing machine and be more likely to crumble into dust and flakes. Oxidative detritiation techniques may also induce flaking similar to that observed in codeposited layers in TFTR after exposure to air [86]. Detachment of blister caps on metallic surfaces may also be contribute to dust. Global estimates of dust production assume that some fraction of material sputtered from plasma facing surfaces ends up as dust (the remainder being vaporized in the plasma). An conservative estimate for the ITER generic site safety report [107] used 30% of the vaporized, sputtered or eroded material and is shown in Fig. 2.

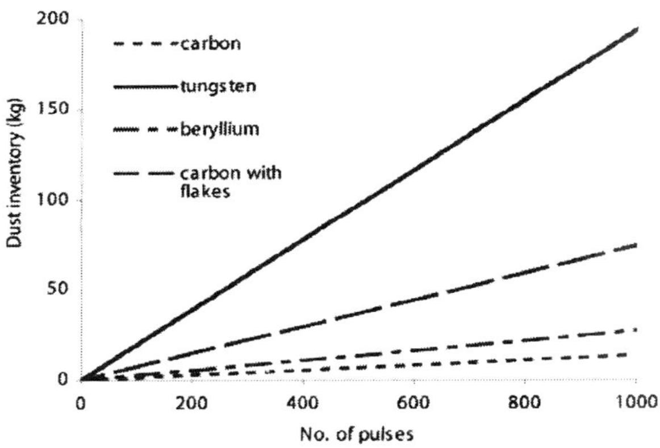

Fig. 2. Estimated dust production rate for ITER FDR as a function of the number of 400s pulses (fraction of vaporized, eroded, or sputtered material: 0.3).

3.4. *In-situ dust measurements*

Besides the 'archeological' type of dust study discussed above, we describe here 'in-situ' study of dust in fusion devices.

3.4.1. *Camera observations of dust in plasmas*

Incandescent particles are frequently observed 'flying' in tokamak plasmas and these provide a route for the transport of impurities (Sec. 5). Particle tracks were observed in TEXTOR that followed the magnetic field lines, indicating that the particles were charged [83]. The recent development of fast cameras has enabled the details of the dust particle motion to be revealed. Debris 'flying' at 100 m/s was observed after a TFTR disruption [108]. A 4500 frames /s camera observed dust moving at 10-50 m/s in TRIAM-1M plasmas and the amount of dust increased with the duration of the discharge [89]. The open geometry of NSTX is particularly suited to observations of dust, and multiple cameras were used in stereoscopic views to obtain detailed 3D trajectories [109]. The cameras had framing rates up to 68,000 fps with pixel arrays 128×128 or larger and used near-infrared and neutral carbon filters to reduce the background light and enhance the visibility of the incandescent dust particle. Particles were most often born in the divertor region during events such as ELMs or disruptions. Particles born on the midplane were most often deflected by the plasma boundary and remained outside the scrape off layer. The dynamics of the dust trajectories could be quite complex exhibiting a large variation in both speed (10-200 m/s) and direction. Some particles had constant velocities or exhibited various degrees of acceleration or deceleration. Abrupt reversals in direction were sometimes observed while some of the larger particles are seen to break apart during mid-flight. 3D trajectories of the dust particles have been derived from measurements of dust trajectories taken simultaneously from two observations points with two fast cameras (Fig. 3) and these have been compared to modeling predictions (Sec. 5).

3.4.2. *Laser scattering techniques*

Laser scattering is a commonly used technique for the detection of dust, having many research and practical applications. However it has been applied successfully only a few times in magnetic confinement fusion devices [110–112]. The primary difficulties are the relatively low density of dust in existing magnetic confinement devices during normal plasma operation, and the small detection volumes commonly used in laser scattering systems. Work on JIPPT-IIU [110] found evidence of light scattering by dust using a laser installed primarily for use as a Thomson scattering diagnostic. Recent work on the FTU and DIII-D tokamaks, using the Nd:YAG lasers and detection optics installed as Thomson scattering diagnostic systems, has demonstrated that sufficient data can be accumulated over months of operation to allow statistical analysis. This analysis leads to important conclusions about the quantity and size distribution of dust, and allows correlation of the prevalence of dust with plasma operating pa-

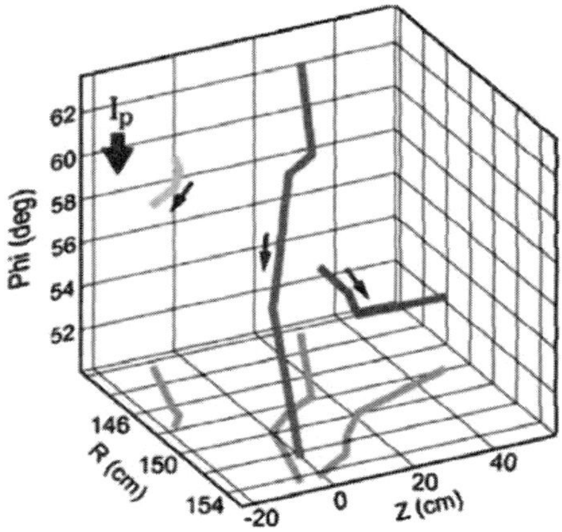

Fig. 3. Example trajectories for particles near the midplane of NSTX. The separatrix is located at R = 148 cm for this discharge. Two of the particles vanish at the separatrix and one crosses to the outboard side. An arrow points in the direction of motion showing a drift in the direction of the plasma current (From ref. [109])

rameters [111–114].

The Thomson scattering system on DIII-D [Fig. 4(a)] has multiple observation points along a laser beam passing through the main plasma and into the upper scrape-off layer (SOL), each with a detection channel centered at the laser wavelength. Dust is observed as a sudden jump of the amplitude of the signal in these detection channels. There is also a set of detection channels along a second laser path that passes through the region of the lower divertor. On DIII-D dust has been observed in the region of the lower divertor and outer SOL during plasma operation. No dust is observable inside the last closed flux surface. Figure 4 shows the average density of dust in the vicinity of the upper SOL as a function of distance from the typical separatrix location along the laser beam path. These data have been accumulated over a nine-month operational period. Shown is the average over all shots, and over selected shots with a dominantly lower or upper divertor configuration. The sharp decrease in the SOL dust density as the separatrix is approached from the far SOL is most likely due to the destruction of the dust by the ever more intense plasma.

Averaging over all shots, the prevalence of dust is strongly correlated with increasing plasma-heating power from neutral beam injectors [113,114]. This correlation is suggestive of edge localized modes (ELMs) as a source of the dust observed during plasma operation. ELMs are a

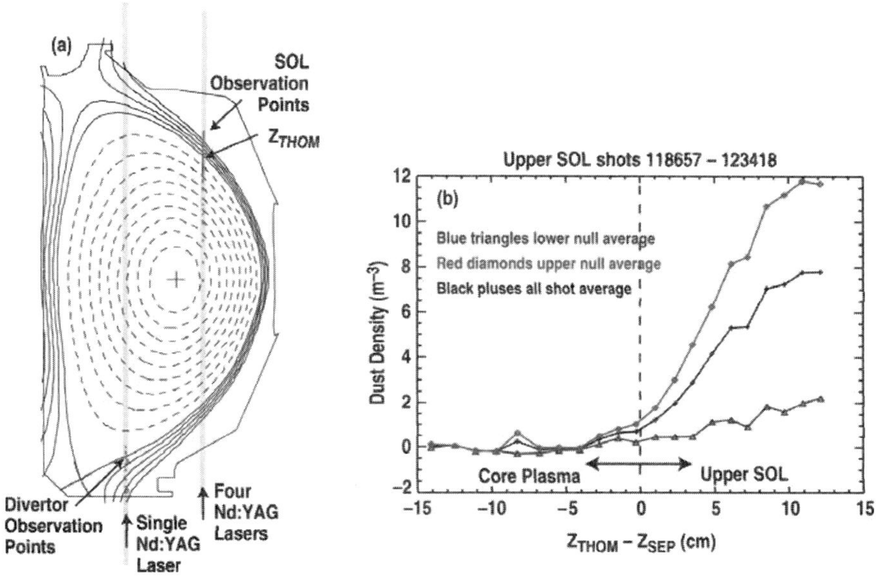

Fig. 4. (a) Schematic of the Thomson scattering diagnostic on DIII-D as related to dust detection. (b) The density of dust as measured by laser scattering in the upper SOL of DIII-D plasmas as a function of the distance along the laser beam, relative to the average location of the magnetic separatrix (ZTHOM). Positive values along the x axis correspond to the SOL, and negative values to the core plasma. The plusses are averaged over all shots, the diamonds over upper null dominated shots, and the triangles to lower null dominated shots. A small background level due to neutron and radiation impact onto the scattered light detectors has been subtracted. (From ref. [113].)

source of impulsive heating of plasma facing surfaces. Their frequency and amplitude are correlated with the injected beam power. This correlation is confirmed by the observation that the ELM-free QH-mode, which typically has relatively high injected beam power, has a dust observation rate a factor of five lower than ELMing phases of the same discharges. The laser scattering data from FTU and JIPPT-IIU indicate that disruptions are also an important source of dust in a tokamak. These observations point to impulsive heat loading of the plasma facing materials as an important mechanism for dust generation in present day tokamaks.

The size distribution of the dust has been modeled from the observed distribution of the pulse height of the scattered laser light [113,114]. The sensitivity of the detection system is calibrated in situ using Rayleigh scattering from argon gas admitted into the vacuum vessel at a pressure of about 2 Torr. The pulse height distribution is measured over three orders of magnitude using two detection channels with different sensitivities at each spatial location. Initial modeling [111] assumed the Rayleigh approximation for a perfectly conducting sphere for the scattering strength. In this approximation the scattering strength increases with the dust particle radius to the power of 6. Direct determination of the size distribution from the observed pulse height distribution is not possible because the laser beam has a variable intensity across its radius. Within the Rayleigh approximation, the size distribution was measured from a=55 nm to a= 248 nm, with an average radius of 95 nm. The measured and modeled pulse height distributions are compared in Fig. 5. The Rayleigh approximation is questionable for particles larger than 150 nm. Recently detailed modeling using the more general Mie scattering formula [115], and including a model for ablation of the particle by the very intense laser pulse, indicates a larger average radius for the dust. Using the index of refraction for graphite at room temperature, an average radius of 166 nm is obtained. Using the refractive index measured for amorphous C:H codeposited films from the TEXTOR tokamak, an average radius of 172 nm is found. An important difference in the results using the Mie formulation is a weaker dependence on scattering strength with particle radius at the large radius end of the distribution, resulting in larger particles playing a more important role in the dust volume distribution.

To date the successful applications of laser scattering to the diagnosis of dust in magnetic confinement devices have come from the use of systems designed for Thomson scattering and far from optimized for dust detection. The experience from DIII-D suggests that improved detection of the very rare events at large particle radius would be of value. A straightforward change to the system that would improve the overall detection rate would be an increase in the laser beam waist at the location of the detection regions by a factor of three or more, from the present 3 mm to $>$ 10 mm. A concomitant modification of the detection optics would also be required. Such an increase would not only provide an order of magnitude increase in the detection volume, and therefore the detection rate, but also would reduce the incident laser light intensity by an order of magnitude and eliminate the issue of destruction of the particles. Unfortunately such a change is not compatible with a primary mission of the diagnostic system, which is to obtain electron density and temperature profiles at high spatial resolution in the vicinity of the magnetic separatrix. Improved interpretation of the dust particle size distribution from the scattered laser light pulse

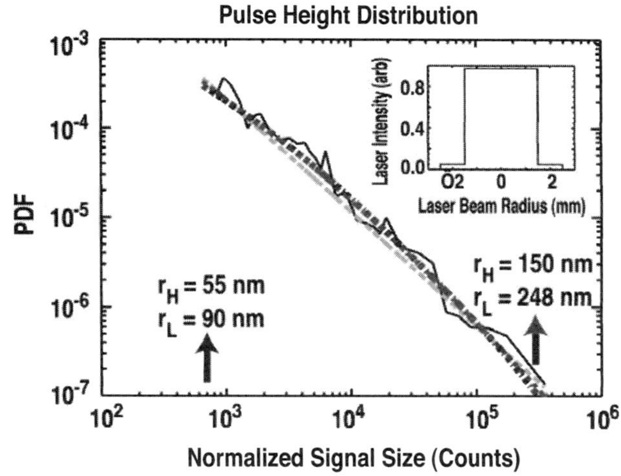

Fig. 5. The measured pulse height distribution of scattered laser light from dust (solid black line) is compared to models of the pulse height distribution based on three dust size probability distribution functions: log normal (red), normal (blue), and 1/a (yellow). The Rayleigh approximation (small particle size) for the scattering strength is assumed. Over the region of validity of the data, all three fits yield an average radius of 95 nm. (From ref. [111].)

height distribution would be achievable by the production of uniform laser intensity across the waist of the beam.

In summary, dust diagnosis in tokamaks by laser scattering has already made key contributions to the understanding of dust and its production during plasma operation. One example is the importance of impulsive heat loading as a source of dust. On DIII-D, the relation of dust production to ELMs has been identified, and on JIPPT-IIU and FTU dust following disruptions was observed. As the interest in dust production in tokamaks increases, these successes will provide important guidance to the design of improved diagnostic systems based on light scattering.

3.4.3. *In-situ measurement of dust on PFC surfaces*

Measurements of the dust inventory in next-step devices will be necessary to demonstrate that they are in compliance with safety limits. It will not be possible to wait until a maintenance period to apply the dust collection procedures of Sec. 3.1. Such procedures would need remote handling anyway in an activated machine. In this section we describe techniques for *in situ* measurement of dust on PFC surfaces.

The size of dust particles is comparable to the wavelength of infrared light and black body emission from dust particles deviates from the Plank function. Observations of thermal emission from Tore Supra limiters revealed dust particles with spectral emissivity falling off with the square of the wavelength [116]. Flakes were identifiable by their fast cool down times. Laser induced breakdown has been proposed for in-situ analysis of dust and loosely attached films on plasma facing surfaces [117]. However quantitative estimation of the mass of dust by these techniques has yet to be demonstrated.

A capacitive diaphragm microbalance has been adapted for surface dust measurements [118,119]. The cumulative mass of dust flakes of film growth on the surface of a diaphragm is determined by the change of capacitance caused by its deflection relative to a fixed plate. A sensitivity of 500 μg/cm^2 and a dynamic range of 10^3 has been demonstrated.

A novel device to detect dust particles that have settled on a remote surface has recently been developed in the laboratory [120,121]. Two closely interlocking grids of conductive traces on a circuit board were biased to 30 - 50 V. Test particles, scraped from a carbon fiber composite tile, were delivered to the grid by a stream of nitrogen. Miniature sparks appeared when the particles landed on the energized grid and created a transient short circuit. Typically the particles vaporized in a few seconds restoring the previous voltage standoff. The transient current flowing through the short circuit created a voltage pulse that was recorded by standard nuclear counting electronics. Tests showed a clear correlation between the recorded counts and particle concentration, especially at finer grid spacing. The device worked well in both atmosphere and vacuum environments. The sensitivity has been enhanced by more than an order of magnitude to 142 counts/μm/cm^2 in vacuum by the use of ultrafine grids. The response to particles of different size categories was compared and the sensitivity, expressed in counts/real density (μg/cm^2) of particles, was maximal for the finest particles (Fig. 6) [122]. This is a favorable property for tokamak dust which is predominantly of micron scale. Larger particles produce a longer current pulse, providing qualitative information on the particle size. A large area (50 $\times$ 50 mm) detector was recently demonstrated [123]. A similar device has been applied to HT-7 [124].

The dust limits for ITER are currently under revision, but it seems highly likely that the dust inventory will approach the safety limit at some point and dust removal will then be necessary for continued plasma oper-

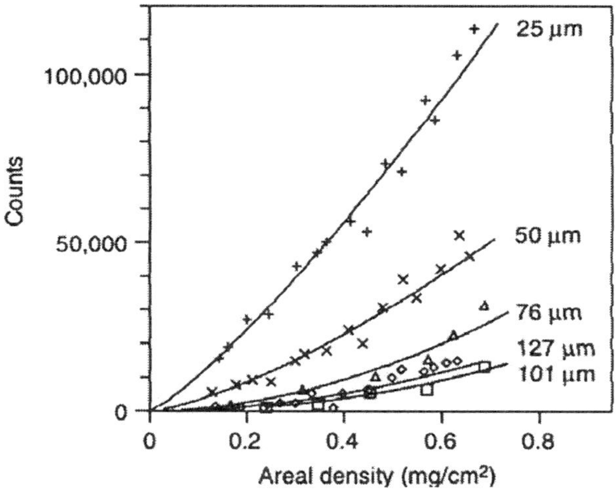

Fig. 6. Response of electrostatic dust detector showing an increase in counts for the finer grids by more than one order-of-magnitude. The grid spacing is listed on the right. The lines are a second order polynomial fit to the data. Reproduced with permission from ref. [122].

ations. Dust can also settle on first mirrors used in diagnostics critical to assure machine safety. This dust will need to be removed or the mirrors cleaned by some means before plasma operations can continue. For example the ITER divertor dome contains mirrors used for diagnosing plasma detachment that are less than 1 m from the strike point. The challenge is outlined in ref. [118] and some potential methods such as a vibratory conveyor, photo-cleaning and a liquid wash and flush are described. An electrostatic dust removal system was investigated in ref. [125].

Dust particles impinging on the electrostatic dust detector described above [120] create a short circuit between the traces, however this short circuit is temporary suggesting that the device may be useful for the removal of dust from specific areas. The fate of the dust particles has been tracked by measurements of mass gain/loss. Heating by the current pulse caused up to 90% of the particles to be ejected from the grid or vaporized [8]. A mosaic of these devices based on nanoengineered traces on a low activation substrate such as SiO_2 could be envisaged for remote inaccessible areas in a next-step tokamak. This mosaic would both detect conductive dust settling on surfaces in these areas and could ensure that these surfaces remained substantially dust free. The above techniques and others yet to be developed are candidates for measurement and removal of surface dust but resources are needed to adapt them to the harsh radiological environment and constrained geometry of next-step devices.

4. Basics of dust-plasma interactions

In this section, we discuss some of the basic physics associated with dust-plasma interaction.

4.1. *Charging*

We consider a single dust grain immersed in a stationary plasma consisting of electrons and singly charged ions, having masses m_e and m_i, temperatures T_e and T_i, and unperturbed densities n_e and n_i, respectively. We assume spherical symmetry and a perfectly conducting, spherical dust grain of radius r_d located at the center of the system and collecting plasma particles. Under typical laboratory conditions and in the absence of electron emission from the grain (this case is known as primary charging), the higher mobility of the electrons with respect to the ions results in a negatively charged dust grain. As the negative charge builds up on the dust grain, the resulting electric field acts against further electron collection and in favor of ion collection such that eventually a dynamical equilibrium is reached where the sum of the plasma currents to the grain is zero (floating condition). For the present purposes, we neglect the effect of an external magnetic field so that plasma particles move only in response to the electric field present in the system due to the charged grain. Collisions with the neutral gas are also neglected.

Current collection by a body in a stationary plasma is a classic problem of plasma physics, originally discussed in the context of probes. There are typically two approaches to the problem. One is a fluid approach known as the Allen-Boyd-Reynolds (ABR) theory [126,127] or cold-ion model. It basically assumes $T_i = 0$ so that the ions move only radially, while the electron density follows the Boltzmann relation. This theory was shown to be in good agreement with some early experimental work on spherical probes [128]. The second approach is the kinetic orbital motion (OM) theory [129–132]. It requires the solution of the nonlinear Poisson equation, taking into account the full particle trajectories and the possibility of potential barriers for the particle motion (see the review article by Allen [128]). In the following, we present a brief overview of a simplified version of the OM theory, known as the orbital motion limited (OML) theory [133]. The basic principle of both OM or OML is that a particle traveling in a central field of force conserves angular momentum and energy. For a plasma particle starting at infinity (outside the Debye sphere determined by the charged grain) with impact parameter b_c and velocity v and grazing the dust surface with velocity v_g, we have

$$m_s b_c v = m_s r_d v_g = \text{const} \tag{25}$$

and

$$\frac{m_s}{2} v^2 = \frac{m_s}{2} v_g^2 + q_s \phi_d = \text{const}, \tag{26}$$

where m_s is the mass of a plasma particle (the subscript $s = e, i$ labels electrons or ions), q_s is its charge ($q_s = \pm e = \pm 1.6 \times 10^{-19}$ C with e the elementary charge) and $\phi_d = \phi(r_d)$ is the electrostatic potential on the grain surface. The critical assumption of OML (as opposed to OM) is that for every energy range some plasma particles can graze the grain surface. In this sense, therefore, the plasma currents are orbital motion limited. It is clear that, for a given velocity, decreasing the impact parameter $b < b_c$ means that the plasma particle will hit the grain. Therefore, the collision cross section is given by

$$\sigma_s = \pi b_c^2 = \pi r_d^2 \left(1 - \frac{2 q_s \phi_d}{m_s v^2}\right), \tag{27}$$

where we have used expressions (25) and (26). Furthermore, the distribution function of the plasma particles far from the grain is assumed Maxwellian:

$$f_s = n_s \left(\frac{m_s}{2\pi k_B T_s}\right)^{3/2} \exp\left(-\frac{m v^2}{2 k_B T_s}\right) \tag{28}$$

with k_B the Boltzmann constant. Finally, the charging current to the grain by plasma collection is

$$I_s = q_s \int_{v_0}^{\infty} v \sigma_s f_s(\mathbf{v}) d\mathbf{v} \tag{29}$$

where the limit of integration v_0 is $v_0 = 0$ for the plasma species attracted by the grain ($q_s \phi_d < 0$) and $v_0 = \sqrt{2 q_s \phi_d / m_s}$ for the plasma species repelled by the grain ($q_s \phi_d > 0$). The latter value of v_0 simply corresponds to $v_g = 0$ in Eq. (26). The integral (29) is readily evaluated to obtain the electron current to the grain

$$I_e = -\pi r_d^2 e n_e \sqrt{\frac{8 k_B T_e}{\pi m_e}} \exp\left(\frac{e \phi_d}{k_B T_e}\right), \quad \text{if } \phi_d < 0; \tag{30}$$

$$I_e = -\pi r_d^2 e n_e \sqrt{\frac{8k_BT_e}{\pi m_e}} \left(1 + \frac{e\phi_d}{k_BT_e}\right), \quad \text{if } \phi_d \geq 0; \tag{31}$$

and the ion current

$$I_i = \pi r_d^2 e n_i \sqrt{\frac{8k_BT_i}{\pi m_i}} \left(1 - \frac{e\phi_d}{k_BT_i}\right), \quad \text{if } \phi_d < 0; \tag{32}$$

$$I_i = \pi r_d^2 e n_i \sqrt{\frac{8k_BT_i}{\pi m_i}} \exp\left(-\frac{e\phi_d}{k_BT_i}\right), \quad \text{if } \phi_d \geq 0. \tag{33}$$

In this simple OML framework, the equilibrium grain charge is obtained by balancing the ion and electron currents to the grain

$$I_e = I_i. \tag{34}$$

Notice that Eq. (34) is an equation for ϕ_d dependent only on mass and temperature ratios and not on the grain radius. In certain applications, ions have a finite streaming speed and the ion distribution function is better approximated by a drifting-Maxwellian. This modifies the ion current and we refer the reader to Refs. [37,134] for the ion current in this case.

Due to its simplicity, the OML theory is widely used to estimate the charging of dust grains in a plasma, although some questions on its validity have been raised [135]. In fact, it can be shown that the OML theory is strictly valid only when there is no absorption radius [132] (which is related to the presence of potential barriers for the particle motion) or, alternatively, when the shielding potential around the grain decreases more slowly than $1/r^2$ [135]. Moreover, Allen and collaborators [135] have shown that this condition is violated for typical dusty plasma scenarios and therefore OML is internally inconsistent. Later, Lampe [136] revisited this objection and concluded that the OML theory is an excellent approximation for small dust radius, since the effect of potential barriers for the plasma particles becomes negligible.

So far we have discussed the primary charging case. However, it is important to recognize that in many conditions electron emission from the dust grain must be included in the charging process. Some examples of such cases include: a meteoroid falling through the Earth's atmosphere is heated through collisions and emits electrons thermionically [137]; dust grains in nebulae photoemit electrons due to the radiation field of a nearby star [138]; dust grains in a plasma environment containing very energetic plasma particles emit electrons by particle impact (this can be the case for a spacecraft in the Earth environment [134]). In the context of fusion applications, the two most important dust emission processes are thermionic and secondary electron emission, while photoemission can generally be neglected. Delzanno and collaborators [139–141] have investigated the charging and shielding of electron-emitting dust grains and showed that, in addition to the dust grains becoming positively charged, the shielding potential around them is modified and can have a potential well.

The usual treatment of thermionic emission starts from the Sommerfeld model of a metal and assumes a Maxwellian distribution function of the emitted electrons [142,143]:

$$f_{\rm th} = \frac{2m_e^3}{h^3} \exp\left(-\frac{\frac{1}{2}m_e u^2 + W}{k_B T_d}\right), \tag{35}$$

where u is the microscopic velocity, h is Planck's constant, W is the thermionic work function and T_d is the dust bulk temperature. Consequently, the thermionic electron current is [142,143]:

$$I_{\rm th} = \frac{16\pi^2 r_d^2 e m_e k_B^2 T_d^2}{h^3} \exp\left(-\frac{W}{k_B T_d}\right), \quad \text{if } \phi_d < 0; \tag{36}$$

$$I_{\rm th} = \frac{16\pi^2 r_d^2 e m_e k_B^2 T_d^2}{h^3} \left(1 + \frac{e\phi_d}{k_B T_d}\right) \exp\left(-\frac{e\phi_d + W}{k_B T_d}\right), \quad \text{if } \phi_d \geq 0. \tag{37}$$

Equation (36) is the well-known Richardson-Dushman expression for the thermionic flux.

Secondary emissions by electron or ion impact with the grain are generally distinguished. Secondary emission by ion impact is neglected as it becomes important only for ion energies above several keV [144]. In the case of emission by electron impact, the yield of secondary electrons (the ratio of emitted to incident electrons) δ is a function of the dust material as well as the kinetic energy of the incident electrons E. Typically, this function peaks at energy $E_m = 300 - 2000$ eV while the maximum yield δ_m is of order unity for metals and semiconductors and of order $2 - 30$ for insulators [145]. (For carbon $\delta_m = 1$ and $E_m = 250$ eV.) This function is commonly approximated by the Sternglass formula [145]

$$\delta(E) = 7.4\delta_m \frac{E}{E_m} \exp\left(-2\sqrt{\frac{E}{E_m}}\right), \tag{38}$$

where the dependence on the angle of incidence of the primary electrons has been neglected. [Chow and collaborators [146] have shown that δ can be enhanced with respect to Eq. (38) if the dust size is comparable to the penetration depth of the bombarding electrons.] The distribution function of the emitted electrons is well approximated by a Maxwellian with temperature T_{se} in the range $T_{\text{se}} \sim 1-5$ eV [147]. The emitted electron current is [15]

$$I_{\text{se}} = \frac{8\pi^2 r_d^2 e}{m_e^2} \int_0^\infty E\delta(E) f_e(E - e\phi_d) dE, \quad \text{if } \phi_d < 0; \tag{39}$$

$$I_{\text{se}} = \frac{8\pi^2 r_d^2 e}{m_e^2} \left(1 + \frac{e\phi_d}{k_B T_{\text{se}}}\right) \exp\left(-\frac{e\phi_d}{k_B T_{\text{se}}}\right) \int_{e\phi_d}^\infty E\delta(E) f_e(E - e\phi_d) dE, \quad \text{if } \phi_d \geq 0, \tag{40}$$

where f_e is the distribution function of the incident electrons. From Eq. (28) we have

$$f_e(E - e\phi_d) = n_e \left(\frac{m_e}{2\pi k_B T_e}\right)^{3/2} \exp\left(-\frac{E - e\phi_d}{k_B T_e}\right). \tag{41}$$

Finally, the dynamic charging equation for a spherical dust grain immersed in a plasma, including thermionic and secondary emission, is:

$$\frac{dQ_d}{dt} = I_e + I_i + I_{\text{th}} + I_{\text{se}}. \tag{42}$$

The grain charge Q_d is related to the surface potential of the grain by $Q_d = C\phi_d$ where C is the grain capacitance. Typically, $C = 4\pi\varepsilon_0 r_d(1 + r_d/\lambda_D)$ is used, with ε_0 the permittivity of free space. The latter is consistent with a Debye-Huckel shielding potential around the grain with a characteristic screening length given by λ_D. Often in dusty plasmas the term $1 + r_d/\lambda_D \simeq 1$ since usually $r_d \ll \lambda_D$.

4.2. *Heating and ablation*

Plasma particles collected during the charging process deposit their kinetic energy on the dust grain and heat it. This is particularly important in a fusion device due to the severe plasma conditions present there and indeed the dust grains are heated to such high temperatures that their destruction can occur. In fact, dust destruction can be effectively used as a diagnostic tool for magnetic field mapping in tokamaks [148,149].

The heat flux on a spherical dust grain due to the direct impact of plasma particles is

$$\Gamma^s = \int_{v_0}^\infty \sigma_s v \left(\frac{m_s}{2} v^2 - q_s\phi_d\right) f_s d\mathbf{v} \tag{43}$$

with the same limits of integration as given in Eq. (29). The calculation of integral (43) for each species requires tedious algebra with no conceptual difficulty, therefore here we only report the results. The heat flux associated with ion direct impact (source) is

$$\Gamma^i = k_B T_i \left(\frac{2 - \dfrac{e\phi_d}{k_B T_i}}{1 - \dfrac{e\phi_d}{k_B T_i}} - \frac{e\phi_d}{k_B T_i} \right) \frac{I_i}{e}, \qquad \text{if } \phi_d < 0; \tag{44}$$

$$\Gamma^i = 2k_B T_i \frac{I_i}{e}, \qquad \text{if } \phi_d \geq 0; \tag{45}$$

while the heat flux associated with electron impact (source) is

$$\Gamma^e = 2k_B T_e \frac{|I_e|}{e}, \qquad \text{if } \phi_d < 0; \tag{46}$$

$$\Gamma^e = k_B T_e \left(\frac{2 + \dfrac{e\phi_d}{k_B T_e}}{1 + \dfrac{e\phi_d}{k_B T_e}} + \frac{e\phi_d}{k_B T_e} \right) \frac{|I_e|}{e}, \qquad \text{if } \phi_d \geq 0. \tag{47}$$

(The ion heat flux is modified if the ion distribution function is a drifting-Maxwellian. We refer the reader to Ref. [37] for the modified expression in this case.) The heat flux due to the thermionically emitted electrons (sink) is

$$\Gamma^{\text{th}} = 2k_B T_d \frac{I_{\text{th}}}{e}, \qquad \text{if } \phi_d < 0; \tag{48}$$

$$\Gamma^{\text{th}} = k_B T_d \left(\frac{2 + \dfrac{e\phi_d}{k_B T_d}}{1 + \dfrac{e\phi_d}{k_B T_d}} + \frac{e\phi_d}{k_B T_d} \right) \frac{I_{\text{th}}}{e}, \qquad \text{if } \phi_d \geq 0. \tag{49}$$

while the heat flux due to the secondary emitted electrons (sink) is

$$\Gamma^{\text{se}} = 2k_B T_{\text{se}} \frac{I_{\text{se}}}{e}, \qquad \text{if } \phi_d < 0; \tag{50}$$

$$\Gamma^{\text{se}} = k_B T_{\text{se}} \left(\frac{2 + \dfrac{e\phi_d}{k_B T_{\text{se}}}}{1 + \dfrac{e\phi_d}{k_B T_{\text{se}}}} + \frac{e\phi_d}{k_B T_{\text{se}}} \right) \frac{I_{\text{se}}}{e}, \qquad \text{if } \phi_d \geq 0. \tag{51}$$

The radiated heat flux from the dust grain (sink) is

$$\Gamma^{\text{rad}} = 4\pi r_d^2 \epsilon_d \sigma_B \left(T_d^4 - T_{\text{wall}}^4\right), \tag{52}$$

with $\sigma_B = 5.67 \times 10^{-8}$ W/m^{2}K^4 and typically $T_{\text{wall}} = 290$ K. A corrective factor ϵ_d is needed when $\lambda_{rad.} \sim r_d$ [38].

Furthermore, given the high temperature reached by the dust grain in fusion devices, the heat lost in the sublimation of the dust material must be taken into account:

$$\Gamma^{\text{sub}} = \frac{\Lambda_s}{\mu}|\dot{m}_d|, \tag{53}$$

where Λ_s is the latent heat of sublimation, μ is the molecular weight and $|\dot{m}_d|$ is the rate at which the dust mass sublimates. In order to calculate $|\dot{m}_d|$, we use the Clausius-Clapeyron equation which governs liquid/gas equilibria [137]

$$n = n_0 \frac{T_0}{T_d} \exp\left[\frac{\Lambda_s}{R}\left(\frac{1}{T_0} - \frac{1}{T_d}\right)\right], \tag{54}$$

where n is the vapour density of the dust material, $R = 8.31$ J/mol·K is the gas constant and n_0 and T_0 are reference numbers. By assuming that the evaporating molecules escape with velocity

$$v_{\text{esc}} = \sqrt{\frac{2k_B T_d}{\pi \mu_d}}, \tag{55}$$

with μ_d the mass of the dust molecules, the sublimation rate is readily obtained:

$$|\dot{m}_d| = 4\pi r_d^2 \mu_0 \sqrt{\frac{2k_B T_d}{\pi \mu_d}} n_0 \frac{T_0}{T_d} \exp\left[\frac{\Lambda_s}{R}\left(\frac{1}{T_0} - \frac{1}{T_d}\right)\right]. \tag{56}$$

For carbon, $\mu_d = 12$ amu, $\Lambda_s = 7.15 \times 10^5$ J/mol, $\mu = 12 \times 10^{-3}$ kg/mol, $n_0 = 1.83 \times 10^{24}$ m^{-3} and $T_0 = 4000$ K.

Finally, assuming the dust grain is heated uniformly, the heat balance equation becomes

$$m_d c_d \frac{dT_d}{dt} = \Gamma^i + \Gamma^e - \Gamma^{\text{th}} - \Gamma^{\text{se}} - \Gamma^{\text{rad}} - \Gamma^{\text{sub}} \tag{57}$$

where c_d is the specific heat of the dust material (for carbon $c_d = 750$ J/kg K) and m_d is the dust mass. Moreover, in order to properly account for the effect of mass loss due to sublimation, the following equation for the dust size variation must also be included in the model:

$$4\pi r_d^2 \rho_d \frac{dr_d}{dt} = -|\dot{m}_d| \tag{58}$$

where ρ_d is the dust density (for carbon $\rho_d = 2.25 \times 10^3$ kg/m^3).

4.3. *Forces*

The determination of the forces acting on the dust grain is crucial to understand its dynamics in the plasma. A comprehensive description of all the forces relevant to dusty plasmas could be the subject of a whole chapter in a book and here we limit ourselves to a brief summary of the most important forces acting on a single dust grain. These are:

- The gravitational force:

$$\mathbf{F}_g = m_d \mathbf{g} = \frac{4}{3}\pi r_d^3 \rho_d \mathbf{g}, \tag{59}$$

  where $\mathbf{g}$ is the gravitational acceleration.
- The electromagnetic force:

$$\mathbf{F}_{EB} = Q_d \left(\mathbf{E}_0 + \mathbf{U}_d \times \mathbf{B}_0\right), \tag{60}$$

  where $\mathbf{E}_0$ and $\mathbf{B}_0$ are the electric and magnetic field and $\mathbf{U}_d$ is the dust velocity. In typical dc discharges the dust grains are levitated in the sheath above the cathode where the electric force balances the gravitational force.
- Neutral drag. The neutral drag is the momentum exchange due to collisions of the dust grain with the neutral background gas. The typical regime of interest for dusty plasmas is the kinetic regime which corresponds to a Knudsen number (the ratio of the neutral mean free path to the dust radius) much larger than unity [15,150]. For specular reflection collisions (where the normal component of the velocity of the neutral is reversed after collision) and assuming a Maxwellian distribution function for the neutral gas one recovers the Epstein formula [151]

$$\mathbf{F}_{nd} = \frac{8}{3}\sqrt{2\pi} r_d^2 m_n n_n V_{th,n} \left(\mathbf{v}_n - \mathbf{U}_d\right) \tag{61}$$

  in the limit $s = |\mathbf{v}_n - \mathbf{U}_d| / \sqrt{2} V_{th,n} \ll 1$. This limit is common for dusty plasmas. Here m_n is the mass, n_n is the density, $V_{th,n}$ is the thermal velocity and $\mathbf{v}_n$ is the velocity of the neutral gas. It is worth pointing out that Draine and Salpeter [34] have obtained an approximate expression for the neutral drag force (specular reflection collisions)

$$\mathbf{F}_{nd} = \frac{8}{3}\sqrt{2\pi} r_d^2 m_n n_n V_{th,n} \sqrt{1 + \frac{9\pi}{64} s^2} \left(\mathbf{v}_n - \mathbf{U}_d\right), \tag{62}$$

accurate to within 1% for all s.

- Ion drag. Similar to the neutral drag, the ion drag is due to collisions between the dust grain and ions. The ion drag is usually very important for applications with a strong ion flow (e.g. in the sheath) or when gravity is not as important (e.g. in microgravity experiments). It consists of two parts: the collection force $\mathbf{F}^{\text{coll}}$ is associated with ions that are collected by the grain, while the orbital force $\mathbf{F}^{\text{orb}}$ is associated with ions that are scattered (and not collected) by the electric field of the grain. Thus

$$\mathbf{F}_{id} = \mathbf{F}^{\text{coll}} + \mathbf{F}^{\text{orb}}. \tag{63}$$

We assume cold, mono-energetic ions since in typical dusty plasma experiments ions entering the sheath have a velocity several times the thermal velocity [150]. We also consider a static grain.
The collection force is easily calculated in light of the considerations described in Sec. 4.1. An ion collected by the dust grain will deposit its momentum on it with the collection cross section given by Eq. (27). Therefore we conclude that [150]

$$\mathbf{F}^{\text{coll}} = \pi r_d^2 n_i m_i v_i \left(1 - \frac{2e\phi_d}{mv_i^2}\right) \sigma^{\text{orb}} \mathbf{v}_i. \tag{64}$$

The scattering cross section is given by [152]

$$\sigma^{\text{orb}} = 4\pi b_{90}^2 \int_{b_c}^{\lambda_D} \frac{b}{b_{90}^2 + b^2} db \tag{65}$$

where b is the impact parameter and

$$b_{90} = r_d \frac{e\phi_d}{m_i v_i^2} \tag{66}$$

is the impact radius for a 90 degree deflection. The lower limit of integration in Eq. (65) is b_c in order to exclude from the calculation those ions collected by the dust grain. The upper limit of integration is a cut-off to avoid divergence of the integral. Physics considerations [15,150] suggest the plasma Debye length, λ_D, as a plausible choice for the cut-off. This implies that ions with an impact parameter larger than λ_D do not contribute to the momentum exchange due to the screening of the plasma. The orbital drag force is therefore [150]

$$\mathbf{F}^{\text{orb}} = 2\pi b_{90}^2 n_i m_i v_i \ln\left(\frac{b_{90}^2 + \lambda_D^2}{b_{90}^2 + b_c^2}\right) \mathbf{v}_i. \tag{67}$$

A widely used formula accounting for thermal effects is given by Barnes *et al.* [29]:

$$\mathbf{F}^{\rm coll} = \pi r_d^2 n_i m_i v_s \left(1 - \frac{2e\phi_d}{mv_s^2}\right) \mathbf{v}_i, \tag{68}$$

$$\mathbf{F}^{\rm orb} = 2\pi b_{90s}^2 n_i m_i v_s \ln\left(\frac{b_{90s}^2 + \lambda_D^2}{b_{90s}^2 + b_c^2}\right) \mathbf{v}_i, \tag{69}$$

where $v_s = \sqrt{v_i^2 + 8k_BT_i/(\pi m_i)}$ is the mean speed (a combination of directed and thermal speed) and $b_{90s} = r_d e\phi_d/(m_i v_s^2)$. Clearly expressions (64) and (67) coincide with (68) and (69) in the limit $v_i \gg \sqrt{8k_BT_i/(\pi m_i)}$.

It is worth pointing out that recently Khrapak and collaborators [30] have revisited the calculation of the ion drag force. Their argument is that the upper integration limit in Eq. (65) should be chosen according to the condition that the ion distance of closest approach be equal to λ_D. This is to account for ions with $b \gtrsim \lambda_D$ which are strongly deflected and enter the Debye sphere of the grain. The resulting, modified $\sigma^{\rm orb}$ was shown to be in much better agreement [than Eq. (65)] with self-consistent numerical simulations of the cross section for elastic ion scattering for a point-like grain [153]. In the superthermal limit ($v_i \gg \sqrt{k_BT_i/m_i}$), Khrapak et al. recover the results reported above while in the subthermal limit ($v_i \ll \sqrt{k_BT_i/m_i}$) results are shown for a shifted Maxwellian ion distribution [Eq. (11) of Ref. [30]]. Furthermore, Hutchinson [31] has recently critically compared PIC simulation results for the ion drag with the analytical formula provided by Khrapak *et al.* [30], showing that in certain limits the analytical formula underestimates the drag force by as much as a factor of 2. Hutchinson [31] then provided an approximate analytical formula by fitting the numerical results. See Ref. [31] for details.

Here we have not discussed the electron drag force due to collision of the dust grain with electrons. The electron drag is usually small compared to the ion drag due to the small electron-to-ion mass ratio but there are instances where it can be important [154].

Finally, the grain dynamics is described by Newton's law including all the relevant forces:

$$\frac{d\mathbf{x}_d}{dt} = \mathbf{U}_d \tag{70}$$

$$\frac{d}{dt}(m_d \mathbf{U}_d) = \mathbf{F}_g + \mathbf{F}_{EB} + \mathbf{F}_{nd} + \mathbf{F}_{id}, \tag{71}$$

where $\mathbf{x}_d$ is the dust grain position.

4.4. *Dust spin*

Charged dust grains can rotate and spin due to their interactions with photons and particles of the surrounding gas. The angular frequency of such rotation and spinning can reach a rather large value ($\sim 10^4 - 10^6$ s^{-1} for thermal dust grains and $\approx 10^9$ s^{-1} for superthermal grains in astrophysical surroundings [155–157]. It has been thought that dust particle spin could be responsible for the production of cosmic radiation [158] in our galaxy. Sensitive observations of variations in the microwave sky brightness have revealed 14-90 MHz microwave emission that is correlated with 100 μm thermal emission from interstellar dust [159]. Marchant [160] reported that astronomers have discovered microwave emissions from the milky way. To explain the findings, astrophysicists had speculated that tiny dust grains with an uneven distribution of electric charge on the grain surface would be set spinning by collisions with the plasma particles that raced past. This would make them radiate at the same frequency as the rate of dust particle spinning. The radiation originates from space dust that fills the voids between stars. Sato [161] has presented a short review of preliminary experiments on the spinning motion of negatively charged spherical fine particles in low-temperature magnetized plasmas. He emphasized the paramagnetic behavior of charged particles in the presence of external magnetic fields. It should be noted that in a hot magnetized plasma of fusion devices, the ablation of dust, which exerts a quite large force on a charged dust grain [38,162], can also cause spinning of a dust particulate.

In laboratory and astrophysical plasmas dust grains can be unevenly charged. Hence, they will acquire an asymmetric electric dipole moment, which will affect the dust particle dynamics in a magnetized plasma containing equilibrium sheared plasma flows and turbulent fluctuations. Following Ref. [163], we shall discuss a possible mechanism for spinning of a charged dust particle in a magnetized plasma. The latter will have the cross-field motion due to the $\mathbf{E} \times \mathbf{B}$ drift of the plasma particles. Since $\mathbf{v}_p \times \mathbf{E} \neq 0$, where $\mathbf{v}_p \equiv \mathbf{V}_{\mathbf{E}\times\mathbf{B}} = \mathbf{E} \times \mathbf{B}/B^2$, the plasma flow would cause asymmetry of dust grain charging, and thereby establishing an electric dipole $\mathbf{D}$, which is not aligned along the direction of the electric field $\mathbf{E}$, viz. $\mathbf{D} \times \mathbf{E} \neq 0$. Subsequently, interactions of the electric dipole with the electric field would produce a torque due to which a charged dust grain will spin up. We emphasize that both laminar (quasi-stationary) and turbulent electric fields will contribute to the torque. Since most of the magnetized plasmas are far away from thermodynamic equilibrium, turbulent electric fields will play a dominant role in dust grain spinning up process that is described herein. We assume that the electron gyroradius is much larger than both the dusty plasma Debye radius λ_D and the effective dust particle radius r_d. We also take $\mathbf{E} \cdot \mathbf{B} = 0$ and assume that the dust particle is made of insulating material. In such a situation, the evolution of the electric dipole $\mathbf{D}$ and the angular velocity of dust grain rotation $\boldsymbol{\Omega}$ are governed by [163]

$$\frac{d\mathbf{D}}{dt} = -\nu_q (\mathbf{D} - \mathbf{D}_p) + \boldsymbol{\Omega} \times \mathbf{D}, \tag{72}$$

and

$$\frac{d\boldsymbol{\Omega}}{dt} = \frac{\mathbf{D} \times \mathbf{E}}{I_d} - \nu_\Omega \boldsymbol{\Omega}, \tag{73}$$

where ν_q is the effective dust charging frequency, ν_Ω is the effective damping frequency of the dust grain rotation caused by interactions of a charged dust with surrounding gas and plasma, $I_d \sim \rho_d r_d^5$ is the moment of inertia of the dust grain, ρ_d is the dust mass density, and $\mathbf{D}_p$ is the equilibrium electric dipole corresponding to the plasma flow with the velocity $\mathbf{v}_p$. We note that $\boldsymbol{\Omega}$ and $\mathbf{B}$ are parallel to each other, and that $\mathbf{E}$ can be non-stationary in our analysis.

For rather small drift (in comparison with the sound speed $C_s = \sqrt{\gamma Z k_B T_e / m_i}$, where T_e is the electron temperature and m_i is the ion mass), we have

$$\mathbf{D}_p = \hat{D} \mathbf{E} \times \mathbf{B}, \tag{74}$$

where $\hat{D}$ is a normalization constant. Based on the result from Ref. [164], we find that

$$\hat{D} = f \frac{\pi r_d^2 k_B T_e}{e B^2 C_s} \left(\frac{r_d}{\lambda_D} + 3 \right), \tag{75}$$

where f is a form factor (e.g. $|f| \sim 1$). As an illustration, we consider the plasma parameters at tokamak edges, with the plasma number density $n_i \sim 10^{13} - 10^{14}$ cm^{-3} and the temperature $T_e \sim T_i \sim 10$ eV, and find that $\nu_q \sim \omega_{pe}/(1 + 3\lambda_d/r_d)$ (e.g. Ref. [164]) is quite large and the dipole reaches its equilibrium value instantaneously. Then, substituting $\mathbf{D} \approx \mathbf{D}_p$ in Eq. (73), we have

$$\frac{d\Omega}{dt} = \frac{\hat{D}E^2 B}{I_d} - \nu_\Omega \Omega, \tag{76}$$

which dictates that the driving term is proportional to E^2, so that even fluctuating turbulent electric field spectrum would result in continuous increase of dust particle angular rotation frequency.

For a tokamak edge plasma, where the plasma number density is usually much higher than the gas density, we have $\nu_\Omega \sim (m_i n_i/\rho_d)(C_s/r_d)$ (e.g. Ref. [164]). However, for $n_i \sim 10^{13} - 10^{14}$ cm^{-3}, $T_e \sim T_i \sim 10$ eV, $r_d \sim 1$ μm, and $\rho_d \sim 1$ g/cm^3, we find that $\nu_\Omega \sim 1$ s^{-1}, while the lifetime of dust particle in fusion plasma (lifetime is determined by either collisions with the confining wall or evaporation), τ_d, is of the order of $10^{-2} - 10^{-3}$ s [38,162]. Therefore, the characteristic rotation frequency Ω_d, a dust particle acquires while going through edge plasmas can be estimated as [163]

$$\Omega_d \sim \pi |f| \left(\frac{r_d}{\lambda_D} + 3 \right) \frac{< E >^2 \rho_s}{\rho_d r_d^3} \tau_d, \tag{77}$$

where $< E >^2$ is the averaged square of the electric field, $\rho_s = C_s/\omega_{ci}$ is the ion sound gyroradius, and $\omega_{ci} = eB/m_i$ is the ion gyrofrequency. For strong turbulence in tokamak edges, we estimate $< E >^2 \sim (k k_B T_e/e)^2$, where k is the characteristic wave number of the turbulence (usually $k\rho_s \sim 0.1 - 0.01$). Hence, Eq. (77) yields

$$\Omega_d \sim 4\pi^2 |f| \frac{m_i n_i}{\rho_d} \frac{C_s}{r_d} \left(\frac{r_d}{\lambda_D} + 3 \right) \frac{\lambda_D^2}{r_d^2} k^2 \rho_s^2 \omega_{ci} \tau_d, \tag{78}$$

which gives $\Omega_d \sim 10^5$ s^{-1} for $k\rho_s \sim 0.03$, $n_i \sim 10^{13-14}$ cm^{-3}, $T_e \sim T_i \sim 10$ eV, $r_d \sim 1\mu$ m, $\rho_d \sim 1$ g/cm^3, $\tau_d = 10^{-3}$ s, and $B = 10^4$ G.

4.5. *Application of the PIC method to simulate dust-plasma interaction*

As noted above, dust particles immersed in plasmas interact with the plasma species by capturing ions and electrons onto their surface (the actual physical process develops at the quantum level but for the present purposes, it is assumed that the charge just remains on the surface). Particles can also be emitted from the dust surface via a combination of the processes of photo, thermionic, secondary and field emission mentioned above.

The theoretical foundation of the study of (electrostatic) charging processes of objects immersed in plasmas is the Boltzmann-Poisson system:

$$\begin{cases} \dfrac{\partial f_s}{\partial t} + \mathbf{v} \cdot \dfrac{\partial f_s}{\partial \mathbf{x}} + \dfrac{q_s}{m_s}(\mathbf{E} + \mathbf{v} \times \mathbf{B}) \cdot \dfrac{\partial f_s}{\partial \mathbf{v}} = \mathrm{St}\{\mathrm{f_s}\} \\ \epsilon_0 \dfrac{\partial \mathbf{E}}{\partial \mathbf{x}} = \displaystyle\sum_s \int f_s d\mathbf{v} \end{cases} \tag{79}$$

governing the evolution in the phase space $(\mathbf{x}, \mathbf{v})$ of the distribution function f_s of particles of the charged species s.

Initial and boundary conditions need to be supplemented. The initial conditions can be chosen from any arbitrary self-consistent equilibrium, usually a uniform drifting Maxwellian. For non-emitting dust, zero outflow boundary conditions at the surface of an object are applied: $f_s(\mathbf{x}, \mathbf{v}, t) = 0$ for $\mathbf{v} \cdot \mathbf{n} > 0$ where $\mathbf{n}$ is the outward normal. More general boundary conditions can be applied to handle dust emission [139]. The electric field at the boundary between the dust and the plasma need to satisfy the following condition [165]:

$$\epsilon_d \, \mathbf{E} \cdot \mathbf{n}\big|_{in} + \epsilon_0 \, \mathbf{E} \cdot \mathbf{n}\big|_{out} = \sigma \tag{80}$$

where σ is the surface charge density and ϵ_d is the dielectric constant of the material composing the dust. Although the system (79) with its boundary conditions might appear simple, it has no general exact analytical solution. The difficulty resides in the nonlinear coupling between the motion of the plasma particles and the electric field $\mathbf{E}$ generated by the plasma itself. Approximate solutions or numerical solutions are required. The theoretical work on the approximate analytical solutions was reviewed above. The following section will discuss the numerical solution of the Boltzmann-Poisson model in the presence of dust.

4.5.1. *Direct simulation with the Particle-In-Cell method*

The Boltzmann-Poisson system (79) with its initial and boundary conditions can be solved with the Particle-in-Cell (PIC) method [166,167]. The phase space is sampled with a limited number of computational particles. The Poisson's equation is discretized on a computational grid where the charge of the particles is collected. A complete description of the PIC method can be found in the classic textbooks [166,167] and is not repeated here.

The boundary conditions relative to the dust-plasma interaction can be imposed most effectively with the immersed boundary method. The immersed boundary method has been developed for fluid simulations [168] and has been extended to plasma simulations [169]. In the study of dust charging, one needs to absorb any ion or electron that hits the surface of the dust particle. Furthermore, the proper boundary conditions for the electric field must be imposed. This task can be very complex for general shapes of the dust particles in multiple dimensions, if traditional methods are used. The immersed boundary method eliminates the need for a detailed knowledge of the position of the interfaces. It replaces the interface conditions with appropriate contributions to the field and particle equations. The interaction of the dust particle with the plasma is described with the immersed boundary method.

The application of the immersed boundary method in PIC codes is described in Ref. [169,170] . In the present work, we summarize the immersed boundary as implemented in the PIC code DEMOCRITUS for dusty plasma simulations [170].

The dust particle is represented by motionless computational particles (object particles) with properties suitable to describe the macroscopic properties of the dust. Dust plasma interface conditions are treated with the immersed boundary method in two steps.

First, we assign to the object particles a susceptibility χ_p that can be interpolated to the vertices of the grid $\mathbf{x}_v$ to obtain a grid susceptibility:

$$\chi_v = \sum_p S_{vp}\chi_p \,, \tag{81}$$

where S_{vp} are the linear assignment weights [170,171]. The grid susceptibility is used to alter the Poisson equation:

$$\mathcal{D}_{cv}(1+\chi_v)\mathcal{G}_{vc'}\phi_{c'} = \rho_c \,, \tag{82}$$

where the potential ϕ and the charge density ρ are defined on the cell centers $\mathbf{x}_c$ and repeated indexes are summed. The operators $\mathcal{D}_{cv}$ and $\mathcal{G}_{vc}$ are a difference approximation of the divergence and gradient, respectively. As discussed in detail elsewhere [169,172], Eq. (82) is solved everywhere, including in the interior of the dust particle. The term $(1+\chi_v)$ gives an approximation to the correct interface conditions for the electric field. For dielectric dust, χ_v is simply the susceptibility of dielectric material; for conductor dust, the susceptibility can be set to a large number to impose a uniform potential within the conducting dust.

Second, the object particles exert a friction on the plasma particles, via a slowing property μ_p that is interpolated to the grid, as in Eq. (81), to produce a grid quantity μ_v used to introduce a damping term to the equation of motion of the plasma particles:

$$\frac{d\mathbf{v}_p}{dt} = \sum_v \mathbf{E}_v S_{vp} - \mathbf{v}_p \sum_v S_{vp} \mu_v \,. \tag{83}$$

The second term in Eq. (83) can be as big as desired to stop the plasma particles on the surface of the dust. The damping term is zero everywhere outside the region occupied by the dust. Equation (82) and Eq. (83) allow one to treat the field and particle boundary conditions on the surface of the dust.

4.5.2. *Adaption strategies*

The use of nonuniform or adaptive grids with the immersed boundary method increases the accuracy of the field description (Poisson equation). However, the description of the phase space (Boltzmann equation) remains poor if a uniform space distribution of computational plasma particles is used. In the following section this limitation is removed.

Grid adaptation Grid adaptation can be achieved by grid refinement (i.e. adding more grid points) in some selected areas or by grid motion (i.e. moving grid points to regions of interest from regions of lesser interest). In the first case, the adaptive mesh refinement (AMR) method [173] is obtained. In the second case, the moving mesh adaptation (MMA) method [174] is obtained.

In dusty plasma physics problems the regions of interest are readily identified: the area immediately surrounding a dust particle. We have recently proposed a new approach [175] to variational grid adaptation [174] based on the minimization of the local truncation error defined above. The method can be constructed starting from the following equidistribution theorem proven in Ref. [175]

THEOREM: *In a optimal grid, defined as a grid that minimizes the local truncation error according to the minimization principle*

$$\int_{\mathbb{V}} | e | \, d^N x \,, \tag{84}$$

the product of the local truncation error in any cell i by the cell volume V_i (given by the Jacobian $J = \sqrt{g}$) is constant:

$$e_i V_i = \text{const} \tag{85}$$

The equidistribution theorem is applied approximately solving the following Euler-Lagrange equations:

$$g^{ij} \frac{\partial}{\partial \xi^i} \left(|e| \frac{\partial x^i}{\partial \xi^j} \right) = 0 \tag{86}$$

Alternatively, an exact approach is to use the Monge-Ampère approach to grid equidistribution [176,177]. This approach creates a grid where $|e_i|V_i$ is constant. Note that the equations above are identical to the equations used by the Brackbill-Saltzman variable diffusion method [174]. The primary innovation is that the monitor function is now directly linked with the local truncation error instead of being left undefined. In the typical implementations of the Brackbill-Saltzman method, the monitor function is defined heuristically by the user.

4.5.3. *Particle adaptation*

Particle rezoning is needed to increase the number of particles in regions where high accuracy is required, and to reduce the number of particles where lower accuracy can be tolerated. The primary effect of increasing the number of particles is to reduce the variance of the statistical description of the distribution function. In a PIC simulation this increases the accuracy defined as typical in Monte Carlo methods, i.e. as the variance of the simulation.

Particle rezoning must be in effect throughout the calculation to constantly keep the local required accuracy. In multiple-length scale problems, the region of interest can move, and particle rezoning must follow the motion to keep the focus where it is needed. The approach followed here is to use adaptive grids to follow the evolution of the system [174,175] and particle rezoning to keep the number of particles per cell constant. This approach leads to finer grid spacing in the region of interest and, automatically, to a higher density of computational particles in that region.

The problem of particle rezoning can be formulated [178] as the replacement of a set of N particles with position $\mathbf{x}_p$, velocity $\mathbf{v}_p$, charge q_p, and mass m_p, with a different set of N' particles with position $\mathbf{x}_{p'}$, velocity

$\mathbf{v}_{p'}$, charge $q_{p'}$, and mass $m_{p'}$. The criterion for replacement is the *equivalence* between the two sets, defined as the requirement that the two sets must represent the same physical system, with a different accuracy. This generic definition of equivalence between two sets is given practical bearing by specifying two rules for equivalence.

Two sets of particles are considered equivalent if [178]:

(1) the two sets are indistinguishable on the basis of their contributions to the grid moments;
(2) the two sets of particles sample the same velocity distribution function.

The first criterion concerns the moments of the particle distribution used to solve the field equations. The moments are defined at the grid points $\mathbf{x}_g$ as

$$M_g = \sum_p S\left(\mathbf{x}_g - \mathbf{x}_p\right) q_p \mathbf{F}(\mathbf{v}_p) \,, \tag{87}$$

where S is the assignment function [167,171]. In general, when nonuniform grids are used, $\mathbf{x}$ is the natural coordinate, i.e. the system of coordinates where the spacing between consecutive points is uniform and unitary in all directions [174]. The function $\mathbf{F}$ of the particle velocity characterizes the moment. In explicit electrostatic codes, only the charge density is required:

$$\rho_g = \sum_p S_g(\mathbf{x}_p) q_p \tag{88}$$

derived from (87) using $\mathbf{F}(\mathbf{v}_p) = 1$ and using a short notation for $S_g(\mathbf{x}_p) = S(\mathbf{x}_g - \mathbf{x}_p)$. Electromagnetic and implicit codes [179] require higher order moments like the current density

$$\mathbf{J}_g = \sum_p S_g(\mathbf{x}_p) q_p \mathbf{v}_p \tag{89}$$

and the pressure tensor

$$\Pi_g = \sum_p S_g(\mathbf{x}_p) q_p \mathbf{v}_p \mathbf{v}_p \,. \tag{90}$$

The first criterion requires the two sets of particles to give the same moments relevant to the field equations. Note that if this criterion is satisfied exactly total energy and momentum are also automatically conserved. The second criterion is more difficult to apply in a quantitative fashion. In previous work [178,180], it has been proposed to use the χ^2 test or the Kolmogorov and Smirnov test to verify that the particle distribution is preserved. In practice, this is not easily achieved.

In fluid PIC codes, the first criterion is the only one to be applied, and general schemes for particle rezoning can be derived [181]. In kinetic PIC codes, the computational particles sample the real plasma velocity distribution, and the second criterion must also be imposed. In the kinetic case the choices are more limited. For this reason, a simpler approach is followed [178,180].

To increase the number of particles per cell, a given particle is split in two or more new particles displaced in space but all sharing the same speed. The weights and displacements can be chosen to conserve exactly the grid moments, and the velocity distribution is not altered because all the particles have the same velocity.

To decrease the number of particles, the splitting operation can be inverted to coalesce two particles into one. The difficulty is that, in general, it is impossible to find two particles with the same velocity. For this reason, particles with different velocity have to be coalesced. To minimize the perturbation of the velocity distribution, the particles to be coalesced must be chosen with similar velocity. An alternative approach is to coalesce three particles into two, which allows one to conserve both energy and momentum [182].

We refer the reader to a previous technical description of the details of the implementation of the particle adaptation algorithms [178].

4.6. *Examples of dust particle charging*

The PIC approach has been applied to numerous problems. To name a few, the direct charging of one spherical dust in a uniform plasma [183], in a plasma with a net relative flow between the dust and the plasma species [170,181,184,185], in a magnetized plasma [186,187] and considering non spherical particles. The PIC method can also effectively consider emission from the dust [139] and collision processes in the plasma between electrons, ions and neutral particles (molecules or atoms) [188]. To show the type of results possible with the PIC method, below we report a typical result.

If no secondary emission or photoemission is present, the equilibrium charge on the dust particle must be negative to repel the more mobile electrons and attract the ions to achieve a balance of electron and ion currents. We consider here the case where a plasma with an ion to electron temperature ratio $T_e/T_i = 20$ and ion to electron mass ratio $m_i/m_e = 1836$ is drifting relative to a spherical dust particle of radius $a/\lambda_{De} = 0.4$, where

λ_{De} is the electron Debye length. The relative velocity w is expressed by the Mach number $M = wm_i^{1/2}/(kT_e)^{1/2} = 10$. The system is simulated using a cylindrical coordinate system with the vertical axis along the direction of the plasma flow and centered in the center of the spherical dust particle. In this configuration, the azimuthal coordinate is invariant, and the problem is 2D axisymmetric.

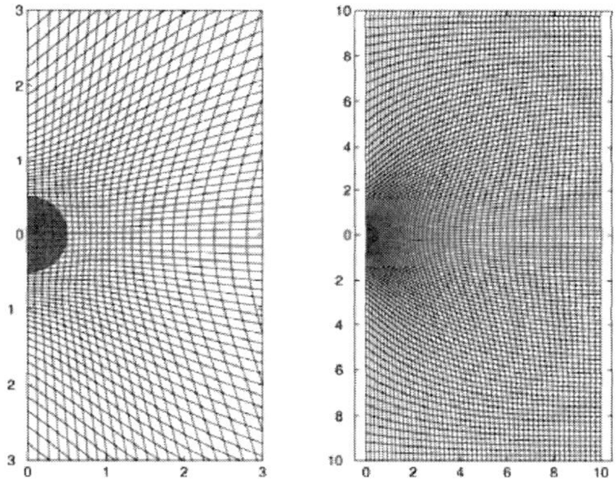

Fig. 7. Initial setup of a dust charging simulation. The dust particle is represented by material computational particles with appropriate dielectric properties for the immersed boundary method. An adaptive grid is used to resolve the small sub-Debye scale dust particle. The blow up of the region near the dust (left) and the full grid (right) are shown.

Figure 7 shows the configuration of the grid and of the dust particle for the problem considered here. Note that a nonuniform (but constant in time) grid is used to describe better the sheath around the dust particle. The distance of the dust particle from the boundaries is 10 λ_{De}. The plasma species are initially loaded according to a drifting Maxwellian distribution with a downward vertical net flow velocity corresponding to a Mach number $M = 10$. To reach an equilibrium, particles that flow out of the lower boundary are replaced by particles injected at the top boundary [170].

Figure 8 shows the history of the net charge accumulated on the dust particle. In this case, particle rezoning was used to ensure the accuracy of the calculation. The particles are loaded, initially, with a constant number of particles per cell, leading to a higher concentration around the dust particle where the cells are smaller. However, the plasma flow tends to empty the region around the dust reducing the accuracy. Splitting the particles moving

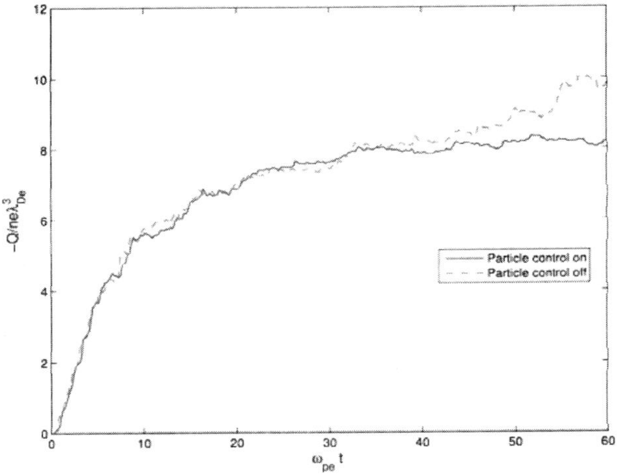

Fig. 8. Evolution of the charge collected by the dust particle. Two runs are shown, both have uniform grids but one has also particle control (solid line) and the other has no particle control (dashed).

toward the dust and coalescing the particles moving away from it is desirable to keep the number of particles per cell and the accuracy constant.

If the calculation is repeated without particle rezoning, the accuracy worsens in time as the region around the dust becomes less populated. Two effects lead to decrease accuracy around the dust particle: the particles originally present are in part captured by the dust and in part just simply flow away according to their average downward velocity of Mach $M = 10$. The new particles that replace them are flowing from regions of larger cells and are less numerous leading to a decrease of accuracy.

As a result of the decrease in accuracy, the dust particle does not reach a steady state in the run without particle rezoning.

5. Dust dynamics, transport, and impact on tokamak plasma performance

5.1. *Electromagnetic acceleration of dust cloud*

Consider acceleration of a dust cloud due to electromagnetic field in quasineutral plasmas, that is, for a uniformly charged dust cloud with dust density n_d, one has

$$m_d n_d \frac{d\mathbf{U}_d}{dt} = Q_d n_d \mathbf{E}, \tag{91}$$

where the electric field is determined from the Poisson equation

$$\nabla \cdot \mathbf{E} = \frac{1}{\epsilon_0}(eZ_i n_i - en_e + Q_d n_d), \tag{92}$$

where Z_i is the ion charge state. Eq. (92) reveals that the electrons and ions are coupled with a dust particle through the space charge electric field, and their motions are governed by, respectively,

$$0 = -en_e\left(\mathbf{E} + \mathbf{v}_e \times \mathbf{B}\right) - \nabla P_e, \tag{93}$$

and

$$0 = eZ_i n_i\left(\mathbf{E} + \mathbf{v}_i \times \mathbf{B}\right) - \nabla P_i, \tag{94}$$

where $\mathbf{v}_e(\mathbf{v}_i)$ is the electron (ion) fluid velocity and $P_e(P_i)$ is the electron (ion) thermal pressure. In Eqs. (93) and (94) we have neglected the electron and ion inertial forces, since both the electrons and ions are light particles in comparison with the dust grains. From Eqs. (91)-(94), we then obtain

$$n_d m_d \frac{d\mathbf{U}_d}{dt} = \epsilon_0 \mathbf{E} \nabla \cdot \mathbf{E} + \mathbf{J} \times \mathbf{B} - \nabla P, \tag{95}$$

where $\mathbf{J} \approx e(Z_i n_i \mathbf{v}_i - n_e \mathbf{v}_e)$ is sum of the ion and electron current densities, and $P = P_e + P_i$ is sum of the electron and ion thermal pressures. Noting that the dust current density is rather insignificant, in view of the low dust density and dust speed in comparison with those of the bulk plasma electrons and ions, we have from Ampère's law

$$\mathbf{J} = \frac{1}{\mu_0} \nabla \times \mathbf{B}. \tag{96}$$

Hence, Eq. (95) yields the dust grain acceleration

$$\frac{d\mathbf{U}_d}{dt} = \frac{\epsilon_0}{n_d m_d} \mathbf{E} \nabla \cdot \mathbf{E} - \nabla \left(\frac{P}{n_d m_d} + \frac{B^2}{2\mu_0 n_d m_d} \right) + \frac{1}{\mu_0 n_d m_d} (\mathbf{B} \cdot \nabla) \mathbf{B}. \tag{97}$$

Since at the plasma surface the tangential component of the space charge electric field is zero and there is no space charge electric field inside the plasma surface [190], therefore there remains only the normal (to the plasma surface) component of the space charge electric field $E = -\partial\phi/\partial n$ at every point, where ϕ is the electrostatic potential and n stands for the normal component, that contributes to acceleration of a dust particle normal to the plasma surface.

Furthermore, in the SOL region near divertor in tokamaks, the plasma pressure and magnetic field gradients are rather week, and acceleration of a dust particle normal to the plasma surface is therefore governed by

$$\frac{d\mathbf{U}_d}{dt} \sim \nabla_n \frac{\mathbf{U}_d^2}{2} = \frac{\epsilon_0}{n_d m_d} \nabla_n E^2, \tag{98}$$

which can be integrated once to obtain the dust speed U_d normal to the plasma surface

$$U_d = \sqrt{\frac{2\epsilon_0}{n_d m_d}} E. \tag{99}$$

It should be stressed that acceleration of a charged dust grain in our investigation is solely caused by the gradient of the electric field intensity involving the normal component of the space charge electric field at the plasma surface.

As an application, we observe that micron-size carbon dust particles (say $r_d = 10$ μm) will have $n_d m_d = 9.3 \times 10^{-7}$ g/cm^3 (if $n_d = 10^2$ cm^{-3}), so that the dust speed U_d, deduced from Eq. (99), for $E = 100$ V/cm will be of the order of one meter per second. Higher velocity dust particles have been observed in DIII-D tokamak [191]. Therefore, electric field alone may not be sufficient to accelerate dust to tens of meter or higher speeds. This result is consistent with both the numerical calculations (Sec. 5.4) and laboratory experiments (Sec. 5.6) described later in this section. Furthermore, it should be mentioned that recently very high-speed dust particles (speed of the order of ten kilometer per second) seems to be detected in the Frascati Tokamak Upgrade (FTU) [192] experiment. The latter also reports the formation of craters due to the impact of such high-speed dust particles onto the FTU wall. The tokamak wall erosion by high-speed charged dust particles ought to be avoided in the next generation fusion reactors.

5.2. *Key ingredients of numerical study of dust transport in fusion devices*

Force analysis on micron-size dust grains in fusion plasmas has shown that dust is mainly accelerated by drag force [35,193,194] associated with strong plasma flows. Such flows exist in edge regions due to plasma recycling processes. In the so-called magnetized sheath region near surfaces, magnetic field lines are very oblique with respect to the surfaces. The thickness of the magnetized sheath is on the order of ion gyro-radius, which for tokamak plasma is about 10^{-2} cm and much larger than a typical dust size $\sim 1\ \mu$m. Both flow along and across (diamagnetic plasma flow) the magnetic field lines can drag dust into motion. Plasma flows at the order of local sonic speed. In the direction perpendicular to the surface, the component of the drag force can be balanced by the repulsive electric force due to the sheath electric field (we assume here that dust grains are negatively charged), the component of drag force parallel to the surface is not balanced and causes continuous acceleration of the grains [35].

Dust grains can be accelerated to speeds of a few hundred meters per second in a matter of milli-seconds due to plasma flow [35,193,194]. These high-speed dust grains can therefore travel distances comparable to a tokamak major radius before they ablate/evaporate [35,193–195]. Thus, dust can play important roles both in transporting eroded wall material and in contaminating hot plasma with impurities. Other interesting and important features of dust dynamics in modern tokamaks are related to the peculiarity of magnetic-field topology, which causes recycling plasma flows in the outer and the inner divertor legs to go in opposite toroidal directions. These opposite flows can push dust grains in opposite toroidal directions [193]. High speed dust, opposite motion of dust in the inner and the outer divertors were observed with fast cameras in MAST tokamak [196].

Even though the perpendicular force balance between the drag and the electric force can be established initially, the synergy of parallel acceleration of the grain and small 'imperfections' of the surface (which always exists in fusion devices!), resulting in the fluctuations of sheath electric field, will couple the parallel motion with the perpendicular motions of the grain and transfer kinetic energy from the parallel to the perpendicular direction [35,193,194]. In some regimes, dust motion becomes stochastic (e.g. see Fig. 9 taken from Ref. [193]). As a result, the perpendicular energy of the dust grain can increase by so much and dust can penetrate through sheath potential and hit the surface. Assuming that dust charging is only due to the electron and the ion currents to the grain, we find a critical perpendicular velocity ($V_{crit.}$) for grain penetration through the sheath

$$U_d^{\perp} \gtrsim V_{crit.} \equiv \Lambda \sqrt{\frac{T_e}{\mu_d} \frac{\lambda_*}{r_d}}, \tag{100}$$

where μ_d is the averaged mass of atoms in the dust grain, $\lambda_* = \sqrt{\frac{T_e \mu_d}{4\pi^2 \rho_d}}$ is the effective Debye length, ρ_d is the mass density of dust material, and $\Lambda \sim 3$ is a constant [193]. For a micron-size carbon dust and $T_e = 10$ eV, we find $V_{crit.} \sim 1$ m/s, which is significantly smaller than a few $\times$ 100 m/s that the dust grain can pick up at tokamak edges.

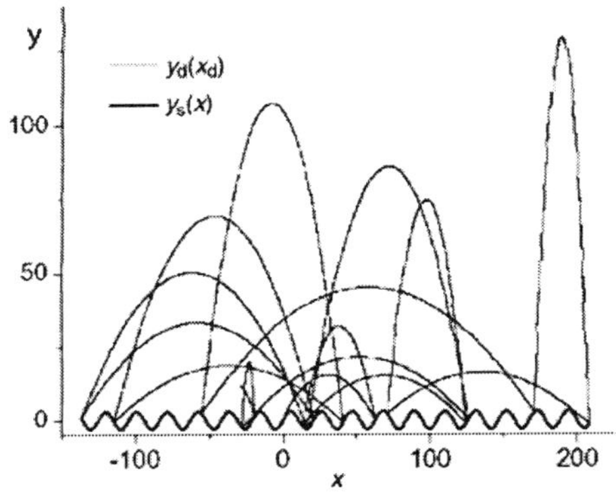

Fig. 9. Trajectory of dust, y_d, moving through the sheath and recycling region over corrugated surface [symbolized by y_s(x)]. Both the perpendicular (y) and the tangential (x) coordinates are normalized by the local ion gyro-radius. (From ref. [193].)

Even though acceleration of dust to high speeds is primarily caused by drag force due to plasma flow, the magnitude of drag force can vary significantly due to large variations in plasma parameters and dust sizes at the tokamak edge. Other forces (e.g. gravity) can be important in regions which are relatively far away from the separatrix and where plasma density is low and where the drag is reduced. In addition, dust charging, heating, and ablation rate also strongly depend on plasma parameters. Therefore, simultaneous tracking of dust dynamics, transport statistics of a large number of dust grains, and variation of plasma parameters in a tokamak geometry is only possible numerically.

5.3. *Computer codes for studying dust dynamics and transport in fusion devices*

Two computer codes have been developed for studying dust dynamics, transport statistics of a large number of dust grains in tokamaks, DUSTT [38,189] and DTOKS [197]. Both codes track the motion of individual dust grains in three-dimensional (3D) tokamak geometry taking into account dust charging, dust momentum balance, energy balance, and dust ablation. However, both DUSTT and DTOKS assume that plasma parameters are toroidally symmetric. The plasma parameters are specified on 2D grids (in the poloidal cross section) from 2D edge plasma transport codes UEDGE [198] and SOLPS [199] respectively. The expressions for charging currents, forces (due to plasma and neutral drags, gravity, and electric field), and heat fluxes associated with plasma and neutral fluxes to the grain in DUSTT are based on the OML theory for spherical grains (e.g. see Ref. [38,154,189,200]. The ablation model takes into account chemical and physical sputtering of the dust particle by plasma, radiation enhanced sublimation and thermal evaporation. Radiation energy loss from the grain accounts for the correction associated with finite ratio of the radiation wavelength to the grain size. Since most of the dust grains are very quickly accelerated to the velocities significantly higher than $V_{crit.}$, sheath electric field is neglected in simulating the dust-wall collisions. However DUSTT code provides possibility of bouncing of the slower grains by the sheath without colliding with the wall. To describe the interactions of dust grain with the surfaces the restitution coefficients and other parameters are used (some of them are found from the simulations with the code LS-DYNA [201]. Further details of the models incorporated into code DUSTT can be found in [38,189] and in Ref. [202] where DUSTT were used for the analysis of different grain materials on dust-plasma interactions for both uniform slab and DIII-D plasmas. Similar models of dust-plasma interactions were used in Ref. [37] for data interpretation of dust grain motion in a flowing plasma with a high-speed camera (slab geometry and uniform plasma parameters were assumed). Code DTOKS uses different models for dust forces, charging, and energy balance than that in Ref. [38,189] (e.g. plasma drag force utilized in DTOKS [197] resembles the supersonic plasma flow limit from [200]).

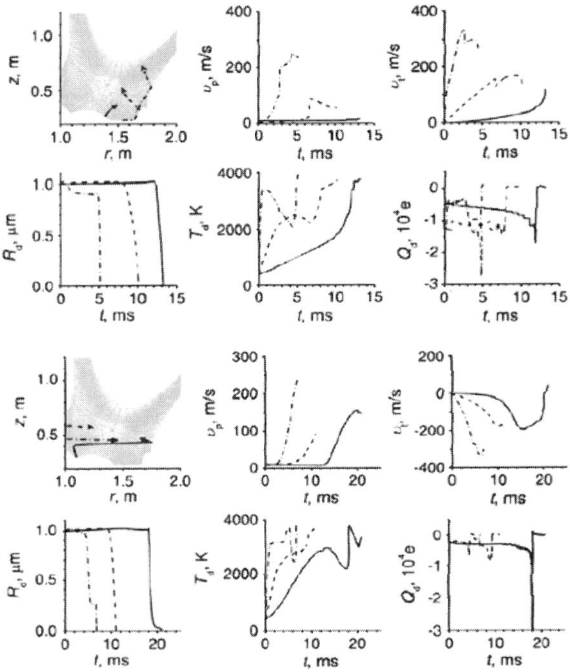

Fig. 10. Trajectories and time dependencies of size, temperature, charge, poloidal and toroidal velocities of carbon dust particles with initial radius 1 μm launched into outer (a) and inner (b) divertor legs of DIII-D tokamak with initial normal velocity 10 m/s. (from Ref. [189]).

5.4. *Computer simulations of the motion of individual dust grains in tokamaks*

Computer simulations of dust particle dynamics in tokamaks confirm theoretical estimates and show that the grains can be accelerated (mainly in the toroidal direction) up to few 100 m/s by plasma flows and, as a result, dust grains are able to penetrate rather deeply into hot plasma regions before they are evaporated (see Fig. 10 taken from Ref. [189]). The dynamics of dust particles in a typical scrape-off layer (SOL) is governed by both the ion drag and the particles' inertia. Dust inertia prevails in low density plasmas of the private flux and far SOL regions, while the ion drag is dominant in dense plasma regions close to the separatrix.

Due to the toroidal geometry, dust grain inertia (resulting in effective centrifugal acceleration) causes rather frequent dust-wall collisions at the outer side of the torus (see Fig. 10a). Dust-wall collisions play an important role in conversion of the high toroidal dust velocity into the poloidal velocity. We note that high-speed dust-wall collisions may also play an important role in dust production by impact-induced failure of wall material. As one can see from Fig. 10b, at the inner side of the torus, inertia "assists" grain penetration toward the core region. It is noticeable that heating and sublimation/erosion of dust accompanied by positive charging due to thermionic emission is strongly enhanced as the dust particles enter hot plasma regions near separatrix where the temperature of dust grain increases significantly.

From Fig. 10 we also see that, in consistency with both theoretical expectations and experimental observations, dust grains launched from the outer and inner divertor legs acquire opposite toroidal velocities. One can see it even more clearly from Fig. 11, where toroidal projections of trajectories of dust grains from Fig. 10 are shown.

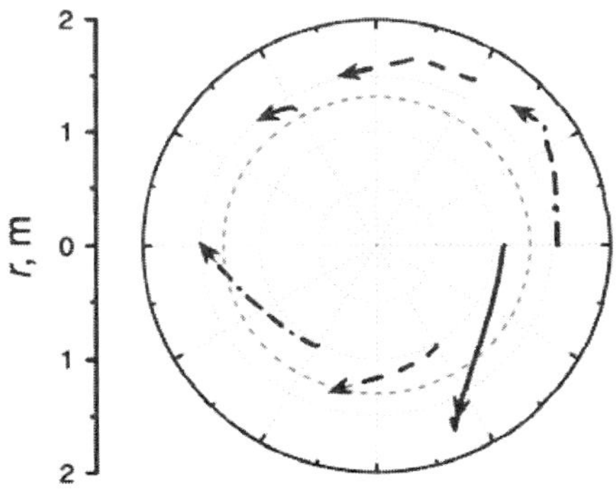

Fig. 11. Top view of the trajectories of dust particles from Fig. 10. The dashed circle divides the inner and the outer divertor leg regions (from ref. [189]).

Dust can penetrate deeply toward the core not only for DIII-D-like tokamaks, but also for ITER, even though edge plasmas in ITER are denser and hotter than in DIII-D. As an example, in Fig. 12 (taken from Ref. [38]), one can see the trajectories of carbon gust grains launched into inner divertor of ITER. Scoping studies aimed at identification of the parameters that have the most pronounced impact on dust dynamics in tokamak plasmas have shown that dust-wall interactions play one of the most important roles.

The behavior of dust particles consisting of various materials (Li, Be, B, Fe, Mo, and W) related to fusion plasmas was simulated with the DUSTT code in Ref. [202]. In this case, the bidirectional phase transition between

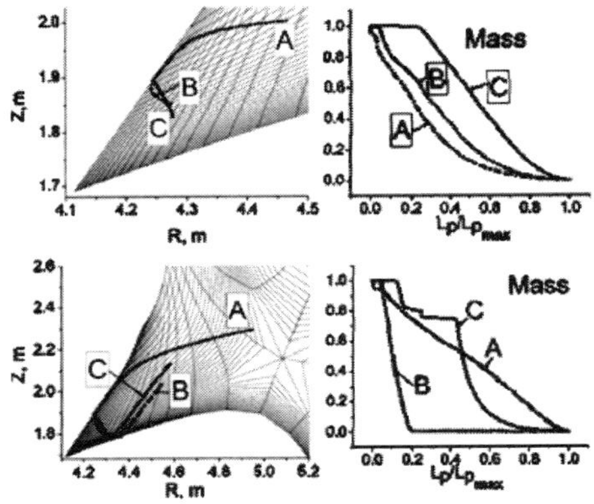

Fig. 12. Trajectories of dust particles launched from the separatrix strike point at inner divertor plate in ITER are shown for initial dust radius 1 μm upper left panel and 10 μm bottom left panel. Trajectories A, B, and C correspond to initial dust velocity of 1, 10, and 100 m/s respectively. The correspondent variation of mass of these particles along the trajectory is displayed in the right panels as a function of the normalized poloidal length Lp of the trajectories shown on the left panels (from Ref. [38]).

solid and liquid states of the dust grain should be taken into account (see Ref. [202] for details). The examples of trajectories for dust consisting of different materials calculated for DIII-D tokamak plasma are shown in Fig. 13 (taken from Ref. [202]). All particles are launched from the same location (close to the outer separatrix strike point) with initial speed 10 m/s. The normalized masses and velocities of these particles can be seen in Fig. 14 (taken from Ref. [202]).

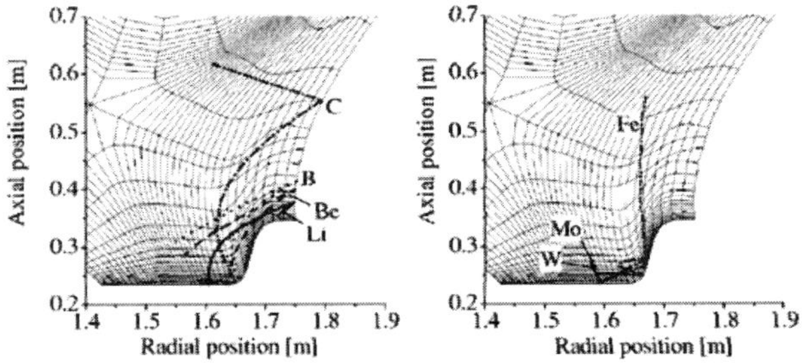

Fig. 13. Trajectories of dust particles consisting of different materials in a DIII-D plasma (from ref. [202]).

As we can see from Fig. 13 and Fig. 14, the lighter dust particles (Li,

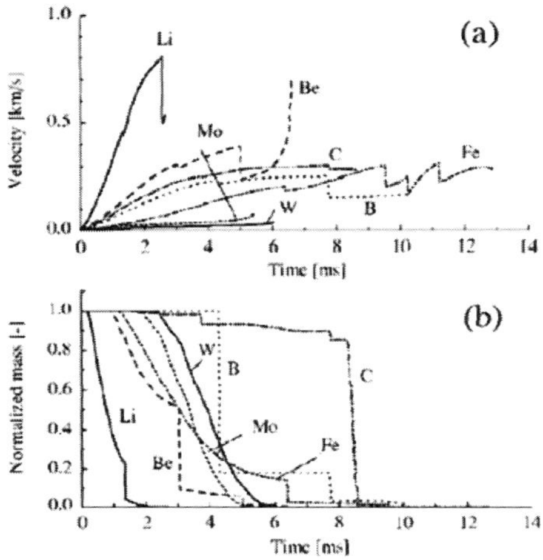

Fig. 14. The velocities (a) and normalized masses (b) of particles from Fig. 13 (from ref. [202]).

B, Be, and C), in comparison with the heavier ones (Mo, W), are more mobile and easier to accelerate by plasma flow, and as a result, the lighter particles travel longer distances and can spread over the divertor region and penetrate deeper toward the core plasma region before being evaporated. The Fe dust grains acquire the largest speed and can travel the longest distance due to the combination of relatively long lifetime and their moderate weight.

5.5. *Statistics of dust in fusion plasmas and impact of dust on plasma performance*

Simulations of a large number of particle trajectories with the DUSTT code allow us to obtain statistically averaged profiles of dust parameters in fusion devices. Fig. 15 (taken from Ref. [189]) displays the profiles of statistically averaged density (a), poloidal velocity (b), toroidal velocity (c), radius (d), temperature (e), and charge (f), obtained from the DUSTT/UEDGE modeling for a typical L-mode DIII-D discharge when the sputtered wall material is converted into dust. Fig. 15 (a and d) shows that both dust density and averaged radius are strongly reduced in the regions with a relatively dense and hot plasma (i.e. regions adjacent to the core and divertor legs). In hot plasma regions, the dust temperature is so high that dust becomes incandescent (Fig. 15e). For such incandescent particles, strong thermionic

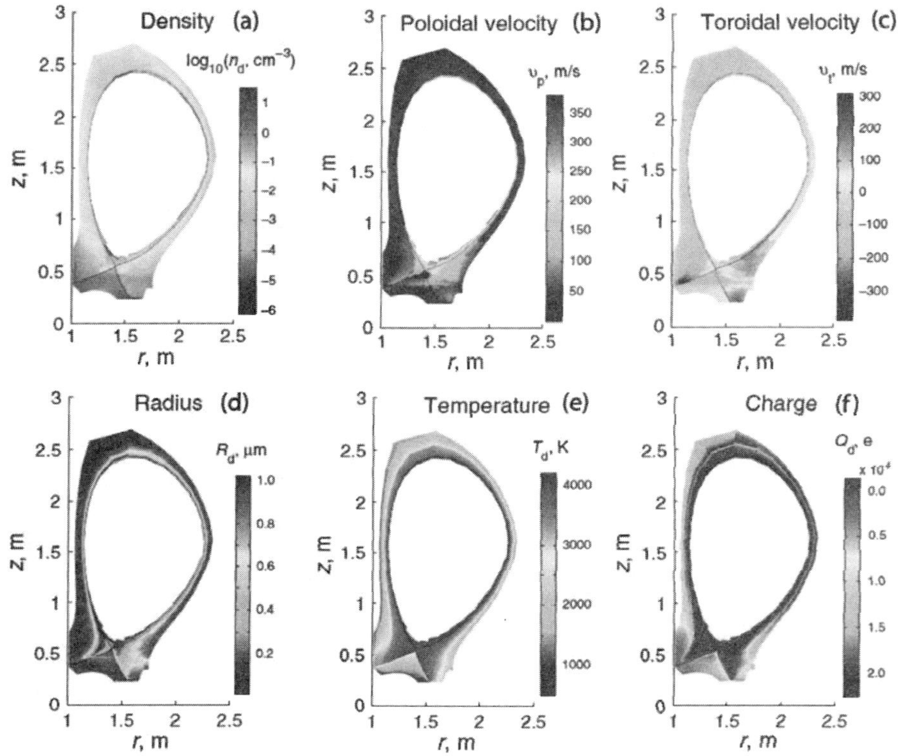

Fig. 15. The profiles of averaged dust parameters calculated with DUSTT for DIII-D plasma (from Ref. [189]).

emission results in the positive values of the dust charge (Fig. 15f). From Fig. 15 (b and c), the ensemble average speed of dust particles, being accelerated by the plasma-dust friction/drag force, reaches a few 100 m/s. While most dust particles have opposite toroidal velocities in the inner and outer divertor regions, one also can see a jet of dust particles originated at the inner separatrix leg and transported into the outer leg region keeping negative toroidal velocity (Fig.15c). The simulated dust profiles predict much deeper penetration of dust ablated carbon atoms toward the core in comparison with directly sputtered carbon atoms from the wall.

Using the DUSTT code in conjunction with the UEDGE package, the effect of dust on the DIII-D divertor plasma parameters via deeper penetration of neutral impurity toward the hot core plasma can be evaluated (see Fig. 16 taken from Ref. [201]). First, only the plasma profiles for wall-sputtered impurities are calculated with UEDGE. Next, the DUSTT code is used to obtain the profiles of neutral carbon generation due to dust ablation based on the conversion coefficient of sputtered impurity into dust (ξ_d). Last, these profiles are used as the fixed volumetric neutral carbon source in UEDGE and new steady-state plasma profiles are calculated. In Fig. 16, the observed change of T_e profiles with increasing ξ_d, highlights strong outer leg detachment due to dust and deep impurity penetration that enhance radiation loss.

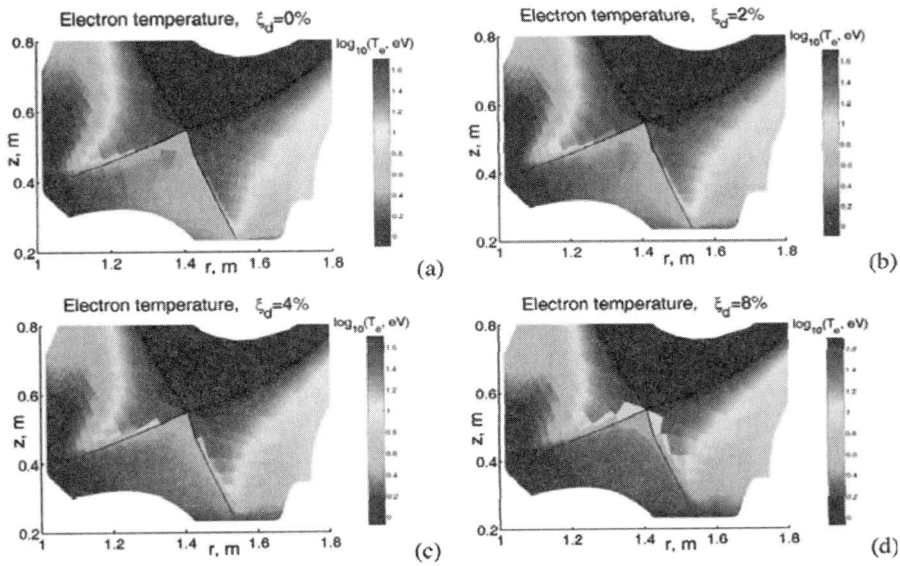

Fig. 16. The results of UEDGE/DUSTT modeling of dust impact on divertor plasma detachment in DIII-D tokamak (from Ref. [201]).

To conclude, in recent years, a significant progress has been made in the understanding of dust dynamics and transport in magnetic fusion plasmas. The processes playing the most important roles in dust dynamics and transport were identified. Rather sophisticated 3D codes, which can track dust in tokamak plasma and provide both the motion of individual grains and statistics of an ensemble of dust particles, were developed and used for dust studies in existing tokamaks and ITER. Good qualitative agreements were found among theoretical predictions, numerical modeling, and experimental observations of the magnitude and directions of dust motion in the tokamak divertor region. Benchmark of numerical codes with experiments is in progress (e.g. see Ref. [109] for a preliminary comparison of dust grain trajectories recorded by multiple fast cameras with DUSTT modeling, another comparison of the laboratory measurement with theory is described in Sec. 5.6). The work for full coupling of dust and plasma transport codes (e.g. DUSTT and UEDGE) makes it first steps.

5.6. *Experiments on dust motion relevant to fusion*

Experiments with dust particles have been extensively performed in weakly ionized gases (the ionization fraction $\lesssim 10^{-5}$) [15,17–23,25,26,28]. In these dusty plasmas, the ion temperature is similar to the neutral temperature($\sim$ 0.025 eV, except at externally biased electrodes), electrons have a temperature of a few eV, and electron density is in the range of 10^{14}-10^{15} m^{-3}. Dust particles acquire negative charges due to larger electron mobility over ion mobility. It has been observed that dust charging is more effective in radio-frequency (rf) driven plasmas (with rf frequencies in MHz range) than direct-current (dc) discharges, partly because variation of the electric fields allows better coupling of rf energy with ionization process. In fact the first systematic observations of dust in laboratory plasma have been reported for a rf plasma reactor used in plasma etching [61,62]. In dusty plasma experiments, dust particles of known size, conductivity (either conducting or dielectric), composition [graphite, Al_2O_3 or plastic (melamine formaldehyde or MF)], are deliberately released into the discharge. Dust usually moves at very 'slow' speeds of a few mm/s. In these weakly ionized plasmas, dust trajectories are strongly affected by electric force, gravity, friction of neutral atoms and ion drag. In particular, ion drag is negligible for larger grains but it becomes important for small grains ($\lesssim 1$) μm and strongly coupled ions [203].

We describe here experiments on dust motion in plasmas that are similar to that of magnetic fusion in density and temperatures. At the edge of a fusion device, plasma density is 5 to 6 orders of magnitude higher than dusty plasmas mentioned above [15,17–23,25,26,28]. In addition, ion temperature is comparable with electrons. Heating of the grains by the electron and ion is significant. At elevated dust temperatures, thermionic emission has become substantial in the balance of plasma currents to the grain, resulting significant grain potential change. For grains at elevated temperatures, their electric potential can even become positive relative to the local plasma potential [139]. We have found that dust acceleration due to the plasma drag force (ion drag primarily) is 3 to 5 orders of magnitude higher than the gravitational acceleration g [36,204,205], therefore, we conclude that plasma drag is the dominant force for dust motion and dynamics in fusion, consistent with the analysis above (Sec. 5.1) and the numerical modeling (Sec. 5.4).

Two types of experiments are reported here [204,205]. In both cases, the Debye length was much shorter than the sizes of dust grains.

In the first type of experiments [206], ions and electrons had similar temperature around 10 eV. The plasma density was $0.5 - 5 \times 10^{19}$ m^{-3}. Dust was stored in a reservoir and released into plasma by a piezo-electric shaker [207]. The grains were made of a kind of fluorescent powder with a density ~ 0.1 g/cm^3. The dust grains were induced to emit visible light by plasma UV light, allowing imaging of dust trajectories by two high-speed cameras with filters and adequate optics. Dust trajectories were observed to be along the direction of plasma flow, as shown in Fig. 17.

The size of the observed grains was not known a priori. The grains diameters were inferred from the parameters of the imaging system: $r_d = \sqrt{d_e^2 - d_s^2}/2M$, where $d_s = 2.44(1+M)\lambda f^{\#}$ is the diffraction-limited spot diameter for the lens set-up with $f^{\#}$, M is the magnification, and d_e is the diameter of a pixel. A 10-nm-wide pass-band filter had a center wavelength of $\lambda = 561$ nm which coincided with the peak emission of dust material. The lens magnification was M $= 0.1$, $f^{\#} = 4$, and $d_e = 6.7$ μ. Most dust particles had radii in the range of 100 to 200. Therefore, individual dust particle radius were about 10 to 20 times the plasma Debye radius.

The total plasma drag exerted on dust ($\mathbf{F}$) has two components, corresponding to the collection force $\mathbf{F}^{coll}$ and the long range force $\mathbf{F}^{orb}$ in Eq. (63). As explained in ref. [37,204], we can approximate the magnitude of the drag force by the collection force

$$F \sim F^{coll} = 2\pi r_d^2 n_i k_B T_i \xi w \tag{101}$$

where $w = U/\sqrt{(}2k_B T_i/\mu_i)$ is the plasma flow speed (U) normalized to the ion thermal speed and ξ= 1.1 to 1.5 depends on models for ion-dust surface impact. Most of the observed dust accelerations (Fig. 18) were in

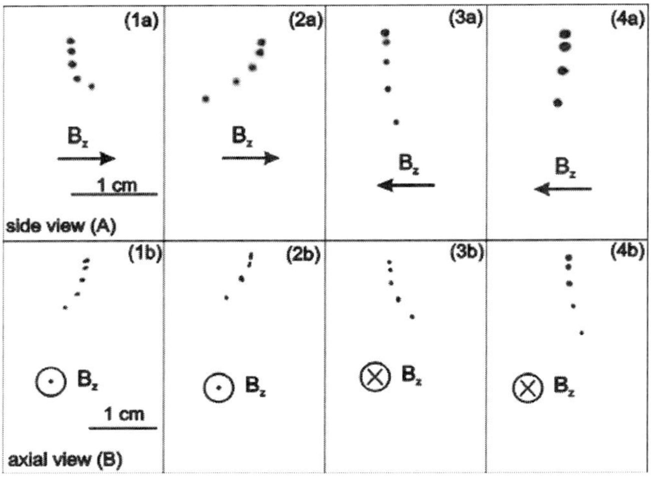

Fig. 17. In terms of the directions of dust motion, four types (1 to 4) of dust trajectories are observed. Each image (with scales similar to (1a) or (1b)) is for the trajectory of the same dust particle inside a plasma. Each pair of the top and bottom images is from the two cameras for the same dust. The top frames (a frame) are from camera A (with a r-z view in a cylindrical geometry), and the bottom frames are from camera B (with a r-θ view in a cylindrical geometry). Each dust trajectory consists of four to five pearl-like dots because the cameras are set up for five short exposures with fixed delay between exposures. The directions of the dust motion are found to be determined by the directions of plasma flow (measured by Mach probes). Directions of the external Bzs are indicated.

good agreement with the Eq. (101).

In the second type of experiments, motion of carbon grains has been studied in a supersonic plasma flow ejected from a compact coaxial gun [207]. The main characteristics of the plasma produced in this coaxial gun are: short pulse discharge ($\approx 400\ \mu$s), ion density of the order of 10^{22} m^{-3} and low ion temperature of only 1-3 eV. The flow speed measured by two photodiodes placed 0.9 m apart was found to be in the range 25-60 km/s. The plasma density has been determined by streaked spectroscopy from the Stark broadening of D_α line, while the ion temperature has been estimated from the angle of plasma expansion relative to the flow speed. The results are presented in Table 2.

A high speed camera (DiCam Pro) equipped with a fish eye lens 16 mm $f/4$ and set for exposures of 50 to 500 ns has been employed to visualize the plasma plume launched in a large vacuum tank, shown in Fig. 19.

Graphite and diamond powders with radii $r_d = 0.5 - 22$ and $20 - 30\ \mu$m and densities of 2.25 and 3.52 g/cm^3, respectively, have been used.

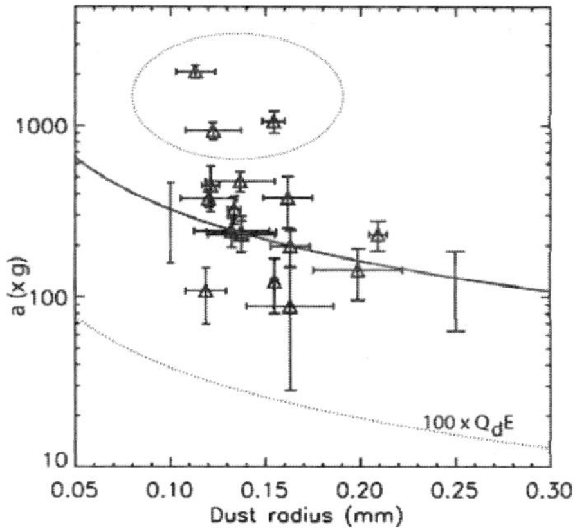

Fig. 18. A collection of experimentally measured dust acceleration in azimuthal direction as a function of dust radius (triangles). Electric force (multiplied by a factor of 100) cannot exceed the dashed line (and therefore is too small). The solid line results solely from direct ion impact (collection component of the drag force) using experimental plasma density ni, Ti, and plasma flow. The error bars for the theoretical curve give the upper and the lower limits for theoretical estimates. A few cases of underestimates are highlighted by the oval (from Ref. [204]).

Voltage (kV)	v_f (km/s)	n_i $(\times 10^{22}\mathrm{m}^{-3})$	Ti (eV)
6	26 ± 0.5	$0.2 - 1.5 \pm 0.1$	1.3 ± 0.3
8	38 ± 1	$0.5 - 2.2 \pm 0.1$	1.7 ± 0.4
10	56 ± 2	$0.5 - 3.1 \pm 0.2$	2.8 ± 0.7

Table 2. Experimentally determined plasma flow speed, ion density and temperature for different discharge voltages.

Dust was imaged from about 2 m with a telescope lens (500 mm $f^{\#}/4$). The capabilities of the imaging system were limited by the resolution and magnification of the optical set-up. Thus the smallest dust size that could be resolved was $r_d \approx 5$ μm. For each shot the camera was operated with different delays relative to the plasma ignition time, between $350 - 3 \times 10^3$ μs. The exposure time was set for 2 to 16 μs. The speed of each grain was inferred by using the time-of-flight technique, from the ratio between the length of the grain's trace and the exposure time of the image. The images of Fig. 20 show diamond (a) and graphite (b) grains located at 1.6 m downstream from the gun muzzle. The exposure times are 4 μs for image (a) and 12 μs for image (b). The speed of the pictured diamond grains is therefore in the range $0.8 - 1.5$ km/s. The graphite grains have

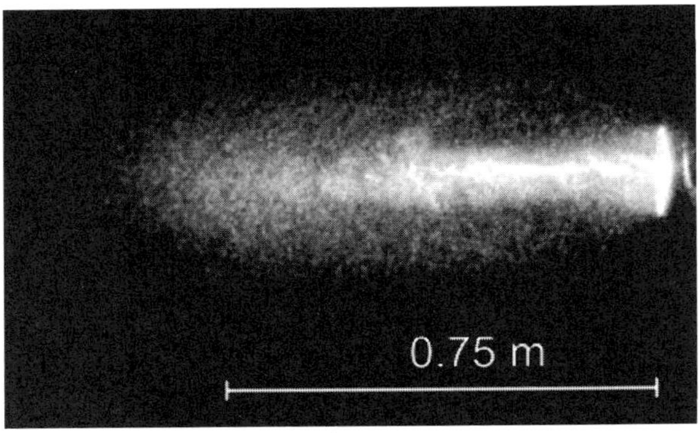

Fig. 19. Images of plasma jet for a 10 kV shot and an exposure of 50 ns from t=0 of the discharge.

similar speeds, between 0.4 and 1.5 km/s, however their longer traces are compensated by the longer image exposure. A distribution obtained by tracking a large number of dust traces (1.94×10^3) is shown in Fig. 21. It is clearly seen from the peaks that the average dust speed increases with the the plasma density and flow speed as expected.

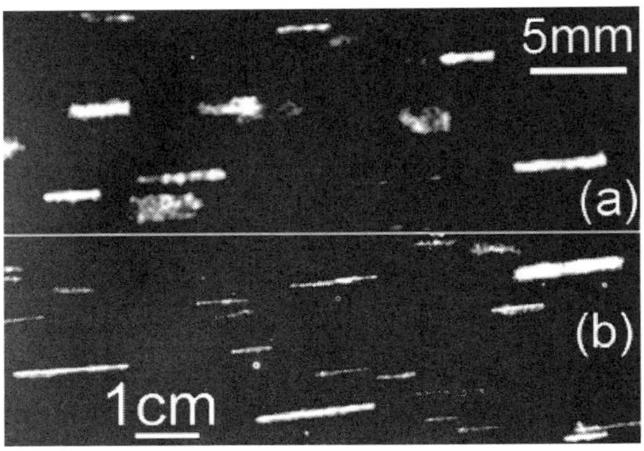

Fig. 20. Images of self-glowing diamond grains (a)graphite grains (b) flying at 1.6 m from the gun muzzle. The exposure time is 4 μs and 12 μs, for (a) and (b).

The pictures of Fig. 20 show also traces of different thickness, which differ qualitatively from each other: on one hand, there are well resolved traces with different widths situated in the focusing region of the lens, while on the other hand there are traces with a blurred aspect and an apparent thickness almost comparable with their length. In this later case, the traces are thought to be situated at the very edge of the field of view and cannot be clearly resolved. While in the first case, the trace widths are correlated with the grain sizes, in the second case the trace appearance is only an optical effect.

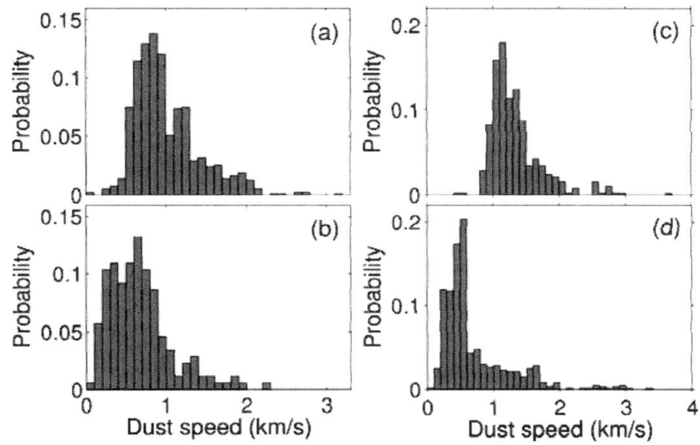

Fig. 21. Dust speed obtained from the traces of 1.94×10^3 flying grains. Probability of observed diamond dust speed for 10 kV shots (a) and 8 kV shots (b); probability of observed graphite dust speed for 8 kV shots (c) and 6 kV shots (d).

Lorentz force was found to be unimportant in the second experiment [205], again, dust acceleration were due to the plasma drag force ($2 \times 10^5 - 9 \times 10^6$ m/s^2).

6. Summary

Dust physics in magnetic fusion is an interesting subject of growing importance on the next generation of magnetic fusion devices. Dust production within fusion devices scales up with the increase in plasma energy content and fusion energy output. Dust can be radioactive due to tritium and neutron activated elements. Dust accumulation could raise the level of tritium content inside a fusion machine with time if left unaddressed. Toxicity, flammability are other dust properties of concern. From the perspective of fusion energy production, one of the most important problems is *can dust be controlled in ways so that it will not become a severe problem for magnetic fusion reactors*? Dust dynamics, small size, and a large quantities of dust make the control and removal of dust a very challenging problem. No proven solution exists at this time.

Dust grains are formed due to plasma-wall interactions in magnetic fusion devices. Physical sputtering, chemical sputtering, and heating due to plasma disruption or edge localized modes (ELMs) are the primary dust production mechanisms. Dust may be produced during the intensive heating of the wall by disruption of ELMs. Dust may also be produced from vapors of atoms and small molecules close to the wall through redeposition, condensation, and coagulation. Redeposition happens on the wall. A nucleation step is necessary for condensation and coagulation to happen away from the wall. Frameworks for nucleation, condensation and coagulation have been established in understanding of aerosol and colloids. Reexamination of these processes can be helpful and still necessary in understanding dust formation in fusion, which may shed light on reducing dust production in the future.

Diagnosis and management of dust particles in tokamaks is a "housekeeping" area that falls in-between traditional plasma diagnostics and plasma technology fields. However, it is critical to the safe operation of next step devices. Particles have been extensively collected from current tokamaks during maintenance periods and analyzed. Because of the concerns with safety, radioactivity from tritium or activated metals, toxicity, and/or chemical reactivity with steam or air, the in-vessel particle inventory must be measured and controlled to assure compliance with regulatory limits. However measuring the dust inventory is a challenge in existing machines let alone one with the radiological environment and scale of ITER.

Proposed methods for dust inventory measurement in ITER fall into three classes: (i) local in-vessel measurement of dust on surfaces and extrapolation to the total inventory. Here the difficulties are hard to access surfaces (e.g. gaps between castellations in hot plasma facing components) and the reliability of the extrapolation. (ii) Measurement of the total erosion in the vessel as an upper bound on the maximum amount of dust. This is challenging for the rather low limit of 6 kg for dust on hot surfaces. (iii) Simulation of an accident situation by the injection of small amounts of inert gas to directly measure mobilizable dust. In July 2007 a Design Change Request DCR-106 'In Vacuum Vessel Dust Control System' was formulated and an ITER task force formed to address these issues. It is highly desirable to have a mockup facility to examine various dust technologies (diagnosis, cleaning, etc.) experimentally.

Basic properties of dust in plasmas, such as charging, drag forces, heating have been understood or can be understood using the existing framework, largely due to the extensive research and significant advances in dusty plasmas, which is less dense and colder compared with plasmas for magnetic fusion. Except for plasmas with Maxwellian or drift-Maxwellian particle distributions, however, most of the dust properties have to be calculated numerically. We have described an approach to the direct simulation of the interaction of dust particles with the ions and electrons comprising the plasma. The approach is based on using computational particles to represent not only the plasma species but also the dust particle themselves. While the plasma species are treated with the usual PIC approach, the dust particles are treated with a special type of computational particles that interact with the fields and the plasma particles via the immersed boundary method. Further improvements have been demonstrated by using grid adaptation and particle population control. Using small grid cells near the dust particles, the rapid field variation around the dust can be captured accurately and by increasing the population of computational plasma particles near the dust, the accurate statistics can be represented.

Understanding of dust dynamics and transport in magnetic fusion plasmas has made significant progress. For example, plasma drag force has been identified both theoretically and experimentally to play a very important role in dust dynamics. Rather sophisticated 3D codes now exist and are used to track dust in tokamak plasma. These codes can provide both motion of individual grains and statistics of an ensemble of dust particles. Good qualitative agreements were found between theoretical prediction, results of numerical modeling, and experimental observations of the magnitude and directions of dust motion in tokamak divertors. Benchmarking of numerical codes by experimental observations will be possible in the near future. The work for full coupling of dust and plasma transport codes (e.g. DUSTT and UEDGE) is in progress. The main focus for further development of theory and modeling of dust dynamics and transport seems to be on dust-wall interactions, dust generation mechanisms, coupling of dust and plasma transport codes, and on more detail comparison of numerical results with experimental data.

Although we have mostly discussed dust as problematic or a potential threat to magnetic fusion devices, however, dust introduced in small and controlled quantities and imaged in real-time at high resolution show great potential as local plasma diagnostics which can complement or confirm spectroscopic or electrical measurements. Micron size dust particles are easy to procure and handle. However, in order to be a suitable diagnostic tool for the hot plasma core, the dust grains need to be injected at high speeds into a fusion plasma. Various methods are available for dust acceleration. It has been demonstrated that acceleration by plasma jet is at least one order of magnitude more effective than by compressed gas, and can results in speeds of a few km/s. This technique has several advantages: it can use any type of gas and can simultaneously work with hundreds of dust grains with sizes from 1 to a few tens of microns.

Finally, we would like to point out that most of the measurements, experiments, understanding and analysis of dust in magnetic fusion are based on understanding and properties of 'conventional wall materials', such as carbon, beryllium and tungsten. Compared with the relatively less frequent discussion of liquid lithium as the first wall material for fusion, 'advanced wall materials' with engineered nano-structures are discussed even less frequently. However, some preliminary results have shown the great promise of such unfamiliar types of materials [45]. Indeed, eventual solution of the dust problem or 'elimination' of dust altogether from fusion device may come from advances in nanomaterials research, which may require us to reassess the plasma-wall interaction physics for nanomaterials.

References

1. G. Federici, C. H. Skinner, J. N. Brooks, *et al.*, *Nucl. Fusion* **41**, 1967 (2001).
2. G. Janeschitz, *et al.*, *J. Nucl. Mater.* **290-293**, 1 (2001).
3. U. Samm in R. E. H. Clark and D. H. Reiter (eds.) *Nuclear Fusion Research*, Chap. 1, Springer, Berlin (2004).
4. S. J. Piet, A. Costley, G. Federici, F. Heckendorn, and R. Little, *Proc. 17th IEEE-NPSS Symp. Fusion Eng.* 167 (1998).
5. J. P. Sharpe, D. A. Petti, and H. -W. Bartels, *Fusion Eng. Design* **63-64**, 153 (2002).
6. J. Winter, Plasma Phys. Control. Fusion **46**, B583 (2004).
7. Radiological source terms, in: ITER Technical Basis, ITER EDA Documentation Series No. 24, IAEA, Vienna (2002).
8. C. V. Parker, C. H. Skinner, A. L. Roquemore, *J. Nucl. Mat.* **363-365**, 1461 (2007).
9. A. Loarte, B. Lipschultz, A. S. Kukushkin, *et al.* Nucl. Fusion **47**, S203 (2007).
10. B. Lipschultz, B. Labombard, E. Marmar, M. Pickrell, J. Terry, R. Watterson, S. Wolfe, *Nucl. Fusion* **24**, 977 (1984).
11. J. Wesson, *Tokamaks*, third ed. (Clarendon Press, Oxford, 2004).
12. R. J. Trumpler, *Publ. Astron. Soc. Pac.* **42**, 214 (1930).
13. B. T. Draine, in D. Pfenniger (ed.) *The Cold Universe:Saas-Fee Advanced Course 32*, (Springer, 2003).
14. P. Goldreich and W. R. Ward, *Astrohphys. J.* **183**, 1051 (1973).
15. P. K. Shukla and A. A. Mamun, *Introduction to Dusty Plasmas* (IoP Pub-

lishing, 2002).
16. T. G. Northrop, *Phys. Scr.* **45**, 475 (1992).
17. R. L. Merlino and J. A. Goree, *Phys. Today* **57**, 32 (2004).
18. J. H. Chu and Lin I, *Phys. Rev. Lett.* **72**, 4009 (1994).
19. H. Thomas, G. E. Morfill, V. Demmel, J. Goree, B. Feuerbacher, and D. Mohlmann, *Phys. Rev. Lett.* **73**, 652 (1994).
20. Y. Hayashi and K. Tachibana, *Jpn. J. Appl. Phys.* **33**, L804 (1994).
21. V. E. Fortov, A. G. Khrapak, V. I. Molotkov, O. F. Petrov, *Phys. Uspekhi* **47**, 447 (2004); Phys. Rep. **421**, 1 (2005).
22. A. Bouchoule (Ed.), *Dusty Plasmas, Physics, Chemistry and Technological Impacts in Plasma Processing*, Wiley & Sons, New York (1999).
23. O. Ishihara, *J. Phys. D. : Appl. Phys.* **40**, R121 (2007).
24. N. N. Rao, P. K. Shukla, and M. Y. Yu, *Planet. Space Sci.* **38**, 543 (1990).
25. A. Barkan, R. L. Merlino, and N. D'Angelo, *Phys. Plasma*, **2**, 3563 (1995).
26. J. B. Pieper and J. Goree, *Phys. Rev. Lett.* **77**, 3137 (1996).
27. G. E. Morfill, H. M. Thomas, U. Konopka, H. Rothermel, M. Zuzic, A. Ivlev and J. Goree, *Phys. Rev. Lett.* **83**, 1598 (1999).
28. O. Arp, D. Block, A. Piel, and A. Melzer, *Phys. Rev. Lett.* **93**, 165004-1, (2004).
29. M. S. Barnes, J. H. Keller, J. C. Forster, J. A. Oneill, D. K. Coultas, *Phys. Rev. Lett.* **68**, 313 (1992).
30. S. A. Khrapak, A. V. Ivlev, G. E. Morfill, and H. M. Thomas, *Phys. Rev. E* **66**, 046414 (2002).
31. I. H. Hutchinson, *Plasma Phys. Control. Fusion* **48**, 185 (2006).
32. G. Lapenta, *Phys. Rev. Lett.* **75**, 4409 (1995).
33. M. J. Barnes, I. P. Williams, and A. S. Asebiomo, *Mon. Not. R. Astron. Soc.* **130**,63 (1965).
34. B. T. Draine and E. E. Salpeter, *Astrophys. J.* **231**, 77 (1979).
35. S. I. Krasheninnikov, Y. Tomita, R. D. Smirnov, and R. K. Janev, *Phys. Plasma.* **11**, 3141 (2004).
36. Z. Wang, C. M. Ticos, L. A. Dorf, and G. A. Wurden, *IEEE Trans. Plasma Sci.* **34**, 242 (2006).
37. C. M. Ticos, Z. Wang, G. L. Delzanno, and G. M. Lapenta, *Phys. Plasma.* **13**, 103501 (2006).
38. A. Yu. Pigarov, S. I. Krasheninnikov, T. K. Solobeva, and T. D. Rognlien, *Phys. Plasma.* **12**, 122508 (2005).
39. S. I. Girshick, C. P. Chiu, and P. H. McMurry, *Plasma Chem. Plasma Proc.* **8**, 145 (1988).
40. A. C. Zettlemoyer (Ed.), *Nucleation*, Marcell Dekker, New York (1969).
41. I. M. MacKinnon, S. M. L. Hamblyn, and B. G. Reuben, Proc. 3rd Intern. Symp. Plasma Chem., Limoges, France (1977).
42. R. Behrisch, *Sputtering by Particle Bombardment*, Springer, Berlin, Vol. I, (1981) Vol. II (1983); and R. Behrisch and K. Whittmaack, Vol. III (1991).
43. J. Roth, in R. E. H. Clark and D. H. Reiter (eds.) *Nuclear Fusion Research*, Chap. 9, Springer, Berlin (2004).
44. W. Eckstein, C. Garcia-Rosales, J. Roth, W. Ottenberger, *Sputtering data*,

Max-Planck-Institute fur Plasmaphysik, Report IPP **9/82** (1993).
45. A. Misra, M. J. Demkowicz, X. Zhang, and R. G. Hoagland, *JOM* Sept. p.62 (2007).
46. J. Roth, C. Garcia-Rosales, *Nucl. Fusion* **36**, 1647 (1996), with Corrigendum, *Nucl. Fusion* **37**, 897 (1997).
47. F. P. Bundy, *Physica* A **156**, 169 (1989).
48. M. Volmer and A. Weber, *Z. Phys. Chem (Leipzig)* **119**, 277 (1926).
49. R. Becker and W. Doring, Ann. Phys. **24**, 719 (1935).
50. J. Zeldovich, J. Exp. Theor. Phys. **12**, 525 (1942).
51. L. D. Landau and E. M. Lifshitz, *Statistical Physics*, 3rd ed. (Reed Educational and Professional Publishing, 1980).
52. S. L. Girshick and C. P. Chiu, *Plasma Chem. Plasma Proc.* **9**, 355 (1989).
53. S. K. Girshick and C. Chiu, *J. Chem. Phys.* **93**, 1273 (1990).
54. S. K. Girshick, *J. Chem. Phys.* **94**, 826 (1991).
55. M. Volmer, *Kinetik der Phasenbildung* (Theodor Steinkopff, Dresden, 1939).
56. J. P. Hirth and G. M. Pound, *Condensation and Evaporation: Nucleation and Growth Kinetics* (Pergamon, New York, 1963).
57. T. Takagi, *Cluster beam deposition and epitaxy*, Parker Ridge, Noyes Publications, New York (1988).
58. G. Scoles, *Atomic and Molecular Beam Methods*, Vol. 1 and 2, Oxford University Press, New York, (1988).
59. H. Pauly, *Molecule and Cluster Beams I*, Springer, New York (2000).
60. I. Langmuir, G. Found and A. F. Dittmer, *Science*, **60**, 392 (1924).
61. G. S. Selwyn, J. Singh, R. S. Bennet, *J. Vac. Sci. Technol.* A **7**, 2758 (1989).
62. G. S. Selwyn, J. E. Heidenreich, K. L. Haller, *Appl. Phys. Lett.* **57**, 1876 (1990).
63. S. K. Friedlander, *Smoke, Dust and Haze* (Oxford University Press, New ork, 2000), 2nd ed.
64. see for example, R. Balescu, *Statistical dynamics* (Imperial College Press, London, 1997).
65. M. von Smoluchowski *Phys. Zeits* **17**, 557 (1916); ibid, 585 (1916).
66. S. Chandrasekhar, *Rev. Mod. Phys.* **15**, 1 (1943).
67. V. S. Safronov, *Evolution of the protoplanetary cloud and formation of the Earth and the planets*, (Nauk), NASA TT-F-677 (1972).
68. G. Lapenta, *Phys. Script.* **57**, 476 (1998).
69. A. V. Ivlev, G. E. Morfill, and U. Konopka, *Phys. Rev. Lett.* **89**, 195502 (2002).
70. L. Spitzer, Jr., *Physical Processes in the Interstellar Medium*, p. 209, John Wiley & Sons, New York (1978).
71. J. H. Oort and H. C. van der Hulst, *Bull. Astron. Inst. Neth.* **10**, 187 (1946).
72. J. Winter, *Phys. Plasma.* **7**, 3862 (2000).
73. S. V. Gorman, W. J. Carmack, P. B. Hembree, *Fus. Technol.* **34**, 745 (1998).
74. W. J. Carmack *et al. Fus. Eng. Des.* **51-52**, 477 (2000).
75. J. P. Sharpe *et al. J. Nucl. Mater.* **313-316**, 455 (2003).
76. W. J. Carmack *et al. Fus. Eng. Des.* **39-40**, 477 (1998).
77. A. T. Peacock *et al. J. Nucl. Mater.* **266-269**, 423 (1999).

78. N. Bekris *et al.* *J. Nucl. Mater.* **337-339**, 659 (2005).
79. J. P. Sharpe *et al.* *J. Nucl. Mater.* **337-339**, 1000 (2005).
80. J. P. Sharpe, W. J. Carmack and D. A. Petti *Fus. Technol.* **39**, 1066 (2001).
81. J. Winter., *Plasma Phys. Control Fusion* **40**, 1201 (1998).
82. J. Winter *et al.*, *J. Nucl. Mater.* **266-269**, 228 (1999).
83. M. Rubel *et al.*, *Nucl. Fus.* **41**, 1087 (2001).
84. W. J. Carmack *et al.*, INEL report. 'Analysis and characterization of TFTR Tokamak Dust' ITER/US/98/TE/SA-8 April 30, (1998).
85. R. A. Anderl and R. J. Pawelko 'BET Surface Area and Tritium Analyses for TFTR Carbon Dust Samples' Sandia Report SAND99-8222 February (1999).
86. C. H. Skinner, C. A. Gentile, M. M. Menon, and R. E. Barry, *Nucl. Fusion* **39**, 1081 (1999).
87. Ph. Chappuis *et al.*, *J. Nucl. Mater.* **290-293**, 245 (2001).
88. J. P. Sharpe *et al.*, *Fus. Tech.* **39**, 1061 (2001).
89. K. Sasaki *et al.*, *J. Nucl. Mater.* **363-365**, 238 (2007).
90. J. P. Sharpe, M. Bourham and J. G. Gilligan, *Fus. Tech.* **34**, 634 (1998).
91. J. P. Sharpe, B. J. Merrill, D. A. Petti, M. A. Bourhan, J. G. Gilligan, *J. Nucl. Mater.* **290-293**, 1128 (2001).
92. L. Khimchenko *et al.* 'Study of erosion products in experiments simulating ELMs and disruptions in ITER on plasma gun QSPA-facility', *Proc.21 Fusion Energy Conf.* Chengdu. China, EX4-5Ra (2006). http://www-pub.iaea.org/MTCD/Meetings/FEC2006/ex_4-5ra.pdf
93. A. Hassanein and I. Konkashbaev, *J. Nucl. Mater.* **313**, 664 (2003).
94. Y. S. Cheng, Y. Zhou, C. A. Gentile, C. H. Skinner, *Fus. Sci. & Tech.* **41**, 867 (2002).
95. J. Winter, V. E. Fortov and A. P. Nefedov, *J. Nucl. Mater.* **290-293**, 509 (2001).
96. C. H. Skinner, C. A. Gentile, L. Ciebiera and S. Langish, *Fus. Sci. & Tech.* **45**, 11 (2004).
97. R. A. Causey, W. L. Chrisman, J. Crawford, J. M. Gahl, *J. Nucl. Mater.* **210**, 333 (1994).
98. C. H. Skinner *et al.*, *J. Vac. Sci.* **A14**, 3267 (1996).
99. A. Nagy *et al.*, Proceedings of the 17th IEEE/NPSS Symposium on Fusion Engineering San Diego, October 6-10, 1997, p321, IEEE, Piscataway, NJ, USA (1998); Princeton University, Plasma Physics Laboratory Report # PPPL-3273 (Nov. 1997).
100. H. Yoshida *et al.*, *J. Nucl. Mater.* **337-339**, 604 (2005).
101. W. J. Carmack *et al.*, *Fus. Tech.* **34**, 604 (1998).
102. K. Niwase *et. al*, *J. Nucl. Mater.* **179-181**, 214 (1991).
103. C. Arnas *et. al*, *J. Nucl. Mater.* **337-339**, 69 (2006).
104. P. Roubin *et. al*, *J. Nucl. Mater.* **337-339**, 990 (2005).
105. Y. Raitses, C. H. Skinner, F. Jiang and T. S. Duffy, 'Raman Spectroscopy of Carbon Dust Samples from NSTX,' submitted to Plasma Physics Reports (2007).
106. C. H. Skinner and G. Federici, *Phys. Scr.* **T 124**, 18 (2006).
107. ITER Generic Site Safety Report v.3 (1994) p. III-54 G 84 RI 3 01-07-13

R1.0.
108. R. J. Maqueda and G. A. Wurden, *Nucl. Fus.* **39**, 629 (1999).
109. A. L. Roquemore, N. Nishino, C. H. Skinner, C. Bush, R. Kaita, R. Maqueda, W. Davis, A. Yu Pigarov, and S. I. Krasheninnikov, *J. Nucl. Mater.* **363-365** 222 (2007).
110. K. Narihara et al., *Nucl. Fusion* **37**, 1177 (1997).
111. W. P. West, B. D. Bray, and J. Burkat, *Plasma Phys. and Control. Fusion* **48**, 1661 (2006).
112. E. Giovannozzi et al., Proc. 33rd EPS Conf. on Plasma Phys. Rome, June 2006 ECA Vol. 30I, P-2.093 (2006).
113. B. D. Bray and W. P. West, *J. Nucl. Mater.* **363-365**,107 (2007).
114. B. D. Bray and W. P. West, Proc. IAEA Conf. on Fusion Energy, Chengdu, China, 2006; *J. Nucl. Mater.* **363-365**, 107 (2007).
115. R. D. Smirnov et al., "Laser-dust interaction and dust size distribution measurements on DIII-D," accepted for publication, *Phys. Plasmas* (2007).
116. R. Reichle *et al.*, *J. Nucl. Mater.* **290-293**, 701 (2001).
117. G. T. Razdobarin *et al.*, *Fus. Sci. & Technol.* **41**, 32 (2002).
118. G. F. Counsell and C. H. Wu, *Phys. Scr.* **T91**, 70 (2001).
119. G. F. Counsell *et al.*, *Rev. Sci. Instrum.* **77** 093501 (2006).
120. A. Bader, C. H. Skinner, A. L. Roquemore, and S. Langish, *Rev. Sci. Instrum.* **75**, 370 (2004).
121. C. H. Skinner, A. L. Roquemore, A. Bader and W. R. Wampler, *Rev. Sci. Instrum.* **75**, 4213 (2004).
122. C. Voinier, C. H. Skinner and A. L. Roquemore, *J. Nucl. Mater.* **346**, 266 (2005).
123. C. H. Skinner, R. Hensley and A. L. Roquemore, submitted to Plasma Physics Reports (2007).
124. B. L. Ling, X. D. Zhang, A. Ti and X. Gao, *J. Nucl. Mater.* **363-365**, 1446 (2007).
125. Y. Oda *et al.*, *J. Fus. Energy* **16**, 231 (1997).
126. J. E. Allen, R. L. F. Boyd, P. Reynolds, *Proc. Phys. Soc. London* **70**, 297 (1957).
127. R. V. Kennedy, J. E. Allen, *J. Plasma Phys.* **67**, 342 (2002).
128. J. E. Allen, *Phys. Scr.* **45**, 497 (1992).
129. I. B. Bernstein, I. N. Rabinowitz, *Phys. Fluids* **2**, 112 (1959).
130. Ya. L. Al'pert, A. V. Gurevich, L. P. Pitaevskii, *Space Physics with Artificial Satellites*, Plenum Press, New York (1965).
131. J. G. Laframboise, *The theory of spherical and cylindrical probes in a collisionless, Maxwellian plasma at rest.* Institute for Aerospace Studies, University of Toronto, Report No. 100, 1966.
132. R. V. Kennedy, J. E. Allen, *J. Plasma Phys.* **69**, 485 (2003).
133. H. M. Mott-Smith, I. Langmuir, *Phys. Rev.* **28**, 727 (1926).
134. E. C. Whipple, *Rep. Prog. Phys.* **44**, 1197 (1981).
135. J. E. Allen, B. M. Annaratone, U. De Angelis, *J. Plasma Phys.* **63**, 299 (2000).
136. M. Lampe, *J. Plasma Phys.* **65**, 171 (2001).

137. G. Sorasio, D. A. Mendis, M. Rosenberg, *Planet. Space Sci.* **49**, 1257 (2001).
138. D. E. Osterbrock, G. J. Ferland, *Astrophysics of Gaseous Nebulae and Active Galactic Nuclei*, University Science Books, Sausalito, California (2006).
139. G. L. Delzanno, G. Lapenta, M. Rosenberg, *Phys. Rev. Lett.* **92**, 035002 (2004).
140. G. L. Delzanno, A. Bruno, G. Sorasio, G. Lapenta, *Phys. Plasmas* **12**, 62102 (2005).
141. G. L. Delzanno, G. Lapenta, *Phys. Rev. Lett.* **94**, 175005 (2005).
142. M. S. Sodha, *Br. J. Appl. Phys.* **14**, 172 (1963).
143. M. S. Sodha, S. Guha, *Adv. Plasma Phys.* **4**, 219 (1971).
144. N. Meyer-Vernet, *Astron. Astrophys.* **105**, 98 (1982).
145. H. Bruining, *Physics and Applications of Secondary Electron Emission* (Pergamon, London, 1954).
146. V. W. Chow, D. A. Mendis, M. Rosenberg, *J. Geophys. Res.* **98**, 19065 (1993).
147. C. K. Goertz, *Rev. Geophys.* **27**, 271 (1989).
148. Z. Wang, G. A. Wurden, *Rev. Sci. Instrum.* **74**, 1887 (2003).
149. Z. Wang, G. A. Wurden, *Rev. Sci. Instrum.* **75**, 3436 (2004).
150. T. Nitter, *Plasma Sources Sci. Technol.* **5**, 93 (1996).
151. P. S. Epstein, *Phys. Rev.* **23**, 710 (1924).
152. J. A. Bittencourt, *Fundamentals of Plasma Physics* (Oxford, Pergamon, 1986).
68, 313 (1992).
153. M. D. Kilgore, J. E. Daugherty, R. K. Porteous, D. B. Graves, *J. Appl. Phys.* **73**, 7195 (1993).
154. S. A. Khrapak, G. E. Morfill, *Phys. Rev. E* **69**, 066411 (2004).
155. M. Harwitt, *Astrophysical Concepts* (Springer, New York, 1988), p. 405.
156. L. Spitzer, *Physical Processes in the Interstellar Medium* (Wiley, New York,1978), p. 182.
157. J. Mahmoodi *et al.*, *Phys. Rev. Lett.* **84**, 2626 (2000); D. D. Tskhakaya *et al.*, JETP **93**, 301 (2001).
158. A. Ferrara and R. J. Dettmar, *Astrophys. J.* **427**, 155 (1994).
159. D. T. Draine and A. Lazarian, *Astrophys. J.* **494**, L19 (1998).
160. J.Marchant, *New Scientist* **160**, 11 (2001).
161. N. Sato, in *New Vistas in Dusty Plasmas*, eds. L. Boufendi *et al.* (AIP Conf. Proceedings No. 799, AIP, New York, 2005), p. 97.
162. S. I. Krasheninnikov *et al.*, *Phys. Plasma.* **11**, 3141 (2004)
163. S. I. Krasheninnikov, V. I. Shevchenko, and P. K. Shukla, *Phys. Lett. A* **361**, 133 (2007).
164. N. Sato, *et al.*, *Phys. Plasma.* **8**, 1786 (2001).
165. J. D. Jackson, *Classical Electrodynamics*, (Wiley, New York, 1998).
166. C.K. Birdsall and A.B. Langdon, *Plasma Physics Via Computer Simulation*, (Taylor & Francis, 2004).
167. R.W. Hockney and J.W. Eastwood, *Computer simulation using particles*, (Taylor & Francis, 1988).
168. FC Hoppensteadt and CS Peskin, *Modeling and simulation in medicine and*

the life sciences, (New York: Springer-Verlag, 2002).
169. G. Lapenta, F. Iinoya and J. U. Brackbill, *IEEE Trans. Plasma Sci.* **23**, 769 (1995).
170. G. Lapenta, *Phys. Plasma.* **6**, 1442 (1999).
171. C. De Boor, *A practical guide to splines*, (Springer, 1978).
172. D. Sulsky and J. U. Brackbill, *J. Computat. Phys.* **96**, 339 (1991).
173. M.J. Berger and J. Oliger, *J. Computat. Phys.* **53**, 484 (1984).
174. J.U. Brackbill, *J. Computat. Phys.* **108**, 38 (1993).
175. G. Lapenta, *J. Computat. Phys.* **193**, 167 (2004).
176. C. J. Budd and J. F. Williams, *J. Phys. A: Math. Gen.* **39**, 5425 (2006).
177. G. L. Delzanno, L. Chacón, J. M. Finn, Y. Chung, G. Lapenta, *An optimal robust equidistribution method for two-dimensional grid generation based on Monge-Kantorovich optimization*, in preparation (2007).
178. G. Lapenta, *J. Computat. Phys.* **181**, 317 (2002).
179. H. X. Vu and J. U. Brackbill, *Comput. Phys. Comm.* **69**, 253 (1992).
180. G. Lapenta and J. U. Brackbill, *J. Computat. Phys.* **115**, 213 (1994).
181. G. Lapenta and J.U. Brackbill, *Comput. Phys. Comm.* **87**, 139 (1995).
182. D. J. Cooperberg and V. Vahedi and C. K. Birdsall, *15th International Conference on the Numerical Simulation of Plasmas*, Valley Forge (1994).
183. S.J.Choi and M. J. Kushner, *IEEE Transactions on Plasma Science*, **22**, 138 (1994).
184. J. W. Manweiler, T. P. Armstrong and T. E. Cravens, *J. Plasma. Phys.*, **63**, 269 (2000).
185. I. H. Hutchinson, *Plasma Phys. Control. Fusion* **44**, 1953 (2002).
186. L. Patacchini and I. H. Hutchinson *Plasma Phys. Control. Fusion* **49**, 1719 (2007).
187. L. Patacchini and I. H. Hutchinson and G. Lapenta, *Phys. Plasma.* **14**, 062111 (2007).
188. B. Rovagnati, M. Davoudabadi, G. Lapenta and F. Mashayek, *J. Appl. Phys.* **102**, 073302 (2007).
189. R. D. Smirnov, A. Yu. Pigarov, M. Rosenberg, S. I. Krasheninnikov, and D. A. Mendis, *Plasma Phys. Control. Fusion* **49** 347 (2007); A. Yu. Pigarov, R. D. Smirnov, S. I. Krasheninnikov, T. D. Rognlien, M. Rosenberg, and T. K. Soboleva, *J. Nucl. Mater.* **363-365** 216 (2007).
190. L. D. Landau, E. M. Lifshitz, and L. P. Pitaevskii, *Electrodynamics of Continuous Media* (Butterworth-Heinemann, Oxford, 1996), p. 2; N. L. Tsintsadze *et al. ibid.* **14**, 073703 (2007).
191. D. L. Rudakov *et al.*, *J. Nucl. Mat.* **363-365**, 227 (2007).
192. C. Castaldo *et al.*, Nucl. Fusion **47**, L5 (2007).
193. S. I. Krasheninnikov and T. K. Soboleva, *Plasma Phys. Control. Fusion* **47** A339 (2005).
194. S. I. Krasheninnikov, T. K. Soboleva, Y. Tomita, R. D. Smirnov, and R. K. Janev, *J. Nucl. Mat.* **337-339** 65 (2005).
195. J. D. Martin, M. Coppins, G. F. Counsell, *J. Nucl. Mat.* **337-339** 114 (2005).
196. G. F. Counsell and F. Lott, Private Communication, (2004).
197. J. D. Martin, M. Coppins, G. F. Counsell, "Modelling dust grain trajectories

in MAST and ITER using plasma backgrounds from B2SOLPS5.0", 33rd EPS Conference on Plasma Phys. Rome, 19 - 23 June 2006, ECA Vol. 30I, O-4.010 (2006).

198. T. D. Rognlien, J. L. Milovich, M. E. Rensink, *et al.*, *J. Nucl. Mater.* **196-198** 34 (1992).

199. R. Schneider, X. Bonnin, K. Borrass, D. P. Coster, H. Kastelewicz, D. Reiter, V. A. Rozhansky, and B. J. Braams, *Contrib. Plasma Phys.* **46**, 3 (2006).

200. I. H. Hutchinson, *Plasma Phys. Control. Fusion* **45**, 1477 (2003).

201. S. I. Krasheninnikov, R. D. Smirnov, D. J. Benson, A. Yu. Pigarov, T. K. Soboleva, T. D. Rognlien, M. Rosenberg, V. I. Shevchenko, P. K. Shukla, and D. A. Mendis, "Modeling of dust-particle dynamics, transport, and impact on tokamak plasma performance", 21st IAEA Fusion Energy Conference, Chengdu, China, October 16-21, 2006 Paper TH/P6-18.

202. Y. Tanaka, A. Yu. Pigarov, R. D. Smirnov, S. I. Krasheninnikov, N. Ohno, and Y. Uesugi, *Phys. Plasma.* **14**, 052504 (2007).

203. S. A. Khrapak, A. V. Ivlev, G. E. Morfill, S. K. Zhdanov, *Phys. Rev. Lett.* **90**, 225002 (2003).

204. Z. Wang, C.M. Ticos, G. A. Wurden, *Phys. Plasma.* **14**, 103701 (2007).

205. C. M. Ticos, Z. Wang, G. A. Wurden, J. L. Kline, D.S. Montgomery, L. A. Dorf, and P. K. Shukla, submitted to PRL (2007).

206. Z. Wang, P. D. Beinke, C. W. Barnes, M. W. Martin, E. Mignardot, G. A. Wurden, S. C. Hsu, T. P. Intrator, and C. P. Munson, *Rev. Sci. Instrum* **76**, 033501 (2005).

207. C. M. Ticos, Z. Wang, L. A. Dorf, and G. A. Wurden, Rev. Sci. Instrum. **77**, 10E304 (2006).

SHORT WAVELENGTH BALLOONING MODE IN TOKAMAKS

A. HIROSE and N. JOINER

Plasma Physics Laboratory, University of Saskatchewan
Saskatoon, Saskatchewan, S7N 5E2, Canada

A skin size ($k_\perp \lesssim \omega_{pe}/c$) plasma mode characterized by a dispersion relation $\omega \simeq ck_\perp k_\parallel / k_{De}$ (k_{De} the electron Debye wavenumber), adiabatic ions, and $\omega \ll k_\parallel v_{Te}$, in a uniform plasma is destabilized in the tokamak geometry by a modest electron temperature gradient η_e and ballooning parameter α_e. When unstable, a large electron thermal diffusivity emerges because of the cross-filed wavelength much longer than that of the conventional electron temperature gradient (ETG) mode.

1. Introduction

The role played by plasma turbulence having cross-field wavelengths of the order of the electron skin depth c/ω_{pe} in anomalous transport has long been noticed in the past. Mikhailovskii [1] made a detailed study of the so-called universal drift mode due to the intrinsic electron Landau damping and found that the maximum growth rate

$$\gamma_{\max} \simeq \frac{c_s}{4\sqrt{\pi}L_n},$$

is reached in the wavenumber regime

$$k_\perp \gtrsim \frac{\omega_{pe}}{c},$$

where $c_s = \sqrt{T_e/M_i}$ is the ion acoustic speed, L_n the density gradient scale length. Simple mixing length estimate for particle diffusivity yields

$$D \simeq \frac{c_s}{4\sqrt{\pi}L_n}\left(\frac{c}{\omega_{pe}}\right)^2. \tag{1}$$

As is well known, the universal drift mode in slab geometry with magnetic shear is absolutely stable [2]. Subsequent studies revealed that in tokamaks, the toroidicity induced drift mode may be unstable [3] and the diffusivity may prevail with some toroidicity corrections. Ohkawa [4] proposed an electron thermal diffusivity in tokamaks,

$$\chi_e = \frac{v_{Te}}{qR}\left(\frac{c}{\omega_{pe}}\right)^2, \tag{2}$$

where v_{Te}/qR is the electron transit frequency with v_{Te} the electron thermal velocity and qR the connection length of the helical magnetic structure. In the formula, the skin depth plays the role of decorrelation length in the radial direction and the transit frequency plays the role of decorrelation rate. It is noted that the derivation of the Ohkawa diffusivity was heuristic. The presence of skin size instability (or turbulence) was assumed but the origin of the turbulence was left unexplained. One feature of the thermal diffusivity which has relevance to experimental observation is that it gives a somewhat natural explanation for the empirical Alcator type scaling law for the energy confinement time [5],

$$\tau_E \propto n, \tag{3}$$

where n is the plasma density. More recently, it is becoming clear that the lower edge of the unstable $k_\perp$ spectrum of the electromagnetic electron temperature gradient (ETG) mode is in the regime $k_\perp \simeq \omega_{pe}/c$ and the presence of skin size plasma turbulence appears to be responsible for the anomalous electron thermal diffusivity.

In the past, electromagnetic dispersion relations involving the skin depth c/ω_{pe} have been considered largely in the limit of $\omega > k_\parallel v_{Te}$ (electron transit frequency). For example, in the slab geometry, the well known dispersion relation for the electromagnetic drift mode

$$\omega = \frac{(1+\eta_e)\,\omega_{*e}}{1+(ck_\perp/\omega_{pe})^2}, \tag{4}$$

emerges in the limit $\omega > k_\parallel v_{Te}$. Here $\eta_e = d\,(\ln T_e)\,/d\,(\ln n_0)$ is the electron temperature gradient relative to the density gradient. In tokamaks, this is modified as

$$\frac{\omega - \omega_{*e}\,(1+\eta_e)}{\omega - \omega_{De}} - \frac{\eta_e \omega_{*e}\omega_{De}}{(\omega - \omega_{De})^2} + \left(\frac{ck_\perp}{\omega_{pe}}\right)^2 = 0, \;\; \omega > k_\parallel v_{Te}, \tag{5}$$

but this dispersion relation predicts stable modes unless $\eta_e < 0$ (electron temperature profile opposite to that of the density). Here, ω_{De} is the electron magnetic drift frequency due to the magnetic curvature. In the same limit $\omega > k_\parallel v_{Te}$, the dispersion relation of the predominantly electrostatic

ETG mode is subject to a small electromagnetic correction involving the electron skin depth [6 Kim-Horton]

$$\tau+\frac{\omega_{*e}}{\omega}+\left((k_\perp\rho_e)^2-\frac{\omega_{De}}{\omega}\right)\left(1-\frac{\omega_{*ep}}{\omega}\right)=\frac{\left(k_\parallel v_{Te}\right)^2}{\omega^2}\left(1-\frac{\omega_{*ep}}{\omega}\right)\frac{k_\perp^2}{k_\perp^2+\left(\omega_{pe}/c\right)^2} \tag{6}$$

To circumvent this problem, it has been suggested [7] that the nonlinear Doppler shift $\mathbf{k}\cdot\mathbf{v}_E$ due to the $E\times B$ drift $\mathbf{v}_E$ may exceed the electron transit frequency in strong turbulence, and skin size electromagnetic turbulence may exist *nonlinearly* being driven by shorter wavelength electrostatic ETG mode.

Finally, in a recent fully kinetic analysis of electromagnetic ETG mode in tokamaks [8], the following electron thermal diffusivity,

$$\chi_e=\frac{qv_{Te}}{\sqrt{RL_{Te}}}\left(\frac{c}{\omega_{pe}}\right)^2\sqrt{\beta_e}, \tag{7}$$

has been found where q is the safety factor, β_e is the electron beta factor and L_{Te} is the electron temperature gradient scale length. The ratio between Eq. (2) and (6) is

$$q^2\sqrt{\frac{R}{L_{Te}}}\sqrt{\beta_e},$$

which is of order unity in tokamaks.

In this lecture, stability of toroidal drift modes in the wavelength regime $k_\perp \simeq \omega_{pe}/c$ is reviewed. In particular, it is shown that tokamak discharges are linearly unstable against an electromagnetic mode (ballooning mode) characterized by adiabatic ions $(k_\perp\rho_i)^2 \gg 1$ (ρ_i the ion Larmor radius) and adiabatic electrons with a phase velocity parallel to the magnetic field smaller than the electron thermal speed $\omega \lesssim k_\parallel v_{Te}$. In these limits, a simple hydrodynamic ballooning mode emerges which is symbolically described by the following dispersion relation,

$$\left(\omega-\omega_{*e}\right)\left(\omega-\omega_{De}\right)+\eta_e\omega_{*e}\omega_{De}-\frac{\left(\omega-\omega_{*e}\right)^2}{1+\tau}=\frac{c^2k_\parallel k_\perp^2 k_\parallel}{k_{De}^2}, \tag{8}$$

where $\tau = T_e/T_i$, $k_{De} = \omega_{pe}/v_{Te}$ is the electron Debye wavenumber, and $k_\perp(k_\parallel)$ is the wavenumber perpendicular (parallel) to the ambient magnetic field. The mode is destabilized by a modest electron temperature gradient η_e and electron ballooning parameter defined by

$$\alpha_e=q^2\frac{R}{L_n}\left(1+\eta_e\right)\beta_e. \tag{9}$$

The mode is intrinsically electromagnetic (because it is a ballooning mode, although there is no resemblance to the ideal MHD ballooning mode), and is not a result of correction to electrostatic modes such as the familiar electron temperature gradient (ETG) mode. The right hand side of Eq. (8) may be approximated by

$$\frac{c^2 k_\parallel k_\perp^2 k_\parallel}{k_{De}^2} \simeq \left(\frac{ck_\theta}{\omega_{pe}}\right)^2 v_{Te}^2 k_\parallel^2, \tag{10}$$

which indicates the electron transit mode and skin depth are intimately related. It is noted that the Debye screening factor $(k_\perp/k_{De})^2 \ll 1$ manifests itself in the dispersion relation even though charge neutrality holds.

2. Review of the Kinetic Ballooning Mode (KBM)

Since the electron ballooning mode to be discussed in this lecture is closely related to the kinetic ballooning mode that has been revealed in [9], we briefly review this mode first. In the analysis, ideal MHD approximation is avoided and the ion density perturbation is evaluated numerically according to the gyro-kinetic formula,

$$\begin{aligned} n_i &= -\frac{e\phi}{T_i} n_0 + \int \frac{\omega + \hat{\omega}_{*i}\left(v^2\right)}{\omega + \hat{\omega}_{Di}\left(v\right)} J_0^2\left(\Lambda_i\right) f_{Mi} d\mathbf{v} \frac{e\phi}{T_i} n_0 \\ &= -\frac{e\phi}{T_i}\left(1 - I_i\right) n_0, \end{aligned} \tag{11}$$

where

$$I_i = \int \frac{\omega + \hat{\omega}_{*i}\left(v^2\right)}{\omega + \hat{\omega}_{Di}\left(v\right)} J_0^2\left(\Lambda_i\right) f_{Mi} d\mathbf{v},$$

is the non-adiabatic part of the ion density perturbation with

$$\hat{\omega}_{*i}\left(v^2\right) = \omega_{*i}\left[1 + \eta_i\left(\frac{v^2}{v_{Ti}^2} - \frac{3}{2}\right)\right], \quad \omega_{*i} = \frac{cT_i}{eB^2}\left(\nabla \ln n_0 \times \mathbf{B}\right)\cdot\mathbf{k}, \tag{12}$$

$$\hat{\omega}_{Di}\left(\mathbf{v}\right) = \frac{Mc}{eB^3}\left(\frac{1}{2}v_\perp^2 + v_\parallel^2\right)\left(\nabla B \times \mathbf{B}\right)\cdot\mathbf{k}, \tag{13}$$

being the energy dependent ion diamagnetic drift frequency and velocity dependent ion magnetic drift frequency, respectively. Eq. (11) is subject to $\omega \gg k_\parallel v_{Ti}$ as appropriate to the ballooning mode which is essentially a destabilized Alfven mode.

For electrons, we assume $\omega \ll k_{\|} v_{Te}$. In this limit, perturbed electron density and parallel current are

$$\begin{aligned} n_e &= \int f_e d\mathbf{v} \\ &= \int \left(\phi - \frac{\omega - \hat{\omega}_{*e}\left(v^2\right)}{\omega - \hat{\omega}_{De}\left(\mathbf{v}\right) - k_{\|} v_{\|}} \left(\phi - \frac{v_{\|}}{c} A_{\|} \right) \right) f_{Me} d\mathbf{v} \frac{e}{T_e} n_0 \\ &\simeq \left(\phi - \frac{\omega - \omega_{*e}}{ck_{\|}} A_{\|} \right) \frac{e}{T_e} n_0, \end{aligned} \tag{14}$$

and

$$\begin{aligned} J_{\|e} &= -e \int v_{\|} \left[\phi - \frac{\omega - \hat{\omega}_{*e}\left(v^2\right)}{\omega - \hat{\omega}_{De}\left(\mathbf{v}\right) - k_{\|} v_{\|}} \left(\phi - \frac{v_{\|}}{c} A_{\|} \right) \right] f_{Me} d\mathbf{v} \frac{e}{T_e} n_0 \\ &\simeq \frac{n_0 e^2}{k_{\|} T_e} \left[\left(\omega_{*e} - \omega\right) \phi + \frac{\left(\omega - \omega_{*e}\right)\left(\omega - \omega_{De}\right) + \eta_e \omega_{*e} \omega_{De}}{ck_{\|}} A_{\|} \right]. \end{aligned} \tag{15}$$

Then from charge neutrality condition $n_i = n_e$ and parallel Ampere's law

$$\nabla_{\perp}^2 A_{\|} = -\frac{4\pi}{c} J_{\|e}, \tag{16}$$

we obtain the following mode equation,

$$c^2 \frac{k_{\|} k_{\perp}^2 k_{\|}}{k_{De}^2} = -\frac{\left(\omega - \omega_{*e}\right)^2}{1 + \tau\left(1 - I_i\right)} + \left(\omega - \omega_{*e}\right)\left(\omega - \omega_{De}\right) + \eta_e \omega_{*e} \omega_{De}, \tag{17}$$

where

$$k_{\|} = -\frac{i}{qR} \frac{\partial}{\partial \theta}, \tag{18}$$

$$k_{\perp}^2 = k_{\theta}^2 \left[1 + \left(s\theta - \alpha \sin\theta\right)^2 \right], \tag{19}$$

$$\omega_{Dj}\left(\theta\right) = \frac{cT_j}{eBR} k_{\theta} \left[\cos\theta + \sin\theta \left(s\theta - \alpha \sin\theta\right) \right]. \tag{20}$$

The mode equation has been analyzed in [9] and the existence of the kinetic ballooning mode outside the stability boundary of the ideal ballooning mode demonstrated, including the regimes of the so-called second stability and negative shear.

In short wavelength limit $\left(k_{\perp} \rho_i\right)^2 \gg 1$, ions become adiabatic,

$$n_i = -\frac{e\phi}{T_i} n_0, \; I_i \to 0,$$

and the dispersion relation reduces to [8]

$$c^2 \frac{k_\| k_\perp^2 k_\|}{k_{De}^2} = -\frac{(\omega - \omega_{*e})^2}{1+\tau} + (\omega - \omega_{*e})(\omega - \omega_{De}) + \eta_e \omega_{*e} \omega_{De}. \quad (21)$$

Ions constitute neutralizing background in this case and instability, if any, is due to electron diamagnetic current coupled to the magnetic drift, *i.e.*, ballooning effect.

3. Local analysis

In a uniform plasma $\omega_{*e} = \omega_{De} = 0$, the dispersion relation

$$(\omega - \omega_{*e})(\omega - \omega_{De}) + \eta_e \omega_{*e} \omega_{De} - \frac{(\omega - \omega_{*e})^2}{1+\tau} = \frac{c^2 k_\| k_\perp^2 k_\|}{k_{De}^2}, \quad (22)$$

reduces to the known form [10],

$$\omega^2 = \frac{1+\tau}{\tau} \left(\frac{c k_\| k_\perp}{k_{De}} \right)^2 = \left(\frac{c k_\perp}{\omega_{pe}} \right)^2 k_\|^2 \frac{T_e + T_i}{m_e}. \quad (23)$$

The conditions of adiabatic ions $(k_\perp \rho_i)^2 \gg 1$ and adiabatic electrons $\omega < k_\| v_{Te}$ impose the range in the cross field wavenumber $k_\perp$ such that

$$\frac{1}{\rho_i} < k_\perp \lesssim \frac{\omega_{pe}}{c}, \quad (24)$$

where c/ω_{pe} is the collisionless electron skin depth. This is possible if the plasma β factor exceeds the electron/ion mass ratio, $\beta \gg m_e/m_i \simeq 3 \times 10^{-4}$, which is well satisfied in tokamaks. However, the dispersion relation in Eq. (22) pertinent to tokamaks is not subject to $ck_\perp < \omega_{pe}$. The electron acoustic mode $\omega = k_\| \sqrt{(T_e + T_i)/m_e}$ and the skin depth c/ω_{pe} are intimately related in this mode and the skin depth naturally appears as a characteristic scale length. The mode can be destabilized in toroidal geometry through the ballooning effect or through electron Landau damping when α_e is subcritical.

The quadratic dispersion relation in Eq. (22) may be solved if the norm of the parallel gradient $k_\|$ is specified. As a rough estimate, we assume $k_\| \simeq 1/(2qR)$. Then the root is given by

$$\frac{\omega}{\omega_{*e}} = 1 - \frac{1+\tau}{2\tau}(1 - 2\varepsilon_n) + i \frac{\gamma}{\omega_{*e}}, \quad (25)$$

where γ/ω_{*e} is the normalized growth rate

$$\frac{\gamma}{\omega_{*e}} = \frac{1+\tau}{2\tau} \sqrt{\frac{4\tau}{1+\tau} \left(2\varepsilon_n \eta_e - \frac{\varepsilon_n (1+\eta_e)}{2\alpha_e} \right) - (1 - 2\varepsilon_n)^2}. \quad (26)$$

The condition for instability is given

$$\alpha_e \left[8\varepsilon_n \frac{\tau\eta_e}{1+\tau} - (1-2\varepsilon_n)^2 \right] > 2\varepsilon_n \frac{\tau}{1+\tau} (1+\eta_e) . \tag{27}$$

The source of instability is in the interchange drive term $(1+\eta_e)\,\omega_{*e}\omega_{De}$ due to the combination of unfavorable magnetic curvature and electron pressure gradient. The mode described by Eq. (22) may thus be called an electron ballooning mode. When compared with the ideal MHD ballooning mode symbolically described by

$$\omega\,(\omega+\omega_{*i})\,(k_\perp\rho_i)^2 = (k_\perp\rho_i)^2\,(k_\parallel V_A)^2 - (1+\eta_e)\,\omega_{De}\omega_{*i} - (1+\eta_i)\,\omega_{Di}\omega_{*i}, \tag{28}$$

where V_A is the Alfven speed, the role of stabilizing Alfven frequency term $k_\perp\rho_i k_\parallel V_A$ in MHD ballooning mode is played by the modified electron transit frequency $(ck_\perp/\omega_{pe})\,k_\parallel v_{Te}$. As is well known, the growth rate of the ideal MHD ballooning mode is practically independent of the ion finite Larmor radius parameter $k_\perp\rho_i$ since $\omega_{De}\omega_{*i} \propto (k_\perp\rho_i)^2$, while the growth rate of the electron ballooning mode is proportional to $k_\perp$.

The condition for the instability given in Eq. (27) is for hydrodynamic ballooning mode and may be relaxed if kinetic effects (electron kinetic resonance) are implemented. Fully kinetic analysis has revealed that the instability persists even in electrostatic limit although the growth rate is small [8]. The compressive magnetic perturbation $B_\parallel$ has little influence on the instability [11].

In the mode described by Eq. (23) for a uniform plasma, energy equipartition holds between the magnetic energy and thermal potential energy. They are out of phase and the sum of the two energy forms is constant, consistent with the general constraint on energy relationship in plasma waves [12]. The magnetic energy density associated with the wave is

$$\begin{aligned} U_m &= \frac{1}{8\pi}k_\perp^2 A_\parallel^2 = \frac{1}{8\pi}\frac{(1+\tau)^2 c^2 k_\parallel^2 k_\perp^2}{\omega^2}\phi^2 \\ &= \frac{1}{8\pi}\tau\,(1+\tau)\,k_{De}^2\phi^2 = \frac{1}{8\pi}(1+\tau)\,k_{Di}^2\phi^2, \end{aligned} \tag{29}$$

while the potential energy density is

$$\begin{aligned} U_p &= \frac{1}{2}n_0 T_i \left(\frac{n_i}{n_0}\right)^2 + \frac{1}{2}n_0 T_e \left(\frac{n_e}{n_0}\right)^2 \\ &= \frac{1}{8\pi}k_{Di}^2\phi^2 + \frac{1}{2}\frac{n_0 e^2}{T_e}\left(\phi - \frac{\omega}{ck_\parallel}A_\parallel\right)^2 \\ &= \frac{1}{8\pi}(1+\tau)\,k_{Di}^2\phi^2, \end{aligned} \tag{30}$$

in agreement with the magnetic energy density. Here the charge neutrality relationship $(1+\tau)\, ck_\parallel \phi = \omega A_\parallel$ has been substituted. It is noted that the dispersion relation is independent of electron and ion masses and thus no kinetic energy is involved in the wave described by Eq. (22).

4. Nonlocal analysis

In order to confirm destabilization of the mode by the ballooning effect in a more rigorous manner, a fully kinetic, electromagnetic integral equation code [13] has been employed to find the mode frequency and growth rate. We consider a high temperature, low β tokamak discharge with eccentric circular magnetic surfaces. Trapped electrons are ignored for simplicity. Also, the magnetosonic perturbation ($\mathbf{A}_\perp$) is ignored in light of the low β assumption and we employ the two-potential (ϕ and $A_\parallel$) approximation to describe electromagnetic modes. As in the preceding section, the basic field equations are the charge neutrality condition (subject to $k^2 \ll k_{De}^2$)

$$n_i(\phi,\, A_\parallel) = n_e(\phi,\, A_\parallel), \tag{31}$$

and the parallel Ampere's law,

$$\nabla_\perp^2 A_\parallel = -\frac{4\pi}{c} J_\parallel(\phi,\, A_\parallel), \tag{32}$$

where the density perturbations are given in terms of the perturbed velocity distribution functions f_i and f_e by

$$n_i = \int f_i d\mathbf{v}, \quad n_e = \int f_e d\mathbf{v}, \tag{33}$$

and the parallel current by

$$J_\parallel = e \int v_\parallel (f_i - f_e) d\mathbf{v}. \tag{34}$$

The perturbed distribution functions f_i and f_e can be found from the gyrokinetic equation in the form

$$f_i = -\frac{e\phi}{T_i} f_{Mi} + g_i(v,\theta) J_0(\Lambda_i), \tag{35}$$

$$f_e = \frac{e\phi}{T_e} f_{Me} + g_e(v,\theta) J_0(\Lambda_e), \tag{36}$$

where $g_{i,e}$ are the nonadiabatic parts that satisfy

$$\left(i\frac{v_\parallel(\theta)}{qR}\frac{\partial}{\partial\theta} + \omega + \widehat{\omega}_{Di} \right) g_i = (\omega + \widehat{\omega}_{*i}) J_0(\Lambda_i) \left(\phi - \frac{v_\parallel}{c} A_\parallel \right) \frac{e}{T_i} f_{Mi}, \tag{37}$$

$$\left(i\frac{v_\parallel(\theta)}{qR}\frac{\partial}{\partial\theta}+\omega-\widehat{\omega}_{De}\right)g_e=-(\omega-\widehat{\omega}_{*e})J_0(\Lambda_e)\left(\phi-\frac{v_\parallel}{c}A_\parallel\right)\frac{e}{T_e}f_{Me}. \quad (38)$$

Here, θ is the extended poloidal angle in the ballooning space, J_0 is the Bessel function with argument $\Lambda_{i,e}=k_\perp v_\perp/\omega_{ci,e}$,

$$k_\perp^2=k_\theta^2\left[1+(s\theta-\alpha\sin\theta)^2\right],$$

$$\widehat{\omega}_{Dj}=2\varepsilon_n\omega_{*j}\left[\cos\theta+(s\theta-\alpha\sin\theta)\sin\theta\right]\left(\frac{1}{2}\hat{v}_\perp^2+\hat{v}_\parallel^2\right),$$

and qR is the connection length. For circulating particles, g_j ($j=i,e$) can be integrated as

$$v_\parallel>0,\qquad g_j^+=-i\frac{e_jf_{Mj}}{T_j}\int_{-\infty}^{\theta}d\theta'\frac{qR}{|v_\parallel|}e^{i\beta_j}(\omega-\widehat{\omega}_{*j})J_0(\Lambda_j')\left(\phi(\theta')-\frac{|v_\parallel|}{c}A_\parallel(\theta')\right), \quad (39)$$

$$v_\parallel<0,\qquad g_j^-=-i\frac{e_jf_{Mj}}{T_j}\int_{\theta}^{\infty}d\theta'\frac{qR}{|v_\parallel|}e^{-i\beta_j}(\omega-\widehat{\omega}_{*j})J_0(\Lambda_j')\left(\phi(\theta')+\frac{|v_\parallel|}{c}A_\parallel(\theta')\right), \quad (40)$$

where

$$\beta_j(\theta,\theta')=\int_{\theta'}^{\theta}\frac{qR}{|v_\parallel|}[\omega-\widehat{\omega}_{Dj}(\theta'')]d\theta''.$$

Substitution of perturbed distribution functions into charge neutrality and parallel Ampere's law yields

$$\nabla^2\phi=-4\pi\sum_{j=i,e}e_j\left(-\frac{e_j}{T_j}\phi+\int\left[g_j^+(\theta)+g_j^-(\theta)\right]J_0(\Lambda_j)d\mathbf{v}\right), \quad (41)$$

$$\nabla_\perp^2A_\parallel(\theta)=-\frac{4\pi}{c}\sum_{j=i,e}e_j\int v_\parallel\left[g_j^+(\theta)-g_j^-(\theta)\right]J_0(\Lambda_j)d\mathbf{v}, \quad (42)$$

where $\int d\mathbf{v}=2\pi\int_0^\infty v_\perp dv_\perp\int_0^\infty dv_\parallel$. This system of inhomogeneous integral equations can be solved by employing the method of Fredholm in which the integral equations are viewed as a system of linear algebraic equations [14]. In the numerical code, the velocity space integration is executed using Gauss-Hermite approximation.

Figure 1 shows the β_e dependence of the normalized eigenvalue ω/ω_{*e} (frequency ω_r/ω_{*e} and growth rate γ/ω_{*e}) when $ck_\theta/\omega_{pe}=0.3$, $T_e=T_i$, $L_n/R=0.3$, $\eta_e=2$, $s=1$, $q=2$, $m_i/m_e=1836$ (hydrogen). The growth rate increases rapidly with the electron pressure (β_e) indicating that the

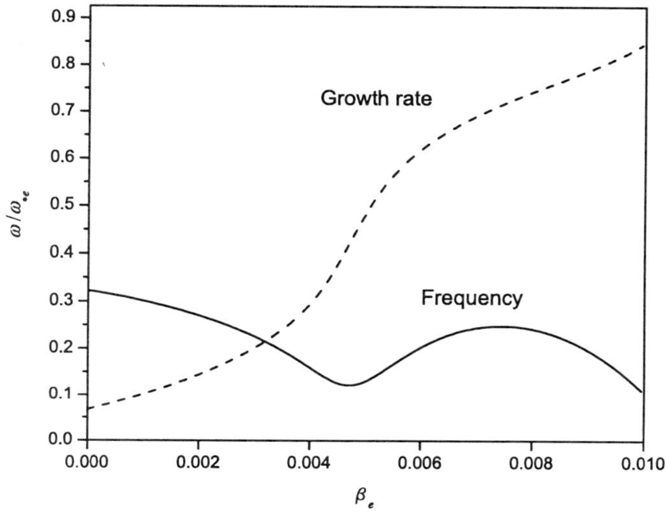

Fig. 1. The normalized mode frequency ω_r/ω_{*e} and growth rate γ/ω_{*e} as functions of the electron beta β_e when $ck_\theta/\omega_{pe} = 0.3$, $L_n/R = 0.2$, $\eta_e = 2$, $q = 2$, $s = 1$.

instability is indeed driven by the ballooning effect. The growth rate found in the numerical analysis qualitatively agrees with the analytic expression presented in Eq. (26).

The mixing length electron thermal diffusivity $\chi_e = \gamma/k_\perp^2$ normalized by the Ohkawa diffusivity

$$\chi_{Ohkawa} = \frac{v_{Te}}{qR}\left(\frac{c}{\omega_{pe}}\right)^2, \tag{43}$$

is shown in Fig. 2 as a function of $k^2 = (ck_\theta/\omega_{pe})^2$ when $L_n/R = 0.2$, $\eta_e = 2$, $s = 1$, $q = 2$, and $\beta_e = 0.4$ %. The maximum diffusivity occurs at $ck_\perp/\omega_{pe} \simeq 0.3$.

In light of the analytic dispersion relation found in this study, it may be concluded that a tokamak discharge can be strongly unstable in the wavelength regime $k_\perp \simeq \omega_{pe}/c$ which is the lower end of unstable $k_\perp$ spectrum. The growth rate is of the order of $\gamma \simeq \sqrt{\eta_e\omega_{*e}\omega_{De}}$. Therefore, for the electron thermal diffusivity, the following estimate emerges,

$$\chi_e \simeq \frac{\sqrt{\eta_e\omega_{*e}\omega_{De}}}{k_\perp^2} = \frac{qv_{Te}}{\sqrt{RL_T}}\left(\frac{c}{\omega_{pe}}\right)^2\sqrt{\beta_e}, \tag{44}$$

where L_T is the temperature gradient scale length. The proportionality of the diffusivity to the safety factor, $\chi_e \propto q$, stems from the condition of most active thermal transport, $\gamma \simeq k_{\|} v_{Te}$, which yields $k_{\perp} \propto 1/q$ [15].

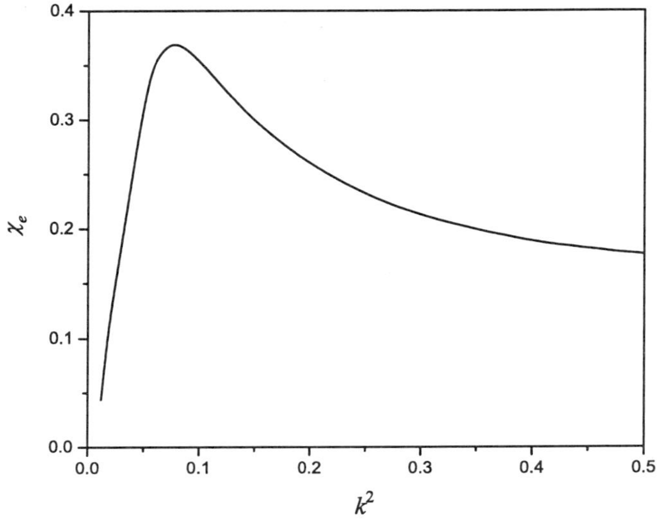

Fig. 2. χ_e/χ_{Ohkawa} vs. $k^2 = (ck_\theta/\omega_{pe})^2$ when $L_n/R = 0.2$, $\eta_e = 2$, $s = 1$, $q = 2$, and $\beta_e = 0.004$.

5. Conclusions

The local dispersion relation

$$(\omega - \omega_{*e})(\omega - \omega_{De}) + \eta_e \omega_{*e} \omega_{De} - \frac{(\omega - \omega_{*e})^2}{1+\tau} = \frac{c^2 k_{\|} k_{\perp}^2 k_{\|}}{k_{De}^2},$$

describes the short wavelength electron ballooning mode subject to the adiabatic electron response $\omega \lesssim k_{\|} v_{Te}$. The mode is intrinsically electromagnetic while the conventional ETG mode is subject to $\omega > k_{\|} v_{Te}$ for which electromagnetic effects appear only as a small correction. Because of the long wavelength nature of the instability ($c/\omega_{pe} \gg \rho_e$), large electron thermal transport emerges even in simple mixing length estimate. The following formula for the electron thermal diffusivity has been found,

$$\chi_e = \frac{q v_{Te}}{\sqrt{R L_T}} \left(\frac{c}{\omega_{pe}} \right)^2 \sqrt{\beta_e}.$$

In the previous investigations [3], it was proposed skin size plasma turbulence may exist if the nonlinear Doppler shift $\mathbf{k} \cdot \mathbf{v}_{E\times B}$ ($\mathbf{v}_{E\times B}$ being the

$E \times B$ drift) exceeds the electron transit frequency $\mathbf{k} \cdot \mathbf{v}_{E\times B} > k_{\parallel} v_{Te}$. The main finding in the present investigation is that the skin depth manifests itself even in the adiabatic limit and governs the lower end of the k spectrum of the ETG mode. It is noted that in the limit of large ballooning parameter α_e, the dispersion relation reduces to

$$(\omega - \omega_{*e})(\omega - \omega_{De}) + \eta_e \omega_{*e} \omega_{De} - \frac{(\omega - \omega_{*e})^2}{1+\tau} = 0,$$

which resembles that of the electrostatic ETG mode in the limit $\omega > k_{\parallel} v_{Te}$,

$$\tau - \frac{\omega_{*e} - \omega_{De}}{\omega - \omega_{De}} + \frac{\eta_e \omega_{*e} \omega_{De}}{(\omega - \omega_{De})^2} = 0.$$

In summary, an electromagnetic ballooning instability having cross-field wavelengths of the order of electron skin depth has been identified analytically and confirmed with an integral equation code which is fully kinetic and electromagnetic. In tokamaks, the mode is destabilized by a modest electron ballooning parameter α_e. A large electron thermal diffusivity emerges because of the long wavelength nature of the instability.

Acknowledgments Helpful comments provided by P. K. Shukla are acknowledged with gratitude. This research has been supported by the Natural Sciences and Engineering Research Council of Canada and by Canada Research Chair Program.

References

1. A. B. Mikhailovskii, in *Reviews of Plasma Physics*, Vol. 3, (Consultants Bureau, New York, 1967), p. 159.
2. D. W. Ross and S. M. Mahajan, Phys. Rev. Lett. **40**, 324 (1978).
3. C. Z. Cheng and L. Chen, Phys. Fluids **23**, 1770 (1980).
4. T. Ohkawa, Phys. Lett. **67A**, 35 (1978).
5. P. R. Parker, *et al.*, Nucl. Fusion **25**, 1127 (1985).
6. J. Y. Kim and W. Horton, Phys. Fluids B **4**, 3194 (1991).
7. W. Horton, Phys. Reports **192**, 1 (1990).
8. A. Hirose, Plasma Phys. Controlled Fusion **42**, 145 (2007).
9. A. Hirose, L. Zhang, and M. Elia, Phys. Rev. Lett. **72**, 3993 (1994).
10. A. B. Mikhailovskii, Sov. Phys.-JTP **37**, 1365 (1967).
11. N. Joiner and A. Hirose, submitted to Phys. Plasmas.
12. J. F. Denisse and J. L. Delcroix, in *Plasma Waves*, (Interscience Publishers, New York, 1963), p. 52.
13. M. Elia, Ph.D. thesis, University of Saskatchewan (2000).
14. G. Rewoldt, *et al.*, Phys. Fluids **25**, 480 (1982).
15. A. Hirose, *et al.*, Nucl. Fusion **45**, 1628 (2005).

EFFECTS OF PERPENDICULAR SHEAR SUPERPOSITION AND HYBRID IONS INTRODUCTION ON PARALLEL SHEAR DRIVEN PLASMA INSTABILITIES

T. KANEKO* and R. HATAKEYAMA

Department of Electronic Engineering, Tohoku University,
6-6-05 Aoba, Aramaki, Aoba-ku, Sendai 980-8579, Japan
** E-mail: kaneko@ecei.tohoku.ac.jp*

Parallel and perpendicular plasma flow velocity shears are independently controlled and superimposed in fully-ionized collisionless magnetized plasmas using a modified plasma-synthesis method with concentrically three-segmented electron and ion emitters. The fluctuation amplitude of the drift wave which has an azimuthal mode number $m = 3$ is observed to increase with increasing the parallel shear strength in the absence of the perpendicular shear. When the perpendicular shear is superimposed on the parallel shear, the drift wave of $m = 3$ changes into that of $m = 2$. Furthermore, the parallel shear strength required for the excitation of the drift wave becomes large with a decrease in the azimuthal mode number. Introduction of hybrid ions, i.e., superposition of two kinds of positive ions, is found to cause unexpected stabilization of the drift wave.

1. Introduction

Investigations of sheared plasma flows parallel and perpendicular to magnetic-field lines have widely been performed in connection with the excitation or the suppression of plasma fluctuations and turbulences, which are considered to generate energetic particles and induce cross-field transport in fusion-oriented[1–3] and space[4] plasmas. However, both the parallel and perpendicular flow shears often coexist in most of the large confinement devices and in the space environment, and cannot be controlled independently. As a result, it is difficult to understand the effects of each flow shear on the turbulences in real situations.

In basic laboratory experiments, on the other hand, a number of works on each of the flow shears have been reported. In the parallel shear case, some experimental investigations related to parallel-shear driven instabilities, such as the D'Angelo mode,[5–7] ion-acoustic,[8,9] and ion-cyclotron[10–13] instabilities, have been performed in various situations. In these experiments, it is to be noted that the instabilities are excited by a slight amount of parallel shear in the presence of the parallel electron current which often exists in conventional configurations such as Q machine experiments,[14] and therefore it is very difficult to experimentally evaluate the precise value of the parallel shear, which leads to difficulty in understanding the effects of the parallel shear on these instabilities.

In the perpendicular shear case, many experimental investigations on the relation between the perpendicular flow shears and the instabilities have been performed,[15–19] and it has recently been reported that the net ion flow shear which is determined based on both $E \times B$ drift and diamagnetic drift is important for stabilizing the drift-wave instability.[20] In such experiments, however, there exists neutral gas and collisional effects on the fluctuations may have to be considered. Thus, it is necessary to carry out experiments in a fully-ionized plasma from the viewpoint of making a comparison with the results obtained in collisionless fusion-oriented plasmas. Although some basic experiments using a fully-ionized collisionless plasma have been done,[21–23] it is difficult to actively change the sign of the radial electric field, or the sign of the perpendicular flow shear in these experimental configurations.

Therefore, it is required to develop novel plasma sources that enable us to actively control the parallel and perpendicular flow velocities and their shears, and to realize the controlled superposition of the parallel and perpendicular flow shears in magnetized plasmas. Furthermore, an experimental scope of the shear-modified instability should be extended to a more general case, namely, effects of several kinds of positive and negative ions (hybrid ions) on the instability should be taken into account because the actual plasmas often contain the hybrid ions.

Thus, the aim of the present work is to independently control and superimpose the parallel and perpendicular flow shears in the basic plasma device with concentrically three-segmented electron and ion emitters,[24] and to carry out laboratory experiments on the low-frequency instability excited and suppressed by the superimposed flow shears in fully-ionized collisionless magnetized plasmas, containing the two kinds of positive ions as the hybrid ions.

2. Experimental setup

Experiments are performed in the Q_T-Upgrade machine of Tohoku University. We attempt to modify a plasma-synthesis method with an electron

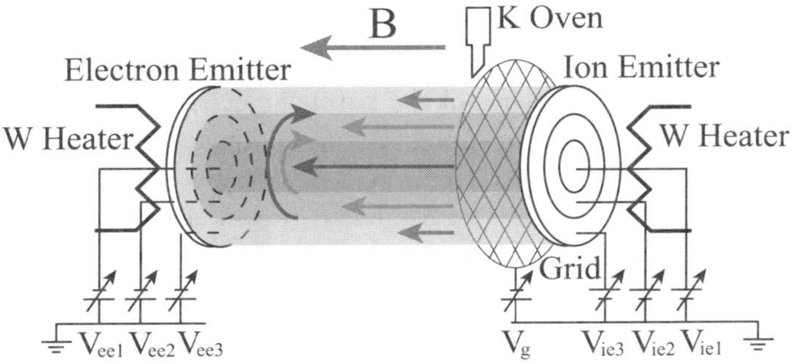

Fig. 1. Schematic of experimental setup for superposition of parallel and perpendicular flow velocity shears.

(e^-) emitter using a 10-cm-diameter tungsten (W) plate and a potassium ion (K^+) emitter using another W plate, which are oppositely located at the machine ends as shown in Fig. 1. The potassium atoms are evaporated using an oven (K oven) and are sprayed on the surface of the ion emitter. The collisionless plasma is produced when the surface-ionized potassium ions and the thermionic electrons are generated by the spatially separated ion and electron emitters, respectively, and are synthesized in the region between these emitters. A negatively biased stainless (SUS) grid, the voltage of which is typically $V_g = -60$ V, is installed at a distance of 10 cm from the ion emitter surface. Since the grid reflects the electrons flowing from the electron emitter side, the electron velocity distribution function parallel to the magnetic fields is considered to become Maxwellian. Here, the axial position z is defined as the distance from the Sus grid ($z = 0$ cm) toward the electron emitter.

Both the emitters are concentrically segmented into three sections with the outer diameters of 2 cm (first electrode), 5.2 cm (second electrode), and 10 cm (third electrode), each of which is electrically isolated. When each section of the electron emitter is individually biased, the radially-different plasma potential, or radial electric field is expected to be generated even in the fully-ionized collisionless plasma. This electric field causes the E×B flows and flow shears perpendicular to the magnetic-field lines.[25,26] Voltages applied to the electrodes set in order from the center to the outside are defined as $V_{ee1}, V_{ee2}, V_{ee3}$, respectively. On the other hand, the parallel K^+ flow with radially different energy, i.e., the parallel K^+ flow shear, is generated when each section of the segmented ion emitter is individually biased ($V_{ie1}, V_{ie2}, V_{ie3}$) at a positive value above the plasma potential that is determined by the bias voltage of the electron emitter. Therefore, these parallel and perpendicular K^+ flow velocity shears can be superimposed by controlling the bias voltage of the ion and electron emitters independently. In the present experiment, V_{ee3} and V_{ie3} are always kept at 0 V which corresponds to the machine-chamber potential. A small radially movable Langmuir probe and an electrostatic energy analyzer are used to measure radial profiles of plasma parameters and ion energy distribution functions parallel to the magnetic fields, respectively. A background gas pressure is less than 10^{-6} Torr.

3. Experimental results

3.1. *Generation and control of ion flow velocity shears*

Figure 2 shows radial profiles of (a) plasma potential ϕ and (b) plasma density n_p of the synthesized plasma, which are measured at $z = 60$ cm for $V_{ie1} = V_{ie2} = 0$ V and $V_{ee1} \simeq V_{ee2} = -2$ V. Here, the dotted lines in Fig. 2 indicate the boundaries of the segmented ion-emitter electrodes. When the bias voltages V_{ee1} and V_{ee2} are set to be the same value, ϕ ($\simeq -5.0$ V) are almost uniform radially within the second electrode, the flat region of which corresponds to the diameter of the heated electron emitter. On the

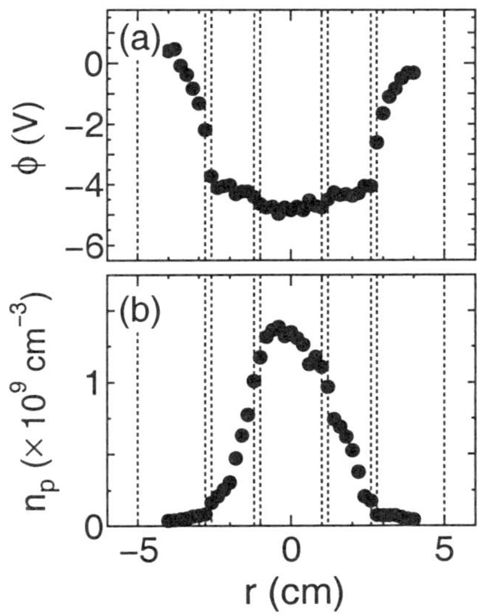

Fig. 2. Radial profiles of (a) plasma potential ϕ and (b) plasma density n_p at $z = 60$ cm for $V_{ie1} = V_{ie2} = 0$ V, $V_{ee1} = V_{ee2} = -2$ V, and $V_g = -60$ V.

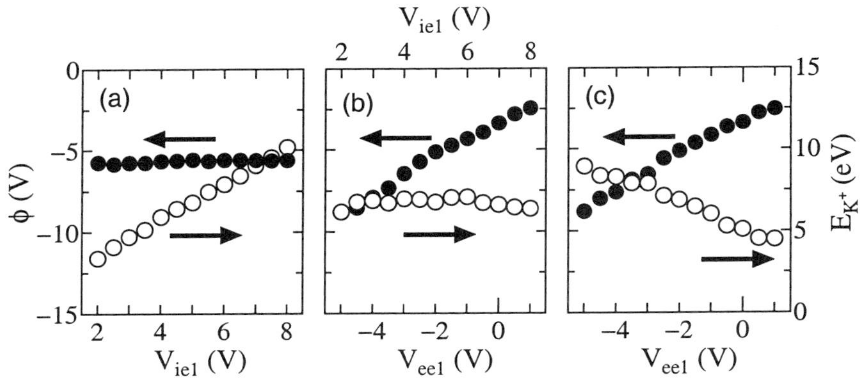

Fig. 3. Plasma potential ϕ (closed circles) and K^+ flow energy E_{K^+} (open circles) as functions of (a) V_{ie1}, (b) V_{ie1} and V_{ee1}, and (c) V_{ee1}. $r = 0$ cm, $V_{ie2} = 5$ V, $V_{ee2} = -2$ V.

other hand, n_p is about 10^9 cm^{-3} at the radial center and the plasma is produced almost within the second electrode, gradually decreasing toward the outside. In this plasma, electron T_e and ion T_i temperatures are around 0.2 eV and their profiles are almost uniform in the radial direction.

We demonstrate the independent control of the parallel and perpendicular K^+ flow velocity shears and the superposition of these shears. Figure 3 shows the plasma potential ϕ (closed circles) and the K^+ flow energy E_{K^+} (open circles) at the radial center $r = 0$ cm of the plasma column as functions of V_{ie1} and/or V_{ee1}, where V_{ie2} and V_{ee2} are fixed at 5 V and -2 V, respectively. When V_{ie1} is changed at constant $V_{ee1} \simeq -2$ V [Fig. 3(a)], the K^+ flow energy is found to increase in proportion to V_{ie1}, while the plasma potential is almost constant at $\phi = -5.5$ V. Since the K^+ flow energy and the plasma potential in the second electrode region are confirmed to have constant values of 7 eV and -5.5 V, respectively, only the parallel flow velocity shear can be generated in the boundary region between the first and second electrodes by changing V_{ie1}.

When V_{ie1} and V_{ee1} are simultaneously changed keeping the bias-voltage difference $V_{ie1} - V_{ee1}$ constant [Fig. 3(b)], on the other hand, the K^+ flow energy does not change, while the plasma potential is found to increase in proportion to V_{ee1}. This result denotes that the parallel shear does not change as far as $V_{ie1} - V_{ee1}$ is constant, and the radial plasma potential difference, i.e., the perpendicular flow velocity shear can be controlled by the bias voltages of the electron emitter.

Since the parallel and perpendicular shears are now able to be controlled independently, we attempt to superimpose these shears. Figure 3(c) presents the plasma potential and the K^+ flow energy as a function of V_{ee1} at constant $V_{ie1} = 5$ V. In this case, the plasma potential is directly changed by V_{ee1}, and the K^+ flow energy is also changed by V_{ee1}, because the bias-voltage difference $V_{ie1} - V_{ee1}$ decreases with an increase in V_{ee1} for the fixed V_{ie1}. Based on these results, the superposition of the parallel and perpendicular flow velocity shears is realized by controlling the V_{ie1} and V_{ee1} simultaneously. These parallel and perpendicular shears are found to give rise to several types of low-frequency instabilities. Here, we concentrate on the drift-wave instability which is excited in the density gradient region around $r = 1.0 \sim 1.5$ cm.

3.2. *Parallel shear driven low-frequency instabilities*

The normalized fluctuation amplitudes $\tilde{I}_{es}/\bar{I}_{es}$ obtained from frequency spectra of an electron saturation current I_{es} of the probe as a function of V_{ie1} (the parallel shear strength) are presented for $V_{ie1} > V_{ie2}(= -0.8$ V) in Fig. 4, together with the frequency spectra of I_{es} in the inset. When $\Delta V_{ie}(\equiv V_{ie2} - V_{ie1})$ is almost zero, the fluctuation is not excited like the case for $V_{ie1} < V_{ie2}$. Once $|\Delta V_{ie}|$ exceeds a certain threshold, the fluctuation with a frequency of about 6.5 kHz is observed to grow as $|\Delta V_{ie}|$ becomes large. Furthermore, the fluctuation is confirmed to be localized in the density gradient region which is a different radial position from the

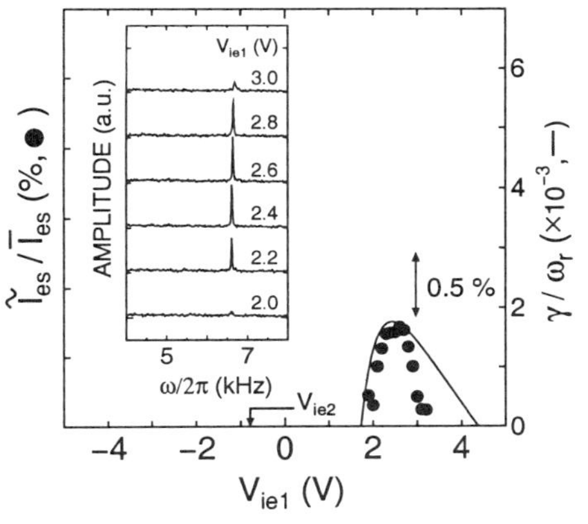

Fig. 4. Frequency spectra of electron saturation current and normalized fluctuation amplitudes $\tilde{I}_{es}/\bar{I}_{es}$ as a function of V_{ie1} for $V_{ie1} > V_{ie2}(= -0.8$ V) at $r = -1.0$ cm. Theoretical growth rate γ of the drift-wave instability is also plotted (solid line).

shear region. When $|\Delta V_{ie}|$ further increases, $\tilde{I}_{es}/\bar{I}_{es}$ attains to a maximum value and gradually decreases after that.

Since the fluctuation is observed to be localized in the large density gradient region, this fluctuation appears to be the drift-wave instability excited by the parallel shear,[27,28] which has never been observed so far in other experiments. In order to identify the parallel-shear excited drift-wave instability, the theoretical growth rate γ as a function of V_{ie1}, which is derived in Ref. [27], is calculated using the plasma parameters experimentally obtained, and is plotted in Fig. 4 (solid line). The experimental results are in good agreement with the theoretical curves and it turns out that the growth rate changes into positive when V_{ie1} exceeds the threshold and gradually increases with an increase in V_{ie1}. This means that the ion-frame phase velocity of the fluctuation becomes large due to the presence of the parallel shear, and thus, the effect of the ion Landau damping on the waves is reduced. For larger V_{ie1}, however, the growth rate saturates and gradually decreases with a further increase in V_{ie1}. When the phase velocity exceeds the ion flow velocity, i.e., the relative electron-ion drift velocity in the ion frame, due to the increase in V_{ie1}, the effect of the inverse electron Landau damping in the ion frame is reduced, which leads to the stabilization of the waves.[27]

3.3. *Superposition of perpendicular and parallel shears*

Since the parallel shear is found to excite the drift wave, we next focus on the effects of superposition of the perpendicular and parallel shears[29,30] on the drift wave. Figure 5(a) shows a contour view of normalized fluctuation amplitudes $\tilde{I}_{es}/\bar{I}_{es}$ obtained from frequency spectra of the electron saturation current I_{es} of the probe as functions of parallel V_{ie1} and perpendicular V_{ee1} shears for $V_{ie2} = 1$ V and $V_{ee2} = -2$ V. A schematic model of the parallel and perpendicular shears introduction is shown in Fig. 5(b), where black arrows described at the ordinate axis mean the parallel ion flow velocity and solid curves described at the abscissa axis mean the radial potential profiles, which are controlled by V_{ie1} and V_{ee1}, respectively, corresponding to the variation of the parallel and perpendicular flow velocity shears. Here, horizontal and vertical dotted lines in Fig. 5 denote the situations in the absence of the parallel and perpendicular shears, respectively, which are confirmed by the actual measurements of the ion flow energy and the space potential.

In the case that the perpendicular shear is not generated at $V_{ee1} \simeq -1.8$ V, the fluctuation amplitude of the drift-wave instability is observed

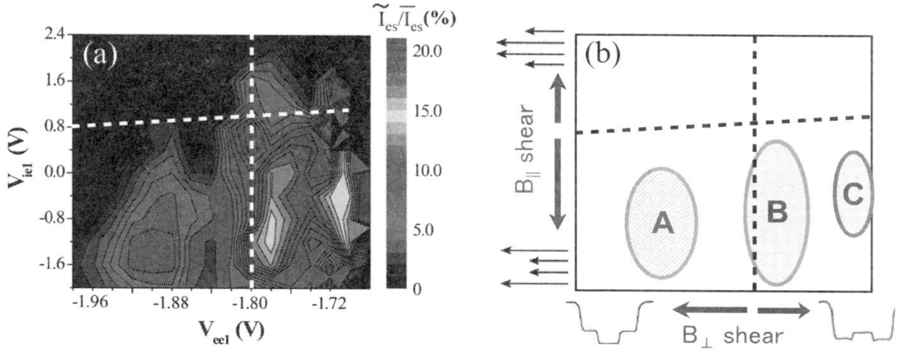

Fig. 5. Contour views of normalized fluctuation amplitudes as functions of V_{ie1} and V_{ee1}. $r = -1.0$ cm, $V_{ie2} = 1$ V, $V_{ee2} = -2$ V.

to increase with increasing the parallel shear strength by changing V_{ie1} to the negative value from 1.0 V, but the instability is found to be gradually stabilized when the shear strength exceeds the critical value, which is the same as the results described in Fig. 4.

When the perpendicular shear is superimposed on the parallel shear, the drift wave excited by the parallel shear for $V_{ie1} \simeq -1$ V and $V_{ee1} \simeq -1.8$ V is found to be modified and finally be suppressed by the sufficiently large perpendicular shear independently of the sign of the perpendicular shear. Figure 6 shows normalized fluctuation amplitudes $\tilde{I}_{es}/\bar{I}_{es}$ as a function of V_{ee1} for $V_{ee2} = -2$ V and $\Delta V_{ie} = 2$ V at $r = -1.0$ cm. Three characteristic fluctuation peaks can be observed at $V_{ee1} = -1.9$ V, -1.8 V, and -1.7 V, which are defined as fluctuations A, B, and C, respectively, as described

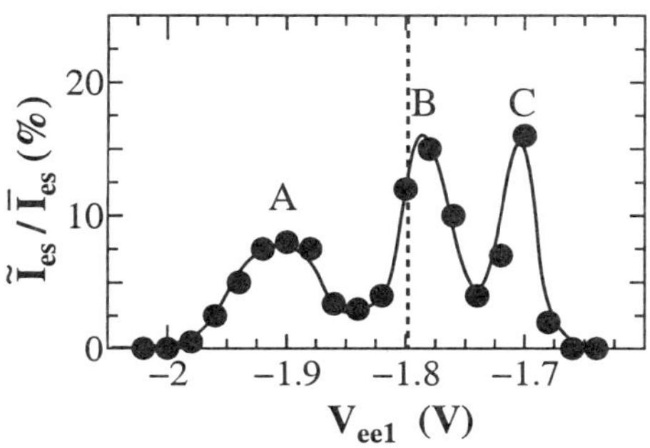

Fig. 6. Normalized fluctuation amplitudes $\tilde{I}_{es}/\bar{I}_{es}$ as a function of V_{ee1} for $V_{ee2} = -2$ V, $\Delta V_{ie}(\equiv V_{ie2} - V_{ie1}) = 2$ V at $r = -1.0$ cm.

also in the schematic model [Fig. 5(b)].

To readily identify the azimuthal component of each fluctuation's wavevector, we measure 2-dimensional (x, y) profiles of fluctuation phase in the plasma-column cross section. The phase is measured with reference to a spatially fixed Langmuir probe located at an axial distance of 26 cm from the 2-dimensionally scanning probe. Figure 7 presents the 2-dimensional phase profiles for (a) fluctuation A and (b) fluctuation B, while the 2-dimensional phase profile for fluctuation C can not be obtained because the fluctuation is like a turbulence and the coherent structure is not detected. Since the phase difference θ between the 2-dimensional probe and the reference probe is plotted as $\sin\theta$, the phase of $+\pi/2$ and $-\pi/2$ relative to the reference probe are indicated by 1.0 (red) and -1.0 (blue), respectively. Green corresponds to the phase difference of zero and π. In the case of the small perpendicular shear strength, i.e., fluctuation B [Fig. 7(b)], the azimuthal mode is found to be $m = 3$. On the other hand, in the presence of the relatively large perpendicular shear, i.e., fluctuation A [Fig. 7(a)], the azimuthal mode changes into $m = 2$. The perpendicular shear is found to modify the azimuthal mode number depending on its strength.

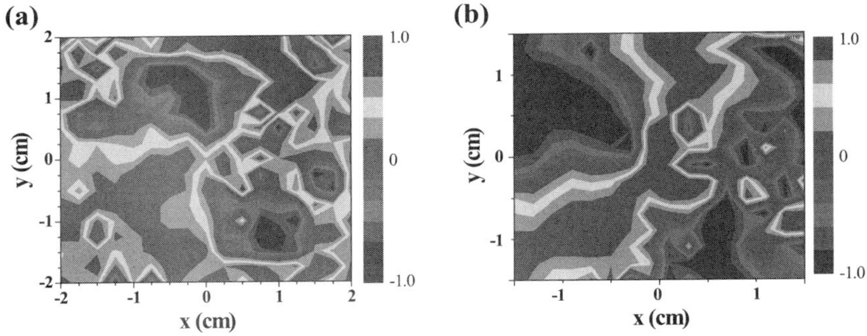

Fig. 7. 2-dimensional profile of fluctuation phase θ which is plotted as $\sin\theta$ for (a) fluctuation A ($V_{ee1} \simeq -1.9$ V) and (b) fluctuation B ($V_{ee1} \simeq -1.8$ V). $V_{ie1} = -1$ V, $V_{ie2} = 1$ V, $V_{ee2} = -2$ V.

For these two kinds of drift waves, we measure the dependence of fluctuation amplitudes on the parallel shear strength. Figure 8 gives normalized fluctuation amplitude $\tilde{I}_{es}/\bar{I}_{es}$ as a function of V_{ie1} for $V_{ee1} \simeq -1.8$ V (closed circles) and -1.9 V (open circles), which are obtained from Fig. 5(a). A dotted line in Fig. 8 denotes the bias voltage of the second electrode V_{ie2}, i.e., the absence of the parallel shear. In this experimental condition, the drift

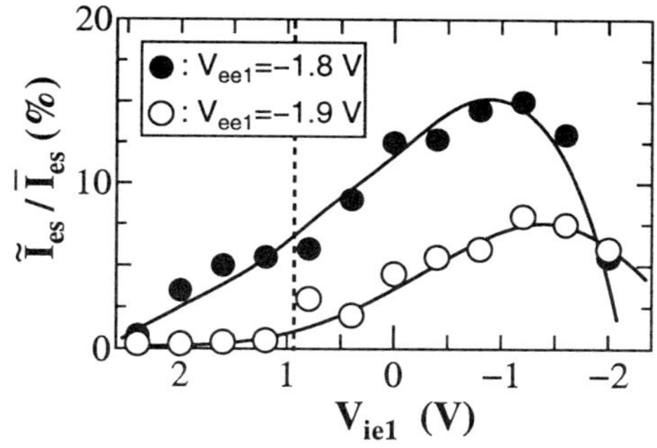

Fig. 8. Normalized fluctuation amplitudes $\tilde{I}_{es}/\bar{I}_{es}$ as a function of V_{ie1} for $V_{ie2} = 1$ V at $r = -1.0$ cm.

wave is excited even in the absence of the parallel shear for $V_{ee1} \simeq -1.8$ V due to the relatively large parallel ion flow (current).[14]

With an increase in the parallel shear strength, it is found that $m = 3$ mode ($V_{ee1} \simeq -1.8$ V) first excited and $m = 2$ mode ($V_{ee1} \simeq -1.9$ V) needs the strong parallel shear strength to be excited.

These phenomena can be explained by the theoretical calculation of the growth rate of the drift wave as shown in Fig. 9. The parameters for the calculation of the growth rate use the experimentally obtained values in this experimental condition. The growth rate for $m = 3$ is larger than that for $m = 2$, which is consistent with the experimental observations in Fig. 8, assuming that saturated mode amplitude is approximately proportional to

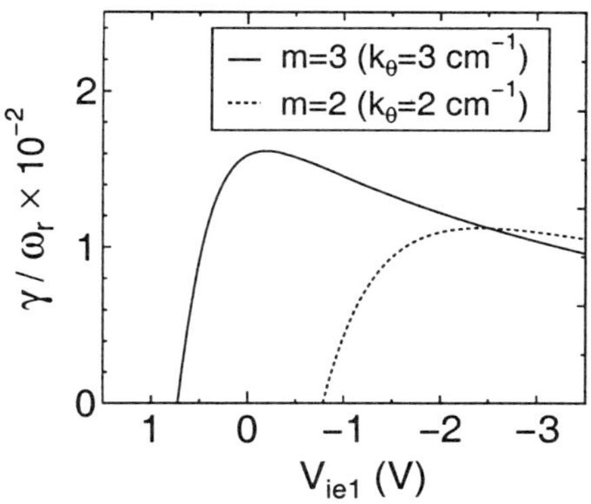

Fig. 9. Predicted dependence of growth rate γ of drift-wave instability on V_{ie1}, where azimuthal mode number m is varied.

the growth rate. Furthermore, the parallel shear strength ΔV_{ie} yielding the peak growth rate is found to decrease for increasing the azimuthal mode number m, i.e., the azimuthal wave number k_θ. This can be understood by noticing that the threshold value of shear parameter σ^2 for instability can be held constant by decreasing the parallel shear strength ΔV_{ie} as k_θ increases.[27,28] The growth rate of the parallel-shear excited drift wave is found to sensitively depend on the azimuthal mode number. Therefore, the superposition of the parallel and perpendicular shears can affect the characteristics of the drift wave through the variation of the azimuthal mode number.

3.4. *Introduction of hybrid ions*

Effects of hybrid ions (two kinds of positive ions) introduction on the parallel shear excited drift wave are important for the purpose of clarifying the exciting mechanism of the drift wave in the actual space and fusion plasmas. Therefore, we equip a new oven which evaporates cesium (Cs) in addition to the conventional oven for potassium (K) as shown in Fig. 10. Here, the bias voltage of the electron emitters are set to realize the absence of the perpendicular shear. To control the density of K and Cs ions, we change the oven temperatures which vary the amount of K and Cs atoms evaporated and sprayed on the surface of the ion emitter (W hot plate).

Figure 11 presents time sequences of (a) oven temperatures and (b) electron density and temperature. Since a time interval between the neighboring measurement points is about 30 min, the total measurement time is about 5 hours. At first, the Cs ion plasma is generated keeping the Cs oven temperature 180 °C (Time = 1-2). Increasing the K oven temperature up to

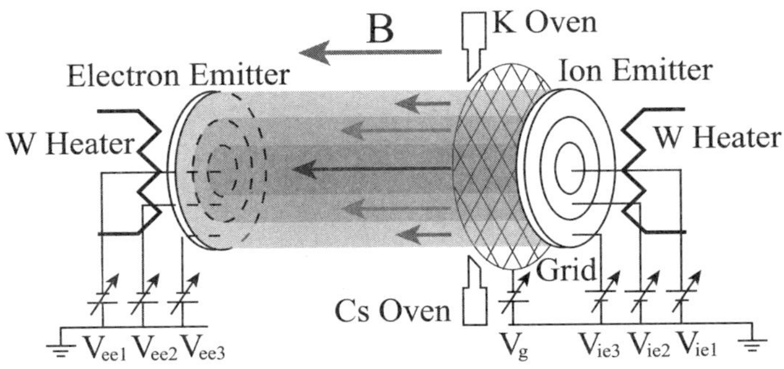

Fig. 10. Schematic of experimental setup for introduction of hybrid ions.

300 °C results in the superposition of the K ions to the Cs ion plasma (Time = 3-7), which is reflected in an increase in the electron density as shown in Fig. 11(b). Finally, the Cs oven temperature is decreased, resulting in

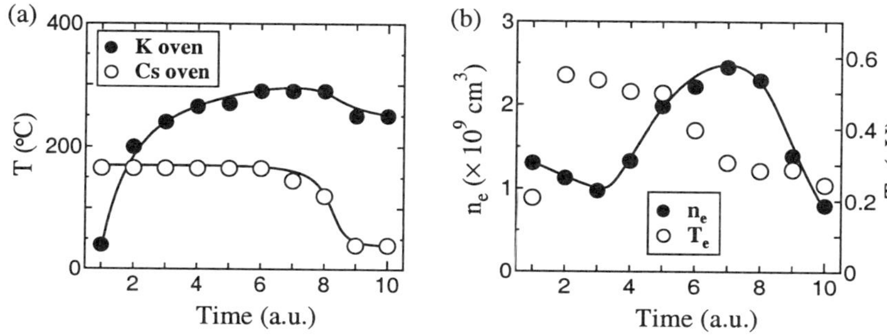

Fig. 11. Time sequences of (a) oven temperature and (b) electron density and temperature.

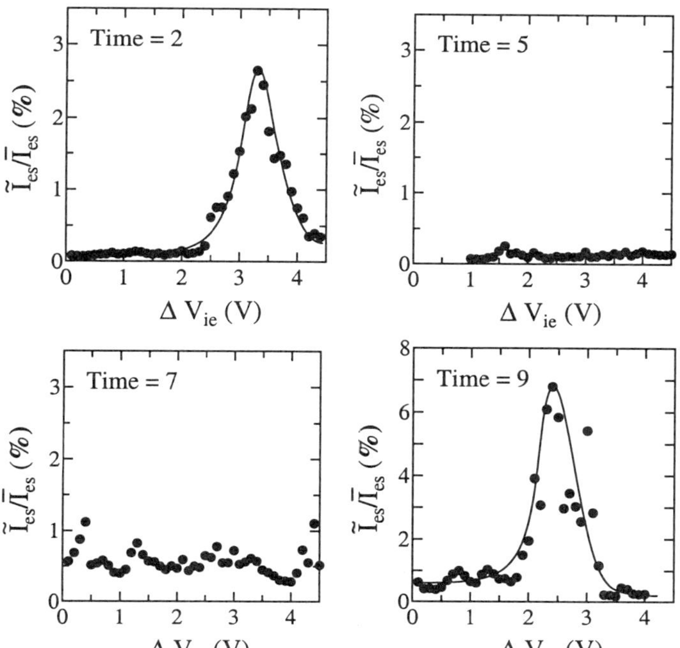

Fig. 12. Normalized fluctuation amplitudes $\tilde{I}_{es}/\bar{I}_{es}$ as a function of ΔV_{ie} for the fixed $V_{ie2} = 1$ V at $r = -1.0$ cm.

the decrease in the Cs ion density and only the K ion plasma is generated (Time = 8-10).

Using this time sequence, normalized fluctuation amplitudes $\tilde{I}_{es}/\bar{I}_{es}$ as a function of ΔV_{ie} for the fixed $V_{ie2} = 1$ V are presented in Fig. 12, where the figures for Time=2 and 9 correspond to the only Cs ion and the only K ion plasmas, respectively, and those for Time=5 and 7 show the case of superposition of the Cs and K ions in the plasma. In the case of the individual ion plasma (Time=2 and 9), the drift wave is excited with an increase in the parallel shear and gradually suppressed by the larger shear in the same manner as in Fig. 4. However, there is a difference in ΔV_{ie} yielding the peak fluctuation amplitude. Since ΔV_{ie} can control the ion flow energy parallel to the magnetic field lines, the radial difference in the ion flow velocity, i.e., the parallel velocity shear, varies depending on the mass number of ions even when ΔV_{ie} is the same value. Based on the results in Fig. 12, it is found that the Cs ions with large mass number need the large ΔV_{ie} to realize the same velocity difference as the K ions and to excite the drift wave.

In the case that these Cs and K ions are superimposed, on the other hand, the fluctuation amplitudes are almost independent of the parallel shear and are unexpectedly suppressed. Although this suppression by the superposition of the hybrid ions was reported in the case of an ion acoustic wave, where the parallel shear is not included, there is no report on the parallel shear excited drift wave modified by the hybrid ions and the theoretical analysis of these phenomena are now under investigation.

4. Conclusions

The independent control of parallel and perpendicular flow velocity shears in magnetized plasmas is realized using a modified plasma-synthesis method with segmented plasma sources. The ion flow velocity shear parallel to the magnetic-field lines is observed to destabilize the drift-wave instability depending on the strength of the parallel shear. On the other hand, when the perpendicular shear is superimposed on the parallel shear, the drift wave of $m = 3$ is found to change into that of $m = 2$, and the instability is suppressed for strong perpendicular shears. The superposition of these shears can affect the characteristics of the drift wave through the variation of the azimuthal mode number. Furthermore, introduction of hybrid ions, i.e., superposition of two kinds of positive ions, causes unexpected stabilization of the drift wave. The mechanism of the stabilization is under investigation.

References

1. R. J. Groebner, K. H. Burrell and R. P. Seraydarian, *Phys. Rev. Lett.* **64** 3015 (1990).
2. K. Ida, *Plasma Phys. Control. Fusion* **40**, 1429 (1998).
3. X. Garbet, C. Fenzi, H. Capes, P. Devynck and G. Antar, *Phys. Plasmas* **6**, 3955 (1999).
4. D. A. Roberts, M. L. Goldstein, W. H. Matthaeus and S. Ghosh, *J. Geophys. Res.* **97**, 17115 (1992).
5. N. D'Angelo and S. V. Goeler, *Phys. Fluids* **9**, 309 (1966).
6. T. An, R. L. Merlino and N. D'Angelo, *Phys. Lett. A* **214**, 47 (1996).
7. J. Willig, R. L. Merlino and N. D'Angelo, *Phys. Lett. A* **236**, 223 (1997).
8. E. Agrimson, N. D'Angelo and R. L. Merlino, *Phys. Rev. Lett.* **86**, 5282 (2001).
9. C. Teodorescu, E. W. Reynolds and M. E. Koepke, *Phys. Rev. Lett.* **88**, 185003 (2002).
10. E. P. Agrimson, N. D'Angelo and R. L. Merlino, *Phys. Lett. A* **293**, 260 (2002).
11. C. Teodorescu, E. W. Reynolds and M. E. Koepke, *Phys. Rev. Lett.* **89**, 105001 (2002).
12. E. Agrimson, S.-H. Kim, N. D'Angelo, and R. L. Merlino, *Phys. Plasmas* **10**, 3850 (2003).
13. M. E. Koepke, M. W. Zintl, C. Teodorescu, E. W. Reynolds, G. Wang and T. N. Good, *Phys. Plasmas* **9**, 3225 (2002).
14. R. Hatakeyama, M. Oertl, E. Märk and R. Schrittwieser, *Phys. Fluids* **23**, 1774 (1980).
15. M. Yamada and D. K. Owens, *Phys. Rev. Lett.* **38**, 1529 (1977).
16. A. Komori, K. Watanabe and Y. Kawai, *Phys. Fluids* **31**, 210 (1988).
17. A. Mase, J. H. Jeong, A. Itakura, K. Ishii, M. Inutake and S. Miyoshi, *Phys. Rev. Lett.* **64**, 2281 (1990).
18. W. E. Amatucci, D. N. Walker, G. Ganguli, J. A. Antoniades, D. Duncan, J. H. Bowles, V. Gavrishchaka and M. E. Koepke, *Phys. Rev. Lett.* **77**, 1978 (1996).
19. M. Yoshinuma, M. Inutake, R. Hatakeyama, T. Kaneko, K. Hattori, A. Ando and N. Sato, *Phys. Lett. A* **255**, 301 (1999).
20. M. Yoshinuma, M. Inutake, K. Hattori, A. Ando, T. Kaneko, R. Hatakeyama and N. Sato, *Trans. Fusion Technol.* **39**, 191 (2001).
21. G. I. Kent, N. C. Jen and F. F. Chen, *Phys. Fluids* **12**, 2140 (1969).
22. D. L. Jassby, *Phys. Rev. Lett.* **25**, 1567 (1970).
23. M. E. Koepke, W. E. Amatucci, J. J. Carroll III and T. E. Sheridan, *Phys. Rev. Lett.* **72**, 3355 (1994).
24. T. Kaneko, Y. Odaka, E. Tada and R. Hatakeyama, *Rev. Sci. Instrum.* **73**, 4218 (2002).
25. H. Tsunoyama, T. Kaneko, E. Tada, R. Hatakeyama, M. Yoshinuma, A. Ando, M. Inutake and N. Sato, *Trans. Fusion Sci. Technol.* **43**, 186 (2003).
26. R. Hatakeyama and T. Kaneko, *Phys. Scripta* **T107**, 200 (2004).
27. T. Kaneko, H. Tsunoyama and R. Hatakeyama, *Phys. Rev. Lett.* **90**, 125001

(2003).
28. T. Kaneko, E.W. Reynolds, R. Hatakeyama, and M.E. Koepke, *Phys. Plasmas* **12**, 102106 (2005).
29. T. Kaneko, K. Hayashi, R. Ichiki, and R. Hatakeyama, *Trans. Fusion Sci. Technol.* **51**, 103 (2007).
30. M.E. Koepke and E.W. Reynolds, *Plasma Phys. Control. Fusion* **49**, A145 (2007).